Batch Processes

CHEMICAL INDUSTRIES

A Series of Reference Books and Textbooks

Consulting Editor

HEINZ HEINEMANN
Berkeley, California

ADDITIONAL VOLUMES IN PREPARATION

Batch
Processes

Ekaterini Korovessi
E.I. du Pont de Nemours and Company
Wilmington, Delaware, U.S.A.

Andreas A. Linninger
University of Illinois
Chicago, Illinois, U.S.A.

CRC Press
Taylor & Francis Group
Boca Raton London New York

CRC Press is an imprint of the
Taylor & Francis Group, an **informa** business
A TAYLOR & FRANCIS BOOK

CRC Press
Taylor & Francis Group
6000 Broken Sound Parkway NW, Suite 300
Boca Raton, FL 33487-2742

First issued in paperback 2019

ISBN-13: 978-0-8247-2522-8 (hbk)
ISBN-13: 978-0-367-39207-9 (pbk)
Library of Congress Card Number 2005051435

Library of Congress Cataloging-in-Publication Data

Korovessi, Ekaterini.
    Batch processes / Ekaterini Korovessi, Andreas A. Linninger.
        p. cm. -- (Chemical industries ; v. 107)
    Includes bibliographical references and index.
    ISBN 0-8247-2522-0 (alk. paper)
    1. Process control--Data processing. 2. Chemical process control--Data processing. I. Linninger, Andreas A. II. Title. III. Series.

TS156.8.K65 2005
660'.281--dc22   2005043727

Visit the Taylor & Francis Web site at
http://www.taylorandfrancis.com

and the CRC Press Web site at
http://www.crcpress.com

Preface

The inherent dynamic nature of batch processes allows for their ability to handle variations in feedstock and product specifications and provides the flexibility required for multiproduct or multipurpose facilities. They are thus best suited for the manufacture of low-volume, high-value products, such as specialty chemicals, pharmaceuticals, agricultural, food, and consumer products, and most recently the constantly growing spectrum of biotechnology-enabled products.

The last 5 to 10 years have witnessed a renewed interest in batch processing technologies by manufacturing businesses that is driven by a number of factors:

- Globalization and strong competition have resulted in reduction of the high margins that have long been enjoyed, especially by the pharmaceutical and high-value specialty chemicals industries.
- The number of differentiated specialty chemicals and biochemicals introduced in the market has increased.
- Businesses must remain compliant with the growing number of regulatory requirements on emission and waste minimization resulting from environmental concerns.

Reduced time to market, lower production costs, and improved flexibility are all critical success factors for batch processes. In response to these industrial needs, batch processes have also recently attracted the attention of the academic world.

The aim of this book is to provide an inclusive review of the wide-ranging aspects of design, development, operations, and control of batch processes. The development of systematic methods for the synthesis and conceptual design of batch processes offers many challenges and has not received considerable attention, with the exception of chemical route selection and solvent selection.

The unsteady nature and flexibility of batch processes pose challenging design and operation problems. Traditional approaches to the design of batch unit operations include short-cut estimation methods, rules of thumb, and design by analogy. Although not necessarily optimal, they are valuable techniques during process development. Especially in the case of solid–liquid and solid–solid separation unit operations, it is not uncommon to hear manufacturers say that their equipment can be properly fitted to a particular task only on the basis of some direct laboratory and pilot plant work. Recent fundamental research is seeking to develop models for unit operations involving solids. In the biochemical and pharmaceutical sectors, batch fermentation and separation of optically pure chiral compounds are examples of research areas that are experiencing a rapid development of new technologies and supporting design methods in both industry and academia.

Batch process scheduling is important for maximization of facility utilization and production rates while meeting product market demands. It is a mature yet still active area of research. Powerful optimization techniques and solutions are available today that are opening up new opportunities for planning and supply chain management, an area of utmost interest for large manufacturing businesses.

Design software tools for the modeling of single-unit operations, such as batch reactors and distillation columns, have been available for a number of years. Significant advances that have taken place in the field of process development, plant operation, and information management have resulted in the recent emergence of tools that model the behavior of the entire batch process and are attempting to address the challenging task of automation of manufacturing batch processes. This book presents the latest of these technologies, shows how they are currently being implemented, and discusses their advantages as well as potential future improvements.

Minimizing safety hazards and designing inherently safe batch processes are crucial for the batch processing industry and have been extensively covered in other publications, which is why we have omitted the subject from this book. In addition, topics such as toxicology and special formulations are not covered here due to their specificity to individual industry sectors.

It is not possible today for any individual to be an expert in all or even many aspects of batch processing. With the exception of very small companies or simple problems, interdisciplinary project teams are required to address and manage the complexities and challenges of batch processes across their life-cycles — from conception through manufacturing. The need for such collaboration is evident in this book. The contributors represent a blend of academic and industrial backgrounds with considerable experience in research, process design, development, operation, and control of batch processes.

The intended audience for this book primarily includes practitioners and researchers in the batch process industries. They may choose to read the introductory comments of each chapter in order to gain an overview of the field, or they may use it as a reference book for specific needs. In either case, they will benefit from the up-to-date coverage of all aspects of batch processing and the combined expertise of the contributing authors. For academic and industrial researchers alike, portions of the book should provide perspectives on the practical context of their work and stimulate research activity in the understudied aspects of batch processes. In recent years, chemical engineering students have been increasingly exposed to batch processing. Although we have not designed the book to be used as a textbook, we hope that many of the ideas presented here can effectively be brought into the classroom.

Contributors

Matthew H. Bassett
Dow AgroSciences LLC
Indianapolis, Indiana

Dennis Bonné
CAPEC, Department of Chemical
 Engineering
Technical University of Denmark
Lyngby, Denmark

Andreas Cruse
Uhde GmbH
Dortmund, Germany

Sean M. Dalziel
Theravance, Inc.
San Francisco, California

James R. Daniel
Department of Foods and Nutrition
Purdue University
West Lafayette, Indiana

Urmila Diwekar
Vishwamitra Research Institute
Clarendon Hills, Illinois

Thomas E. Friedmann
Solae Denmark A/S
Aarhus C, Denmark

Rafiqul Gani
CAPEC, Department of Chemical
 Engineering
Technical University of Denmark
Lyngby, Denmark

Lars Gregersen
COMSOL A/S
Søborg, Denmark

Priscilla J. Hill
Dave C. Swalm School of Chemical
 Engineering
Mississippi State University
Mississippi State, Mississippi

Sten Bay Jørgensen
CAPEC, Department of Chemical
 Engineering
Technical University of Denmark
Lyngby, Denmark

Ki-Joo Kim
U.S. Environmental Protection
 Agency
Cincinatti, Ohio

Ekaterini Korovessi
Central Research and Development
E.I. du Pont de Nemours and
 Company
Wilmington, Delaware

Andreas A. Linninger
Departments of Chemical and
 Bioengineering
University of Illinois at Chicago
Chicago, Illinois

Andrés Malcolm
Department of Chemical Engineering
University of Illinois at Chicago
Chicago, Illinois

Wolfgang Marquardt
Lehrstuhl für Prozesstechnik
RWTH Aachen
Aachen, Germany

Conor M. McDonald
Microsoft Corporation
Dublin, Ireland

Jan Oldenburg
BASF Aktiengesellschaft
Ludwigshafen, Germany

Irene Papaeconomou
CAPEC, Department of Chemical
  Engineering
Technical University of Denmark
Lyngby, Denmark

Martin Schlegel
BASF Aktiengesellschaft
Ludwigshafen, Germany

Karl D. Schnelle
Dow AgroSciences LLC
Indianapolis, Indiana

Nilay Shah
Centre for Process Systems
  Engineering
Imperial College London
London, United Kingdom

Keith G. Tomazi
Tyco/Healthcare/Mallinckrodt
St. Louis, Missouri

John Villadsen
Department of Chemical Engineering
  and Biocentrum-DTU
Technical University of Denmark
Lyngby, Denmark

Table of Contents

Part I

Batch Processing General Overview

1 Introduction

Ekaterini Korovessi and Andreas A. Linninger

The inherent dynamic nature of batch processes allows for their ability to handle variations in feedstock and product specifications and provides the flexibility required for multiproduct or multipurpose facilities. They are thus best suited for the manufacture of low-volume, high-value products, such as specialty chemicals, pharmaceuticals, agricultural, food, and consumer products, and most recently for the constantly growing spectrum of biotechnology-enabled products.

This book presents the latest batch processing technologies, how they are currently being implemented, and discusses advantages, limitations, and potential future improvements.

The book is divided into three sections. Part I presents an overview of the batch processing industries. The unsteady nature and flexibility of batch processes, which make them so attractive, pose challenging design and operation problems. Process and equipment design issues in batch processing are covered in Part II, and Part III deals with the management of batch process operations.

A description of the structure and contents of the book is included in this chapter. Chapter 2 begins with an introductory discussion of the distinct characteristics of batch processes and batch dynamics and proceeds with a review of the industries that use predominantly batch processing. The subsequent discussion of the various stages of the lifetime of a drug, from conception through manufacturing, focuses on the interactions between the parallel tracks of product development and batch process development and the challenges of their optimization under the constraints of the Food and Drug Administration (FDA) regulatory approval process. Use of the Merrifield process for solid-phase peptide synthesis is presented as a case study that is discussed in some detail and is used to introduce the use of various batch unit operations and demonstrate the flexibility of batch processing. The diversity of the specialty chemicals industries is exemplified in the section of the chapter that introduces specific equipment, critical variables, varying quality specifications, and operational practices for the high-purity chemicals and cosmetics industrial sectors. Selected examples of the use of batch processing in the food industry are presented in the last section of Chapter 2.

Chapter 3 covers some recent academic work addressing the problem of synthesis of batch processes. The solution strategies of the mathematically formulated batch synthesis problem can be based on heuristic, mathematical programming, or hybrid approaches. The discussion focuses on a framework of methods and tools that are required by the various solution methodologies for the problem formulation, generation of alternative feasible solutions, and ultimately evaluation of these alternatives and the arising challenges. Simple illustrative examples complement the chapter.

3

The remaining chapters in Section II provide detailed descriptions of some of the most important batch unit operations, including operation overview, design, and, in some cases, simulation and control. Traditional approaches to the design of batch unit operations include short-cut estimation methods, rules of thumb, and design by analogy. Although not necessarily optimal, they are valuable techniques during process development. In the biochemical and pharmaceutical sectors, batch and fed-batch fermentation is a research area experiencing rapid development of new technologies and supporting design methods in both industry and academia. Kinetics and mass transfer considerations during the design of batch bioreactors are discussed in detail in Chapter 4. Theoretical studies of batch distillation, one of the oldest and most widely used unit operations in the batch industries, have advanced to state-of-the-art, computer-aided design techniques that are based on rigorous mathematical models. A complete review of batch distillation is presented in Chapter 5. Especially for the case of solid–liquid and solid–solid separation unit operations, it is not uncommon for manufacturers to fit their equipment specifically to a particular task only after some direct laboratory experimentation and pilot plant work. Recent fundamental research has sought to develop models for unit operations involving solids. Crystallization is often the final purification step for pharmaceuticals and specialty chemicals. Despite the importance of batch crystallization on product quality, methodologies for the design of batch crystallizers are still in developmental stages. The fundamental concepts of batch crystallization from solution and the theory of crystallization kinetics are presented in Chapter 6. A design methodology is proposed and a few examples are used to illustrate the presented design principles. Chapter 7 addresses practical aspects of the development of batch crystallization and the associated downstream solid–liquid separations while at the same time considering the integration and interdependency of these technologies. The chapter brings together a compilation of valuable important considerations, approaches, and resources required by practitioners of crystallization and solid–liquid separations. It also serves as a guide for laboratory and pilot testing, as well as for scale-up aspects.

Minimization of waste generation can most effectively be addressed during the design of batch processes. Chapter 8 summarizes the waste sources and pollution concerns in the batch pharmaceutical and specialty chemical industries and provides an overview of the various pollution control methodologies. The impact of environmental regulations and regulatory incentives for pollution prevention is discussed, and guidelines for selection of pollution prevention strategies are presented. The chapter includes a comprehensive list of either commercially available or academic software tools that is not limited to computer-aided pollution prevention but also covers batch process modeling.

Recent research activities aimed at the use of rigorous mathematical modeling and optimization for the design and optimization of batch processes are discussed in Chapter 9. The chapter presents an overview of mathematical techniques for the optimal design of batch processes of predetermined fixed structure as well as batch processes having a structure subject to optimization. The incorporation of uncertainties in the optimization problems is also considered. Finally, the concept of dynamic real-time optimization and control is presented. The management of batch processes involves decision making at all levels of the plant operation hierarchy. Such decisions include scheduling, planning, monitoring, control, and supply-chain management. Chapter 10 introduces batch process production scheduling and campaign planning problems, outlines approaches to their solution, and discusses the practical implementation of these approaches. Both problems are information intensive, and solution methodologies should be able to handle uncertainty in the input data. The importance and challenges of integrating information management systems with production scheduling and campaign planning are emphasized. Chapter 11 provides an overview of advanced monitoring and control concepts for batch processes and presents recent advances in the development of methods for the practical implementation of batch process control. The purpose of supply-chain management, the subject of Chapter 12, is to deal effectively with external strategic changes, such as globalization, as well as operational uncertainties, such as demand fluctuations, with the ultimate goal of driving down the overall supply-chain costs. The authors discuss typical categories of supply-chain problems: supply-chain infrastructure design, supply-chain planning, and supply-chain execution and control. Scenario-based approaches for dealing with uncertainties during the strategic supply-chain network design are presented. The chapter also discusses the main elements of supply-chain planning — namely, management of demand, inventories, production, and distribution, as well as recent academic work on the topic of dynamic supply-chain simulation. The latter is an emerging area aiming at the effective execution and control of supply-chain management decisions.

The authors of the individual chapters have included extensive bibliographies to serve as resources for more detailed information regarding the topics discussed or briefly mentioned in the chapter.

2 Batch Processing Industries

Keith G. Tomazi, Andreas A. Linninger, and James R. Daniel

CONTENTS

2.1 INTRODUCTION

Batch chemical processing is still widely used and offers advantages over continuous operations in certain cases. These cases include manufacturing operations in which flexibility is required in either the rate or mix of products, in cases where the throughput is very small or the chemistry is not applicable to continuous operations, and in cases where lot integrity must be maintained. The selection of a batch vs. continuous operation is based on many factors (see Table 2.1). A large-scale manufacturing process, such as the refining of petroleum, involves many continuous operations. A small-scale process, such as the production of a highly specialized chemical reagent, may be mostly done in batch operations. A heuristic rule suggests that products with annual requirements of more than 500,000 kg are more economically efficient when done in continuous operations.[1]

Certain products have a demand that changes over time or have seasonal variability. A manufacturer of these products is faced with either scaling back production during periods of slack demand or holding product in inventory. This type of variability is well suited to batch manufacturing, as it is a simple matter to adjust production by either operating the facility on fewer shifts or manufacturing a different product that is in higher demand. Continuous unit operations do not usually have much operating flexibility. For example, a continuous distillation column has an efficient range of production that is set by the hydraulic limitations of flooding at the high end and weeping at the low end. Operations outside these limits are not feasible. In multiproduct facilities, each product is manufactured in campaigns that are sized to use the equipment efficiently. The campaign size represents a trade-off between the cost of changing over from one product to another and the size of the inventory created during the campaign.

Products that have repeated process steps may also require less equipment to produce in a batch process. For example, a process that isolates solids in several steps may be less expensive to manufacture if one set of filtration and drying equipment is reused, or shared between these steps. A multistep batch synthesis may be able to share the same reactor between each synthesis step. A continuous process will normally require dedicated equipment for each process step.

TABLE 2.1
Selection of Batch vs. Continuous Processes

Reasons for Batch Operations	Reasons for Continuous Operations
Small volume of production (production typically <500,000 kg/yr)	Large volume of production
Variability in production rate	Steady production rate
Reuse of equipment (shared equipment)	Dedicated-use equipment (single product use)
Multiproduct operation	Single-product operation
Process variables subject to adjustment (uncertainties in the reactivity or potency of raw materials)	Invariable process conditions (minor uncertainties in the reactivity or process is sufficiently robust)
Many isolation steps	Few purification steps
Lot integrity required	Lot integrity arbitrary or not required

Certain process chemistries, such as the synthesis of large organic molecules, are sensitive to conditions that may be beyond the control of the operator. As we will see later in this chapter, the synthesis reactions of polypeptides do not proceed at the same rate for each step. As a result, the operator must be able to easily adjust the reaction conditions to compensate for the change in reaction rate. It may not be feasible to adjust the processing time of continuous operation steps outside of a narrow range because of the effect on subsequent steps. For example, unusually long processing time for one step may cause the process to run empty on downstream operations.

Processes that have a large number of isolation steps, particularly in the isolation and drying of solids, are more expensive to process in continuous operations. For example, continuous filtration equipment is typically more expensive than batch filtration equipment (rotary vacuum filters or belt filters have moving parts, while Neutsch filters or plate-and-frame filters do not.)

Finally, certain types of processes that require a rigorous maintenance of batch integrity (such as the production of drug products) may be more easily done in batch operations, where a certain number of batches is blended together to make a finished lot of raw material. It is a straightforward task to document each lot of raw material that makes up each lot of finished goods.

2.2 BATCH DYNAMICS

Batch chemical manufacturing is by nature a dynamic process. For example, a batch chemical process may involve charging reactants to a reactor, stirring the reactants while heating to a desired temperature, adding the remaining reactants, holding at a specified temperature for a period of time, then cooling to a final temperature over a certain profile to produce a crystalline suspension. The process chemistry may be strongly affected by the dynamic steps of heating and cooling. For example, the specific cooling profile of a crystallization process (holding at a specified temperature for a period of time then reducing the temperature at a certain linear rate) may be selected to produce a desired particle size distribution of the product. However, the dynamic nature of batch processing may be difficult to scale up to larger batch sizes, as the rate of heating or cooling is proportional to the surface area shared by the vessel and the batch, while the amount of energy liberated or consumed by a reaction is proportional to the mass (or volume) of the batch.

Certain batch operations may be less expensive to do than continuous operations. For example, a ternary system consisting of a low-boiling impurity, an intermediate product, and a high-boiling impurity may require two continuous distillation columns in series but only one batch column. The continuous columns may be operated in such a manner that the first column produces a distillate stream consisting of the low-boiling impurity and the desired product and a bottoms stream consisting of the high-boiling impurity. The second column has as its feed stream the distillate from the first column and produces a distillate stream of the low-boiling impurity and a bottoms stream of the desired product.

On the other hand, the batch column may be operated to produce successive distillate cuts of the low-boiling impurity and the desired product, with a pot residue of the high-boiling impurity. Although batch chemical processing is widespread throughout many different facets of the chemical processing industry, this chapter focuses on specific examples of the pharmaceutical industry and the food industry.

2.3   INDUSTRIES THAT USE BATCH PROCESSING

Batch and continuous operations are not absolutely restricted by industry; however, high-volume industries such as petroleum refining typically use continuous operations, while low-volume industries or industries using specialized types of chemistry typically emphasize batch operation over continuous operations. Industries in which batch operations are more common include pharmaceutical, specialty chemical, and food. These industries are quite diverse, encompassing a very large number of products. The specialty chemical industry is particularly diverse and difficult to categorize. Two examples of specialty chemical products are discussed here: high-purity chemicals and cosmetic pigments. The remainder of this chapter discusses batch processing in the pharmaceutical industries and introduces operational practices in the specialty chemical and household chemical industries. It concludes with a brief overview of batch processes in the food industry.

2.4   BATCH PROCESSING IN THE PHARMACEUTICAL INDUSTRY

The pharmaceutical industry had U.S. sales in 2002 of $192 billion. Research and development expenses for that year were $32 billion.[2] The pharmaceutical industry employs a number of special batch unit operations in the manufacturing of products. Many of the new products being developed are compounds that are active in very small doses, have process chemistries that use these special batch unit operations, or both. Batch manufacturing techniques are commonly used in these products. A description of the life-cycle of a drug product and a specific example of a synthetic biochemical product are presented below.

2.4.1   LIFE-CYCLE OF A DRUG PRODUCT

An innovator company may devote many years of research and many hundreds of millions of dollars to discovering and developing a new drug. On average, the total timeline required to navigate a medication from the researcher's laboratory to the patient is about 10 to 15 years.[3] DiMasi et al.[4] found that the average cost of developing a new drug (new molecular entity) is $403 million in actual dollars, and $802 million if the time between discovery and marketing is accounted for (the product may not be sold until approval has been granted). Despite these efforts, the vast majority of all new candidate drugs that are discovered fail to reach the marketplace.

Figure 2.1 shows the typical life-cycle of a new drug product. Pharmaceutical manufacturing can be broadly classified into two phases: (1) *product development*, the drug discovery and clinical testing phase; and (2) *process-development*, the synthesis of batch operations, scale up, and industrial production phase. It can be inferred from Figure 2.1 that, after molecular discovery, drug development is the critical time-to-market path. Patents may be filed for promising new drugs right after molecular discovery. The total patent life of a new drug is approximately 20 years, and the clock starts ticking right after drug discovery, several years before the product goes to market. After the patent of a drug expires, other companies can copy its formula and sell generic versions.

A candidate drug originates from knowledge of the pharmacology and biochemistry of the disease or medical condition to be treated. A candidate molecule is proposed and then studied under laboratory conditions (discovery). Some preliminary animal testing (preclinical testing) may be done to establish toxicity and efficacy. The innovator may file to patent the drug candidate. At the same time, early development work (batch process synthesis) is done on the chemical process to manufacture the drug. Numerous routes to synthesize (or extract the drug from natural sources) may be evaluated and discarded due to process economics, quality, safety, or environmental concerns (sequence selection → route-selection → recipe development). A final process is then devised, and an impurity profile is obtained. Formal *preclinical studies* to demonstrate the safety and biological activity of the drug in laboratory and animal studies are conducted. An *Investigational New Drug Application* (INDA) for the candidate drug is filed with the Food and Drug Administration (FDA) in order to conduct clinical studies in humans. After approval of the INDA, *clinical trials* (phases I to III) are

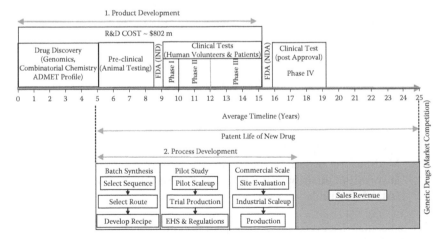

FIGURE 2.1 Life-cycle of a new drug product (total patent life = 20 years).

performed in order to test the efficacy of the drug on human subjects. At the same time, process development by pilot testing, Environmental Health and Safety (EHS) studies, and trial productions are set up. The process is then filed for New Drug Approval (NDA) at the FDA. When the FDA approves the NDA, the new medicine becomes available for physicians to prescribe. The FDA-approved drug manufacturing recipe is finally scaled up to an industrial production scale in order to meet market demands. The pharmaceutical company must continue to submit periodic reports to the FDA, including any cases of adverse reactions and appropriate quality-control records. The next sections describe in detail the activities involved in each of the phases (i.e., product and process development phases).

2.4.1.1 Pharmaceutical Product Development

Figure 2.2 illustrates the following stages involved in development of a new drug: (1) discovery, (2) preclinical trial, (3) Investigational New Drug Application (INDA), (4) clinical trials, (5) New Drug Application (NDA) for FDA approval, and (6) follow-up evaluations (post-approval clinical trials). Each of these stages is discussed in the following paragraphs. (See Table 2.2.)

2.4.1.1.1 Drug Discovery

Pharmaceutical companies constantly engage in the process of discovering new chemical substances with desired pharmacological properties. Drug discovery and development are creative, complex, and highly regulated processes.[5] The process of drug discovery and development is time consuming and expensive, and the rate of success is low. Only five in 5000 compounds that enter preclinical testing make it to human testing. Finally, only one of these five tested in people is approved. Therefore, pharmaceutical companies invest significant resources in new technologies (e.g., recombinant DNA, genomics, macromolecule synthesis) and developing alliances with research organizations to improve the discovery rate of new chemical entities and molecules. Drug discovery usually involves the following three steps:

1. Selection and validation of target drugs (enzymes, regulatory proteins, or other bioactive molecules) that are related to the disease of interest
2. Design of appropriate biochemical and biological assay models to screen libraries of compounds (from natural resources or synthesis labs) that interact with the selected biological targets in order to discover lead compounds
3. Modification of lead structures in order to optimize efficacy and the adsorption, distribution, metabolism, excretion, and toxicology (ADNET) profile

After molecular discovery, the critical path in time-to-market is clinical testing. Table 2.2 highlights the major activities involved in clinical testing during the drug development phase.

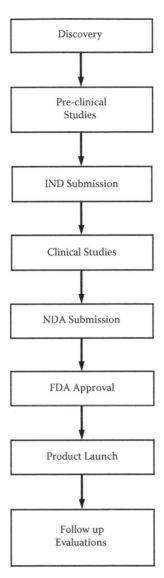

FIGURE 2.2 Development process for a new drug.

2.4.1.1.2 *Preclinical Studies*

A pharmaceutical company conducts laboratory and animal studies to determine the biological activity of the compound against the targeted disease, and it evaluates the compound for safety. These tests take approximately 3.5 years.

TABLE 2.2
Drug Development: Clinical Trials

—	Preclinical Testing	File IND at FDA	Phase 1	Phase II	Phase III	File NDA at FDA	FDA	Total Test	Phase IV
Years	3.5	—	1	2	3	—	2.5	12 total	—
Test population	Laboratory and animal studies	—	20–80 healthy volunteers	100–300 patient volunteers	1000–3000 patient volunteers	—	—	—	—
Purpose	Assess safety and biological activity	—	Determine safety and dosage	Evaluate effectiveness; look for side effects	Verify effectiveness; monitor adverse reactions from long-term use	—	Review process and approval	—	Additional post-marketing tests as required by FDA
Success rate	5000 compounds evaluated	—	Five enter trials	—	—	—	One approved	—	—

Source: Adapted from Wierenga, D.E. et al., *Phases of Product Development*, Pharmaceutical Manufacturers Association, Office of Research and Development (http://www.allp.com/drug_dev.htm), accessed September 11, 2003.

2.4.1.1.3 Investigational New Drug Application

After completing preclinical testing, the company files an INDA with the FDA to begin testing the drug in humans. The INDA contains information on the results of previous research, including any toxic effects found in animal studies and how the drug is thought to work in the human body.

2.4.1.1.4 Clinical Studies

Clinical trials consists of three phases:

- In phase I, studies are done to establish the human toxicity of the candidate drug and related impurities and to determine dosage requirements. The studies also determine how a drug is absorbed, distributed, metabolized, and excreted, as well as the duration of its action. These tests involve about 20 to 80 normal, healthy volunteers and take about a year to complete.
- In phase II, a candidate drug that passes the toxicology studies must then be evaluated for dosage, efficacy, and side effects of short-term use among a larger patient population of approximately 100 to 300 volunteer patients (people with the targeted disease). This study takes approximately 2 years.
- In phase III, if the short-term results are satisfactory, the efficacy of the drug and its safety over the long term are then verified on a patient population of several thousand volunteers. Phase III studies gather precise information on the effectiveness of the drug for specific indications, determine whether the drug produces a broader range of adverse effects than those exhibited in the small study populations of phase I and II studies, and identify the best way to administer and use the drug for the purpose intended. This phase lasts about 3 years and usually involves 1000 to 3000 patients in clinics and hospitals.

2.4.1.2 New Drug Application

The regulatory approval process for drugs is normally established by law in each country. Regulation of the manufacture and distribution of drugs in the United States is administered by the FDA. Regulatory authority of the FDA was granted under the Federal Food, Drug, and Cosmetic Act of 1938. The FDA recognizes two broad classes of drugs: new and generic. The regulatory approval process is more demanding for new drugs. For a new drug to be marketed in the United States, an NDA must be submitted to the FDA. The NDA contains the following information:[5]

- Documentation of investigations showing that the drug is safe and effective for the intended use
- A complete list of components of the drug
- A full statement of the composition of the drug

- A full description of the methods, facilities, and controls for the manufacturing, processing, and packaging of the drug
- Samples of the drug and components
- Specimens of labels proposed for the drug

An NDA may be tens of thousands of pages long. The innovator may be required to include all clinical data, including adverse events (side effects) in the NDA. The FDA may audit the innovator's proposed manufacturing facility and inspect the manufacturing unit, quality control labs, warehouses, and other facilities. The FDA may audit the manufacturing and testing procedures and documentation, validation studies of the analytical methods and key process parameters, documentation of the processing and packaging equipment for the process, and control of the storage and handling of labels. All drugs must be manufactured under the current good manufacturing practices (GMPs) as established by current regulations. Documentation of the manufacturing and testing is extremely important; each lot of drug produced must have written records that document the steps followed during manufacturing and each lot of raw material and each batch or lot of intermediate used. All testing, including intermediate and final lot testing, must also be documented.

2.4.1.3 Approval

The innovator may begin selling the drug when approval has been granted. The innovator normally begins stockpiling the drug in advance of the date of approval in order to generate revenue from the product as quickly as possible. This means that the manufacturing facility must be ready some months prior to the expected approval date. All equipment must be installed, validated, and cleaned. All employees must be hired and trained, and all procedures must be written, reviewed, and approved. Only a small number of candidate drugs achieve approved status. Kaitin and Cairns[6] reported that during the period of 1999, 2000, and 2001 the total number of new drugs approved by the FDA was 86: 35 in 1999, 27 in 2000, and 24 in 2001. Of these 86 drugs, 82 were New Chemical Entities (NCEs). The NCEs had a mean clinical phase time of 5.5 years, a mean approval phase time of 1.4 years, and a mean total development time of 6.9 years. Figure 2.3 shows the average clinical time and approval time for biopharmaceutical drugs for the last two decades.[7]

2.4.1.4 Post-Approval Clinical Studies (Phase IV)

Experimental studies and surveillance activities are also undertaken after a drug is approved for marketing. Clinical trials conducted after a drug is marketed (referred to as phase IV studies in the United States) are an important source of information on as yet undetected adverse outcomes, especially in populations that may not have been involved in the premarketing trials (e.g., children, the elderly, pregnant women), as well as the long-term morbidity and mortality profile of the

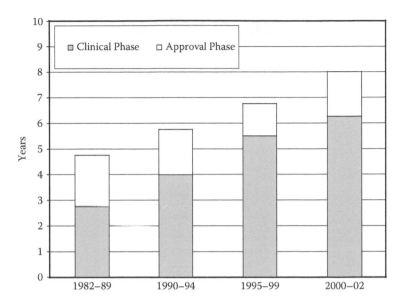

FIGURE 2.3 Typical clinical phase and approval phase time for biopharmaceutical drugs. (Adapted from Tufts Center for the Study of Drug Development, *Tufts CSDD Outlook: 2004*, http://csdd.tufts.edu, 2004.)

drug. Regulatory authorities can require companies to conduct phase IV studies as a condition of market approval. Companies often conduct post-marketing studies even in the absence of a regulatory mandate.

2.4.1.5 Pharmaceutical Process Development

Figure 2.4 outlines the typical work flow of process development in a pharmaceutical company. Pharmaceutical process development primarily consists of three stages: (1) batch process synthesis (sequence selection → route selection → recipe development → environmental health and safety measures), (2) pilot studies (pilot scale-up → trial production → regulatory compliance), and (3) industrial production (site evaluation → industrial scale-up → production). The different process stages are developed parallel to product testing after drug discovery. Batch synthesis is done parallel to preclinical animal testing, pilot studies are done during clinical trials, and industrial production begins after FDA approval. The next sections discuss the various stages of the drug development process.

2.4.1.5.1 Batch Synthesis (Preclinical Process Development)

Development of the manufacturing process for the drug candidate closely follows the regulatory approval process for the drug. During the first stage of process development, the pharmaceutical company evaluates the economics of the candidate drug and estimates a budget for the manufacturing process and the cost of

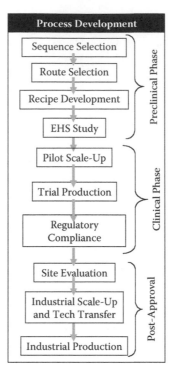

FIGURE 2.4 Typical process development work flow in a pharmaceutical company.

the new manufacturing plant. Various routes to produce the drug are proposed and evaluated in terms of the expected yield, expected raw material and labor costs, and indirect costs of the manufacturing facility. In addition, the proposed processes are evaluated in terms of process safety and hygiene and compliance with environmental regulations. Demonstrations of the manufacturing process are tested via bench-scale runs, and impurities are identified and characterized. The most promising route is selected, and product specifications are established. Small-scale laboratory experiments are conducted to establish operating conditions. At this point, the INDA is filed, and phase I clinical studies begin.

2.4.1.5.2 Pilot Study (Process Development During Clinical Studies)

The development of the manufacturing process is well underway by the time the phase I studies are complete. The process development effort becomes focused on scaling up to the proposed manufacturing scale when the phase I study begins. Pilot plant studies are conducted in order to obtain the engineering data required for design of the manufacturing facility. Analytical methods are developed, and in-process specifications are determined. Key process variables, operating conditions, and analytical methods (for in-process and finished lot testing) are validated to ensure quality control throughout the entire process. Cleanout procedures are developed and validated. Documentation of the manufacturing process (which includes batch records and work instructions) is completed. All of these tasks must be completed prior to submitting the NDA. Documentation of these activities becomes part of the NDA.

2.4.1.5.3   Industrial Production (Process Development After NDA)

After the NDA is submitted, the capital project to build the manufacturing facility (or to make modifications to an existing shared facility) is written and approved. Detailed engineering designs are completed and documented. Equipment is purchased, installed, tested, and qualified. Installation, operational, and performance qualifications are conducted and documented. The FDA makes an on-site inspection and notes any deficiencies that must be corrected.

2.4.2   OPTIMIZATION

2.4.2.1   Challenges

The main challenge in the pharmaceutical product/process development work flow is getting the new product to market quickly and cost effectively. While it continues to cost approximately $800 million and take 10 to 15 years to bring a new drug to market (see Figure 2.1), 90% of drugs produce revenues of less than $180 million, making it difficult to justify such a large research and development (R&D) investment.[8] The long product-development cycle from patent to market can take more than a decade, which means pharmaceutical companies often find that their products have only a few years to be revenue producers before the exclusive formula goes public. The two factors governing the profitability of a drug are (1) market exclusivity or available patent life for the product, and (2) market demand for the product.

2.4.2.1.1   Marketing Exclusivity

The innovator normally has 20 years of patent protection for a novel drug under current U.S. law; however, a significant portion of that period is consumed by the development and clinical trials of the drug. As a result, only about 8 to 10 years of marketing exclusivity are available. During this period, the innovator faces competition from existing, but different, drugs. After the patent expires, generic copies of the drug may be sold by other companies, and the innovator must now compete on price with essentially identical drugs. A generic manufacturer does not face as stringent approval requirements as the innovator and may obtain approval of an Abbreviated New Drug Application (ANDA), which requires information identical to that on the NDA for the new drug except that safety and efficacy data are not required. The generic manufacturer must, however, show that the effects of the generic drug are similar to those of the innovator drug (typically, similar bioequivalence). As a result of the regulatory status of generic drugs, the risk assumed by the generic manufacturer is relatively low (e.g., the drug is recognized as safe and effective, although the generic manufacturer must show bioequivalence). On the other hand, the generic manufacturer must compete on price; however, the generic manufacturer has more time to devote to optimizing the manufacturing process and as a result may be able to devote resources to develop a more economical manufacturing process than the innovator.

2.4.2.1.2 Demand for Product

The innovator normally faces an increasing demand for the new drug after approval has been granted and the drug wins acceptance in the marketplace. In the early years of the life of the product, the innovator may experience considerable pressure to expand the manufacturing capacity of the drug; however, if the market share is eventually eroded by generic manufacturers or by new innovator drugs, the demand for the original drug will decline. The original innovator now faces pressure to reduce manufacturing costs, which may include elimination of slack production time. Batch processing offers such flexibility during periods of decreasing demand. Continuous equipment may normally be "turned down" to a somewhat limited extent; however, batch operations may be operated at reduced capacity simply by making fewer batches. This can be done by reducing the number of batches produced, by reducing the hours of operation, or by shutting done some parallel trains of equipment. If the drug is produced in a multiproduct manufacturing facility, the scheduled time for the drug may be reduced and the equipment used to manufacture new drugs or other existing drugs. In other words, the product mix that is produced may be adjusted to meet changing market demands.

2.4.2.2 Optimization of Product Development

Improvements in the drug development process can dramatically reduce the total cost of new drugs. Table 2.3 shows the impact of drug development time and clinical success rates on total cost.[9] A company can save approximately $200 million by reducing the development time by 41% or by increasing the success rate by 31%.

2.4.2.2.1 Optimal Drug Discovery

The key innovation lies in the drug-discovery phases in which pharmaceutical companies invest significant resources in new technologies (e.g., recombinant DNA, genomics, macromolecule synthesis) and developing alliances with research organizations to increase the discovery rates of new chemical entities/molecules. The pharmaceutical industry has begun employing the newest technologies available, e.g., high throughput screening, biochips, combinatorial synthesis, and biochemoinformatics to accelerate the drug discovery and development process and reduce the cost of drug development.

TABLE 2.3
Impact of Drug Development Time on Total Cost

Action by Drug Manufacturer	Impact on Cost	
	$100 Million	**$200 Million**
Reduce time by	18.9%	41.3%
Or improve clinical success rates by	25.2–25.6%	30.4–31.7%
Or cut out-of-pocket preclinical costs by	29.8%	59.6%

Note: Current average cost of drug development is approximately $802 million.

Source: DiMasi, J.A., *Pharm. Econ.*, 20(53), 1–10, 2002.

2.4.2.2.2   Optimal Clinical Trials

In phase I, the manufacturing process is frozen at this time. Process changes that alter the impurity profile may require additional clinical studies to establish the toxicity of the drug due to the revised process. Laws vary by country, but a manufacturer may find that all impurities present in amounts greater than 0.1% must be characterized (the chemical structure identified), and impurities present in amounts greater than 0.15% may require toxicology studies. Shifts in the relative amounts of known impurities are cause for further investigation and corrective action, while the presence of new impurities may be unacceptable without additional clinical studies. As a result, the innovator faces a trade-off between the amount of time and expense required for process development leading to a true optimum manufacturing process (with the risk that the candidate drug might not pass the approval process) and a shortened development and approval time using a workable process. In many cases, this trade-off leads to batch chemical processing. Most laboratory and pilot plant work begins with batch operations, as batch operations are typically easier to set up on the bench than continuous operations. The innovator may also be constrained by process chemistry that is not amenable to continuous processing. For a discussion of some of the optimization techniques that may be employed during this stage, such as multi-objective optimization, and stochastic optimization (to obtain the value of additional research), see Diwekar[10] and Chakraborty and Linninger.[11,12]

Under certain conditions (such as the treatment of a fatal illness), phase II and phase III studies may be combined. The clinical studies must be based on good science. For example, the effectiveness studies may be double-blind experiments, in which neither the patient nor the clinician knows if the patient receives the candidate drug, a placebo, or an approved drug that is known to be effective. In some cases, it is not considered appropriate to administer a placebo (for example, in the investigation of drugs for lethal medical conditions). The candidate drug must have a statistically significant improvement in patient response relative to the placebo and at least a statistically favorable response in comparison with an existing approved drug.

2.4.2.3 Optimal Process Development

Key to achieving an accelerated time-to-market are workflow unification and compression and consolidating the time from discovery through pilot-scale and full-scale manufacturing. Recently, companies have sought to unite and compress the work flow because studies indicate that process optimization that keeps the same synthetic route can yield manufacturing savings of as much as 40%. Process optimization that considers changes in synthetic route (i.e., optimal batch recipes) can result in savings of up to 65%; however, it should be noted that after FDA approval neither the drug nor its operating procedures may be altered without a new drug approval process. Batch operating procedures should, therefore, be optimized at the conceptual level, as later improvements require expensive FDA reapproval. The following case study demonstrates typical operational sequences used in pharmaceuticals manufacturing.

2.4.3 CASE STUDY: SOLID-PHASE PEPTIDE SYNTHESIS

Certain types of chemical reaction syntheses are particularly well suited to batch processing. The synthesis of polypeptides by the Merrifield method is an example of such a process.[13] Peptides are an important class of drugs composed of sequences of amino acids. An amino acid is an organic chemical containing both an amine group ($-NH_2$) and a carboxylic acid ($-COOH$). Examples of amino acids include lysine, glycine, leucine, and aspartic acid. An example of a peptide is human insulin. The peptide bond between successive amino acids occurs between the amine and carboxylic acid and has the structure $-COHN-$. The amino acids form a chain of n amino acids with the following structure: $H_2N-AA_n-COHN-AA_{n-1}-COHN-...-COHN-AA_2-COHN-AA_1-COOH$. The peptide has an amine end (at the last amino acid in the sequence) and a carboxylic acid end (at the first amino acid in the sequence). Peptides are similar to proteins in that they are both composed of sequences of amino acids. The chief distinction between peptides and proteins is that peptides are so small (typically with a molecular weight less than 10,000). Peptides are typically clinically active in doses of micrograms to milligrams; however, some very notable exceptions to this dosage have occurred. The high potency of these drugs typically means that demand is small, on the order of kilograms to hundreds of kilograms per year. Again, some notable exceptions to this demand can be found.

The Merrifield[13] method of synthesizing peptides employs a sequential assembly of the peptide by coupling individual amino acids to form chains of the desired sequence that are linked to a solid-phase support. After synthesis of the molecule is complete, the crude peptide is cleaved from the support. The crude peptide is purified, typically by using liquid chromatography. Additional purification may be done to remove pyrogens by using membrane technology. The final, purified peptide is then isolated from solution, typically by using lyophilization. The sequence of operational steps is depicted in Figure 2.5.

2.4.3.1 The Main Reaction Step

The first step in the synthesis of a peptide is to substitute the C-terminal amino acid to a solid-phase support using a suitable linker. (The C-terminal amino acid contains the free carboxylic acid, while the N-terminal amino acid contains the free amine.) The solid-phase support is typically polystyrene resin beads, crosslinked with divinylbenzene for strength and rigidity. The polystyrene beads are typically about 100 μm in size initially, but the beads grow as the length of the intermediate peptide increases. The amino acids are protected at the amine group in order to improve selectivity of the reaction. Typical protecting groups include Boc (*t*-butyloxycarbonyl) and Fmoc (9-fluorenylmethoxycarbonyl). After the initial substitution, the resin is washed to remove all traces of unreacted amino acid. Each amino acid addition in the synthesis includes the following steps:

1. Deprotect the N-terminal amino acid in the sequence to produce a free amine.
2. Wash the resin.
3. Activate the next amino acid in the sequence.
4. Couple the activated amino acid to the free amine on the peptide fragment.
5. Wash the resin.

Testing is then done to ensure that all free amines have been coupled to the amino acid. This sequence is repeated for each amino acid in the peptide.

2.4.3.1.1 Deprotection

The synthesis process typically occurs in an analog of a stirred-tank reactor. The specific reaction used to deprotect the N-terminal free amine depends on the protecting group. The Boc group is normally removed with a strong acid (such as 50% trifluoroacetic acid), while the Fmoc group is normally removed with a weak base (such as piperidine). This reaction is normally time sensitive. Excessive contact time between strong trifluoroacetic acid and the intermediate peptide may lead to degradation of the peptide, and therefore low yields or high levels of impurities. The typical reaction is to stir a suspension of peptide–resin in the acid or base for a specified amount of time. The solution is then removed from the resin by filtration.

2.4.3.1.2 Washing

The peptide–resin is then washed to remove residual acid or base. The washing may be done by reslurrying the peptide–resin in a clean solvent and then filtering. The wash step may be repeated as needed.

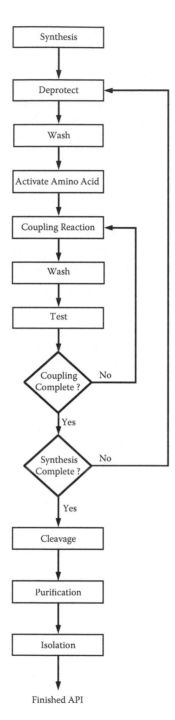

FIGURE 2.5 The Merrifield process for solid-phase peptide synthesis.

2.4.3.1.3 Amino Acid Activation

The next amino acid to be added to the intermediate peptide is then dissolved in solvent and activated using a suitable reagent. The activated amino acid may be in the form of a symmetric anhydride of the amino acid or an active ester. The amino acid may be activated in a second vessel under refrigerated conditions, or activation may be done *in situ* with the resin suspension.

2.4.3.1.4 Coupling

The peptide–resin is suspended in a solvent with stirring, then the activated amino acid is added. The batch is allowed to stir for 1 to 24 hours to allow the reaction to proceed to completion. The batch is then sampled and tested for residual free amine. If the batch fails, the reaction may be continued or repeated as necessary. After the reaction is complete, the batch is filtered and the slurry washed to remove residual amino acid. The process is repeated until the synthesis is complete. Peptides with up to about 40 amino acids are typically synthesized in this manner. A number of innovations in the process technology have increased the speed and scale of the process and reduced solvent usage. Displacement washes are more efficient than slurry washes. The geometry of the reaction may be difficult to scale using conventional stirred-tank reactor technology with integrated filter media, because the resin is very soft and has a tendency to pack during the filtration steps. This means that the resistance to liquid flow through the resin bed may become excessive. As a result, shallow beds are sometimes mandatory to minimize contact time during time-sensitive steps. The properties of the peptide–resin may change during the course of the synthesis. The resin particle size tends to increase as the length of the intermediate peptide increases. This means that the resin volume increases over the course of the synthesis. The amount of required reaction time may change during the course of a synthesis. As the intermediate peptide grows, coupling reactions may take longer to go to completion. In addition, certain reactions may always be difficult. Reactions involving secondary amines are an example of difficult coupling reactions. As a result, the residence time in the reactor may vary during the course of a synthesis batch. Batch reactor technology is sufficiently flexible for this task, as an operator can simply allow the reactor to stir until the reaction is complete or the step is repeated.

2.4.3.1.5 Cleavage

The next phase of the process is to cleave the peptide from the resin support. The exact nature of the process depends, again, on the type of amino acid chemistry used. The linker used in Boc amino acid chemistry must be stable in the strong acid used during the synthesis deprotection steps. As a result, an extremely strong acid, such as liquid anhydrous hydrogen fluoride, is used to cleave the peptide from the resin. The linkers used in Fmoc chemistry are stable in a base but labile in a weak acid. As a result, a weaker acid, such as dilute trifluoroacetic acid, is used. A second treatment in more concentrated acid may be used to remove any additional protecting groups from other potentially reactive sites on the peptide. In either case, the cleavage reaction is done in a stirred-tank reactor with a high degree of temperature control. The cleavage reactions may be exothermic, and peptides normally have very limited stability to moderate (room) temperatures under acidic conditions. As a result, the yield and purity of these reactions are very sensitive to both reaction temperature and reaction time; therefore, the acid must then be immediately neutralized or removed from the peptide as soon as the reaction is completed. This may be done by vacuum distillation or by the addition of another chemical species that is protonated by the acid, such as ether. The crude peptide is then recovered by filtration and is washed to remove residual acid.

2.4.3.1.6 Purification

The third phase of the process is to purify the crude peptide. Typical impurities that may be present include deletion peptides (one or more amino acids is missing), insertion peptides (one or more extraneous amino acids is present), or peptides missing one or more functional groups (such as a carboxylic acid). These impurities are chemically very similar to the desired product and as a result are very difficult to remove. One unit operation that has proved to be successful is high-pressure liquid chromatography (HPLC), which uses a column packed with a stationary phase, a high-pressure pump, and a detector. (In reverse–phase HPLC, the stationary phase consists of porous silica particles with a hydrocarbon ligand attached.) The size of the silica particles may be selected based on a trade-off between the pressure drop required to achieve the required flow of mobile phase and the number of theoretical plates of separation required for the purification. Smaller packing (10-μm average size) may be able to achieve more than 10,000 theoretical plates in a 50-cm bed but at a pressure drop of over 300 psi. The chromatography operation consists of loading an aliquot of peptide solution onto the column, then eluting the peptide with a mixture of organic and aqueous solutions. Separation occurs according to the hydrophobicity of each species. Impurities that are more hydrophobic are more strongly retained on the stationary phase and, as a result, elute at a slower rate than less hydrophobic species. The operator can monitor the separation using a suitable detector and, at the proper time and detector response, collect one or more product fractions that contain the purified peptide. The purification step may be operated in a more efficient manner by slowly increasing the concentration of organic solvent in the mobile phase over time, a process known as gradient chromatography. Certain types of large-scale chromatography technology have been developed to approximate semibatch or continuous operations; however, the typical chromatography described above, especially using gradient elution, is performed as a batch unit operation. Additional purification using high-performance filtration (such as ultrafiltration) may be done to remove pyrogens (fragments of microorganisms that can produce fevers if injected into a patient). An ultrafilter consists of a stirred cell containing a membrane that has a porosity sufficiently large to pass the peptide solution but sufficiently small to retain the pyrogens. The stirred cell prevents polarization (i.e., a layer of peptide from blinding the membrane).

2.4.3.1.7 Isolation

The final step is isolation of the peptide from solution. Several separate steps may be performed in the final isolation. First, any residual organic solvent must be removed from the product fractions by vacuum distillation or diafiltration, in which the peptide is retained on the filter membrane but solvent passes through. Fresh water or a sterile aqueous buffer solution is continuously added to keep the peptide in solution. The final isolation may then be done by lyophilization. The peptide solution is frozen to a low temperature (−50°C), then a vacuum is applied. A typical lyophilizer vacuum is less than 100 mTorr (0.1 mmHg) absolute pressure; however, the pressure is fixed by the operating temperature. Heat is applied to the frozen peptide solution, and the water is removed first by sublimation and then desorption. The lyophilized peptide is then packaged under asceptic conditions. Lyophilization may be done in bulk or in single-dosage vials.

2.4.3.1.8 Summary

The peptide manufacturing process illustrates a case in which many different batch unit operations are used. Process chemistry that involves a repetitious series of heterogeneous reactions and washes and high-performance purification and isolation is well suited to batch manufacturing technology but poorly suited to continuous processing. The typical small demand for peptides also fits with the philosophy of using batch technology for small-volume manufacturing.

One other important regulatory requirement is that written records must be maintained for each lot of drug produced. These records include all raw materials and equipment used and documentation of all process steps used to make the drug. Written records are also maintained of the results of the quality testing of all raw materials, intermediates, and finished goods produced. The written records are necessary in order to rapidly obtain a list of all raw materials and a complete set of manufacturing records in case of an inquiry about a particular lot of a drug. A lot of finished goods that does not have the complete set of required documentation is considered adulterated under the law and is subject to recall from the market. In a batch manufacturing process, records are kept of each batch of drug manufactured (and of each batch of intermediate produced in a multistep manufacturing process, such as the manufacturing of peptides). One or more batches of finished drug are then combined and blended together to make a lot of drug product.

2.5 BATCH PROCESSING IN THE SPECIALTY CHEMICALS AND HOUSEHOLD CHEMICALS INDUSTRIES

The specialty chemical industry is very diverse. Generally, specialty chemicals are designed for specific applications or customers and are typically produced in small volumes. Categories include adhesives, catalysts, coatings and paints, electronic chemicals, industrial gasses, plastic additives, water management chemicals, and lubricants. The total U.S. market for specialty chemicals was $110 billion in 2002.[14] The household nondurables industry includes household products (such as cleaners and detergents) and personal care products (hair-care products, color cosmetics and fragrances, skin-care products, deodorants, oral-care products, shaving products, sun-care products, nail products, and hair colorants). The global sales of this industry are estimated at $150 billion, and U.S. sales were estimated at $75 billion in 2002.[15] This section presents one example for each of these industries: manufacturing high-purity chemicals and manufacturing cosmetics.

2.5.1 HIGH-PURITY CHEMICALS

Potassium sulfate (or sulfate of potash) has a wide variety of uses.[16] A relatively low-purity variety of this chemical may be used in the manufacture of fertilizers; however, higher purity and other special properties (such as specified particle size) are required for uses in construction materials (gypsum wallboard), ordnance, analytical reagents, glass making, and drugs. Low-purity potassium chloride may be used as a fertilizer and for ice control. Higher purity grades may be used in food (as a replacement for sodium chloride), photography supplies, buffer solutions, electrodes, and drugs. Low-purity calcium chloride[17] may be used as a raw material in the manufacture of concrete, glues, and rubber; in ice control; and for sizing fabrics. Higher purity grades and certain particle size material may be used as an antifreeze, for extinguishing fires, as a dehydrating agent, as a preservative, in foods (cheesemaking, canning, and brewing), and in drugs.

To meet higher purity specifications, additional unit operations must be performed. These unit operations normally include purification by recrystallization,[16] classification by particle size, and size reduction. Specific unit operations include the use of stirred tanks for dissolving and crystallizing the product, centrifugation for collecting and washing the product, a forced-air dryer to dry the intermediate product, classification to isolate various particle sizes, and milling to produce the finest crystals or powder. Small- to medium-scale manufacturing of the higher purity grades of these chemicals may be done in batch dissolvers, crystallizers, and centrifuges.

An operator of a chemical plant may have several sources of raw materials that meet certain specifications, and these feedstocks may have a significant variation in availability and price. The operator of the plant may select an individual feedstock for each product or may alter the manufacturing process for each product while using a single feedstock. The latter policy may have economic advantages over the use of multiple raw materials if an inexpensive grade of raw material may be used for all of the products. This reduces the number of raw materials held in inventory, allows the use of a small number of suppliers, and offers flexibility in scheduling the various products.

In order to save raw material costs, manufacturers want to recycle mother liquors from the purification process and off-specification product from the classification and size-reduction steps. One strategy is to start each campaign with all fresh raw material and produce the highest purity product first. As the levels of impurities increase, the production may be shifted to produce products with a lesser purity. When the impurities become too high to produce the least pure product, the mother liquors may be discarded and a new campaign begun.

High-purity solvents may start with an industrial grade feedstock and then be subjected to one or more distillation steps to reduce the levels of impurities to acceptable levels. Demand for individual solvents may be insufficient for dedicated continuous stills, so a train of batch distillation units may be used. The typical product cycle is to distill a batch in three portions: a waste cut containing low-boiling impurities, a product cut that meets the required specifications, and a bottoms cut containing the high-boiling impurities. The intended use of the product places restrictions upon the purity of the product. For example, a low-purity grade of ethanol may be used as a gasoline additive, but a high-purity grade is required as a solvent in drug manufacturing, and a very-high-purity grade is required for use as a mobile-phase solvent in liquid chromatography for analytical chemistry, where the purity must be sufficient to eliminate any extraneous signals in the detector. (Special grades of solvents are available that are free of extraneous substances that may absorb ultraviolet light.)

2.5.2 COSMETICS

Another industry in which batch chemical processing is used due to small volumes and specialized chemistry is the cosmetics industry. The relatively small volumes of individual cosmetic pigment products and the specialized precipitation and pigment-bonding operations favor batch manufacturing. In addition, the flexibility of batch manufacturing makes it relatively easy for a manufacturer to switch products due to seasonal or fashion variations. One widely used type of cosmetic pigment consists of a reflective metal salt precipitated onto a substrate to which a pigment may be bonded. Cosmetic pigments are then in turn used to provide color and reflectance to consumer products such as eye shadow and lipstick. The unit operations for manufacturing cosmetic pigments involve several batch unit operations, including batch reactions for dissolving the metal or metal salt, precipitating the salt onto the substrate, heat treatment to produce the desired reflectance (which may include changing the metal from a hydroxide to an oxide), high-shear dispersion of one or more color pigments into a solvent, bonding of the color pigment onto the substrate, isolation and drying of the cosmetic pigment, and blending various batches of cosmetic pigment to product a finished lot. Blending of individual batches may be done to ensure lot-to-lot consistency. These steps are summarized in Figure 2.6.

The critical variables of the precipitation step include the size and shape of the substrate, the thickness of the metal salt coating on the substrate, and the crystalline form of the metal salt. These variables can have a profound effect upon the appearance of the product; the wrong thickness of salt can lead to the formation of undesirable interference colors, while the wrong crystalline form of the metal salt can lead to a chalky appearance instead of a shiny or glittery

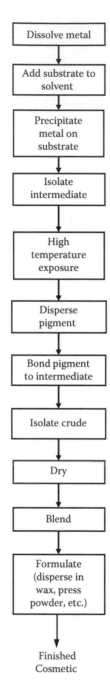

FIGURE 2.6 Flow chart for a cosmetic pigment.

appearance. Certain products also have a heat-treatment step that may be done in a rotary kiln or high-temperature oven. For example, titanium products may be exposed to a temperature of 850°C under mildly oxidizing conditions to convert the hydroxide form of the metal into the dioxide form. The dispersion and pigment bonding steps are also important because the intensity of the color may be affected, or attrition of the pigment during the dispersion step may cause a dull appearance of the color.

Cosmetic pigments are also suitable for batch manufacturing because the demand for individual colors is highly variable. A plant operator may schedule colors in campaigns to meet current demands and may select the order of manufacture within certain groups of colors in order to minimize changeover costs. For example, it is not necessary to perform a thorough cleaning of the equipment when changing from a white pigment to a black pigment, but a very thorough cleaning is necessary when changing from a black pigment to a white pigment. As a result, campaigns may be organized into color groups such as white–yellow–orange–brown or white–blue–black in order to minimize the time and expense of cleaning out between individual colors.

2.6 BATCH PROCESSES IN THE FOOD INDUSTRY

Batch processes are widely used in the food industry. Although sometimes supplanted by continuous processes, batch processes are still very important and sometimes irreplaceable in the preparation of certain food items or ingredients, among them candy[18] and modified starches.[19,20] Batch processing in the food industry has some unit operations, such as panning, whipping (emulsifying), and the pulling of taffy, that are not too often used in other industries, as well as other operations, such as blending, tabletting (including granulation), and lyophilization, that are common in other industries (such as the pharmaceutical industry). As is the case in the pharmaceutical industry, regulations are in place regarding the use of approved ingredients and the cleanliness of the manufacturing facility. Products that could provide media for bacterial growth may require treatments to kill germs (pasteurization), as well as sanitary processing equipment that can be easily sterilized and which avoids stagnant areas where bacteria may grow.

Edwards[18] has pointed out that some batch operations have advantages over continuous operations. For example, the cooking of toffees depends on localized overheating to produce color and flavor (by the Maillard reaction). If localized overheating does not occur, the appropriate color and flavor are not formed. Continuous plants must have special facilities (such as specific residence times in heated vessels) to form color and flavor. The process of making toffee includes dissolving sugars in syrup and water, adding fat and skim milk, formation of an emulsion, cooking (in open saucepans to wiped-film evaporators, depending on the scale of the operation) to reduce the water to the desired level, and shaping the toffee. Shaping may be done on a cooled slab with turning to promote even cooling. Once the toffee has cooled, it may either be cut into sheets or fed batchwise into rollers that form the toffee into a rope. The rope may then be rolled into the desired thickness, cut into the desired size pieces, wrapped, and packaged. A schematic of a toffee process is presented in Figure 2.7.

Amaranth starch, unique because of its microcrystalline starch granules (1 to 3 μm in diameter) and definitely a specialty food ingredient, has been isolated using a batch process.[21] Pearled and unpearled amaranth seed was wet milled in a high-pH, batch-steeping process, and the various parts of the seed were separated by methods similar to those used in corn wet milling. The process produced 98%+ pure starch from both pearled and unpearled starch. Pearled amaranth gave a higher yield of starch (32.6%) than did unpearled amaranth (11.2%). Less germ was recovered from the unpearled amaranth (1.8% vs. 7.3% for pearled seed).

Beverages have long been the subject of batch processes. A patent described the production of ethanol and fermented beverages via a batch process;[22] a fermentable substrate was contacted with yeast cells encapsulated with a porous, semipermeable matrix, in this case an alginate gel. The biochemistry of malting was reviewed in an article that presents the major biochemical reactions occurring during malting.[23] While malting is generally referred to as a batch process, it should be noted that it really is a series of three steps (in some sense continuous), the reaction products of which are the starting materials for subsequent steps. The recovery of beer from surplus yeast was described by Roegener et al.[24] This process employs a filtration process; the recovered beer is of good quality and can be added to freshly brewed beer up to 10% by volume without adverse composition or sensory effects.

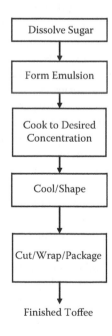

FIGURE 2.7 A toffee process.

Batch processes are common in candy manufacture and chocolate manipulation. A survey of conching systems for chocolate has been published,[25] in which the optimal features of these systems with respect to conching time, energy consumption, and rheological properties are explored. Also, the formation of difficult candy bar materials has been examined,[26] mainly with the aim of looking at various techniques of combining different textures within a single candy bar. Typically, on a small scale, sheeting and cutting performed on slabs comprise the preferred technique. On a larger scale, specifically developed slitting, spreading, and cross-cutting devices are employed.

In the dairy field, hydrolysis of lactose enzymatically using either a batch or continuous process has been detailed.[27] The lactase enzyme can be recycled in these processes, usually by ultrafiltration. The kinetics of yogurt fermentation in a continuous process has been compared to the corresponding batch process.[28] In this study, 45°C milk was inoculated with a mixed culture of *Streptococcus thermophilis* and *Lactobacillus bulgaricus* (3%) at dilution rates of 0, 0.6, 1.2, and 1.5/hr. Yogurts prepared by this method have similar characteristics to those prepared by batch inoculation; however, the continuous nature of this process reduced the manufacturing time by 10 to 15% compared to the traditional batch process. Protein–lipid interactions in processed cheese produced by both batch and extrusion cooker methods have been studied.[29] Adding melting salts and premelted cheese increased the binding of fat in the cheeses, and proteolysis was greater in the extrusion cooker method than in the batch method. Lactose-free milk has been prepared by a batch process.[30] *Saccharomyces fragilis* (40% w/v) was entrapped in 2% calcium alginate and used to completely remove the lactose from milk. Ten grams of immobilized cells removed all the lactose from 100 mL of milk in 3.5 hr. Such an immobilized preparation could be used repeatedly (up to 15 times in this study) without any decrease in enzymatic activity. A way to accelerate the production of stirred yogurt has been described.[31] This process utilizes fed-batch prefermentation and a higher initial concentration of inoculating culture. The quality of the yogurt made by this process is similar to that of batch-made yogurt. When the cooling of raw milk by batch or continuous processes was compared, it was found that the continuous-process milk had slightly lower total bacterial growth and slightly lower hydrolysis and oxidation of milk fat than the batch-process milk.[32] The milks were analyzed after 2 to 6 days for total plate count, psychrotropic bacteria, sensory score, fatty acid degree, and thiobarbituric acid (TBA) value. Also, a batch process has been used in the production of a milk substitute that has a low content of free calcium ions (which destabilize the casein micelles).[33] The composition of this material is 10 to 12% skim milk solids, 5 to 9% whey solids, 7 to 10% lipid, 68 to 80% water, 0.01 to 0.03% carrageenanates, and 0.1 to 3% calcium sequestering agent. The substitute is manufactured by sequential dissolution of the ingredients in water (a series of batch processes). The resulting dispersion shows good thermal stability despite having a large amount of whey protein.

Browning in dried egg whites has also been addressed by way of batch processing.[34] This is a major problem in the manufacture of dried egg whites due to the formation of brown pigments via the Maillard reaction between the reducing monosaccharide D-glucose and primary or secondary amines in the egg white. Usually this problem is treated by enzymatic conversion of the D-glucose to D-gluconic acid, which is not a participant in the Maillard reaction. In this particular approach, eggwhite foam was crosslinked with 0.75% glutaraldehyde in which glucose oxidase and catalase were immobilized. This enzyme preparation was able to remove >95% of the glucose from the egg white in 5 hr and could be reused at least 10 times without loss of enzyme activity.

Improvement of food flavors has also been addressed by batch processing. Shaw and coworkers[35] detailed the use of batch or continuous fluid-bed processes to improve the flavor of navel orange and grapefruit juices by removal of the bitter components. A beta-cyclodextrin polymer (seven alpha-1,4-linked glucoses in a macrocyclic ring) at 1 g polymer per 50 mL juice reduced the levels of limonin, nomilin, and naringin in grapefruit juice and limonin and nomilin in navel orange juice by about 50%. The chelating polymer could be regenerated for further batch use by treatment with aqueous alkali or ethanol. Sensory investigation revealed that the panelists preferred the reduced bitterness juices compared to the control juice. The polymer treatment was quite specific and did not affect the soluble solids, total acid, or vitamin C content of the juices but did reduce the citrus oil level by about 40%. In addition, alpha-cyclodextrin (six alpha-1,4-linked glucoses in a macrocyclic arrangement) was also effective in removing the bitter principals. Similar work was reported by Ujhazy and Szejtli[36] in which they used a cyclodextrin polymer to remove naringin from grapefruit juice in a batch process. Two grams of the polymer would bind >75% of the naringin content of a 80 mg/100 mL dispersion. The polymer could be regenerated by washing with aqueous sodium bicarbonate solution. A small initial loss of binding activity was noticed, but the polymer could be used for up to five regenerations without significant loss of binding. Continuous and semibatch processes for recovering the aroma components of apple juice have been compared.[37] The cost of the semibatch process was found to be an order of magnitude higher than for the continuous process, and it was concluded that perevaporation has the potential to become a viable alternative for the recovery and concentration of food aromas.

Batch processes are employed in the freezing of foods.[38] The Cryomix™ process (a batch process for mixing, coating, and quick freezing food materials) has been reviewed, considering cryogenics generally, and early and current applications, including individual quick freezing (IQF) as applied to fried vegetables, ratatouille, chili con carne, and coated fruits or cakes.

Batch applications have found use in the preparation of gum materials, as well. While not a typical food application, it is near enough to warrant mention. Coating of chewing gum and bubble gum can be achieved by both continuous and batch processes.[39] Small cores of gum material can initially be coated by a batch process before being introduced into the continuous coating drums.

Carbohydrate macromolecules (specialty items compared to, for instance, corn starch) have also been the target of batch processes. Xanthan gum was prepared in a repeated batch process utilizing *Xanthomonas cucurbitae* in sugar-cane juice or synthetic sucrose/salt media.[40] Fermentation in a 10-L fermenter was faster and more complete than in 1-L flasks; that is, the rate of synthesis and percent conversion of sucrose to gum was greater in the fermenter than in the flask. The efficiency of the culture was unchanged in three cycles of 50 hours each. Gum formation was greater in the sucrose/salt medium than in sugar cane juice. Another polysaccharide, pectin, has been extracted from sunflower heads by both batch and continuous processes.[41] Variables investigated in this study were solvent pH and liquid-to-solid ratio (LSR). Dried sunflower heads were ground, sieved, and extracted. Both batch and continuous processes produced a maximum pectin extraction of about 50%; however, the continuous process achieved a higher pectin yield over a broader range of LSR and pH values. The result of this is that the pH could be selected to yield a desired pectin firmness without affecting polymer yields. The lack of sensitivity to LSR gave significant advantage to the continuous process with respect to solvent levels and wastewater treatment costs.

Pasta fillings prepared via batch and continuous processes have been the subject of investigation.[42] This study was conducted to assess the effects of processing type on meat-based fillings for pasta such as tortellini. Among variables examined were total plate counts and total coliform counts, *Escherichia coli*, molds, staphylococci, and *Streptococcus faecalis* in the meat samples. The continuous process had significant advantages over the batch process, and the spices used were a significant source of bacterial contamination. Also, the use of frozen meat improves microbial quality compared to fresh meat.

The use of steam to sterilize chicken breast strips has been reported.[43] Fully cooked chicken breast strips were surface inoculated with *Salmonella* or *Listeria*. After vacuum packaging, the chicken was steam pasteurized at 88°C in either a continuous process (26 to 40 minutes) or a batch process (33 to 41 minutes). For the continuous-process product, a cooking time of 34 minutes was required to achieve a 7-log reduction of microorganisms, while a cooking time of 40 minutes was required for the same log reduction in the batch-process product. A general review of batch processing by Russell[44] addresses the market for batch process equipment, batch control, availability and facility upgrades, combination systems, and component groupings and presents examples of systems for specific food applications.

Use of batch processes for crystallization of glucose or sucrose has also been explored. Kraus and Nyvlt[45] examined the crystallization of glucose in three different sized crystallizers (1 L, 4 L, or 15 m^3) with variable cooling rates. Small glucose crystals grow more slowly that larger ones, and the addition of fructose slightly accelerates the nucleation process. A batch process for the production of large sucrose crystals was described by Bruhns et al.[46] The crystals they produced were >5 mm, preferably >8 mm without adherent threads. Relative motion of the supersaturated solution and temperature are controlled and are important variables in the batch crystallization. The use of both batch and continuous processes to dehydrate maple syrup was described by Rees.[47] Maple syrup (34% moisture) was dehydrated in a batch process in flat pans at 54°C under vacuum or in a continuous process utilizing a vapor separator, crystallizer, dryer, mill, and sizing screen. Physical differences in products from the two processes were minimal; however, microscopic examination showed very different structures for the products of the two processes. Either product would provide the maple industry a new, useful product, but the lower water content and greater density of the continuous-process product made it the more favored of the two dehydrated maple syrups.

In a comparison of tofu made by batch and continuous processes,[48] coagulation of the tofu was carried out using glucono–delta–lactone at 90°C and at 120°C (2 kg/cm^2 pressure in a retort). Scanning electron microscopy was used to identify the tofu structures, and physical properties related to texture were evaluated with a Tensipresser. Onion vinegar has been produced by a two-step batch process system.[49] This process combined the use of a flocculating yeast with a charcoal pellet bioreactor. Red onion juice (67.3 g/L total sugar) was converted to onion alcohol (30.6 g/L ethanol). The operation was stable, and maximum productivity was about 8.0 g/L/hr. A packed-bed bioreactor containing charcoal pellets was then used in a continuous process to convert onion alcohol to onion vinegar. Maximum productivity was about 3.3 g/L/hr, and maximum acetic acid concentration was about 37.9 g/L. This two-step process was operated for 50 days and was competitive with other systems of producing onion vinegar.

2.7 SUMMARY

Batch chemical processing is a flexible and widely used means of manufacturing. Certain types of process chemistry are easily done in batch operations, such as the manufacturing of polypeptides and certain types of food processing. Common pharmaceutical batch unit operations include preparative chromatography, lyophilization, and membrane processes such as ultrafiltration and diafiltration. Chemical reactions that have an uncertain endpoint are conveniently done in batch reactors. Common batch unit operations in the specialty chemical and pigment industries include batch distillation, centrifugation, classification, and high-shear dispersion. Product campaigns may be scheduled in order to maximize the purity of a product or to minimize changeover costs. Batch operations are also widely used in the food industry. Specific examples of batch manufacturing include beverage manufacturing; growing sugar crystals; production of carbohydrates, dairy products, and tofu; and in confection manufacturing. A drug product may experience a period of rapidly increasing demand early in the product life-cycle; the demand may be relatively steady as the product matures and then decline as generic versions or competing new drug products come onto the market. Manufacturing using batch technology and short process development times offers important flexibility to the pharmaceutical industry.

REFERENCES

1. Douglas, J.M., *Conceptual Design of Chemical Processes*, McGraw-Hill, New York, 1988.
2. Frederick, J., New-drug pipeline slows, but sales remain steady, *Drug Store News*, 25, August 18, 2003.
3. Ewalt, D.M., Drug discovery: technology plays an important role, *Inform. Week*, December 15, 2003.
4. DiMasi, J.A., Hansen, R.W., and Grabowski, H.G., The price of innovation: new estimates of drug development costs, *J. Health Econ.*, 22, 151–185, 2003.
5. Beers, D.O., *Generic and Innovator Drugs: A Guide to FDA Approval Requirements*, 4th ed., Aspen Publishers, Englewood Cliffs, NJ, 1995.
6. Kaitin, K.I. and Cairns, C., The new drug approvals of 1999, 2000, and 2001, *Drug Inform. J.*, 37, 357, 2003.
7. Tufts Center for the Study of Drug Development, *Tuft CSDD Outlook: 2004* (http://csdd.tufts.edu), 2004.
8. Manian, B.S. and Hemrajani, R.J., Drug discovery and healthcare in knowledge economy, *Pharma Pulse Express*, April 17, 2003.
9. DiMasi, J.A., The value of improving the productivity of the drug development process: faster times and better decisions, *Pharm. Econ.*, 20(53), 1–10, 2002.
10. Diwekar, U.M., *Introduction to Applied Optimization*, Kluwer, Norwell, MA, 2003.
11. Chakraborty, A. and Linninger, A.A., Plant-wide waste management. 1. Synthesis and multi-objective design, *Ind. Eng. Chem. Res.*, 41, 4591–4604, 2002.
12. Chakraborty, A. and Linninger, A.A., Plant-wide waste management. 2. Decision-making under uncertainty, *Ind. Eng. Chem. Res.*, 42, 357–369, 2003.
13. Merrifield, R.B., Solid phase peptide synthesis. I. The synthesis of a tetrapeptide, *J. Am. Chem. Soc.*, 85, 2149, 1963.
14. O'Reilly, R., *Standard and Poor's Industry Surveys: Chemicals, Specialty*, Vol. 171, Section 2, October 2, 2003.
15. Choe, H., *Standard and Poor's Industry Surveys: Household Nondurables*, Vol. 171, Section 1, December 18, 2003.
16. Searles, J.J., Potash, in *Encyclopedia of Chemical Processing and Design*, Vol. 41, McKetta, J.P., Ed., Marcel Dekker, New York, 1992 pp. 138–181.
17. Reid, K.I.G. and Kust, R., Calcium chloride, in *Encyclopedia of Chemical Technology*, 4th ed., Vol. 4, Kroschwitz, J., Ed., John Wiley & Sons, New York, 1996 pp. 801–812.
18. Edwards, W.P., *The Science of Sugar Confectionery*, Royal Society of Chemistry, Cambridge, U.K., 2000.
19. Rutenberg, M.W. and Solarek, D., Starch derivatives: production and uses, in *Starch: Chemistry and Technology*, Whistler, R.L., BeMiller, J.N., and Paschall, E.F., Eds., Academic Press, New York, 1984, 311–388.
20. Wurzburg, O.B., Ed., *Modified Starches: Properties and Uses*, CRC Press, Boca Raton, FL, 1986.
21. Myers, D.J. and Fox, S.R., Alkali wet-milling characteristics of pearled and unpearled amaranth seed, *Cereal Chem.*, 71, 96–99, 1994.
22. Hsu, W.P., Batch Fermentation Process, U.S. Patent No. 4,659,662, 1987.
23. Sebree, B.R., Biochemistry of malting, *Tech. Q. Master Brewers Assoc. Am.*, 34, 148–151, 1997.

24. Roegener, F., Bock, M., and Zeiler, M., New batch process for recovery of beer from surplus yeast, *Brauwelt*, 143, 505–508, 2003.
25. Anon., A survey of conching systems, *Confection. Manuf. Market.*, 28, 25–26, 29, 1991.
26. Anon., Gentle formation of "difficult" candy bar materials, *Confection. Prod.*, 56, 859–860, 1990.
27. Miller, J.J. and Brand, J.C., Enzymic lactose hydrolysis, *Food Technol. Austral.*, 32, 144, 146–147, 1980.
28. Reichart, O., Kinetic analysis of continuous yoghurt fermentation, *Acta Aliment.*, 8, 373–381, 1979.
29. Blond, G., Haury, E., and Lorient, D., Protein–lipid interactions in processed cheese made by batch and extrusion cooker methods: effect of processing conditions, *Sci. des Aliments*, 8, 325–340, 1988.
30. Rao, B.Y.K., Godbole, S.S., and D'Souza, S.F., Preparation of lactose free milk by fermentation using immobilized *Saccharomyces fragilis*, *Biotech. Lett.*, 10, 427–430, 1988.
31. Otten, J., Verheij, C.P., and Zoet, F.D., Accelerated production of stirred yoghurt by fed-batch prefermentation, *Voedingsmiddelentechmologie*, 28, 15–18, 1995.
32. Guul-Simonsen, F., Christiansen, P.S., Edelsten, D., Kristiansen, J.R., Madsen, N.P., Nielsen, E.W., and Petersen, L., Cooling, storage and quality of raw milk, *Acta Agric. Scand. A*, 46, 105–110, 1996.
33. Wyss, H., Engel, H., and Kuslys, M., Milk Substitute and Process for Its Manufacture, EP (European Patent) 0 832 565 A1, 1998.
34. Mariola, K.Z. and D'Souza, S.F., Removal of glucose from hen egg using glucose oxidase and catalase co-immobilized in hen egg white foam matrix, *J. Food Sci. Technol.*, 31, 153–155, 1994.
35. Shaw, P.E., Tatum, J.H., and Wilson III, C.W., Improved flavor of navel orange and grapefruit juices by removal of bitter components with beta-cyclodextrin polymer, *J. Agric. Food Chem.*, 32, 832–836, 1984.
36. Ujhazy, A. and Szejtli, J., Removal of naringin from aqueous solution with cyclodextrin bead polymer, *Gordian*, 89, 43–45, 1989.
37. Lipnizki, F., Olsson, J., and Tragardh, G., Scale-up of perevaporation for the recovery of natural aroma compounds in the food industry. II. Optimization and integration, *J. Food Eng.*, 54, 197–205, 2002.
38. Vie, J.F., Food freezing, *New Food*, 3, 52–53, 2000.
39. Degady, M. and van Niekerk, M., Continuous coating of chewing gum materials, U.S. Patent 2002/0009517 A1, 2002.
40. Qadeer, M.A. and Shahjahan, B., Microbial synthesis of xanthan gum by repeated-batch proccss, *Pak. J. Sci. Ind. Res.*, 30, 886–889, 1987.
41. Wiesenborn, D.P., Wang, J., Chang, K.C., and Schwarz, J.G., Comparison of continuous and batch processes for pectin extraction, *Ind. Crops Prod.*, 9, 171–181, 1999.
42. Vezzani, E. and Foti, S., Effects of cooling temperature on the microbiological characteristics of fillings for pasta, *Tecnica Molitoria*, 41, 577–581, 1990.
43. Murphy, R.Y., Duncan, L.K., Johnson, E.R., Davis, M.D., and Wolfe, R.E., Thermal lethality of *Salmonella* Seftenberg and *Listeria innocula* in fully cooked and packaged chicken breast strips via steam pasteurization, *J. Food Protect.*, 64, 2083–2087, 2001.

44. Russell, M.J., Process control: emphasis on batch, *Food Eng.*, 60, 53, 56–58, 62, 64, 1988.
45. Kraus, J. and Nyvlt, J., Crystallization of anhydrous glucose. IV. Batch crystallization and secondary nucleation, *Zuckerindustrie*, 119, 407–413, 1994.
46. Bruhns, M., Huwer, T., Kohnke, J., and Zingsheim, O., Batch Process for Manufacture of Sugar Crystals Without Threads and Equipment for Its Implementation, German Patent 101 35 079 A1, 2003.
47. Rees, F.M., Dehydrated maple syrup, *J. Food Sci.*, 47, 1023–1024, 1982.
48. Matsuura, M., Obata, A., and Murao, S., Properties of tofu gel made from continuously produced soymilk, *J. Jpn. Soc. Food Sci. Technol.*, 43, 1042–1048, 1996.
49. Horiuchi, J., Kanno, T., and Kobayashi, M., Effective onion vinegar production by a two-step fermentation system, *J. Biosci. Bioeng.*, 90, 289–293, 2000.
50. Wierenga, D.E. and Eaton, C.R., *Phases of Product Development*, Pharmaceutical Manufacturers Association, Office of Research and Development (http://www.allp.com/drug_dev.htm), accessed September 11, 2003.

Part II

Batch Processing Design Issues

3 Conceptual Design and Synthesis of Batch Processes

Rafiqul Gani and Irene Papaeconomou

CONTENTS

3.1 INTRODUCTION

The conceptual design and synthesis of batch processes can be defined as, given an initial charge, the identification of a set of batch operations (tasks) required to produce a sequence of specified products, subject to operational constraints.[1–3] Each batch process may be defined as a series of operational tasks, such as mixing, reacting, and separating, and within each task is a set of subtasks, such as heating, cooling, charging, and discharging. For example, a batch mixing operation at constant temperature may require a sequence of charging and heating and cooling operational subtasks while a batch separation task (for crystallization) may require a sequence of heating and cooling and discharging operational subtasks. Only a single production line consisting of a network of tasks is considered in this chapter; that is, only the recipe or sequence of tasks necessary for a single production line or a single batch operation is considered here. The design of batch processes involving multiple production lines is usually solved as a planning and scheduling problem (not covered in this chapter) where the batch recipe for the single production line is important starting information.[4]

As in the synthesis of continuous process flowsheets, more than one feasible set of operations might be able to produce the specified products; therefore, an optimal sequence with respect to a defined performance index also must be identified. Also as in synthesis of continuous process flowsheets, the batch operation synthesis problem may be solved by various solution approaches, such as knowledge-based heuristic approaches, mathematical programming approaches, and hybrid approaches. In principle, the knowledge-based heuristic approaches are easy to use and may be able to quickly identify a feasible sequence of batch operations but not necessarily the optimal solution. Mathematical programming techniques, on the other hand, may be able to find the optimal solution from a predefined space of feasible solutions (defined through the models used or a superstructure of candidate solutions). Hybrid approaches combine aspects of heuristic and mathematical programming approaches; they are more flexible and robust and have a wider application range. Essentially, all approaches follow the generate-and-test paradigm, but, while some approaches only generate and test a few alternatives (heuristics), others attempt to generate and test all possible alternatives (enumeration) or only those leading to the optimal solution within a defined search space (mathematical programming). Heuristic approaches are relatively easy to develop and apply, whereas mathematical programming approaches require process plus operational models (an operational model is a sequence of instructions necessary to perform a desired operation with a particular piece of equipment) and solvers that can handle large sets of potentially highly nonlinear equations. Enumeration techniques can suffer from combinatorial explosion and are not practical for problems with too many degrees of freedom. Note that heuristic approaches may be considered as a special-purpose (limited) enumeration approach.

All of the above-mentioned solution approaches require a set of methods and tools, and this chapter focuses primarily on them. These methods and tools can be classified as three types: those that help to generate and organize information (that is, help to formulate the synthesis or design problem), those that help to generate alternatives (that is, generate feasible solutions), and those that help to evaluate alternatives (that is, analyze and verify alternatives). Combinations of all three types of methods and tools are necessary for any of the above-mentioned solution approaches. In this chapter, some of the methods and tools are described within the context of conceptual design and synthesis of batch processes. The methods and tools discussed in this chapter, however, should not be regarded as the best available but as those used to illustrate and highlight important issues in conceptual design and synthesis of batch operations with respect to a single production line. Simple illustrative examples are used to highlight the methods and tools, and two case studies highlight various aspects of the hybrid solution approach.

3.2 SYNTHESIS PROBLEM FORMULATION AND SOLUTION APPROACHES

The synthesis of batch operations can be represented through the following generic mathematical description.

$$F_{OBJ} = \max\{C^T y + f(x,z)\} \tag{3.1}$$

s.t.

$$h_1(x,z) = 0 \quad \text{Product design specifications} \tag{3.2}$$

$$dx/dt = h_2(x,z,t) \quad \text{Process model equations} \tag{3.3}$$

$$h_3(x,z,t) = 0 \quad \text{Operation model equations} \tag{3.4}$$

$$l_1 \le g_1(x,z) \le u_1 \quad \text{Process design constraints} \tag{3.5}$$

$$l_2 \le g_2(x,z,t) \le u_2 \quad \text{Operation design constraints} \tag{3.6}$$

$$l_3 \le By + Cx \le u_3 \quad \text{Logical constraints} \tag{3.7}$$

In the above equations, x represents the vector of continuous variables (e.g., flowrates, mixture compositions, condition of operation), z represents a vector of design variables (e.g., equipment design parameters, operation design parameters), y represents the vector of binary integer variables (e.g., operation task and subtask identities, solvent identity), $h_1(x,z)$ represents the set of equality constraints related to operation (design) specifications (e.g., reflux ratio, reactor temperature, heat addition), $h_2(x,z,t)$ represents the set of process model equations (i.e., mass and energy balance equations), $h_3(x,z,t)$ represents the set of equality constraints related to the operational model (e.g., charge for a specified time period, heat at a constant rate), $g_1(x,z)$ represents the set of inequality constraints with respect to process design specifications, and $g_2(z,t)$ represents the set of inequality constraints with respect to operational constraints. The binary variables typically appear linearly as they are included in the objective function term, and in the constraints (Equation 3.7) to enforce logical conditions. The term $f(x,z)$ represents a vector of objective functions that may be linear or nonlinear, depending on the definition of the optimization problem. For process-operation optimization, $f(x,z)$ is usually a nonlinear function, while for integrated approaches, $f(x,z)$ usually consists of more than one nonlinear function.

The solution to the synthesis problem defined by Equations 3.1 to 3.7 is given in terms of:

- The production recipe
- A list of necessary equipment for each task and subtask (and the equipment design)
- A list of operational instructions (amounts, time, utilities consumption, etc.) for each task and subtask

In principle, what is obtained from the solution of the above problem is a network of tasks and subtasks with their associated design and operational data as well as an operation model (see Figure 3.1).

Many variations of the above mathematical formulations may be derived to represent different batch-operation synthesis problems and their corresponding solution methodologies. Some examples are given below.

- Heuristic approaches solve synthesis problems formulated by Equations 3.2 to 3.7. Although their solutions are feasible, they are not necessarily optimal. Rule-based procedures are employed to generate feasible solutions. The rules guide the user toward promising but not

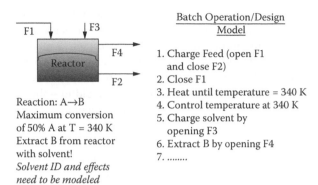

Reaction: A→B
Maximum conversion
of 50% A at T = 340 K
Extract B from reactor
with solvent!
*Solvent ID and effects
need to be modeled*

Batch Operation/Design
Model

1. Charge Feed (open F1
   and close F2)
2. Close F1
3. Heat until temperature = 340 K
4. Control temperature at 340 K
5. Charge solvent by
   opening F3
6. Extract B by opening F4
7. ....

FIGURE 3.1 A solvent-based batch reaction process (process and operation models).

necessarily optimal solutions. Examples of these approaches are provided by Linninger et al.[1] and Papaeconomou et al.[2,3]

- Mathematical programming approaches solve synthesis problems that involve Equation 3.1 plus different combinations and forms of Equations 3.2 to 3.7. For example, the solvent recovery targeting algorithm of Barton et al.[5] employs linear equations for all constraints (and Equation 3.3 represents a steady-state process model). This results in the formulation of a mixed-integer linear programming (MILP) problem. Kondili et al.,[6] on the other hand, solved a nonlinear programming (NLP) problem for fixed-state task networks by employing Equations 3.1 to 3.6 for fixed y.

- Solving all the equations represents an integrated process/product design problem and is usually quite difficult to achieve due to the potential complexity of the model equations. Usually, a superstructure of known networks of tasks and subtasks is specified to restrict the search space and complexity of the process and operational models.

- Hybrid approaches decompose the synthesis problem into subproblems such that each subproblem satisfies a subset of the constraints (Equations 3.2 to 3.7). The objective function plus the remaining constraints form an NLP or a mixed-integer nonlinear programming (MINLP) problem with a well-defined search space. Combining heuristic and mathematical programming approaches produces a hybrid solution approach. The solution approaches developed by Linninger et al.[1] and Papaeconomou et al.[2,3] fall into the hybrid category, as they propose a final optimization step after identifying a small search space where the optimal solution is likely to be found.

Note that for all batch-process synthesis problem formulations, unlike synthesis of continuous processes (usually considering only the steady state), an operational model (Equation 3.3) is needed in addition to the process model. Figure 3.1 illustrates a batch solvent-based, two-phase reactor process, where, in addition to the dynamic model for a two-phase reactor, a model for the batch operation is also needed. For the problem highlighted in Figure 3.1, determination of the operation model requires the use of a number of methods and tools — for example, a tool for solvent selection, a model for liquid miscibility calculations, and a tool to evaluate the reaction kinetics. The conceptual design and synthesis of batch operations requires determination of the operation model for a desired batch process.

Using the heuristic approach, an optimal design may be obtained by ordering all the generated feasible candidates according to the objective function (Equation 3.1) value. Global optimality, however, can only be guaranteed if and only if all possible alternatives were considered in the generation of the feasible set of candidates.

On the other hand, trying to solve all the equations may become too complex if the process and operation models are highly nonlinear and discontinuous. Also, the solution approach is not able to accommodate multiple process and operation models for the same sequence of operation tasks. Although these problem formulations can determine the optimal design, their application range is usually not very large; however, because relatively simple process and operation models are employed in planning and scheduling, problem formulations of this type are common.

3.3 METHODS AND TOOLS FOR KNOWLEDGE GENERATION

Information related to the initial charge, a set of desired products, and a set of operational constraints is usually not enough to formulate and solve problems of conceptual design and synthesis of batch processes. Additional information (knowledge or data) must be generated and analyzed to define a subset of constraints (Equations 3.2, 3.5, 3.6, and 3.7). The following four classes of information (knowledge or data) must be generated when they are not available so the synthesis problem can be formulated with sufficient clarity.

1. *Identify the batch operation tasks that must be performed.* It is necessary to establish as early as possible the tasks necessary to achieve the desired operation. Some of these are obvious; for example, a reaction task is required if a reacting system is specified, or a separation task may be required when products are to be separated from a specific mixture. The objective is to use this information to generate more details on the synthesis problem so the subtasks necessary to achieve the identified main tasks can be identified (see steps 2 through 4). Consider an initial charge of benzene, monochlorobenzene (MCB),

and *ortho*-di-chlorobenzene (DCB), at 1 atm pressure and 300 K temperature, from which benzene needs to be removed. Obviously, this involves a separation task as the initial mixture already contains all the three compounds, which are not known to react under the specified conditions. This problem will be referred to as the benzene separation problem and is discussed further in the text below.

2. *Analyze the initial charge and the desired products.* It is necessary to know the state of the initial charge and products (solid, liquid, gas, or multiphase), their properties (e.g., phase behavior, presence of azeotropes), reacting or nonreacting, and many other variables. This helps to identify the subtasks required for any main operational task. Considering the benzene separation problem, at the specified condition the ternary mixture of benzene, MCB, and DCB is a liquid, none of the binary pairs of compounds form azeotropes, and because of the differences in their vapor pressures separation by batch distillation is feasible for the desired product (benzene). Possible subtasks (for a single stage batch still) could include charge the feed, add heat to the still, and condense and collect the vapor as it exits from the top.

3. *Identify and select the operational variables to be considered.* For each batch operation task and subtask, the sets of operational variables that will define the corresponding batch operation must be selected. For example, in mixing operations, flowrates of the additives and the heating or cooling medium may be important operational variables, while in batch distillation operations heating and reflux flowrate may be important operational variables. For the benzene separation problem, the amount of initial charge will define the size of the equipment, while the product (benzene) identity, recovery, purity, etc., will determine the temperature and pressure of the batch distillation operation and the amount of energy required to achieve the separation task.

4. *Define the operational constraints and the performance index.* Each operational task and subtask may be constrained in terms of the operational variables through specified or identified operational tasks — for example, minimum and maximum temperatures in a reactor, minimum reaction selectivity, or maximum flowrate of a solvent. For the benzene separation problem, the operational constraints could include product purity, temperature of the residue liquid (at a fixed pressure), and total time of operation.

When the necessary information (knowledge) has been generated, it must be organized and represented through modeling frameworks so the methods and tools for solving the synthesis problem and evaluating the batch process (e.g., simulation engine) can use them.

3.3.1 BATCH OPERATION TASK IDENTIFICATION

Three of the most common operational tasks found in many batch processes —
mixing, reaction (single or multiple-phases), and separation (various types) —
are considered below.

3.3.1.1 Liquid Mixing

Mixing (liquid) operations may be necessary in order to prepare the initial charge
for a subsequent reaction or separation as well as in the manufacture of blended
or formulated products. Here, two or more chemicals (or mixtures) are mixed to
form a liquid (single-phase or emulsion) product. The mixture is not reactive and
does not (usually) split into two or more phases. The product is a liquid solution
(or emulsion) with a specified composition and properties.

3.3.1.2 Reaction

Reactors are required whenever the identities of compounds in the initial charge
are different from the identities of the specified products. To produce the appropriate reaction, information related to the reaction chemistry must be generated
or specified. In this chapter, it will be assumed that the reaction chemistry
information is available, but in the case of multiple reactions the reaction sequence
will be considered a design parameter and therefore must be determined by
solving the synthesis problem.

3.3.1.3 Separation

Separation tasks are necessary whenever the number and identity of compounds
in the cumulative products obtained at different times of the batch operation (each
product having different compositions) are the same as the initial charge. Usually,
a sequence of separation-related (operation) tasks must be performed. The separation may be achieved by creating additional phases with differences in properties and compositions. The additional phases may be created by heating or
cooling, by the addition of mass separating agents, or by creating a barrier, among
other means. Batch (separation) operations through heating or cooling, such as
evaporation, crystallization, distillation, and liquid–liquid extraction, are among
the most common. In this chapter, only these separation operations will be
considered. The type of separation task required can be identified through an
analysis of the initial charge and the desired products (see Section 3.3.2.3).

3.3.2 ANALYSIS OF INITIAL CHARGE AND PRODUCTS

The objective here is to generate enough information (knowledge) so steps 3 and
4 can be carried out efficiently. Identification of the batch operation tasks required
to achieve the desired product also implicitly defines the information necessary
for their analysis and which must be generated by the analysis. Note that the

initial charge is used in the discussion below to indicate the amount of material (chemicals) charged to a production line or to a specific task in a production line.

3.3.2.1 Liquid Mixing

In the liquid mixing task, the most important information to be generated is related to the effect of the mixing operation. For example, do the initial charge and additions form a single homogeneous liquid (or emulsion)? If not, is the creation of another phase desirable? Does the mixing operation generate negligible heat of mixing? If not, should the temperature be controlled? In the case of emulsion, what is the critical miceller concentration and how can it be reached? Also, for reactive systems, if the initial charge does not have all the reactants, then the missing reactants will have to be added. Information related to the above questions can be generated by property estimation features in commercial simulators, or specialized software such as iCAS[7] may be used. Heats of mixing data may be obtained through experimentally measured data, retrieved from an appropriate database, or predicted through specialized software. Choice of the appropriate property models is an important first step. Referring to Figure 3.1, it is clear that the reactants plus the reagent must be charged and mixed before the reaction can take place.

3.3.2.2 Reaction

The number and identity of the phases are important and must be analyzed together with the reaction chemistry. The heat of reaction is an important property as is the reaction kinetics (or chemical equilibrium). For multiple reactions systems, the total heat of reaction will identify the need for heating or cooling. If the total heat of reaction is negligible, adiabatic reaction operation is feasible; isothermal reaction operation is recommended when the heat of reaction is not negligible. Identity of the state (phase type and number) of the reacting mixture is important information, and an initial guess can be very quickly made through the pure component boiling points (at the operating pressure) and the melting points. Creation of two phases is considered when it is feasible or necessary to simultaneously react and separate products or byproducts. Referring to Figure 3.1, upper and lower bounds on the reactor temperature could be set to maintain a high yield of the desired product; therefore, necessary tasks during the reaction are to heat or cool until the reaction has been completed. Note that if a solvent is added while the reaction is going on, the solvent effect (and simultaneous separation reaction) will also have to be considered.

3.3.2.3 Separation

Because many separations are caused by the creation of an additional phase, the number and identity of the phases present at the initial point (which can be the discharge from a reaction operation) are important with respect to the type of the separation task, while the behavior of the phases as a function of temperature and pressure provides information related to the products that can be obtained. For vapor–liquid separations, the existence of azeotropes must be determined, while for solid–liquid separations any eutectic points must be identified. For liquid–liquid extraction, a liquid phase split with sufficient differences in compositions in the two phases must be maintained. The selection and effect of adding mass separating agents must also be analyzed. For purposes of illustration, consider an aqueous mixture containing a valuable product that must be recovered or a contaminant that must be removed. The identity of the product or contaminant influences the separation technique to be employed. Consider now that the product or contaminant is phenol. If only a small amount of phenol is present, solvent-based liquid extraction or crystallization may be preferred, while batch distillation or short-path distillation may be preferred if water is present in only a small amount. In the case of solvent-based liquid extraction, a solvent (for example, 2-methylpropyl ester) that selectively dissolves phenol and creates a phase split of the original mixture would be necessary.

In order to identify the required separation tasks, a method for synthesis of process flowsheets, based on thermodynamic insights and developed by Jaksland et al.,[8] can be applied. The method was originally developed for (continuous) process flowsheet synthesis but it is clearly valid for identifying batch (separation) operations as well. According to this method, the mixture to be separated is analyzed in terms of phase state, azeotropes, eutectic points, mutual solubility, and so on. All possible binary pairs for the mixture components are then identified, and for each pair binary ratios of the properties are calculated:

$$B(k,l)_{ij} = \theta(l)_i / \theta(l)_j \qquad (3.8)$$

where $B(k,l)_{ij}$ is the property ratio of binary pair k consisting of compounds i and j, $\theta(l)_i$ is the property l for compound i, and $\theta(l)_j$ is the property 1 for compound j. If the $B(k,l)$ values are much greater than unity for property l and binary pair k, then a separation technique based on exploiting this property difference is feasible (compounds i and j are selected such that $\theta(l)_i > \theta(l)_j$). Detailed examples of application of this algorithm can be found in Jaksland,[9] and a simple example is given here for the benzene separation problem. The mixture contains three compounds, so the binary ratios of properties for the three binary pairs of compounds must be generated and analyzed. Table 3.1 provides a partial list of binary ratios of properties for the three binary pairs. Note that, because the ratios of normal boiling points are relatively larger than 1 and none of the pairs forms azeotropes, separations involving vapor–liquid phases are feasible. For the same reasons, crystallization is also possible but, because of the low temperatures required, may not be economically feasible.

In a similar way, we can take advantage of differences in the compositions of two phases (for example, vapor–liquid). Gani and Bek-Pedersen[10] referred to these differences in compositions as driving forces and exploited these differences to sequence distillation columns and for hybrid separation techniques. The same principle can also be used to identify the feasibility of batch separations. (When

TABLE 3.1
Binary Ratio of Properties for Three Binary Pairs (Benzene Separation Problem)

Binary Pair	Binary Ratio Property Values						
	M_w	T_c	P_c	T_b	T_m	V_m	V_{dw}
Benzene–MCM	1.44	1.12	1.08	1.15	1.22	1.14	1.20
Benzene–DCM	1.88	1.25	1.20	1.28	1.09	1.26	1.39
MCM–DCM	1.31	1.11	1.11	1.12	1.12	1.10	1.10

Binary Pair	Separation Technique			
	Adsorption	Distillation	Crystallization	Membrane-Based Separation
Benzene–MCM	Ratio of V_{dw}	Ratio of T_b	Ratio of T_m ($T < 278$ K)	Ratio of V_m
Benzene–DCM	Ratio of V_{dw}	Ratio of T_b	T_m too low (<227 K)	Ratio of $V_m^{\cdot}$
MCM–DCM	Ratio of V_{dw}	Ratio of T_b	T_m too low (<227 K)	Ratio of V_m

the driving force is zero, no further separation is possible; when the driving force is at a maximum, the separation is easy and low cost.) Calculating the driving forces requires knowing the compositions of two coexisting phases (in equilibrium or not) for binary or multicomponent mixtures. For the benzene separation problem, only the two binary pairs involving benzene need to be investigated. Figure 3.2a and Figure 3.2b show calculated driving force diagrams for the benzene–MCB mixture and residue curves on a ternary benzene–MCB–DCB diagram. See also Section 4.3 for use of driving force diagrams to generate batch distillation operation alternatives.

3.3.3 IDENTIFICATION AND SELECTION OF OPERATIONAL VARIABLES

The variables for specifying a batch operation task must be identified. In liquid mixing, the amounts of chemicals necessary to achieve a blend or formulation with a desired set of properties are important variables. The temperature can also be a variable if the desired properties of the blend or formulated product are dependent on it. For reactors, reaction selectivity, reactor liquid composition, heating and cooling, temperature, and pressure are important variables. Note that not all variables have to be considered; for example, in a single-phase (liquid) batch reactor operation, pressure may be assumed to play an insignificant role, while in a two-phase vapor–liquid (but not solid–liquid) reacting system, pressure plays a very important role. For all reactor systems with significant total heats of reaction, temperature is always very important and upper and lower bounds

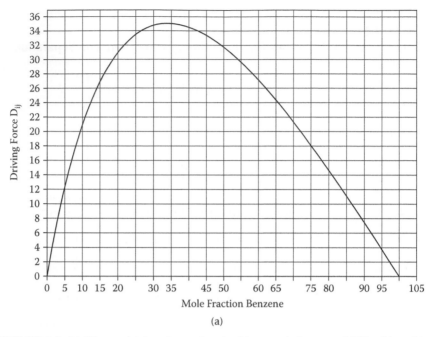

(a)

FIGURE 3.2 (a) Binary driving forces for the binary pair benzene–MCB; (b) residue curves (batch distillation) for the ternary system benzene–MCB–DCB.

are usually specified. For multiple reacting systems, an attainable region diagram of concentration of the product vs. concentration of the key reactant provides useful information related to the selection of operational variables. Figure 3.3 shows the attainable region for a multiple reaction system. Clearly, the best operating point is to maintain the batch operation around the maximum concentration of the product (that is, maintain the operation at the corresponding key reactant concentration). This diagram can also be used to generate the list of subtasks needed to attain the maximum area within the curve, starting from the key reactant concentration of 1 and moving beyond the value corresponding to the maximum product concentration. As these curves are functions of reaction rates (kinetics), changing the reacting temperatures as the concentration of reactant decreases is one option. In the case of separation, the variables depend on the type of operation as well as the identity of the phases. For example, in batch distillation, the amount of heating or cooling and the reflux rate are important; in crystallization, heating and cooling and evaporation and mixing are important.

3.3.4 Definition of Operational Constraints and Performance Index

The operational constraints define the batch operational windows (boundaries). This point is more important for reaction and separation operations than for mixing operations. These boundaries are governed (implicitly) by the sensitivity

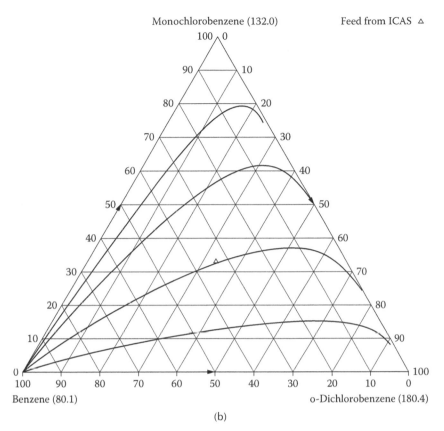

Monochlorobenzene (132.0) Feed from ICAS △

Benzene (80.1)

o-Dichlorobenzene (180.4)

(b)

FIGURE 3.2 (Continued).

of the important properties on the intensive variables (temperature, pressure, and composition). For example, in reactions and separations, the temperature may change the reaction rate and separation driving force, respectively. In the case of multiple reactions, product yield and reaction selectivity provide important performance index criteria, while temperature, pressure, and reactant concentration impose operational constraints. In the case of separation (azeotropic separation or crystallization), temperature and pressure also define distillation and crystallization boundaries and, therefore, the sequence of operations required to obtain a product. The performance index in all cases may be related to the cost of operation or total time required for the operation. In their tool for solvent recovery targeting, Ahmad and Barton[11] used the identified distillation boundaries as operational constraints to predict all possible product sequences achievable for different (azeotropic) mixtures of a given set of components. In order to obtain the optimal product sequence for a given mixture, they used the maximum achievable recovery (of the solvent) in each product cut as the performance index. Linninger et al.[1] have shown how environmental constraints can be incorporated into the synthesis of batch processes to minimize avoidable pollution. In some cases (e.g.,

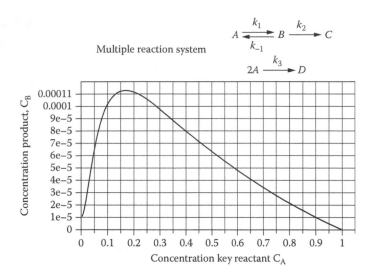

FIGURE 3.3 Attainable regions for a multiple reaction system.

synthesis based on driving forces[10]), near-optimal solutions with respect to both time and cost can be obtained without directly using a performance index because the driving force by definition is inversely proportional to the energy (external medium) cost and directly proportional to ease of operation.

3.3.5 INFORMATION (KNOWLEDGE) REPRESENTATION

The recipe for the manufacture of a product is modeled as a network of tasks and each task as a sequence of subtasks. A task consists of all of the operations performed in a single item of equipment; a subtask consists of a model of one of these operations. Tasks have associated with them requirements for specific types of equipment and selection priorities; thus, the synthesis of batch operations also implies modeling of the batch operations. A modeling framework is needed to represent (organize) the generated information so the synthesis, design, and simulation problems can be formulated and solved and the results analyzed. The modeling task can be decomposed into two distinct activities: modeling fundamental physical behavior and modeling the external actions imposed on this physical system resulting from the interaction of the process with its environment due to disturbances, operation procedures, or other control actions.[4] The model for the physical behavior of the system is the process model (see Equation 3.2; see also Sections 3.3.2 to 3.3.4 and Sections 4.4.1 to 4.4.4). The model for external actions imposed on the batch process (physical system) is the operations model (Equation 3.3), where a process representation framework is usually employed.

3.3.5.1 Process Representation Frameworks

Kondili et al.[6] proposed a representation of the state–task network (STN) that is similar to flowsheet representations of continuous plants but is intended to describe the process itself rather than a specific plant. The distinctive characteristic of STN is that it has two types of nodes: state nodes, representing the feeds, intermediates, and final products, and the task nodes, representing the processing operations that transform material from input states to output states. Circles and rectangles denote state and task nodes, respectively (see Figure 3.4a). Process equipment and its connectivity are not explicitly shown. Other available resources are also not represented.

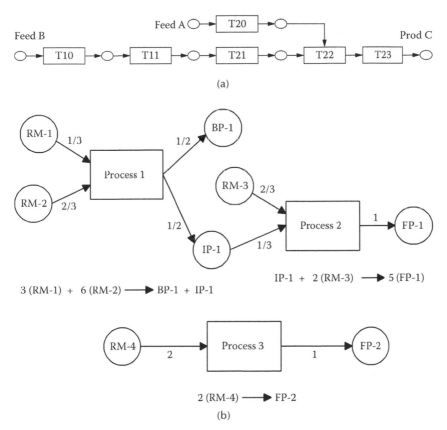

(a)

IP-1 + 2 (RM-3) ⟶ 5 (FP-1)

3 (RM-1) + 6 (RM-2) ⟶ BP-1 + IP-1

2 (RM-4) ⟶ FP-2

(b)

FIGURE 3.4 (a) State-task network representation of chemical processes; (b) PMN describing the processing of two products.

The STN representation is equally suitable for networks of all types of processing tasks: continuous, semicontinuous, or batch. It assumes that an operation consumes material from input states at fixed ratio and produces material for the output state also at a known fixed proportion. The processing time of each operation is known *a priori* and is considered to be independent of the amount of material to be processed; otherwise, the same operation may lead to different states (products) with different processing times. The rules followed when constructing this model are:

- A task has as many input (output) states as different types of input (output) material.
- Two or more streams entering the same state are necessarily of the same material. If mixing of different streams is involved in the process, then this operation should form a separate task.

Barton and Pantelides[12] proposed an alternative representation, the resource–task network (RTN). In contrast to the STN approach, where a task consumes and produces materials while using equipment and utilities during its execution, in this representation a task is assumed only to consume and produce resources. Processing items are treated as though consumed at the start of a task and produced at the end. Furthermore, processing equipment in different conditions can be treated as different resources, with different activities consuming and generating them, which allows a simple representation of changeover activities.

Graells et al.[13] proposed a modeling (framework) environment that employs a continuous time representation for the scheduling of batch chemical processes. For this environment, the process structure (individual tasks, entire subtrains, or complex structures of manufacturing activities) and related materials (raw, intermediate, or final products) are characterized by means of a processing network, which describes the material balance. Manufacturing activities are considered at three different levels of abstraction: process level, stage level, and operation level. At the process level, the process and materials network (PMN) provides a general description of production structures (e.g., synthesis and separation processes) and the materials involved, including intermediates and recycled materials. An explicit material balance is specified for each of the processes in terms of a stoichiometric-like equation relating raw materials, intermediates, and final products (Figure 3.4b). Each process may represent any kind of activity necessary to transform the input materials into the desired outputs.

3.4 METHODS AND TOOLS FOR GENERATING ALTERNATIVES

While the literature on design, planning, and scheduling for multiproduct batch processes is quite extensive (see the review by Reklaitis[14]), literature on the conceptual design and synthesis of batch processes is very limited.[1–3,15–18] In this section, step-by-step algorithms for the generation of batch recipes (route selection) are given for three types of batch operation tasks. When more than one operation task is needed, the corresponding algorithms are repeated in an established sequence. The objective for the synthesis of batch operational sequences is to minimize the operating time and expense required to obtain specified products. The focus may shift from achieving the optimal time to minimizing energy and operating costs. Because of the trade-off between these two objectives, ultimately an optimization problem will have to be formulated and solved where appropriate weights can be given to time and operating costs.

This section describes a systematic (hybrid) methodology developed by Papaeconomou et al.[2,3] for the synthesis of batch operational tasks. The methodology consists of a set of algorithms that generate feasible and near-optimum batch recipes for specified operational and end constraints. The common ground of these algorithms is the existence of a number of constraints that must be satisfied at all times and use of manipulated variables to ensure feasible operation. The algorithms take care of the operational modeling of each operation by identifying the sequence of tasks that must be performed in order to achieve the objectives of the specified operation. This is done with the help of a set of knowledge-based rules (e.g., thermodynamic insights), which are employed to identify the end of each task and determine the next feasible task. Note that the objectives of this synthesis methodology are to identify and define the contents of the operation model. When this has been achieved, any modeling framework can be employed to represent the information. In this way, the modeling and synthesis of batch operations are simultaneously achieved.

3.4.1 Liquid Mixing Task

The synthesis problem here is defined as follows: Given a list of candidate chemicals and their available amounts, determine which chemicals should be mixed and in what amounts in order to achieve a desired formulation or blend. It is assumed that all the candidate chemicals, when mixed, will form a totally miscible liquid (or form an emulsion when added to a specific product). It is assumed that when the chemicals have been mixed for a sufficiently long time, a homogenously mixed, totally miscible liquid will be obtained. The synthesis problem does not consider the time required to achieve this but calculates the time and flowrates of the chemicals that should be mixed to achieve the desired product. Mixing time is not considered; only what to mix and how much to mix are considered.

A graphical, composition-free method developed by Eden et al.[19] is applied here to solve synthesis problems related to mixing. According to this method, the mixing (mass balance) model equations are reduced in terms of property clusters, which are linear functions of properties or their functions. A property cluster (φ) may be defined as a linear function (with respect to composition) of property (θ), as shown below (Eq. 3.9):

$$\phi = \Sigma_i x_i \theta_i \tag{3.9}$$

where x_i is mole fraction of component i, and θ_i is a property (or property function) for component i. The mixing model (mass balance) is written as:

$$My_i = \Sigma_j f_{ij} F_j, \text{ where } i = 1, NC \tag{3.10}$$

where M is the total mixed flowrate, y_i is the mole fraction of component i in the mixed liquid, f_{ij} are the mole fractions of component i in stream j ($j = 1, NS$), F_j are flowrates of feed streams j ($j = 1, NS$), NC is the number of compounds in the mixture, and NS is the number of feed streams being mixed. Multiplying each term in Equation 3.10 with θ_i, summing each term with respect to $i = 1, NC$, replacing them with Equation 3.9, and applying some algebraic manipulations give us the following (for three property clusters):

$$\phi_m = z_1 \phi_1 + z_2 \phi_2 + z_3 \phi_3 \tag{3.11}$$

where the subscripts 1, 2, and 3 represent property clusters 1, 2, and 3, respectively; the subscript m represents the mixture; and z_1, z_2, and z_3 are mole-fraction-like values of property clusters 1, 2, and 3, respectively. Comparison of Equations 3.10 and 3.11 indicates that a composition-based mass balance equation has been transformed into a property-cluster-based (mass) balance equation for the mixing (operation) model. Note that this transformation satisfies Equation 3.10 exactly. This means that the synthesis of mixing operations in terms of how much and which chemical to add in order to produce a desired blend or formulation may now be achieved graphically (or by linear programming) for any number of candidate chemicals; that is, all candidate chemicals may be located on a ternary diagram of property clusters together with the target mixture properties. Finding feasible solutions means simply joining two or more points on the ternary diagram and checking if the line passes through the target point. Because Equation 3.11 satisfies the inverse lever-arm rule, the amounts of each chemical in the target mixture can be calculated from the distances between the target and the chemical. Figure 3.5 illustrates this method by considering replacing six mutually miscible compounds with an environmentally friendlier blend. This means that the mixing operation can be performed as a function of the desired properties on a composition- and compound-free basis, making the control of the batch operation easier.

3.4.2 REACTION TASK

Reaction task synthesis problems identify a network of batch reaction operations where single or multiple reactions may take place and where operational constraints on product yield, selectivity, temperature, or pressure may be imposed. For multiple reactions, particular reactions may be desired while others may be competing reactions that must be suppressed. Different end objectives (e.g., the

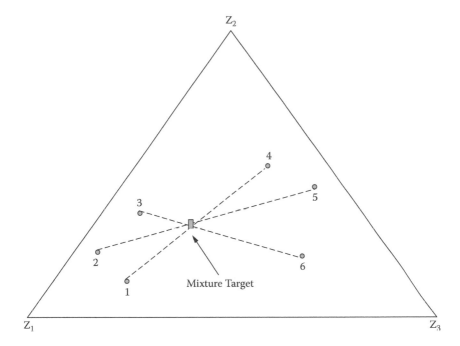

FIGURE 3.5 Composition-free synthesis of mixing/blending operation.

mole fraction of the limiting reactant in the reaction of interest should be as low as possible and the progress of the reaction of interest should be as high as possible) may be specified. At least one of the end constraints has to be satisfied. Additional constraints related to selectivity may also be introduced.

This algorithm[2,3] helps to identify the first operating step (task) based on a check of selectivity or product yield. To generate the initial recipe, rules are employed at all points to identify the end of each task and determine the next feasible task, such as isothermal operation, adiabatic operation, heating, or cooling. The procedure is repeated until the product (end) constraints are met. Figure 3.6 illustrates a flow diagram for the complete algorithm. It is valid for synthesis of single-phase (liquid) as well as multiple-phase (liquid–vapor) batch reactor operations. In the case of single-phase reactor operation, some of the rules and constraints are not necessary (e.g., rule 6 and pressure constraint). Detailed examples of applying this algorithm can be found in Papaeconomou.[2,3]

3.4.3 Separation Task: Batch Distillation

This algorithm is based on the use of driving forces; that is, the operation is driven to always employ the maximum available driving force for the desired separation. The synthesis algorithm also simultaneously predicts the behavior of the batch operations, so a simple simulation model based on driving forces has also been developed and used to solve the synthesis and the simulation problem

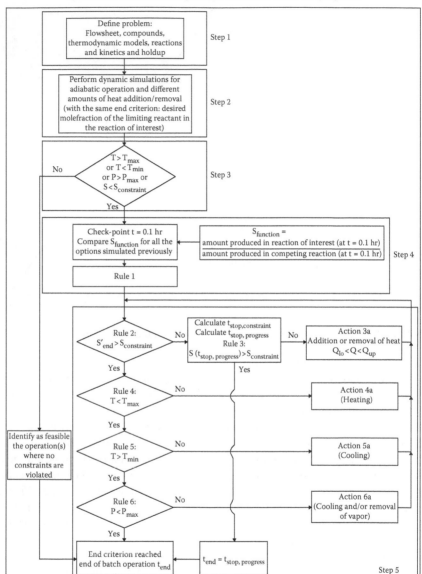

Synthesis of feasible operational sequences for a batch reactor (vapor and liquid present)

FIGURE 3.6 Flow diagram for algorithm for synthesis of batch reactor operations.

simultaneously. The algorithm for this simple model based on driving forces is presented below, followed by an illustrative example.

The algorithm uses a set of simple equations for the distillation column and adapts well-known methods, such as the driving-force approach and the McCabe–Thiele diagram, to quickly find a near-optimum recipe for the separation task. A batch distillation column with negligible holdup in the perforated plates and the condenser, as described in Diwekar,[20] also follows the Rayleigh equation for simple distillation. The overall material balance and material balance for the most volatile component around the complete column give us the following equation:

$$x_D dB = d(Bx_B) = Bdx_B + x_B dB \tag{3.12}$$

Integrating Equation 3.12 leads to the Rayleigh equation:

$$\ln\left(\frac{B}{F}\right) = \int_{x_F}^{x_B} \frac{dx_B}{x_D - x_B} \tag{3.13}$$

In the above equations, B is the amount of product remaining in the still, F is the initial charge, x_D is the instantaneous distillate composition, and x_B is the still composition of the more volatile component.

The overall material balance around the top section gives us:

$$dD \,/\, dt = \frac{V}{R+1} \tag{3.14}$$

In the above equation, D is the amount of distillate, V is the vapor boilup rate, and R is the reflux ratio. For the batch distillation column described here, the entire column section above the still may be considered as a rectifying section. So, the functional relationship between x_D and x_B turns out to be given by the operating line equation in the enriching section of a continuous distillation column:[20]

$$y_j = \frac{R}{R+1} x_{j-1} + \frac{1}{R+1} x_D \tag{3.15}$$

The easiest method for identifying the bottom composition (x_B) is the McCabe–Thiele graphical method, which can be extended to batch distillation by writing the equations for the operating line at different time intervals. In order to calculate x_B with this method, we need to know the number of trays of the column and the reflux ratio; however, the reflux ratio cannot be chosen arbitrarily. If it is too small the separation might not be feasible, and if it is too high the separation might consume excessive energy; therefore, it is necessary to identify the minimum reflux ratio R_{min} that is needed to perform the separation. One way to find R_{min} is to use the driving-force approach. The driving force, as defined by Gani and Bek-Pedersen,[10] is the difference in composition of a component in two coexisting phases; thus, for the case of a batch distillation column (from Equation 3.15), we have:

$$DF = y - x = \frac{-1}{R+1}x + \frac{1}{R+1}x_D \qquad (3.16)$$

The existence of a driving force is what makes the distillation possible. Operating at the largest driving force leads to near-minimum energy expenses.[10] From Figure 3.7 it is obvious that, for a specific feed composition, the largest driving force corresponds to the minimum reflux ratio; however, the minimum reflux ratio can only be supported by an infinite number of plates. Thus, for a specific number of plates a ratio larger than the minimum reflux ratio has to be used. As the composition of the more volatile component in the still moves to the left (decreases), the reflux ratio used (R_1) approaches the minimum value for the corresponding composition ($R_{min,1}$). At that point, a new reflux ratio has to be used (R_1). This process is repeated until the end (product) constraint is reached.

Driving-force diagrams (see Figure 3.7) give users visual, physical insights leading to operation at near-optimum energy costs; moreover, a large driving force, which corresponds to a low reflux ratio, results in a faster separation. This finding is also supported by Equation 3.14, where the lower the R the larger the

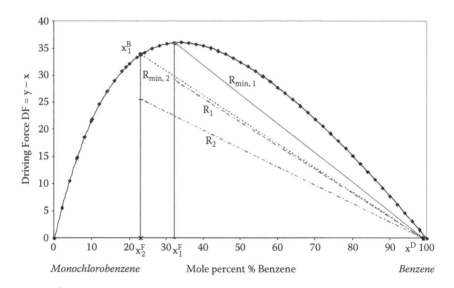

FIGURE 3.7 Driving force diagram for the binary system benzene–MCB.

distillate rate. Two basic modes of operation of a batch distillation column may be considered:

1. Constant reflux and variable product composition
2. Variable reflux and constant product composition of the key component

In this chapter, we are considering the second mode; that is, the reflux ratio is changed at specific intervals of time (note that, during any time interval, the reflux ratio remains constant). Also, we use a constant value for the distillate composition (a time-averaged value that is above a specified minimum). Now, Equation 3.13 becomes:

$$B = F \frac{x_D - x_F}{x_D - x_B} \tag{3.17}$$

And the overall material balance:

$$D = F - B \tag{3.18}$$

In the above equations, F is the initial charge, x_D is the allowed minimum distillate composition, x_F is the initial feed composition, D is the amount distilled, and B and x_B are, respectively, the amount remaining in the still and its corresponding composition by the end of the period operated with a specific constant reflux ratio.

The time for the period of constant reflux ratio can be found from the integration of Equation 3.14:

$$\int_0^T dt = \int_0^D \frac{R+1}{V} dD \Leftrightarrow T_{period} = \frac{R+1}{V} D \tag{3.19}$$

To determine the sequence of periods (subtasks) where at each period the reflux ratio is changed, we can employ an algorithm that identifies the value of the reflux ratio and the time of operation for each subtask (described below).

3.4.3.1 Synthesis Algorithm

The algorithm for synthesis of batch distillation operations is illustrated in Figure 3.8. The objective of the algorithm is to identify in advance the necessary sequence of tasks in order to achieve a number of end objectives for the distillation operation. These objectives are product purity and product yield. Initially, the relationship between the vapor boilup rate and the reflux ratio must be investigated in order to locate operating problems, such as flooding. In this way, when the chosen reflux ratio for a subtask is too high for the corresponding vapor boilup rate, corrective action can be taken.

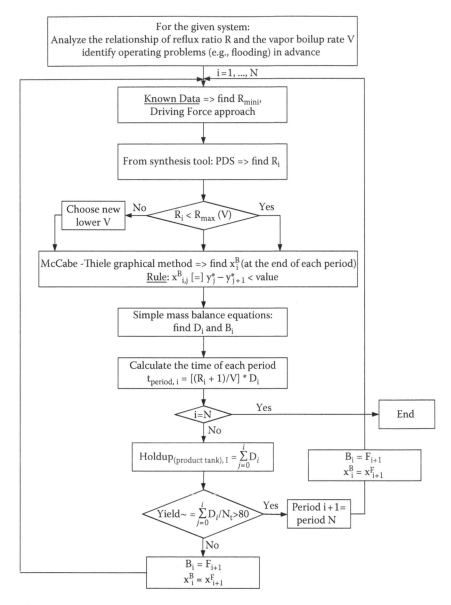

FIGURE 3.8 Schematic presentation of the algorithm for batch distillation.

The first subtask (operation for period 1) is identified using the following procedure, which is then repeated until the end objectives are reached. From the known data, which are the composition of the feed and desired product (distillate) composition, the driving-force approach is employed to find the minimum reflux ratio for the specific feed. The next step is to calculate the reflux ratio (R) that will match the specified number of plates of the column. This is done by performing a special stage-by-stage calculation for a distillation column using a synthesis tool developed by Hostrup[21] (involving sequential calculations of mass balance and phase equilibria starting from the feed condition and going toward the endpoint) and implemented in iCAS.[22] If the chosen reflux ratio is lower than the flooding value for the specific vapor boilup rate, the next step is to determine the bottom composition at the end of this period; otherwise, a new value for the vapor boilup has to be selected first.

The composition of the amount remaining in the still at the end of this period can be found by applying the simple McCabe–Thiele graphical method; however, because the available number of plates might be excessive for the separation, the bottom composition (x_B) is not necessarily identified from the last plate (N) but from plate j, where no significant improvement in the vapor composition occurs between the adjacent plates j and $j + 1$. The amount (D) distilled in this period and the amount (B) remaining in the still, as well as the operating time (end) of this subtask, are calculated from Equations 3.10 to 3.12. The next step, in effect, is designating the bottom composition (x_B) and amount in the bottom (B) as the feed composition (x_F) and feed (F) for the next task. Repeating this procedure leads us to synthesis of a batch recipe for a batch distillation column. The last task is identified as the one where the yield of the product achieved in the previous subtask is above a high value of 80%.

3.4.3.2 Short Example

We want to distill a mixture of methanol and methyl acetate (15% kmol MeOH and 85% kmol MeAc). The objective is to achieve 99%-pure methyl acetate at the bottom of the column, with 95% recovery of the maximum yield. In the case of minimum boiling azeotropic mixtures, the azeotrope is considered to be the first product. The algorithm uses the azeotropic composition (as it is a low-boiling azeotrope) as the distillate specification. The application of the algorithm provides the following sequence of subtasks. The simulation engine in iCAS for batch processes (BRIC) was employed for dynamic simulation to verify the suggested recipe. The operating conditions for each subtask were defined, with the distillate composition being the end constraint for each subtask, except for the last subtask, where the purity of MeAc left in the column was the end constraint. The results (sequence of tasks and subtasks) are shown in Table 3.2.

When this operational sequence was compared with a constant-reflux-ratio operation, it was found that for low values of the reflux ratio the two end objectives could not both be achieved. For high values of the reflux ratio, the operating time was significantly higher than the one achieved by the operational sequence. Compared to operation with a constant reflux ratio where both product objectives are achieved, the time in the generated recipe is 30% faster, as can be seen in Table 3.3. Note that, because this example handles only a binary mixture, there is only one task with several subtasks. For a multicomponent system, more than one task would be sequenced in terms of the highest driving force. The identified

TABLE 3.2
Operational Sequence for Distilling Off the Binary Azeotrope, Leaving Pure MeAc in the Column

Subtask No.	Reflux Ratio	Vapor Boilup Rate	Operating Time (hr)	Simulated Operating Time (hr)	Acetone/MeAc Purity (%)	Acetone/MeAc Recovery (%)
1	4.8	100	0.942	1.176	95.89	—
2	10	100	0.465	0.459	99.05	95.91

TABLE 3.3
Results for Constant Reflux Ratio Operation

Operation	Number of Intervals	Reflux Ratio	Vapor Boilup Rate	Simulated Operating Time (hr)	MeAc Purity (%)	MeAc Recovery (%)
a	1	4.8	100	1.615	99.05	84.90
b	1	8.5	100	2.324	99.05	95.05
c	1	10	100	2.629	99.05	96.90
d	2	[4.8, 10]	100	1.635	99.05	95.91

separation tasks would be ordered in terms of their corresponding driving forces in descending order.

3.4.4 SEPARATION: SOLUTION CRYSTALLIZATION

An algorithm for the operational design in batch crystallization is illustrated in Figure 3.9. The objective of the algorithm is to identify in advance the necessary tasks and their sequence in order to achieve a specified set of objectives for the separation task — namely, the recovery (yield) of solutes. The algorithm uses insights obtained from analyses of solid–liquid equilibrium phase diagrams for the purpose of generating a sequence of feasible and near-optimal operational steps.

The algorithm consists mainly of a repetitive procedure that identifies the nature of each subtask and the corresponding operating conditions. The initial step is the actual generation of data for the phase diagram through tools for knowledge generation (provide a visual picture of the solution behavior). The phase diagram is obtained by drawing the solubility data for a given temperature range with respect to operation of the crystallizers. In the repetitive procedure, the feed mixture is located on the phase diagram and the feasibility of precipitation of the desired solid is checked. This is done by first checking the location of the feed point (if it is in the solid–liquid region, a solute can be made to precipitate)

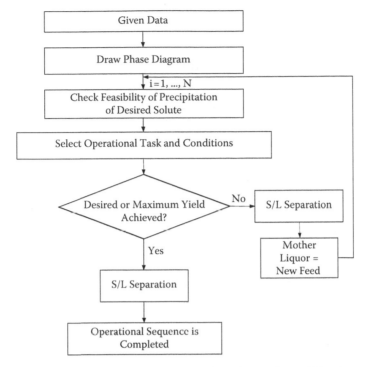

FIGURE 3.9 Schematic presentation of the algorithm for batch crystallization.

and then joining the corresponding solute vertex (on a ternary phase diagram) with the feed point and extending it to meet the saturation curve (calculated at a given temperature), as shown in Figure 3.10. This means that by evaporating (or removing) the solvent at a constant temperature (same as the temperature of the saturation curve), the solute will precipitate as a solid in equilibrium with a liquid solution given by the point of intersection on the saturation curve. Thus, by changing the temperature and adding or removing solvents, it is possible to move from different solid–liquid regions where different solids precipitate. Another way to check for the identity of the solid in the case of multiple solutes is to calculate the solubility index for solute j (SI_j), given by:

$$SI_j = \left[\Pi \left(K_j \right)^{v_i} \right] \Big/ \left[\sum \left(K_j \right)^{v_i} \right] \tag{3.20}$$

If SI_j for solute j at temperature T is >1, then solute j already exists as a solid in a solid–liquid region. If SI_j for solute j at temperature T is <1 but larger than any other solute, then by removing the solvent at constant temperature T solute j can be precipitated as a solid. The exact amount of solvent to evaporate or add can be found from the lever-arm principle, as the material balances can be represented on the phase diagrams in the form of tie lines. Unless the objective is to produce

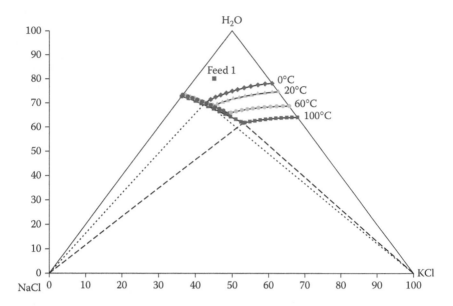

FIGURE 3.10 Phase diagram (salt solubility) at different temperatures.

a solid complex, it is important to avoid regions on the phase diagram where two solutes may precipitate. For this reason, evaporation is always less than the maximum, and dilution is more than the minimum. Knowing the exact location of the slurry, we can compute the product yield using the lever-arm principle. The composition of the mother liquor (saturated liquid solution) in equilibrium with the precipitated solute is identified on the phase diagram and treated as a new feed for the next solid product. The above-described procedure is repeated until the desired (or maximum) yield for the corresponding feed mixture is achieved. It can be noted that each precipitation of a solute may require a number of subtasks. A detailed case study in Section 3.6 highlights all the steps in the generation of a feasible and near-optimal operation model for a batch (crystallization) process.

3.4.5 NETWORK OF TASKS

As shown in Figure 3.1 and Figures 3.4a and 3.4b, most batch processes usually require more than one task. The multiple tasks may be of the same type or of different types. Thus, even for a single production line, the sequences of tasks and their corresponding subtasks must be generated. Essentially, a two-level approach can be used. In the outer level the sequence of tasks is established, and in the inner level the corresponding sequences of subtasks are established.[2,3]

Linninger et al.[1] developed a systematic method for synthesis of batch processes (Process_Synthesizer in BatchDesign_Kit) based on the hierarchical approach proposed by Douglas.[2,3] Linninger et al. proposed the following four steps for their synthesis (conceptual design) procedure:

1. Construction of multistage (or task) plant diagram
2. Hierarchical decomposition and synthesis of conceptual design for each task
3. Integration of the process (task) flowsheet for each task
4. Generation and improvement of alternatives

Application of the synthesis method was used to generate a detailed separation system for the zero-achievable-pollution (ZAP) design of the carbinol production stage.

3.5 METHODS AND TOOLS: VERIFY AND ANALYZE ALTERNATIVES

The effective simulation of batch processes requires representation of the dynamics of the individual batch operations, the decision logic associated with the start and stop of operations, as well as decisions associated with the assignment of equipment and other resources to specific operations as defined through the product recipe. The BATCHES (Batch Process Technologies, Inc.; www.bptech.com) simulation framework accommodates the above-mentioned batch process features and uses advances in combined discrete–continuous dynamic simulation methodology to follow the progress of a batch plant over time. Linninger et al.[1] also provide a Process Assessor tool in their BatchDesign_Kit (BDK) software,[24] which is a computer-aided design environment for the interactive development of processes for manufacturing pharmaceuticals and specialty chemicals.

After generation of a batch recipe, it is important to verify the operational sequence through simulation. For batch operation simulation, the model must be able to simulate the behavior of the process subject to a specific set of operational instructions. Commercial simulators provide this simulation option. The important issue here is to be able to capture all the operational instructions to match the sequence of operations exactly as the synthesis algorithms generate them. In this chapter, we have used the BRIC toolbox in iCAS[7] to illustrate all batch operation simulation. BATCHES is also a useful tool for evaluation of alternatives. For input, it requires the recipe network, equipment network, and set of processing directions (operation model). As output, it gives the states at the end of each task and subtask.

Another useful analysis tool is to identify areas where the operation can be improved; for example, study the effect of substituting one solvent with another or creating a two-phase reactor operation by adding a solvent. Algorithms for finding suitable replacement solvents based on computer-aided molecular design (CAMD) have been developed and are available as a toolbox in iCAS. A review of various types CAMD methods, tools, and applications (especially to solvents for organic synthesis) can be found in Achenie et al.[25] Other tools from iCAS that are useful for conceptual synthesis and design of batch processes are ProPred (pure component property estimation), TML (a thermodynamic model library for calculating phase diagrams), and PDS (a process design studio for calculation of distillation boundaries, residue curves, and batch distillation verification)

3.6 CASE STUDIES

3.6.1 Network of Separation (Crystallization) Tasks

This case study addresses the generation of a batch recipe for crystallization (precipitation) of two salts from an aqueous electrolyte mixture. The synthesis results have been verified through rigorous dynamic simulation, and iCAS software has been used for all calculation steps.

3.6.1.1 Problem Definition

Recover 95% of the dissolved sodium chloride (NaCl) from a mixture of water, NaCl, and potassium chloride (KCl) with a feed composition (on a 100-kg feed basis) of 80% water, 15% NaCl, and 5% KCl. The operating temperature range is 273 to 373 K.

3.6.1.2 Step 1: Generate the Phase Diagrams at Different Temperatures

Note that, even though the main components are water, sodium chloride, and potassium chloride, some extra compounds (double salts and hydrates) must be included in the component list, as they may also precipitate at various temperatures. These extra compounds are the double salts of the two single salts and their hydrates (mono- and dihydrates). Figure 3.10 shows the calculated phase diagrams at different temperatures and the saturation lines for different salts. The position of the feed is in the unsaturated region, which is the upper part of the triangle defined by the solubility curves. Below the solid–liquid equilibrium curves, one or two salts precipitate, depending on the position in the triangular diagram. In the area defined by the NaCl vertex of the triangle, the invariant point, part of the solubility curve, and the NaCl–H$_2$O axis, the only salt precipitating is NaCl. In a similar area on the right side of the triangle, the only salt precipitating is KCl. In the area defined by the NaCl and KCl vertices and the invariant point, both salts precipitate. This is illustrated in Figure 3.11. All the phase diagrams are generated through the electrolyte options of TMS in iCAS.

3.6.1.3 Step 2: Check for Feasibility of the Specified Salt

Let (X_F, Y_F) be the rectangular coordinates of the feed and (X_{inv}, Y_{inv}) the coordinates of the invariant point at a specific temperature, and consider the following cases.

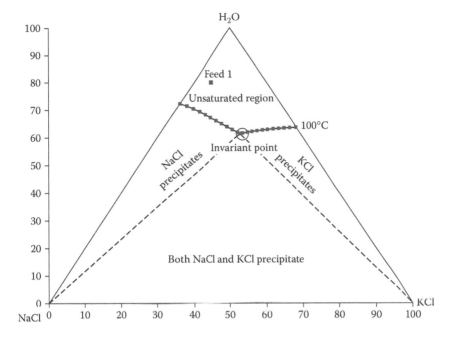

FIGURE 3.11 Phase diagram and location of feed and saturation lines.

a. $Y_F > Y_{inv}$. If

$$X = \frac{50 * (Y_{inv} - Y_F) - X_F * (Y_{inv} - 100)}{100 - Y_F} < X_{inv} \qquad (3.21)$$

- then the salt placed in the bottom left vertex of the triangle will precipitate. If the X described above is greater than X_{inv}, then the salt in the bottom right vertex of the triangle will precipitate. If $X = X_{inv}$, then both salts will precipitate.

b. $Y_F < Y_{inv}$. In addition to the constraint above, the following constraint also has to be satisfied:

$$Y_F > \frac{Y_{inv}}{X_{inv}} * X_F \qquad (3.22)$$

- for the salt in the bottom right vertex of the triangle to precipitate. Otherwise, both salts will precipitate.

3.6.1.4 Step 3: Choose the First (Operation) Crystallization Temperature

For the specific problem, the precipitation of the desired salt (NaCl) is feasible for all the temperatures in the given operating range (constraint 1 is satisfied), so crystallization at any temperature within the specified range will give NaCl. Because the feed lies in the unsaturated region, evaporation (to remove the solvent, water) will have to be applied. As water is removed, the composition of the feed moves away from the H_2O vertex along the line connecting the water vertex and the feed (as shown in Figure 3.12). From Figure 3.12, it can be seen that the area where only NaCl precipitates is larger at temperature $T = 373$ K. At this temperature, evaporation can be more extensive and the resulting slurry has a larger density:

$$\text{Slurry density} = \text{Mass of salts/mass of slurry}$$

$$= \text{Length slurry\_mother liquor/length NaCl\_mother liquor}$$

Therefore, the operating temperature for the crystallization is chosen to be 373 K.

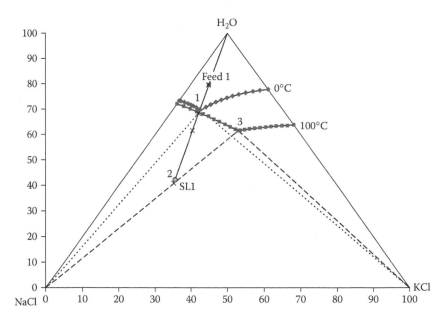

FIGURE 3.12 Location of operation lines for crystallization operation at 373 K.

3.6.1.5 Step 4: Precipitation of Salts (by Evaporation of Solvent)

As mentioned previously, as water is removed, the composition of the feed moves away from the water vertex along the line connecting the water vertex and the feed; however, a maximum evaporation can be achieved without crossing the boundary into the area where both salts precipitate (unwanted). The composition of the slurry on that boundary is found as the intersection of the evaporation line and the line connecting the NaCl vertex and invariant point at $T = 373$ K. The actual slurry composition is found for about 99% of the maximum evaporation.

- *First precipitation operation:* Evaporate feed solution (feed 1). Saturation occurs at point 1, on the curve. Continue evaporation until the slurry composition is at point 2.

- *Amount of solvent evaporated:* Length Feed1_pt.2/length H_2O_pt.2 * 100 kg feed = 65.278 kg H_2O. At point 2, the remaining feed (charge) is $100 - 65.278 = 34.722$ kg slurry. Reaching point 2, we have only removed water. The composition at point 2 is given by:

$$80 - 65.278 = 14.722 \text{ kg } H_2O; \ 15 \text{ kg NaCl}; \ 5 \text{ kg KCl}$$

- Slurry at point 2 is NaCl + liquor at point 3. Point 3 is found as the intersection of the line connecting NaCl vertex and the slurry (point 2) and the solubility curve at $T = 373$ K:

  Slurry density = Length 2_3/length NaCl_3 = 0.3176.
  Solids precipitated = 0.3176 * 34.722 = 11.028 kg NaCl.
  Product yield = 11.028/15 = 73.52% < desired yield (95%).
  Residue at point 3 is $34.722 - 11.028 = 23.694$ kg mother liquor.

Moving from point 2 to point 3 we have only removed NaCl.

- *Second precipitation operation:* Because the desired yield for NaCl was not achieved, the mother liquor from the solid–liquid separation is treated as a new feed. A feasibility check makes it obvious that it is not possible to precipitate NaCl at any temperature for that feed (diluted or condensed); however, the precipitation of KCl is feasible for a wide temperature range and, as can be seen from the phase diagram (Figure 3.13), point 3 lies in the saturated region for KCl so no evaporation is required. The operating temperature for the crystallization of KCl is chosen so the slurry density is the highest. This task is achieved at temperature $T = 273$ K; therefore, the second crystallizer is to operate at $T = 273$ K. Also, because the mother liquor (point 3) from the first

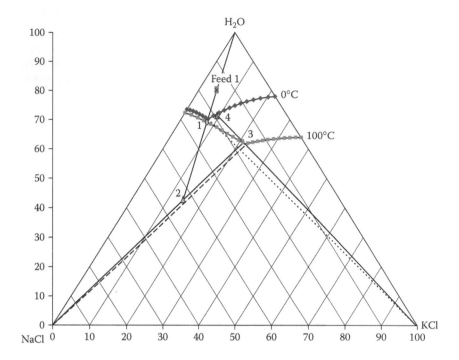

FIGURE 3.13 Cooling for crystallization at 273 K.

crystallizer is in the saturated region for KCl at 273 K, no evaporation
is needed (see Figure 3.13):

> Slurry at point 3 is KCl + liquor at point 4.
> Point 4 is found similarly to point 3.
> Slurry density = Length 3_4/length KCl_4 = 0.1322.
> Solids precipitated = 0.1322 * 23.694 = 3.132 kg KCl.
> Residue at point 4 is 23.694 – 3.132 = 20.562 kg mother liquor.

The mother liquor in equilibrium with the precipitated salt in the second
crystallizer (point 4) is treated as a new feed (see Figure 3.13), for
which, precipitation of NaCl is feasible for a temperature range of 293
to 373 K. For the same reasons as in the first subtask (step 3), the operating temperature is chosen to be T = 373 K. The new feed (point 4) lies
in the unsaturated region and evaporation must be applied. The evaporation ratio is calculated such that the resulting slurry is in equilibrium
with a mother liquor of composition point 3 – ML1. This means that the
resulting slurry has a composition at the intersection of the evaporation
line connecting point 4 and the water vertex and of the line connecting
the NaCl vertex and point 3 (ML1). The amount of solvent evaporated
is:

> Length pt.4_pt.5/length H_2O_pt.5 * 20.562 kg feed (point 4) = 9.209 kg H_2O.
>
> From point 4 to point 5: Remaining feed at point 5 = 20.562 – 9.209 = 11.353 kg slurry.
>
> Slurry density = Length 5_3/length NaCl_3 = 0.2179.
>
> Solids precipitated = 0.2179 * 11.353 =2.474 kg NaCl.
>
> Yield = (11.028 + 2.474)/15 = 90.01% < desired yield = 95%.

- *Repeated sequence of precipitation and heating and cooling.* Repeating the procedure and moving from points 3 to 4 to 5 and again to 3 on the phase diagram (see Figure 3.14) gives us the precipitation of NaCl and KCl at 373 K and at 273 K, respectively. The operational path (model) required to recover 95% of the NaCl from the salt solution is listed in Table 3.4.

3.6.1.6 Step 5: Verification by Simulation

The BRIC toolbox in iCAS has been used to model and simulate the generated sequences of batch operations. BRIC uses the operation model to set up the

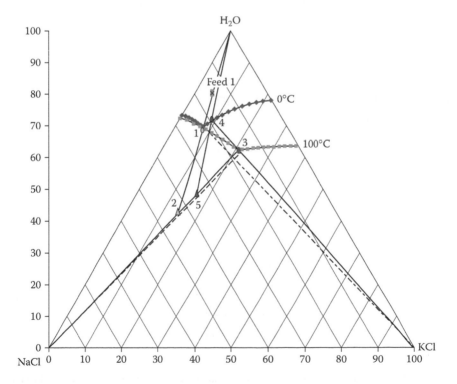

FIGURE 3.14 Second crystallizer operation and location of operating lines.

TABLE 3.4
Sequence of Tasks and Subtasks for Crystallization Case Study

Crystallizer	Temperature (K)	Operational Task	Predicted Product Yield (%)
1. NaCl (product)	373	Evaporation 65.28 kg	73.60
2. KCl (product)	273	Cool + crystallization	62.55
1. NaCl (product)	373	Evaporation 9.21 kg	90.09
2. KCl (product)	273	Cool + crystallization	85.95
1. NaCl (product)	373	Evaporation 3.15 kg	> 95

Note: In between operations with crystallizers 1 and 2, heating and cooling are also needed.

dynamic process model to validate the sequence of batch operations. Figure 3.15 shows a screen shot from iCAS where the model flow diagram is illustrated. CRYST is the dynamic crystallizer model. It receives a charge from stream 1, and through streams 2, 3, and 4 it discharges the vapor, mother liquor and solid, respectively (charging/discharging is achieved by opening and closing of stream valves). Tmixer1 and Tmixer 2 are tanks that collect the mother liquor and precipitated salts. Simulation runs are set up by specifying the time of operation, an evaporation rate, an operation temperature, etc., depending on whether it is necessary to precipitate, heat or cool, or mix. At the end of each run (operation), the mother liquor is placed in the CRYST and a new operation is simulated.

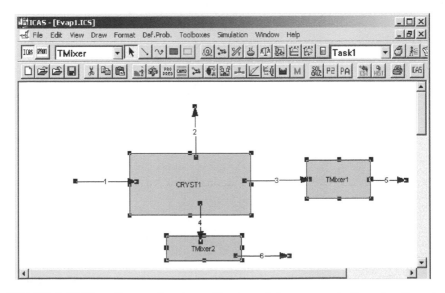

FIGURE 3.15 Flow diagram for the verification of the generated alternative through dynamic simulation.

To reach the first evaporation point, 1 hour of evaporation operation at 3.6234 kmol/hr was needed. At this point, 65.278 kg of solvent (water) was evaporated. Running the crystallization operation at zero vaporization for 0.67 hr yielded exactly the same amount of NaCl (11.028 kg). At this point, the mother liquor was placed back in the crystallizer, which was cooled to 273 K; vaporization was set to zero and precipitation for 0.53 hr yielded 3.132 kg of KCl. Heating to 373 K and operating for 1 hour at an evaporation rate of 0.511 kmol/hr removed 9.21 kg of solvent. Subsequent precipitation at zero vaporization gave a further 3.15 kg of NaCl. Thus, this model allowed us not only to verify the sequence of operations needed to obtain the required amounts of salts but also to calculate the operating times (not estimated by the synthesis algorithm, unlike the batch distillation algorithm). Placing this simulation in a dynamic optimization loop, the cost of operation and the time of operation may be optimized. Note, however, that the operating temperatures and evaporation rates have been chosen (implicitly) for the largest amount of precipitation with the lowest amount of evaporation (as defined by the phase diagrams); thus, the solution obtained from the first simulation is already a very good operation sequence.

3.7   USEFUL REFERENCES FOR FURTHER STUDY

- Separation task selection — Jaksland,[9] Chakraborty and Linninger[17,18]
- Solvent selection and design — Achenie et al.[25]
- Optimization for synthesis and design — Bhatia and Beigler[26]
- Simulation and design — Salomone et al.[27]
- Batch process systems engineering — Rippin[28]
- Computer-aided tools related to batch process — Puigjaner et al.[29]

3.8   CONCLUDING REMARKS

This chapter has highlighted important issues related to the conceptual design and synthesis of batch processes in terms of the methods and tools that can be applied to such problems. The methods and tools discussed in this chapter are not necessarily the best available but are ones that would be able to help the reader in formulating, analyzing, and solving typical conceptual design and synthesis problems. For illustration purposes, application examples have been kept simple in order to provide a visual picture that is easy to understand and appreciate. More realistic and complex problem formulations and solutions can be found in the referenced papers. The main message from this chapter is that, even though the complete synthesis and design of batch operations may be quite complex, the problem can be divided into smaller problems for whose solution systematic and easy-to-use methods and tools exist. A number of the smaller problems require analysis of the available (or generated) information and data, and design decisions are made based on these analyses. Such synthesis and design give us one or more operation models on which to base further studies with respect to planning and scheduling, control, and operational analyses.

NOTATION

B	Cost data for structural (binary) variables; amount remaining in batch distillation still
$B(k,l)_{ij}$	Binary ratio matrix of binary pair k, property l, and compounds i and j
C	Cost data for structural and process (real) variables
D	Amount of distillate
DF	Driving force
F	Initial charge
f_{ij}	Flow rate of compound i in stream j
F_j	Total flow rate of stream j
l_i	Lower limit value for constraint i
M	Total mixed flow rate
M_w	Molecular weight
P_c	Critical pressure
R	Reflux ratio
t	Independent variable
T_b	Normal boiling point
T_c	Critical temperature
T_m	Normal melting point
u_i	Upper limit value for constraint i
V	Vapor boilup rate
V_{dw}	Van der Waal's volume
V_m	Moral liquid volume
x	Vector of continuous variables
x_i	Mole fraction of compound i
X_j	Composition in rectangular coordinates (x-axis) for stream j
y	Vector of binary integer variables
y_i	Mole fraction of compound i in mixture
Y_j	Composition in rectangular coordinates (y-axis) for stream j
z	Vector of design variables
z_i	Mole-fraction-like values for property cluster i

Greek Symbols

q	Property function or property
q	Property cluster

REFERENCES

1. Linninger, A.A., Ali, S.A., Stephanopoulos, E., Han, C., and Stephanopoulos, G., Synthesis and assessment of batch processes for pollution prevention, *AIChE Symp. Ser.*, 90(303), 46, 1994

2. Papaeconomou, I., Gani, R., and Jørgensen, S.B., A general framework for the synthesis and operational design of batch processes, in *Computer-Aided Chemical Engineering*, Vol. 12, Grievink, J. and van Schijdel, J., Eds., Elsevier, Amsterdam, 2002, 289.

3. Papaeconomou, I., Gani, R., Jørgensen, S.B., and Cordiner, J., Synthesis design and operational modelling of batch processes, in *Computer-Aided Chemical Engineering*, Vol. 13, Kraslawski, A. and Turunnen, I., Eds., Elsevier, Amsterdam, 2003, 245.

4. Puigjaner, L., Espuna, A., and Reklaitis, G.V., Frameworks for discrete/hybrid production systems, in *Software Architectures and Tools for Computer-Aided Process Engineering*, Braunshweig, B. and Gani, R., Eds., Elsevier, Amsterdam, 2002, 663.

5. Barton, P., Ahmed, B., Cheong, W., and Tolsma, J., Synthesis of batch processes with integrated solvent recovery, in *Tools and Methods for Pollution Prevention*, Sikdar, S. and Diwekar, U., Eds., Kluwer Academic, The Netherlands, 1999, 205.

6. Kondili, F., Pantelides, C.C., and Sargent, R.W.H., A general algorithm for short-term scheduling of batch operations. 1. MILP Formulation, *Comput. Chem. Eng.*, 17, 211, 1993.

7. Gani, R., *ICAS Documentations*, CAPEC Publ. PEC02-14, Technical University of Denmark, Lyngby, 2003.

8. Jaksland, C.A., Gani, R., and Lien, K., Separation process design and synthesis based on thermodynamic insights, *Chem. Eng. Sci.*, 50, 511, 1995.

9. Jaksland, C.A., Separation Process Design and Synthesis Based on Thermodynamic Insights, Ph.D. dissertation, Technical University of Denmark, Lyngby, Denmark, 1996.

10. Gani, R. and Bek-Pedersen, E., Simple new algorithm for distillation column design, *AIChE J.*, 46, 1271, 2000.

11. Ahmad, B.S. and Barton, P.I., Solvent recovery targeting, *AIChE J*, 45, 335, 1999.

12. Barton, P.I. and Pantelides, C.C., Modelling of combined discrete/continuous processes, *AIChE J.*, 40, 966, 1994.

13. Graells, M., Canton, J., Peschaud, B., and Puigjaner, L., A general approach and tool for the scheduling of complex production systems, *Comput. Chem. Eng.*, 22(suppl.), 395, 1998.

14. Reklaitis, G.V., Progress and issues in computer aided batch process design, in *Foundations of Computer-Aided Process Design*, Sirola, J.J., Grossmann, I.E., and Stephanopoulos, G., Eds., CACHE and Elsevier, Amsterdam, 1990, 241.

15. Faqir, N.M. and Karimi, I.A., Optimal design of batch plants with single production routes, *Ind. Eng. Chem. Res.*, 28, 1191, 1989.

16. Allgor, R.L., Barrera, M.D., Barton, P.I., and Evans, L.B., Optimal batch process development, *Comput. Chem. Eng.*, 20, 885, 1996.

17. Chakraborty, A. and Linninger, A., Plant-wide waste management. 1. Synthesis and multiobjective design, *Ind. Eng. Chem. Res.*, 41, 4591, 2002.

18. Chakraborty, A. and Linninger, A., Plant-wide waste management. 2. Decision making under uncertainty, *Ind. Eng. Chem. Res.*, 42, 357, 2003.

19. Eden, M.R., Jørgensen, S.B., and Gani, R., A new modeling approach for future challenges in process and product design, in *Computer-Aided Chemical Engineering*, Vol. 13, Kraslawski, A. and Turunnen, I., Eds., Elsevier, Amsterdam, 2003, 101.

20. Diwekar, U., *Batch Distillation: Simulation, Optimal Design and Control*, Taylor & Francis, London, 1995.

21. Hostrup, M., Integrated Approach to Computer-Aided Process Synthesis, Ph.D. dissertation, Technical University of Denmark, Lyngby, Denmark, 2002.

22. Gani, R., Hytoft, G., Jaksland, C.A., and Jensen, J.A., An integrated computer aided system for integrated design of chemical processes, *Comput. Chem. Eng.*, 21, 1135, 1997.

23. Douglas, J.M., Process synthesis and waste minimization, *Ind. Eng. Chem. Res.*, 31, 238, 1992.

24. Linninger, A.A., Ali, S.A., Stephanopoulos, E., Han, C., and Stephanopoulos, G., Synthesis and assessment of batch processes for pollution prevention, *AIChE Symp. Ser.*, 90(303), 46, 1994.

25. Achenie, L.E.K., Gani, R., and Venkatasubramanian, V., *Computer-Aided Molecular Design: Theory and Practice*, Elsevier, Amsterdam, 2002.

26. Bhatia, T. and Biegler, L.T., Dynamic optimization in planning and scheduling of multiproduct batch plants, *Ind. Eng. Chem. Res.*, 35, 2234, 1996.

27. Salomone, H.E., Montagna, J.M., and Iribarren, O.A., Dynamic simulation in the design of batch processes, *Comput. Chem. Eng.*, 18, 191, 1994.

28. Rippin, D.W.T., Batch process systems engineering: a retrospective and prospective view, *Comput. Chem. Eng.*, 17(suppl.), 1, 1993.

29. Puigjaner, L., Graells, M., and Reklaitis, G.V., Computer tools for discrete/hybrid production systems, in *Software Architectures and Tools for Computer-Aided Process Engineering*, Braunshweig, B. and Gani, R., Eds., Elsevier, Amsterdam, 2002, 433.

4 Batch Reactors in the Bioindustries

John Villadsen

CONTENTS

ABSTRACT

Stoichiometric relations and kinetic expressions are given for bioreactions as an introduction to the design of bioreactors, both for aerobic and for anaerobic processes. It has been shown that the exchange of mass between a gas phase and the liquid phase reaction medium is an important part of the design, and correlations are given to calculate mass transfer coefficients in different types of tank reactors. Time profiles for biomass concentration are calculated for simple Monod kinetics and for rate expressions that exhibit either product inhibition or substrate inhibition. It is argued that straightforward batch operation is not common practice in the bioindustry, primarily because undesired side reactions can severely curtail the overall yield of the desired product in the typical batch reaction where the substrate concentration is initially high. A variant of standard batch operation, the so-called fed-batch cultivation, is shown to be a good option. The undesirable catabolite-repression that makes it unattractive to operate in strict batch mode is avoided, and the problems concerned with steady-state continuous operation are also circumvented. The design of a fed-batch process for baker's yeast is shown as an example. Time profiles are determined for the increase of biomass concentration and of medium volume, and thereafter the energy requirement for transport of oxygen to the medium is calculated. Due to an increasing superficial gas velocity with increasing medium volume, the power input needed to achieve a given mass transfer coefficient may actually decrease for large reactor volumes.

4.1 INTRODUCTION

Producing bulk chemicals by fermentation-based routes rather than by the classical organic chemistry-based routes is receiving increased attention. The reaction conditions are environmentally friendly. Reaction temperature is low, and the reaction medium is usually a highly diluted aqueous solution of mineral salts to which the carbon-containing substrate is added. Inexpensive carbon-containing substrates such as glucose can have a bulk price of less than 20¢ per kg (at least in the midwestern United States), and glucose is the preferred feed of many industrially used microorganisms.

Right now, large-scale bioprocesses are in the process of being commissioned for production of lactic acid (Cargill-Dow) and propane 1,3-diol (DuPont). Huge quantities of ethanol to be used as a fuel additive are being produced by fermentation. Single-cell protein (SCP; a fish meal substitute and a possible future ingredient in human diets) is produced by microbial oxidation of methane. The first natural-gas based-SCP plants were built in the Soviet Union more than 20 years ago. Today, a much more energy-efficient process is used in Norway,[7] and the product is sold as feed to the salmon-aquaculture industry.

Reactions in which living cells are used to produce desirable chemicals are different from other chemical reactions, most of all because of the autocatalytic nature of the reaction. The rate of reaction is proportional with the biomass concentration, and biomass is one of the products of the reaction. This leads to a negative-order kinetics for the reaction, and a continuous stirred-tank reactor will have a higher productivity than a batch reactor, at least down to very small residual concentrations of the reactants (substrates).

Another feature of bioreactions that favors continuous steady-state operation is the immense complexity of the regulatory network of microorganisms. Even small changes in the environment can lead to the switching-off of the desired pathway in the organism, and the carbon flux is then directed toward undesirable byproducts. An example is the production of industrial enzymes by fermentation of glucose using filamentous fungi. The genes responsible for production of the desired proteins will be downregulated as soon as the glucose concentration in the medium exceeds about 20 mg L^{-1}, which would make it impossible to produce the proteins by batch fermentation; only worthless biomass but no heterologous protein is being produced.

Still, the batch reactor has many advantages, and it is being used for small-scale productions of biochemicals if catabolite repression can be avoided. A variant of batch operation, the so-called fed-batch process is also used in the production of high-volume/low-unit-cost chemicals; hence, this review of the use of batch reactors in the bioindustry focuses on the fed-batch operation.

To understand how bioreactions are designed, it is necessary to discuss rate expressions of various degrees of complexity. This is done in the following sections with an emphasis on the applicability of different rate expressions in different situations. Simple, unstructured kinetics can be used for both steady-state continuous operation and usually also batch operation. Rate expressions that reflect deeper layers of the regulation of mass flow in microorganisms must be used for rapid, transient changes in the environment of the cell. The coupling between mass transfer and the chemical reaction is crucial for the successful operation of bioreactors; consequently, calculation of mass transfer rates is discussed in some detail.

The current debate in the bioindustry revolves around the advantages and disadvantages of batch or steady-state continuous operation. With an increasing understanding of the details of microbial physiology and with the appearance of robust strains that are less prone to degrade to nonproducing "bald" strains by mutation, some of the objections to continuous fermentations are now disappearing. Low-value products must eventually be made in continuous culture to cut down the cost of processing, including manpower. The long preproduction breeding of inoculum is avoided, and (cleaning in place) CIP and reactor sterilization is done less frequently in continuous operation. Even pharmaceuticals are now produced in continuous culture (e.g., human insulin in recombinant yeast), and other high-value products will follow suit. One can, however, predict that the fed-batch variant of batch fermentation will hold its own for a long time. The productivity can be almost as high as in continuous fermentation, and extraction of the desired product from the more concentrated broth of the batch- or fed-batch fermentation is easier than downstream processing of the effluent from a continuous reactor.

4.2 STOICHIOMETRIC CONSIDERATIONS

In contrast to "normal" chemical reactions, the stoichiometry of bioreactions is very complex, and the study of biostoichiometry is a science in itself. *Metabolic flux analysis* is the name given to the analysis, at steady state, of the carbon distribution to the many different metabolic pathways in a given microorganism. One can obtain widely different stoichiometries for the bioreaction at different dilution rates, $D = v/V(h^{-1})$, in a steady-state, continuous culture or in more general terms at different specific growth rates $= \mu(h^{-1})$ (which happens to be equal to D in a steady-state, continuous culture); consequently, the yield coefficient, $Y_{sp} = $ g product (g substrate)$^{-1}$, can depend heavily on the operating conditions of the bioreactor.

The stoichiometry of the bioreaction can be written as a *black box model*:

$$-CH_2O - Y_{sn}NH_3 - Y_{so}O_2 + Y_{sx}X + Y_{sp}P + Y_{sc}CO_2 + Y_{sw}H_2O = 0 \quad (4.1)$$

where biomass X, metabolic product P, and $CO_2 + H_2O$ are the products formed from the substrates glucose, ammonia, and oxygen. In Equation 4.1, the sum of the yield coefficients of the three carbon-containing products is 1 on a C-mol basis; therefore, we will usually work on a one-carbon atom basis for the carbon- and energy-containing substrate and with a formula weight for X based on a one-carbon atom. If nothing else is stated, we will use a standard biomass with formula $X = CH_{1.8}O_{0.5}N_{0.2}$ ($M_x = 24.6$ g DW (C-mol)$^{-1}$).

The coefficients of Equation 4.1 are related not only by a carbon balance but also by a nitrogen, an oxygen, and a hydrogen balance. Usually one is not interested in the small amount of water formed by the reaction in the very dilute aqueous medium, and the last term in Equation 4.1 is often left out. In this case, a combined oxygen + hydrogen balance, a so-called *degree of reduction balance*, is a further constraint on the stoichiometric coefficients. The oxygen demand is given by:[6]

$$Y_{so} = 1/4(4 - 4.20Y_{sx} - \kappa_p Y_{sp}) \qquad (4.2)$$

where κ_p is the degree of reduction of the metabolic product and is calculated from the formula of the product (on a one-carbon atom basis) and with the definition of one redox neutral compound for each element ($\kappa_{H2O} = \kappa_{CO2} = \kappa_{NH3}$ $= 0 \rightarrow \kappa_O = -2$, $\kappa_C = 4$, and $\kappa_N = -3$, when the redox unit is defined as $\kappa_H = 1$). With only one metabolic product (P) and a carbon, nitrogen, and redox balance, all stoichiometric coefficients in Equation 4.1, except Y_{sw}, can be calculated when Y_{sx} is measured.

The flaw of this procedure, as indicated above, is that the stoichiometric coefficients change with the operating conditions. In a cultivation of *Saccharomyces cerevisiae* (baker's yeast), Y_{sp} is practically zero when the specific growth rate (μ) is small (e.g., below 0.20 to 0.25 h^{-1} for normal strains of *S. cerevisiae*), but it increases rapidly (to about 0.5 C-mol (C-mol)$^{-1}$) when μ increases above a threshold value (e.g., 0.25 h^{-1}, corresponding to a volumetric flow $v = 1/4$(h^{-1}) V in a steady-state continuous tank reactor).

In reality the black-box model, Equation 4.1, is obtained by linear combination of several redox and adenosine triphosphate (ATP)-balanced elementary reactions:

$$-(\alpha + 1)CH_2O - 0.2NH_3 + X + \alpha CO_2 + (4(\alpha + 1) - 4.20)/2NADH - \beta ATP = 0 \qquad (4.3)$$

$$-3/2CH_2O + CH_3 O_{1/2} + 1/2CO_2 + 1/2ATP = 0 \qquad (4.4)$$

$$-CH_2O + CO_2 + 2NADH = 0 \qquad (4.5)$$

$$-O_2 - 2 NADH + 2P/O ATP = 0 \qquad (4.6)$$

In Equations 4.3 to 4.6, it is assumed that β moles of ATP are consumed to make one C-mol biomass. The ATP production by conversion of 3/2 C-mol glucose to one C-mol ethanol is 1/2 mol. For each C-mol of biomass produced in Equation

4.3, $(2\alpha - 0.10)$ moles of redox equivalents are formed, and by complete oxidation of glucose (in the TCA-cycle) 2 mol of NADH are produced.

The rate of oxygen consumption (r_{O2}) per C-mol biomass is obtained from Equation 4.6. $\Sigma(r_{NADH})$ is the total production rate of NADH by reactions 4.3 and 4.5, and respiration of the NADH gives rise to a rate of formation of ATP (r_{ATP}), which is equal to $-2r_{O2}\ P/O$, where the P/O ratio is a measure of the efficiency of the respiratory system to generate metabolic energy by oxidation of NADH (or H_2, in a less formal nomenclature). P/O as well as the coefficients α and β in Equation 4.3 are empirical parameters with values that depend on the environment of the microorganism (e.g., the dilution rate in a steady-state continuous culture).

The black box–stoichiometry is obtained (e.g., per 100 C-mol glucose consumed) by a linear combination of the four *metabolic fluxes* (v_i), the rates of the four reactions in Equations 4.3 to 4.6. To find the fluxes, one may use an overall ATP balance as the net rate of metabolic energy production is zero in the steady state, but even the simplest formulation of the metabolic network still has three empirical parameters: α, β, and P/O. These parameters must either be estimated from biochemical data or they must be found by measurements (e.g., of CO_2, the biomass production rate, or the oxygen consumption rate, $-r_{O2}$).

If the desired metabolic product is an amino acid, a large amount of NADPH is required for synthesis of the product. Thus, for production of L-lysine from glucose by *Corynebacterium glutamicum*:

$$-C_6H_{12}O_62NH_3 - 4NADPH + C_6H_{14}O_2N_2 + 2NADH = 0 \qquad (4.7)$$

$$-1/3C_6H_{12}O_6 + 2CO_2 + 4NADPH = 0 \qquad (4.8)$$

$$-O_2 - 2NADH = 0 \qquad (4.9)$$

with the overall reaction:

$$-4/3C_6H_{12}O_62NH_3 + C_6H_{14}O_2N_2 + 2CO_2(+ 4H_2O) = 0 \qquad (4.10)$$

In Equations 4.7 to 4.9, another redox-carrying cofactor, NADPH, is introduced. This is synthesized in the so-called PP-pathway at the expense of glucose that is oxidized to CO_2.

From Equation 4.10 it follows that the maximum yield of L-lysine is 0.75 C-mol per C-mol glucose, but other products are formed together with lysine (first of all biomass), and Equation 4.10 must be coupled with Equation 4.3 and possibly another equation that describes glucose consumption for maintenance of cells without any biomass production.

It should be evident that the correct formulation of a stoichiometric representation of the large number of bioreactions that together lead to formation of a desired product is a major task that must necessarily precede any attempt to set up a set of rate equations, and the rate equations are needed to set up the mass balances that constitute the basis for design of the bioreactor.

4.3 KINETICS OF BIOREACTIONS

Just as is the case for ordinary chemical reactions, the rate expression for a bioreaction is a relation between reactant concentrations (concentrations of substrates and products) and the rate of the reaction. The stoichiometric coefficients of the previous section were determined by measurements of rates at a given set of environmental conditions, including the concentrations of reactants, but each set of measured rates as used in metabolic flux analysis constitutes an isolated package. With a kinetic expression one can extrapolate to new environmental conditions, and if the expression is based on a reasonable theoretical or experimental foundation one obtains reliable estimates of the rates at new conditions without having to do additional experiments.

Rate expressions for bioreactions are grouped into expressions that can be used for *balanced growth* and expressions that are suitable to simulate *unbalanced growth*. Balanced growth is characteristic for continuous, steady-state cultivations, whereas rapid transients can only be reliably simulated using a rate expression for unbalanced growth.

At first it might seem absurd that one has to work with two sets of rate expressions, but the difference between the two lies only in the amount of detail included in the model. It is inconceivable that a single rate expression with a manageable set of parameters would be able to describe at all environmental conditions the growth and product formation of an organism on a given set of substrates. This would disregard the vast amount of biochemical knowledge that is available on mass flow through the metabolic network and the equally important knowledge of the signal flow network that exerts a tight control on cell behavior.

Conversion of substrates to products in a batch reactor is of course in principle a transient process, but for all practical purposes one may regard growth to be balanced in the batch reactor. The *inoculum* is often taken from a breed reactor at balanced growth conditions similar to those that will be experienced in the production reactor. As will be explained below, cell growth and associated product formation will take place in an environment where the substrate concentrations are usually so high that the rate is independent of the substrate concentrations. In the very last part of the cultivation, the relation between substrate concentrations and rate may become more complicated, but this has hardly any influence on the outcome of the process.

The most frequently used rate expression for balanced growth was proposed by Jacques Monod in 1942 and is known as the Monod rate expression:

$$q_x = r_x x = \mu x = \frac{\mu_{max} s}{s + K_s} x \qquad (4.11)$$

Variants of the Monod expression that account for inhibition of the reaction by either the substrate itself or of a metabolic product P are given in Equations 4.12 and 4.13:

$$q_x = \frac{\mu_{max} s}{\dfrac{s^2}{K_i} + s + K_s} \, x \tag{4.12}$$

$$q_x = \frac{\mu_{max} s}{s + K_s} \left(1 - \frac{p}{p_{max}} \right) x \tag{4.13}$$

In all three cases, the rate expression bears a resemblance with kinetic expressions derived for enzymatic reactions. It must, however, be realized that, whereas the corresponding expressions for enzyme kinetics have a definite background in mechanistically derived rate models, all three expressions Equation 4.11 to 4.13 are strictly speaking only data fitters. There is, however, a good reason for their applicability to predict the rate of balanced growth. Equation 4.11 is a mathematical formulation of a verbal model for cell growth that tells us that the activity of the cell to accomplish a certain goal (e.g., to increase its weight) must have an upper limit (μ_{max}) determined by the amount of "machinery" that can be harbored within the cell. Also, the rate of any chemical process is eventually going to be first order in the limiting reactant concentration when this concentration decreases to zero. Hence, the hyperbolic form of $\mu(s)$ in the Monod expression is intuitively correct.

It should be noted that in all rate expressions for bioreactions the *volumetric rate* (q_x) is the product of the *specific rate* (cell activity), $\mu(s)$ or $\mu(s,p)$, and the biomass concentration, x (g L^{-1}). The reaction rate in the real reactor, the cell, is denoted r_i (specifically, μ (g DW (g DW h)$^{-1}$) $= r_x$ for the rate of biomass production) while the rate of production in the vessel (the reactor) in which the cells live is q_i (g i (L medium h)$^{-1}$).

It is *only* at conditions of balanced growth that the influence of the biomass can be expressed in terms of the biomass concentration alone. The *quality* of the biomass does not enter the picture as it must necessarily do in rapid transients where the activity of the cell changes due to a changing biomass composition. Cell *viability* is not considered, although the culture may contain an increasing fraction of unproductive or even dead cells at the end of a batch cultivation. *Mutation*, whereby a production strain reverts to the unproductive wild-type strain, is another phenomenon that can invalidate the assumption of balanced growth.

A simple example of an empirical rate expression where the formation of product during a batch fermentation has a long-term influence on the quality of the cells is the kinetics for ethanol formation by the bacterium *Zymomonas mobilis* proposed by Jöbses et al.[4] Here, the activity of the cells to produce both biomass and the product P = ethanol depends on a cell component (E), which at steady-state conditions will be produced in a minute but constant ratio (e_{ss}) to the cell mass. When the ethanol concentration increases, the rate of formation of this essential biomass component (it could be related to the RNA content of the cell) decreases, and cell growth, as well as the ethanol production rate, decreases. Equations 4.14 and 4.15 are modifications of the kinetics proposed by Jöbses et al.:[4]

$$q_x = f(s)ex; \quad q_p = (Y_{xp}f(s)e + m_p)x; \quad q_s = -(Y_{xs}f(s)\, e + m_s)x; \quad q_e = Y_{xe}f(s)ex$$
$$(4.14)$$

$$Y_{xe} = (k_1 - k_2p + k_3p^2) \tag{4.15}$$

In Equation 4.14, $f(s)e$ is the specific growth rate (μ) of the culture; $f(s)$ is given by the Monod expression, Equation 4.11. A characteristic feature of many design models is that the rates are taken to be proportional ($q_i = Y_{xi}q_x$ for products and $-Y_{xs}q_x$ for substrates). In the above model, this is true for the rate of formation of E while an extra term, accounting for *maintenance* (m_px and $-m_sx$) is added to the growth-related term in the expression for q_s and q_p. An extra feature of the kinetic model is that the volumetric growth rates are based not only on x but also on the tiny fraction e of biomass that constitutes the essential cell component, E. The yield coefficient of E on X is furthermore taken to be a function of the product concentration, p.

It would appear from the form of the simple rate expressions in Equations 4.11 to 4.14 that a change in the value of the substrate concentration (s) would lead to an immediate change of r_i. This would be in analogy with rate expressions typical for chemical reactions. When the hydrogen partial pressure over an ammonia catalyst is changed it takes at most a few seconds to change the coverage of H_2 on the catalyst surface, and thereafter the rate of ammonia production changes to a new value. This is not so for bioreactions. It may take on the order of hours to obtain a new steady-state value of μ after an increase of the dilution rate, despite the fact that the glucose concentration in the tank changes within a few minutes to a level that is perhaps 100 times higher than the level in the previous steady state. As demonstrated by Duboc et al.[1] and by Melchiorsen et al.,[5] catabolism becomes completely decoupled from anabolism during the first part of the transient. Cell growth is slow, but essential enzymes for the production of ATP in catabolism are produced much faster — an intuitively reasonable strategy for any living organism that suddenly experiences a much more benign environment.

The large difference between the time constant for substrate concentration change and the time constant for the growth of cellular components is exactly what makes it obligatory to work with models for steady-state growth (or balanced growth) and with other models that are able to capture the complex structure of the bioreactions in transient growth. The kinetics of Equations 4.14 and 4.15 is a simple version of models that are sufficiently structured to capture long-range effects, also in a transient operation. The influence of a higher ethanol concentration is not direct, as in Equation 4.13, but indirect, via the growth effect on E. One may say that the inhibition is observed after a time delay that depends primarily on the parameters of $f(p)$ and on the two maintenance coefficients m_p and m_s.

When Equations 4.14 and 4.15 are used in a dynamic model for a stirred-tank continuous reactor, very complex oscillatory patterns appear. These are also observed in experimental studies of *Zymomonas mobilis*. Oscillations are not likely to occur in batch cultivation, as all elements of the state vector will change monotonically (sugar is converted to products, but neither biomass nor ethanol can be used to regenerate the sugar).

In the previous discussion it has been implicitly assumed that the growth-limiting substrate (S) can be identified. This is, however, not always easy. Assume that simple Monod kinetics with maintenance substrate consumption is to be studied. At a low dilution rate, much of the substrate is used for maintenance. The observed yield coefficient of biomass is $Y_{sx} = \mu/(-r_s)$, and this is related to the "true" growth yield coefficient Y_{sx}^{true} by:

$$Y_{sx} = \frac{1}{\left(Y_{sx}^{true}\right)^{-1} + \dfrac{m_s}{\mu}} \tag{4.16}$$

For increasing specific growth rate, Y_{sx} and Y_{sx}^{true} approach each other. If the feed composition is taken to be the same at both a high and low specific growth rate, one may easily run into a deficiency of another substrate that enters into a biomass in a constant atomic ratio to, for example, carbon. This happens when μ changes from a low to a higher value during the batch fermentation, as will typically happen in carbon-substrate inhibited kinetics (Equation 4.12). The nitrogen source is a typical representative of other substrates that may become limiting during the cultivation. Thus, in lactic acid fermentation, the nitrogen source is a complex mixture of easily absorbable peptides with two to five carbon atoms and other nitrogen compounds that are much more difficult to digest by the lactic bacteria.

A "clean" cultivation with glucose as the limiting substrate is characterized by a value of the saturation constant K_s in Equation 4.11 of the order 1 to at most 150 mg L^{-1}. The lowest values of K_s are observed for growth of filamentous fungi (used in the production of industrial enzymes). Here, K_s is probably no larger than 1 mg L^{-1}. When cultivating yeast, the saturation constant can be as high as 150 mg L^{-1}. In all cases, the ratio between the initial glucose concentration $s_0 = s$ ($t = 0$) and K_s is likely to be more than 200 to 500, and the batch fermentation operates at the maximum specific growth rate μ_{max} during practically the entire processing time. This is, of course, what gives rise to the linear relationship between $\ln(x/x_0)$ and cultivation time that is characteristic of a successful batch operation. Unfortunately, one does see published values of K_s that are in the range of 0.5 to 2 g L^{-1}. This result, obtained by data fitting of batch fermentation data using Equation 4.11, bears no relation to the true value of K_s, which is probably at least an order of magnitude smaller. What is observed in reality is that another substrate takes over as the limiting substrate. It could be an amino acid, but it could also be a mineral (sulfur or phosphorous) or a vitamin (e.g., B_{12}) that the organism cannot synthesize and which must be supplied with the medium.

Another typical source of errors when fermentation data are used to fit parameters in kinetic expressions is that the influence of a product on the rate is not recognized. Thus, in a cultivation that is product inhibited, the $\ln(x/x_0)$ vs. time profile will have a long tail for large t values. If Equation 4.11 rather than Equation 4.13 is used to fit the kinetic parameters, then a very large value of K_s will be obtained, and this of course is quite wrong. Product inhibition can rather easily be detected by repetition of the batch cultivation with an initial non-zero concentration of the product, whereas the detection of another substrate that becomes growth limiting is a much more difficult task as one may not suspect the nature of the missing substrate; it may even be a trace element present in some batches of the complex substrate used in industrial fermentations but lacking in other batches.

4.4 SUBSTRATES EXTRACTED FROM A GAS PHASE

In aerobic fermentations, oxygen must be supplied by transport from a gas phase that is sparged to the liquid medium. For a batch reactor, the volumetric rate of oxygen consumption $(-q_{O2})$ is given by:

$$-q_{O2} = k_l a(s_1^e - s_1) \qquad (4.17)$$

In Equation 4.17, s_1 and s_1^e are, respectively, the oxygen concentration in the liquid phase at the conditions of the bioreaction and the oxygen concentration that is in equilibrium with the gas phase where the partial pressure of oxygen is π; s_i^e is proportional to π, and values of s_i^e for $\pi = 1$ bar are available in tables for many different fermentation media. Thus, for a fermentation medium with 20 g L^{-1} glucose and 2 to 5 g L^{-1} mineral salts, one can use $s_1^e = 1.16$ mM at 30°C. $k_l a$ is the *mass transfer coefficient* (h^{-1}). It depends on the medium properties (coalescent media giving a lower mass transfer coefficient than noncoalescent media), and it depends on the equipment used to disperse the gas into the liquid. But, for all types of mass transfer equipment, $k_l a$ depends on the specific power input P/V (W m^{-3}) to the reactor volume V. It also depends on the volumetric flow rate v_g (m^3 h^{-1}) of the gas that is dispersed into the liquid. In most correlations for mass transfer the superficial gas velocity u_g (m s^{-1}) is used rather than v_g.

$u_g = v_g/(3600A)$, where A is the cross-sectional area of the reactor (m^2)

$$(4.18)$$

For mechanically stirred reactors, van't Riet[9] collected a large number of data for both small pilot reactors and larger industrial reactors and correlated the data by:

$$k_l a \,(\text{h}^{-1}) = 93.6 \, u_g^{0.5}(P/V)^{0.4} \text{ (where } u_g \text{ in ms}^{-1} \text{ and } P/V \text{ in W m}^{-3}) \quad (4.19)$$

The power input can be provided by means other than by mechanical stirring, and because some of the alternative methods can give the same mass transfer coefficient at a lower power input and at considerably lower investment costs these newer types of mass transfer equipment are gaining popularity. The mass transfer is obtained by circulating the entire mixture of liquid medium and injected gas through a large loop in which a number of static mixers are installed (a 300-m^3 reactor for production of single-cell protein is operating according to this principle[7]), or a small fraction of the total medium volume may be circulated very rapidly through a loop and reinjected into the bulk liquid. In the latter procedure, the gas is fed to the pressurized liquid (up to 10-bar gauge pressure) in the loop, and the gas–liquid mixture is injected into the reaction medium through rotary jet heads.[8]

One may conceive of the Norferm reactor as equipment in which the gas is contacted with the liquid in a more or less plug flow. When the gas is separated from the liquid after having traveled with the medium through the loop, a satisfactory degree of conversion of oxygen (or methane in the SCP reactor) must have been reached. In the system described by Nordkvist et al.,[8] the circulation of liquid and gradual feed of gas to the loop ensure that the reactor works as an almost ideal stirred-tank reactor.

The power input to the liquid flowing in the loop is proportional to v_l^3 if the liquid flow is turbulent and the power input needed to move the gas phase is neglected:

$$P = f v_l^3/0.036 \text{ (where } P \text{ is in W and } v_l \text{ is in m}^3 \text{ h}^{-1}) \quad (4.20)$$

In the Norferm reactor, factor f is a function of the number of static mixers and their construction. In the system described by Nordkvist et al.,[8] f is a function of nozzle diameter d in the rotary jet heads used to dispense the air–liquid mixture from the loop into the bulk liquid. It was found for all the rotary jet heads used and independent of the liquid viscosity that

$$f(d) = (0.1054d^2 + 0.4041d)^{-2} \text{ (where } d \text{ is in mm)} \quad (4.21)$$

In their investigation, a medium volume of $V = 3.4$ m^3 was used, and this was circulated 5 to 10 times per hour through a loop of volume 70 L. The holding time of liquid in the loop was from 7 to 14 sec, and liquid velocities in the range of 1 to 2 m s^{-1} were obtained. Obviously, a very efficient mass transfer was obtained when the gas–liquid mixture was injected into the bulk volume.

The power input (P) to the liquid–gas mixture in the loop is:

$$P = P_0 u_g^{0.60} N^{-3+0.4} = P_0 u_g^{0.6} N^{-2.6} \qquad (4.22)$$

In Equation 4.22, P_0 is the power input to an unaerated liquid flow v_l calculated by Equations 4.20 and 4.21, and N is the number of rotary jet heads used to dispense the circulating liquid plus the gas fed to the loop into the bulk liquid. The power input decreases rapidly with N due to the strong dependence of P on the volumetric rate of liquid flow shown in Equation 4.20.

Correlation of the oxygen mass transfer data obtained with respectively one and two rotary jet heads gives the following relations between $k_l a$, the specific power input, and the superficial gas velocity for a water-like medium:

$$k_l a = 278 \, u_g^{0.631}(P/V)^{0.267} \quad (N = 1) \qquad (4.23)$$

$$k_l a = 440 \, u_g^{0.631}(P/V)^{0.267} \quad (N = 2) \qquad (4.24)$$

In both expressions $k_l a$ is in h^{-1}, u_g in $m \, s^{-1}$ and (P/V) in $W \, m^{-3}$.

The ratio between the two numerical constants (1.58) corresponds closely to the ratio between the total power input to one and two rotary jet heads which is predicted from Equation 4.22. For a given value of P/V, one would expect that $k_l a$ is factor $(2)^{2.6 \cdot 0.267} = 1.62$ higher when two machines rather than one are used to dispense the circulating liquid (plus gas) into the bulk liquid in the tank.

In both mechanically stirred reactors and in reactor configurations with forced-flow circulation of the medium the power input is used to set the liquid into motion and thereby increase the effectiveness of the gas–liquid mixing process that leads to transfer of oxygen from the gas to the liquid phase. The power input used to compress the gas is not included, but it is usually much smaller than the power input to the liquid.

In bubble columns the only power input is that used to compress the gas and disperse it through spargers to the medium; consequently, the mass transfer coefficient depends only on the linear gas velocity (i.e., on the volumetric gas flow rate) and on the liquid properties. For a water-like medium, Heijnen and van't Riet[3] correlated mass transfer data for bubble columns of widely different size (diameters ranging from 0.08 to 11.6 m; height, from 0.3 to 21 m; and u_g, from 0 to 0.3 m s^{-1}). Their correlation is:

$$k_l \, a = 1152 \, u_g^{0.7}(k_l \, a \text{ in } h^{-1} \text{ and } u_g \text{ in } m \, s^{-1}) \qquad (4.25)$$

Application of Equation 4.25 for a typical μ_g value of 0.06 m s^{-1} yields $k_l a = 161$ h^{-1}. For a relatively small extra power input of 2000 W m^{-3} used to circulate liquid through the pressurized loop, one obtains a $k_l a$ value of 358 h^{-1} (Equation 4.23) and 566 h^{-1} (Equation 4.24) when, respectively, one and two rotary jet head machines are used. In order to reach a satisfactory biomass concentration in a batch fermentation, it may become very important to use an efficient but still not excessively energy-demanding method for contacting the gas and the liquid phases.

Experimental determination of the mass transfer coefficient should preferably be done in the actual vessel in which the bioreaction is going to be carried out, although some correlations such as Equation 4.23 and 4.24 are very robust for scale-up as one could just add more rotary jet heads in a larger medium volume, assuming that each machine works in its own part of the total reactor volume.

There are a number of well-proven experimental methods for determination of $k_l a$.[6] The best method appears to be the so-called peroxide method originally proposed by Hickman[2] and now used as standard for large industrial bioreactors by Novozymes, the world's leading producer of industrial enzymes.

A small amount of catalase is added to the medium, and a constant volumetric flow of air is sparged into the reactor. The saturation tension (p^{sat}) is measured on an oxygen electrode positioned in the medium at a position where the absolute pressure is p^t. A constant flow of H_2O_2 to the reactor (M moles of H_2O_2 per hour) is started at time $t = 0$. The peroxide is converted to O_2 and H_2O by the catalase, and the excess oxygen is stripped off by the constant flow of air through the reactor. A steady state is reached after a short time (4 to 5 min for a reactor volume of 3 m³), and the oxygen tension p in the liquid is recorded. ($p/p^{sat} - 1$) $= DOT - 1$ is positive because the oxygen tension in the liquid is higher than p^{sat} as long as the peroxide is degraded. After 20 to 30 min, a large number of steady-state readings of p has been recorded, and the flow of H_2O_2 is stopped, and DOT returns to 1 after 3 to 8 min. The term $k_l a$ is calculated as:

$$k_l a = \frac{M}{2 \, V \, s^{sat} \, p^t (DOT - 1)} \qquad (4.26)$$

In Equation 4.26, s^{sat} (mol m⁻³ bar⁻¹) is the saturation concentration of oxygen in equilibrium with the gas leaving the reactor volume (V); s^{sat} is slightly higher than the saturation concentration when the medium is in contact with air alone due to the release of O_2 to the gas.

Figure 4.1 shows the result of a typical experiment in which v_g (= 40 Nm³ air h⁻¹) is contacted with V (= 3.4 m³ water at 25°C). The constant flow of H_2O_2 is 93.0 mol h⁻¹. In the steady state, $DOT - 1 = 0.5$. $p^t = 1.03$ bar, and $s^{sat} = 0.223$ mol O_2 m⁻³ bar⁻¹ (saturated with water and with 1.14 m³ O_2 released h⁻¹); $k_l a$ is calculated to 96 ± 1.5 h⁻¹.

4.5 MASS BALANCES FOR BIOREACTORS WORKING IN THE BATCH MODE

The mass balances for bioreactors are of course not different from the balances that can be set up for batch reactors working with ordinary chemical reactions.

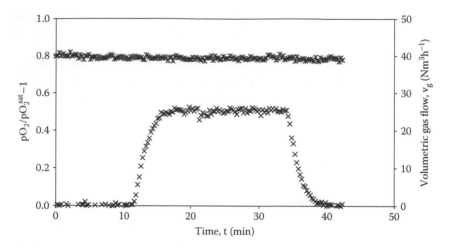

FIGURE 4.1 Typical mass transfer experiment according to the peroxide method. The medium is water, and one rotary jet head (d = 10 mm) is used.

There are, however, a few complicating factors due to the complexity of the kinetic expressions for the bioreactions. When any of the unstructured kinetic models, Equations 4.11 to 4.13, are used in the mass balance the result is quite simple:

$$\frac{dx}{dt} = \mu x \ , \text{ where } x = x_0 \text{ for } t = 0 \tag{4.27}$$

$$\frac{ds}{dt} = -Y_{xs}\mu x \ , \text{ where } s = s_0 \text{ for } t = 0 \tag{4.28}$$

$$\frac{dp}{dt} = Y_{xp}\mu x \ , \text{ where } p = 0 \text{ for } t = 0 \tag{4.29}$$

At first it would appear as though there are three independent mass balances, but the substrate (S), the biomass (X), and the product (P) are coupled through total mass balances if the rates are proportional (i.e., if the yield coefficients are constant):

$$s = s_o - Y_{xs}(x - x_o); \ p = Y_{xp}(x - x_o) \text{ if } p_o = p \ (t = 0) = 0 \tag{4.30}$$

Inserting Equation 4.30 in Equations 4.27 to 4.29 leads to one differential equation in x. It can often be solved analytically by separation of variables, but numerical solution is not a problem when a specific case is being studied. If maintenance substrate consumption and corresponding product formation must be included (and this is necessary for many fermentations, such as lactic acid fermentation), then an analytical solution of Equations 4.27 to 4.29 is not possible, but numerical solution is always possible.

When a specific cell component such as E in the kinetics of Equations 4.14 and 4.15 is included in the kinetic expressions, then the mass balance for E must be included with the three balances, Equations 4.27 to 4.29. Thus, in the case of Equation 4.14:

$$\frac{d(ex)}{dt} = Y_{xe}q_x \quad or \quad \frac{de}{dt} = Y_{xe}\mu - e\mu = r_e - \mu e \qquad (4.31)$$

This equation is independent of the reactor type used in the process. It says that component E must be generated by cell reactions at least as fast as the culture grows (i.e., at least as fast as μx; otherwise, the component will eventually be "washed out" by the growth of the biomass. Because $r_e = Y_{xe}f(s)\,e = Y_{xe}\mu$, r_e will slowly decrease as the ethanol concentration enters the region where Y_{xe} is negative (i.e., between the zeros of the quadratic). Nielsen et al.[6] provide a number of other examples to show how the so-called compartment-structured kinetic models are used in mass balances for different types of bioreactors.

Here, only a few examples are given to show the integrated form of the mass balances for unstructured kinetics and constant yield coefficients. Thus, for the three kinetic models Equations 4.11 to 4.13, the analytical solutions are:

$$\mu_{max}t = \left(1 + \frac{a}{1 + X_0}\right)\ln\left(\frac{X}{X_0}\right) - \frac{a}{1 + X_0}\ln(1 + X_0 - X) = F(X) \quad (4.32)$$

$$\mu_{max}t = b(1 + X_0)\ln\left(\frac{X}{X_0}\right) - b(X - X_0) + F(X) \qquad (4.33)$$

$$\mu_{max}t = (X_{max} - X_0)\left[\frac{1 + X_0 + a}{X_{max}(1 + X_0)}\ln\left(\frac{X}{X_0}\right) - \frac{a}{(1 + X_0)(X_{max} - 1 - X_0)}\right.$$

$$\left.\ln(X_0 + 1 - X) - \frac{X_{max} - 1 - X_0 - a}{X_{max}(X_{max} - 1 - X_0)}\ln\left(\frac{(X_{max} - X)}{(X_{max} - X_0)}\right)\right]$$

$$(4.34)$$

$$X = \frac{x}{s_0 Y_{sx}}, \quad X_0 = \frac{x_0}{s_0 Y_{sx}}, \quad a = \frac{K_s}{s_0}, \quad b = \frac{s_0}{K_i}$$

$$X_{max} = X_0 + \frac{p_{max}}{Y_{sp}s_0} \ (Eq.4.13), \ p_0 = 0)$$

Because, in general, a is of the order of 0.001, the logarithmic dependence of x on t for small t is apparent for the Monod kinetics (Equation 4.32). For $t \rightarrow \infty$, X approaches $1 + X_0$, and the last term in Equation 4.32 dominates, although the numerical constant is small. Equation 4.33 is the corresponding analytical expression for substrate-inhibited kinetics (Equation 4.12). The last term in Equation 4.33 is equal to the expression $F(X)$ in Equation 4.32, but for $X/X_0 \sim 1$, the sum of the first two terms can be approximated by $b(X - X_0)/X_0$, and if b is large it may take quite some time before X moves away from X_0. The last expression, Equation 4.34, is the analytical solution of the mass balances for q_x given by Equation 4.13. The product inhibition makes it impossible to reach the expected value of X for large values of s_0, because the last factor in Equation 4.13 becomes zero at a certain $p = p_{max} = Y_{sp}(s_0 - s)$.

Figure 4.2 shows the solution to the mass balances for the three kinetic expressions, Equations 4.11 to 4.13. In all three figures, $\mu_{max} = 0.4$ h^{-1}, $K_s = 0.2$ g substrate L^{-1}, and $x_0 = 0.1$ g biomass L^{-1}. The yield coefficients are $Y_{sx} = Y_{sp} = 0.5$ g g^{-1}. In Figure 4.2b $p_{max} = 6$ g L^{-1}, and in Figure 4.2c $K_i = 0.5$ g L^{-1}. Each figure has two curves, one for an initial substrate concentration of $s_0 = 5$ g L^{-1} and one for $s_0 = 20$ g L^{-1}. In Figure 4.2a it can be seen that the difference $x(final)$ – x_0 increases by a factor of 4 when s_0 increases from 5 to 20 g L^{-1}. This also happens in the substrate-inhibited case (Figure 4.2c), but for $s_0 = 20$ g L^{-1} it takes forever to start the climb toward $x(final)$. In the product inhibition case, $x - x_0$ is limited by $p_{max}Y_{sx}/Y_{sp} = p_{max}Y_{px}$. For $p_{max} = 6$ g L^{-1} and $Y_{px} = 1$, the maximum difference between x and x_0 is 6 g L^{-1}, as also seen in Figure 4.2b.

The sharp bend of the $x(t)$ curves when the substrate concentration decreases below K_s is particularly noticeable in Figure 4.2a. $K_s/s_0 = a$ is 0.025 and 0.01 for the two s_0 values used in the example. Determination of K_s by batch experiments is clearly impossible, and the only kinetic parameter that can be obtained is the maximum specific growth rate, μ_{max}. In Figure 4.2b, falsification of K_s by product inhibition is particularly noticeable for $s_0 = 20$ g L^{-1}, while the strong substrate inhibition in Figure 4.2c leads to falsification of μ_{max}.

4.6 FED-BATCH OPERATION

As already mentioned in the introduction, a "clean" batch operation is rarely the best choice. Catabolite repression will frequently make it impossible to achieve any satisfactory yield of the desired product on the substrate (glucose), and the process is not economically viable. The *fed-batch operation* is, however, very popular in the fermentation industry, so the necessary mathematical background for design of fed-batch operation is reviewed in the following text.

The reactor is started as a batch, and a suitably large biomass concentration is obtained by consumption of the initial substrate. The rate of product formation is typically low or even zero during the batch cultivation. At a certain time, usually when the substrate level has decreased to a very low level, a feed of (usually very concentrated) substrate is initiated. At the same time, an *inducer* may be added to switch on the metabolic pathways that lead to the desired product. During the

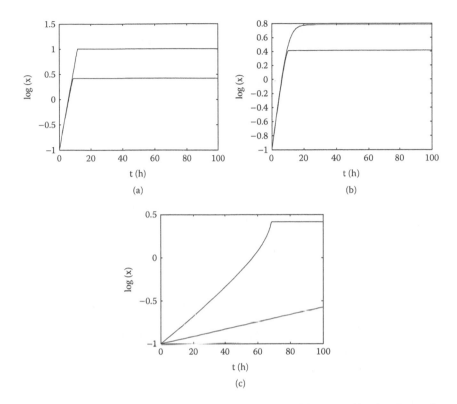

FIGURE 4.2 Time profiles for biomass concentration x for (a) Monod kinetics; (b) product inhibition, and (c) substrate inhibition (Equations 4.11 to 4.13). Batch reactor with $x_0 = 0.1$ g L^{-1} and s_0 either 5 or 20 g L^{-1}; $p_0 = 0$; $Y_{sx} = Y_{sp} = 0.5$ g g^{-1}.

entire fed-batch period, no product is withdrawn from the reactor, and the medium volume keeps increasing. At the end of the fed-batch period, a certain portion of the reactor volume may be withdrawn and sent to downstream processing. New substrate is added, and when the biomass concentration has reached a desired value by consumption of the added substrate a second fed-batch period is started. This is known as a *repeated fed-batch* operation. It will work satisfactorily if the remainder of biomass used to grow up a new culture has not been weakened, perhaps producing the desired product with a low yield. Thus, in penicillin fermentation, one may perhaps withdraw 70% of the reactor volume after the first fed-batch period, which can take up to 400 h. New substrate is added and fresh biomass is formed in batches. A second fed-batch period can be successful, but the high-yielding strains used today in the antibiotics industry are likely to revert to a very poor penicillin-producing strain ("wild-type").

When a time-varying feed stream, $v(t)$, containing one or more of the substrates at a constant concentration (c_f) is admitted to a stirred-tank reactor without withdrawal of a corresponding effluent stream, the mass balances become:

$$\frac{d(\mathbf{c}V)}{dt} = V\frac{d\mathbf{c}}{dt} + \mathbf{c}\frac{dV}{dt} = V\frac{d\mathbf{c}}{dt} + \mathbf{c}v(t) = V\mathbf{q}(\mathbf{c}) + \mathbf{c}_f v(t) \qquad (4.35)$$

The mass balances for the fed-batch operational mode therefore become:

$$\frac{d\mathbf{c}}{dt} = \mathbf{q}(\mathbf{c}) + \frac{v(t)}{V(t)}(\mathbf{c}_f - \mathbf{c}) \qquad (4.36)$$

A striking similarity exists between Equation 4.35 and the mass balances for a stirred-tank continuous reactor. In Equation 4.35, the dilution rate, $D = v(t)/V(t)$, is, of course, a function of time, and the fed-batch model is in principle a purely transient model, although most results can safely be derived assuming that the growth is balanced.

One may visualize the fed-batch operation as a control problem. Subject to certain constraints, it is possible to choose the control function, $v(t)$, such that a given goal is reached. This goal may be defined at the end of the fermentation process, where $V(t)$ has reached a specified value. This endpoint control problem is a classical problem of control theory — for example, choosing s_f and the initial values s_0 and x_0 that characterize the state when the fed-batch process $v(t)$ is initiated so a given state (x,s) is reached in the shortest possible time. The chemical engineering literature abounds with solutions to this kind of problem. Here, two simpler problems will be studied. The control action is applied with the purpose of achieving certain metabolic conditions for the cell culture at every instant during the fermentation. The concept of an instantaneous control action is illustrated in sufficient generality with only one growth-limiting substrate and the biomass as the state vector. To simplify the discussion, maintenance-free kinetics will be used. The two most obvious feed policies are:

- Choose $v(t)$ so $s = s_0$ throughout the fermentation.
- Choose $v(t)$ so $q_x = q_x^0$ throughout the fermentation.

The two policies correspond to fermentation at a constant, specific growth rate and at a constant volumetric rate of biomass production, respectively. Both policies have obvious practical applications. When s is kept at a level below that at which part of the added substrate is converted to undesired products, a large amount of biomass (together with an associated protein that may be the real product) is produced at a reasonably high rate, and a high final biomass concentration can be obtained. In the production of baker's yeast, neither the continuous stirred-tank reactor (one is afraid that the subtle qualities of the yeast that give the optimal leavening properties of the product will be lost in a long fermentation run) nor the classical batch reactor (diversion of glucose to ethanol, which inhibits the growth and represents a considerable loss of substrate) are suitable reactor choices. Fed-batch operation is typically the preferred choice. The constant volumetric rate policy is important if removal of the heat of reaction is a problem or if the capability to supply another substrate (e.g., oxygen) is exceeded when $q_x > q_x^0$.

Calculation of $v(t)$ corresponding to constant $s = s_0$ (which means that μ is constant and equal to μ_0) is quite simple. From Equation 4.35 and with $x_f = 0$:

$$\frac{d(xV)}{dt} = \mu_0 xV \implies xV = x_0 V_0 \exp(\mu_0 t) \tag{4.37}$$

while the substrate balance taken from Equation 4.36 reads:

$$\frac{ds}{dt} = 0 = -Y_{xs}\mu_0 x + \frac{v(t)}{V(t)}(s_f - s) \tag{4.38}$$

or

$$v(t) = \frac{Y_{xs}\mu_0}{s_f - s_0} xV = \frac{Y_{xs}\mu_0}{s_f - s_0} x_0 V_0 \exp(\mu_0 t) \tag{4.39}$$

$V(t)$ is found by integration of Equation 4.39 from $t = 0$, and $x(t)$ by inserting $V(t)$ in Equation 4.37:

$$\frac{V}{V_0} = 1 - bx_0 + bx_0 \exp(\mu_0 t) \tag{4.40}$$

where $b = Y_{xs} / (s_f - s_0)$, and

$$\frac{x}{x_0} = \frac{\exp(\mu_0 t)}{1 - bx_0 + bx_0 \exp(\mu_0 t)} \tag{4.41}$$

Equations 4.39 to 4.41 provide the complete explicit solution to the constant specific growth rate problem. v is seen to increase exponentially with time. The biomass concentration x is a monotonically increasing function of time with an upper limit:

$$1/b = Y_{sx}(s_f - s_0) \quad \text{for} \quad V \to \infty$$

The value of x for a specified V/V_0 is calculated from Equation 4.41 using a value of t obtained by solution of Equation 4.40.

The constant q_x fed-batch fermentation can also be designed quite easily. One would, of course, not voluntarily abandon the constant μ policy that gives the maximum productivity, consistent with the constraint that no byproducts should be formed, unless forced to do so. But, with the increasing biomass concentration x (Equation 4.41) a point may be reached at $t = t^*$ when, for example, the rate of oxygen transfer or the rate of heat removal from the reactor can no longer match the increasing q_x. From that point on, we must work with $q_x = q_0 = q(x^*)$.

In the constant μ period the value of $s = s_0 = s(\mu_0)$ is usually many orders of magnitude smaller than s_f. In the continued fermentation with constant q_x from V^* to V_{final}, the substrate concentration in the reactor decreases even further because μ_x is constant and there would be no reason to continue beyond t^* if x did not increase.

If we assume that the reactor volume keeps increasing exponentially also after t^*, then the transient mass balance for the biomass from $t_1 = t - t^*$ when the constant μ period ends is:

$$\frac{dx}{dt_1} = \mu x - \frac{v(t)}{V(t)} x = q_0 - kx \tag{4.42}$$

The exponential increase of V is given by:

$$V = V^* \exp(kt_1) \tag{4.43}$$

The value of the parameter k will be determined shortly. The substrate balance reads:

$$\frac{ds}{dt_1} = -Y_{xs}\mu x + \frac{v(t)}{V(t)}(s_f - s) = -Y_{xs}q_0 + k(s_f - s) \tag{4.44}$$

Integration of a weighted sum of the substrate and biomass balances yields:

$$x + Y_{sx}s = (x^* + Y_{sx}s_0 - Y_{sx}s_f) \exp(-kt_1) + Y_{sx}s_f \tag{4.45}$$

Because s and s_0 are negligible compared to s_f, Equation 4.45 can be simplified to:

$$x \quad (x^* - Y_{sx} s_f) \exp(-kt_1) + Y_{sx}s_f \tag{4.46}$$

Integration of Equation 4.42 yields another expression for x:

$$x = (x^* - q_0/k) \exp(-k t_1) + q_0/k \tag{4.47}$$

The two expressions for x become identical if:

$$q_0/k = Y_{sx}s_f \text{ or } k = q_0/(Y_{sx}s_f) \tag{4.48}$$

The design of the constant q_x policy is therefore quite explicit; k is chosen according to Equation 4.48 where q_0 as well as $Y_{sx}s_f$ are known. Hence, the time (t_1) to reach V_{final} is calculated from Equation 4.43 and the corresponding x value from Equation 4.47. The approximation in Equation 4.45 does not affect the result.

The constant q_x period can be shown to end with the same biomass concentration as would have been obtained if the constant μ policy could have been maintained until V_{final} was reached, but the processing time $t^* + (t_1)_{final}$ is longer, and hence the productivity is somewhat smaller.

These concepts will be illustrated by a practical example concerned with the design of a baker's yeast production. In the example, it will be assumed that the yeast grows on glucose with NH_3 as nitrogen source. The specific growth rate is given by:

$$\mu = \frac{0.4s}{s + 150 \ (mg \ L^{-1})} \tag{4.49}$$

For μ 0.25 h^{-1} (s 250 mg L^{-1}), the growth is purely respiratory and $Y_{xo} = 0.6836$ mol O_2 (C-mol biomass)$^{-1}$.

It is desired to design an optimal fed batch process starting at the end of a preliminary batch period in which the biomass concentration has increased to $x_0 = 1$ g L^{-1} and the glucose concentration has decreased to $s_0 = 250$ mg L^{-1}. The feed concentration during the fed-batch operation is 100 g glucose L^{-1}. At $t = 0$, the reactor volume is V_0, and the fed-batch process stops when $V = 4V_0$. The temperature is 30°C and the oxygen is fed as air with 20.96% O_2.

Obviously the constant μ policy will select $\mu = \mu_0 = 0.25$ h^{-1}, the largest value of the specific growth rate for which no byproducts are formed. Y_{sx} is calculated from a redox balance:

$$(1 - 1.05Y_{sx}) = Y_{so} = 0.6836Y_{sx} \text{ or } Y_{sx} = 0.5768$$

From Equations 4.39 to 4.41 and with $b = Y_{xs}/(s_f - s_0) = (24.6/30)/(100 \cdot 0.5768) = 0.02114$ L g^{-1}, one obtains the following:

$$v(t) = 0.25 \cdot 0.02114 \cdot 1 \cdot V_0 \exp(0.25 \ t) \tag{4.50}$$

$$V(t) = V_0(1 - b + b \exp(0.25 \ t)) \tag{4.51}$$

$$x(t) = \exp(0.25t) \ V_0/V for \ x_0 = 1 \ g \ L^{-1} \tag{4.52}$$

At the time t_{final} when $V = V_{final} = 4 \ V_0$, one obtains from Equation 4.51 that $(1 - b + b \exp(0.25 \ t)) = 4$ and t_{final} is calculated to 19.84 h. From Equation 4.52, the corresponding x value is determined to be 35.7 g L^{-1}.

Assume that the largest attainable value of $k_l a$ is 650 h^{-1} and that the oxygen tension in the medium needs to be 10% of the saturation value. The volumetric air feed is assumed to be so large that the partial pressure of oxygen in the inlet can be used to determine the saturation concentration of oxygen in the liquid:

$$(q^t)_{max} = 650 \ (1.16 \ 10^{-3} \cdot 0.2096) \cdot 0.9 = 0.1422 \ \text{mol } O_2 \ \text{h}^{-1} \ \text{L}^{-1.}$$

This oxygen uptake can support a volumetric biomass growth rate:

$$(q_x)_{max} = (0.6836)^{-1} \ 24.6 \ (q^t)_{max} = 5.116 \ \text{g L}^{-1} \ \text{h}^{-1}, \text{ corresponding to}$$
$$x = x^* = 20.46 \ \text{g L}^{-1}$$

Solving Equation 4.52 for $x = 20.46$ g L^{-1} yields $t^* = 14.26$ h, and from Equation 4.3 $V = V^* = 1.73 \ V_0$. From Equation 4.48, $k = 5.116/(0.4730 \cdot 100) = 0.1082$ h^{-1}. Thus, from $t = t^*$ ($t_1 = 0$) to t_{final} the reactor volume increases as $V = V^*$ exp(0.1082 t), and for $V = 4V_0$ (i.e., $V/V^* = 2.31$) one obtains $t_1 = 7.73$ h and t_{final} = 22.0 h. The value x_{final} is calculated from Equation 4.47 to be 35.62 g L^{-1}, and apart from the permissible approximation in Equation 4.45 this is the same as the concentration reached at the end of a constant-μ fed-batch fermentation. The increase in production time from 19.8 to 22 h is not large.

The optimal design of a fed-batch fermentation that gives the maximum productivity and yet satisfies the constraint imposed by a limited oxygen transfer should follow the lines illustrated in this example. One should, however, not be misled into believing that this is the overall best production policy. A continuous steady-state fermentation offers far greater productivity.

Let the reactor volume be 4 V_0, as this volume must be available at the end of the fed-batch process. If a continuous production of biomass with $x = 35.62$ g L^{-1} is to be maintained in the reactor then s_f should be $35.62/Y_{sx} = 61.75$ g L^{-1} when the miniscule effluent glucose concentration is neglected. If $D = 0.25$ h^{-1}, the highest D value for which no ethanol is produced, then $q_x = 0.25 \cdot 35.62 = 8.905$ g L^{-1} h^{-1}, which cannot be supported by the available mass transfer coefficient. To obtain $q_x = q_{xmax} = 5.116$ g L^{-1} h^{-1}, the dilution rate must be lower; namely, $D = 0.1436$ h^{-1}. But, still, a much higher volume of glucose can be processed to give $x = 35.62$ g L^{-1}: $v = 4V_0 \cdot 0.1436$ L h^{-1}; in other words, in 22 hours (t_{final}) a total volume of 12.64 V_0 is achieved compared to only 4 V_0 by the optimal fed-batch process.

To complete the design example one needs to calculate the power input required to transfer the required oxygen from the gas phase. Assume that the oxygen partial pressure in the gas leaving the reactor is 0.18 bar, and assume that the gas phase is ideally mixed into the liquid by the use of the efficient rotary jet heads that were described in the section on mass transfer. In the following calculation, the partial pressure of oxygen in equilibrium with the medium will be taken to be the average 0.1948 bar between the inlet partial pressure (0.2096

TABLE 4.1
Calculation of Specific Power Input P/V During Fed-Batch Fermentation of Baker's Yeast

x (g L^{-1})	1	10	20.46	26	32	35.62
V (m^3)	2	2.48	3.45	4.35	6.06	8.0
k_la (h^{-1})	34.16	341.6	699	699	699	699
qO_2 (mol m^{-3} h^{-1})	6.947	69.47	142.1	142.1	142.1	142.1
VqO_2 (mol h^{-1})	13.89	172.3	490	618	861	1137
v_g (m^3 h^{-1})	11.48	142.3	405	511	711	939
u_g (m s^{-1})	0.00259	0.0322	0.0917	0.1156	0.1610	0.2126
P/V (W m^{-3})	501	7259	8952	5197	2366	1226
P (kW)	1.002	18.0	30.9	22.6	14.3	9.81

Note: Until x = 20.46 g L^{-1}, the specific growth rate μ of the culture is constant at 0.25 h^{-1}, and the k_la value needed to sustain a certain biomass concentration increases proportional to x. During the constant q_x operation for larger x values, k_la remains constant at 699 h^{-1}.

bar) and the outlet partial pressure 0.18 bar. The amount of air needed to give a certain oxygen quantity Vq_{O2} mol h^{-1} is $(Vq_{O2})/0.0296$ mol h^{-1}.

In Table 4.1, the mass transfer coefficient k_la that is needed to sustain a given biomass concentration is calculated at different times during the process. Also shown in the table is the volumetric airflow rate (v_g) and the power input P (kW) calculated for one rotary jet head and using Equation 4.23.

It is of course quite unnecessary to use the high mass transfer coefficient (650 h^{-1} when calculating the course of the fed-batch process and now changed to 699 h^{-1} due to the lower oxygen partial pressure [0.1948 bar] used in the calculations of the table) when x is below 20.46 g L^{-1}. At this point, the operation must be switched from a constant μ to a constant q_x mode as a consequence of the upper limit (here, 699 h^{-1}) imposed on k_la.

It is at first surprising that P/V and even P have a maximum for an x value between 1 and 35.6 g L^{-1}, but this result is a consequence of the rapidly increasing value of μ_g when the oxygen demand and hence v_g increases. In a constant diameter reactor, the superficial velocity μ_g increases proportional with V, whereas it increases with $V^{1/3}$ when scale-up is done for a constant height-to-diameter ratio. In the fed-batch design, the aspect ratio, of course, does not change when the medium volume increases.

REFERENCES

1. Duboc, P., von Stockar, U., and Villadsen, J., Simple generic model for dynamic experiments with *Saccharomyces cerevisiae* in continuous culture: decoupling between anabolism and catabolism, *Biotechnol. Bioeng.*, 60, 180–189, 1998.
2. Hickman, A.D., Gas–liquid oxygen transfer and scale-up: a novel experimental technique with results for mass transfer in aerated agitated vessels, *Proc. 6th Eur. Conf. on Mixing*, Pavia, Italy, 369–374, 1988.
3. Heijnen, J.J. and van't Riet, K., Mass transfer, mixing and heat transfer phenomena in low-viscosity bubble columns, *Chem. Engr. J.*, 28, B21–B42, 1984.
4. Jöbses, I.M.L., Egberts, G.T.C., Luyben, K.C.A.M., and Roels, J.A., Fermentation kinetics of *Zymomonas mobilis* at high ethanol concentrations: oscillations in continuous cultures, *Biotechnol. Bioeng.*, 28, 868–877, 1986.
5. Melchiorsen, C.R., Jensen, N.B.S., Christensen, B., Jokumsen, K.V., and Villadsen, J., Dynamics of pyruvate metabolism in *Lactococcus lactis*, *Biotechnol. Bioeng.*, 74, 271–279, 2001.
6. Nielsen, J., Villadsen, J., and Lidén, G., *Bioreaction Engineering Principles*, 2nd ed., Plenum, New York, 2002.
7. Norferm, http://www.norferm.com, 2002.
8. Nordkvist, M., Grotkjær, T., Hummer, J.M., and Villadsen, J., Applying rotary jet heads for improved mixing and mass transfer in a forced recirculation tank reactor system, *Chem. Eng. Sci.*, 58, 3877–3890, 2003.
9. van't Riet, K., Review of measuring methods and results in non-viscous gas–liquid mass transfer in stirred vessels, *Biotechnol. Bioeng.*, 18, 357–364, 1979.

5 Batch Distillation

Ki-Joo Kim and Urmila Diwekar

CONTENTS

5.1 INTRODUCTION

Distillation has been widely accepted for product separation, purification, and waste removal in chemical process industries. Depending on whether the industry is handling petrochemicals, bulk chemicals, specialty chemicals, or pharmaceuticals, the distillation process can be divided into two categories: (1) batch distillation, which is mainly used in specialty chemical, biochemical, and pharmaceutical industries; and (2) continuous distillation, which is primarily implemented in the petrochemical and bulk chemical industries. Figure 5.1a shows a conventional batch distillation column where the feed is initially charged into the reboiler at the beginning of operation. After a total reflux operation (i.e., all condensates are recycled to the column), the distillate is continuously withdrawn while the bottom residue with a high-boiling-temperature component is concentrated, making this a time-varying process. In continuous distillation (Figure 5.1b), the feed is constantly supplied to the column, and the top and bottom products are simultaneously obtained under a steady-state operation. The upper section of the feed point is referred to as the *rectifying section*, as a low-boiling-temperature component is enriched. The lower section is referred to as the *stripping section*, as a low-boiling-temperature component is stripped off.

Batch distillation is the oldest separation process and the most widely used unit operation in the batch industry. Batch distillation is highly preferable to continuous distillation when high-value-added, low-volume chemicals must be separated. It is also widely used in chemical processing industries where small quantities of materials are to be handled in irregularly or seasonally scheduled

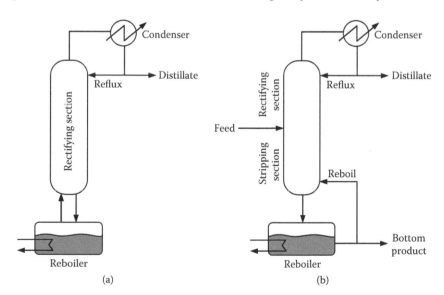

FIGURE 5.1 Types of distillation processes: (a) batch distillation, and (b) continuous distillation.

periods, and it is implemented when the feed composition varies widely from period to period or where completely different feed stocks have to be handled.

Theoretical studies on batch distillation began with a simple distillation still in a laboratory. In this type of distillation, a still is initially filled with a feed mixture, which evaporates and leaves the still in the vapor form. This vapor, which is richer in the more volatile component, is collected in the condenser at the top and accumulated in a receiver. In this operation, no liquid is refluxed back to the still, and no plates or packing materials are present inside the still. This simple distillation still is an example of a batch operation, often referred to as *Rayleigh distillation*[1] because of Rayleigh's pioneering theoretical work in simple distillation. The concept of reflux and the use of accessories such as plates and packing materials to increase the mass transfer converts this simple still into a batch distillation column, as shown in Figure 5.1a. Because this batch column essentially performs the rectifying operation, it is often referred to as a batch *rectifier.*

The most outstanding feature of batch distillation is its flexibility in operation. This flexibility allows one to deal with uncertainties in feed stocks or product specifications. In addition, one can handle several mixtures just by switching the operating conditions of the column. The basic difference between batch distillation and continuous distillation is that in continuous distillation the feed is continuously entering the column, while in batch distillation the feed is charged into the reboiler at the beginning of the operation. The reboiler in batch distillation gets depleted over time, so the process has an unsteady-state nature. A conventional batch column can be operated under the following operating conditions or policies:

- Constant reflux and variable product composition
- Variable reflux and constant product composition of the key component
- Optimal reflux and optimal product composition

Under conditions of constant reflux, the instantaneous composition of the distillate keeps changing because the bottom still composition of the more volatile component is continuously depleted. On the other hand, under variable reflux, the composition of the key component in the distillate can be kept constant by increasing the reflux ratio. The third type of operation, known as optimal reflux, is neither constant nor variable; instead, this type of operation exploits the difference between the two operating modes. Thus, the optimal reflux policy is essentially a trade-off between the two operating modes, and is based on the ability of the process to yield the most profitable operation.

The flexible and transient nature of batch distillation allows us to configure the column in a number of different ways, some of which are shown in Figure 5.2.[2] The column in Figure 5.2a is a conventional batch distillation column, with the reboiler at the bottom and the condenser at the top. A single column can be used to separate several products using the multifraction operation of batch distillation presented in Figure 5.2b. Some cuts may be desired and others may

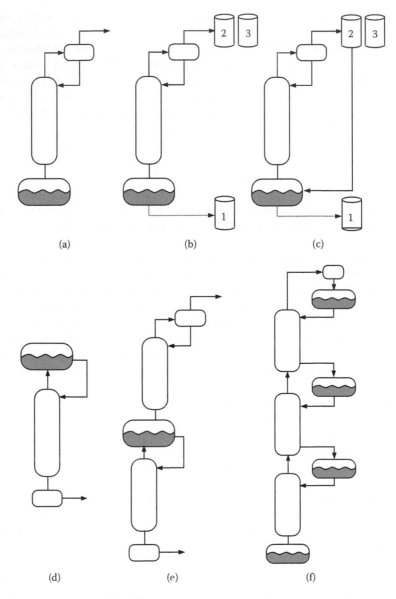

FIGURE 5.2 Examples of ways to configure the batch distillation column.

be intermediate products. These intermediate fractions can be recycled to maximize profits or minimize waste generation. Figure 5.2c shows a periodic operation in which each charge consists of a fresh feed stock mixed with recycled off-specification material from the previous charge. Figure 5.2d represents a stripping column for separating a heavy component as the bottom product where the liquid feed is initially charged into the top still. In 1994, Davidyan et al.[3] presented a batch distillation column that has both stripping and rectifying sections embedded in it (Figure 5.2e). Although this column has not been investigated completely, recent studies demonstrated that it provides added flexibility for the batch distillation operation. Recently, Skogestad et al.[4] described a new column configuration referred to as a *multivessel* column (Figure 5.2f) and showed that the column can obtain purer products at the end of a total reflux operation. These emerging column designs play an important role in separations of complex systems such as azeotropic, extractive, and reactive batch distillation systems. The batch rectifier configuration for such separations may be very restrictive and expensive.

These emerging designs, combined with different possible operating modes similar to the ones described earlier for the rectifier, provide greater flexibility but result in a large number of column configurations. Because of the unsteady-state nature of the operation, embedded in the design problem is the optimal control problem of deciding time-dependent variables such as reflux ratios, reboil ratios, vapor flow rates, and vessel holdups. Given this flexibility, batch distillation poses a difficult synthesis problem involving the selection of optimal column configurations and optimal operating conditions. Complex systems, such as azeotropic, extractive, and reactive batch distillation systems, add another dimension to the synthesis problems as the cuts (fractions) in the multifraction operation can have significantly different characteristics depending on the feed mixture of these systems. The complexity in design, synthesis, and analysis of batch distillation due to the (1) unsteady-state nature, (2) operational flexibility, and (3) emerging column design can only be handled systematically using computer-aided design techniques and recently developed software tools.

This chapter presents a complete review of batch distillation starting from the first analysis in 1902 by Rayleigh to the current state-of-the-art, computer-aided design techniques. The chapter introduces an early theoretical analysis of simple distillation and various operating policies in Section 5.2. Section 5.3 examines the challenges involved in rigorous modeling of batch distillation dynamics and provides a hierarchy of models of varying complexity and rigor. Recent advances in optimal design and control problems are discussed in Section 5.4. Emerging columns, complex systems, and batch synthesis are described in Section 5.5, followed by an overview of available software packages. The last section provides overall conclusions and addresses the direction of future research.

5.2 EARLY THEORETICAL ANALYSIS

This section presents early theoretical analysis of simple distillation, which was first analyzed by Rayleigh.[1] The limitations of simple distillation that led to the development of the batch rectifier are discussed, as is the operational flexibility of batch distillation with regard to the type of operation.

5.2.1 Simple Distillation

The analysis of simple distillation presented by Rayleigh in 1902 marks the earliest theoretical work on batch distillation. Simple distillation, also known as *Rayleigh distillation* or *differential distillation*, is the most elementary example of batch distillation. In this distillation system, the vapor is removed from the still during a particular time interval and is condensed in the condenser. The more volatile component is richer in the vapor than in the liquid remaining in the still. Over time, the liquid remaining in the still begins to experience a decline in the concentration of the more volatile component, while the distillate collected in the condenser becomes progressively more enriched in the more volatile component. No reflux is returned to the still, and no stages or packing materials are provided inside the column; therefore, the various operating approaches are not applicable to this distillation system.

The early analysis of this process for a binary system, proposed by Rayleigh is given below. Let F be the initial binary feed to the still (mol) and x_F be the mole fraction of the more volatile component (A) in the feed. Let B be the amount of compound remaining in the still, x_B be the mole fraction of component A in the still, and x_D be the mole fraction of component A in the vapor phase. The differential material balance for component A can then be written as:

$$x_D \, dB = d(B \, x_B) = B \, dx_B + x_B \, dB, \tag{5.1}$$

giving:

$$\int_F^B \frac{dB}{B} = \int_{x_F}^{x_B} \frac{dx_B}{x_D - x_B}, \tag{5.2}$$

or:

$$\ln\left(\frac{B}{F}\right) = \int_{x_F}^{x_B} \frac{dx_B}{x_D - x_B}. \tag{5.3}$$

In this simple distillation process, it is assumed that the vapor formed within a short period is in thermodynamic equilibrium with the liquid; hence, the vapor composition (x_D) is related to the liquid composition (x_B) by an equilibrium relation of the form $x_D = f(x_B)$. The exact relationship for a particular mixture may be obtained from a thermodynamic analysis depending on temperature and pressure. For a system following the ideal behavior given by Raoult's law, the equilibrium relationship between the vapor composition y (or x_D) and liquid composition x (or x_B) of the more volatile component in a binary mixture can be approximated using the concept of constant relative volatility (α), which is given by:

$$y = \frac{\alpha x}{(\alpha - 1)x + 1}.\tag{5.4}$$

Substitution of the above equation in Equation 5.3 results in:

$$\ln\left(\frac{B}{F}\right) = \frac{1}{\alpha - 1}\ln\left[\frac{x_B(1 - x_F)}{x_F(1 - x_B)}\right] + \ln\left[\frac{1 - x_F}{1 - x_B}\right].\tag{5.5}$$

Although the analysis of simple distillation historically represents the theoretical start of batch distillation research, a complete separation using this process is impossible unless the relative volatility of the mixture is infinite. Therefore, the application of simple distillation is restricted to laboratory-scale distillation, where high purities are not required, or when the mixture is easily separable.

Example 5.1

A mixture of components A and B with 0.6 mole fraction of A and relative volatility of 2.0 is distilled in a *simple* batch distillation column. The feed is 133 mol, and 29.3% of the mixture is distilled. Find the distillate composition (derived from Converse and Gross[5]).

Solution

Because 29.3% of the feed is distilled, the residue amount is 94.031 mol. The bottom composition can be found using Equation 5.5:

$$\ln\left(\frac{94.031}{133}\right) = \frac{1}{2 - 1}\ln\left[\frac{x_B(1 - 0.6)}{0.6(1 - x_B)}\right] + \ln\left[\frac{1 - 0.6}{1 - x_B}\right] \quad \Rightarrow \quad x_B = 0.4793$$

Then the distillate composition ($x_D = y$) can be obtained from Equation 5.4, resulting in a distillate composition of 0.6480.

Because this distillate composition is quite low for separation purposes, simple batch distillation cannot be used in real practice. To obtain products with high purity, multistage batch distillation with reflux has been used. As seen in Figure 5.1a, the batch rectifier is comprised of multiple thermodynamic stages (manifested by internal trays or packing) inside the rectifying section. The feed is normally charged to the reboiler at the beginning of the operation. Although the top products are removed continuously, no bottom product withdrawal occurs in batch distillation, and the reboiler becomes depleted over time. This makes batch distillation an unsteady-state but flexible operation.

5.2.2 OPERATING MODES

The two basic modes of batch distillation are (1) constant reflux and (2) variable reflux, resulting in variable distillate composition and constant distillate composition, respectively. The third operating model of a batch distillation, optimal reflux or optimal control, is neither constant nor variable but is between the two. Similar operating modes are also observed in the emerging batch distillation columns. For example, a stripper can also have three operating modes: (1) constant reboil ratio, (2) variable reboil ratio, and (3) optimal reboil ratio. For a middle vessel column, the combination of the three reflux and three reboil modes results in at least nine possible operating policies. The operating modes of a multivessel column can be derived based on the middle vessel column, but this column configuration requires additional considerations with respect to operating variables such as the holdup in each vessel. The total reflux mode can be also considered especially in the middle vessel and multivessel columns. As these column designs are still under extensive research, early analyses of operating modes are mainly restricted to the batch rectifier and are discussed below.

5.2.2.1 McCabe–Thiele Graphical Method

The difference between simple distillation and batch distillation operations is the relation between the distillate composition (x_D) and the bottom composition (x_B) due to the presence of reflux and column internals. The graphical analysis presented by McCabe and Thiele[6] for continuous distillation provided the basis for analyzing batch distillation operating modes. They suggested a graphical method to calculate this relation using the following procedure. In the McCabe–Thiele method, the overall material balance with no holdup is considered from the condenser to the jth plate. This leads to the following operating equation:

$$y_j = \frac{R}{R+1} x_{j-1} + \frac{1}{R+1} x_D. \tag{5.6}$$

This operating equation represents a line through the point (x_D, x_D) where $y_i = x_{j-1} = x_D$ at the top plate, with a slope of $R/(R+1)$. Starting from this point (x_D, x_D), Equation 5.6 and the equilibrium curve between y_j and x_j can be recursively used from top plate 1 to the reboiler (the reboiler can be considered as the $(N+1)$th plate). This procedure relates the distillate composition (x_D) to the still composition (x_B) through the number of stages.

In the case of batch distillation, however, the still composition (x_B) does not remain constant, as observed in continuous distillation, thus the instantaneous distillate composition (x_D) is also changing. This necessitates using the recursive scheme several times. If this scheme is used while keeping the reflux ratio constant throughout the operation, just like normal continuous distillation, the composition of the distillate keeps changing. This is the constant reflux mode of operation. On the other hand, the composition of the key component in the distillate can be maintained constant by changing the reflux, resulting in the variable reflux mode of operation. The third mode of operation of batch distillation, optimal reflux or optimal control, is designed to optimize a particular perk mode such as maximum distillate, minimum time, or maximum profit functions.

5.2.2.2 Constant Reflux Mode

Smoker and Rose[7] presented the first analysis of the constant reflux operation of a binary batch distillation with no holdup. They used the Rayleigh equation in conjunction with the McCabe–Thiele graphical method to capture the dynamics of the batch distillation column. In their procedure, the relationship between x_D and x_B is recursively determined by the McCabe–Thiele graphical method, then, the right-hand side of the Rayleigh equation (Equation 5.3) is integrated graphically by plotting $1/(x_D - x_B)$ vs. x_B. The area under the curve between the feed composition (x_F) and the still composition (x_B) now gives the value of the integral, which is $\ln(B/F)$. The average composition of the distillate can be obtained from the following equation:

$$x_{D,avg} = \frac{F\, x_F - B\, x_B}{F - B}. \tag{5.7}$$

Although Smoker and Rose presented the calculation method independent of time, time can be introduced through the vapor boilup rate (V) of the reboiler. The resulting equation for determining batch time is given by:

$$T = \frac{R+1}{V}(F - B) = \frac{R+1}{V} D. \tag{5.8}$$

This operation policy is easy to implement and is commonly used.

Example 5.2

We have seen in Example 5.1 that the purity obtained by simple distillation is not satisfactory. Let us add four stages and make this a batch distillation column operating under constant reflux of 1.82. Using the McCabe–Thiele graphical method, find the distillate and still composition when 29.3% of the feed mixture is distilled. What is the average distillation composition? If the feed is 133 mol and the vapor boilup rate is 110 mol/hr, what is the total time required to complete the distillation operation and what is the average distillate composition?

Solution

From Equation 5.3, we have:

$$\ln\left(\frac{B}{F}\right) = \ln\left(\frac{0.707}{1}\right) = -0.3467 = \int_{x_B}^{x_F} \frac{dx_B}{x_D - x_B}.$$

For various values of x_D, operating lines (Equation 5.6) are drawn to obtain the x_B values using the McCabe–Thiele graphical method (see Figure 5.3a). Then, values of x_B vs. $1/(x_D - x_B)$ are plotted in Figure 5.3b, where the area under the curve is equal to the right-hand side of the Rayleigh equation:

$$\int dx_B / (x_D - d_B) \Rightarrow \int \frac{dx_B}{x_D - x_B}$$

The operation is stopped when the integral is –0.3467, which is equivalent to $\ln(B/F)$. From Figure 5.3b, $x_{B,final}$ is 0.4755, which satisfies the above integral. The average distillate composition becomes:

$$x_{D,avg} = \frac{F x_F - B x_B}{F - B} = \frac{133 \times 0.6 - 94.031 \times 0.4755}{133 - 94.031} = 0.9001.$$

Even though the bottom compositions are similar, the distillate composition of batch distillation with multiple stages and reflux is significantly increased from 0.6480 to 0.9001. The time required for the distillation, as given by Equation 5.8, is 0.999 hr.

5.2.2.3 Variable Reflux Mode

In 1937, Bogart[8] presented the first analysis of the variable reflux policy for a binary system. The steps involved in calculating the variable reflux mode are similar to those in the case of the constant reflux mode; however, for variable reflux, the reflux ratio is varied instead of the distillate composition at each step. Moreover, the Rayleigh equation, though valid for the variable reflux condition, takes a simplified form. Because the distillate composition remains constant (remember that we are considering binary systems here) throughout the operation, the Rayleigh equation reduces to the following equation:

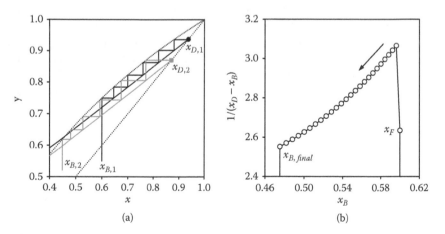

(a)   (b)

FIGURE 5.3 (a) McCabe–Thiele method for plate-to-plate calculations, and (b) graphical integration of the right-hand side of the Rayleigh equation.

$$\frac{B}{F} = \frac{x_D - x_F}{x_D - x_B}. \tag{5.9}$$

The second step is to establish the relation between R and x_B using the McCabe–Thiele graphical method. Several values of R are selected, operating lines are drawn through the fixed point (x_D, x_D) with slope $R/(R + 1)$, and steps are drawn between the operating line and the equilibrium curve to obtain the bottom composition (x_B). This recursive scheme is repeated until the desired stopping criterion is met, thus B and x_B can be found at each value of the reflux ratio. The time required for this operation at a given product purity is calculated by plotting the quantity $(R+1)/V \times \{F(x_D - x_F)\}/(x_D - x_F)^2$ vs. x_B in the following equation and then finding the area under the curve:

$$T = \int_{x_B}^{x_F} \frac{R+1}{V} \frac{F(x_D - x_F)}{(x_D - x_B)^2} dx_B. \tag{5.10}$$

The variable reflux operation policy is commonly used with a feedback control strategy because the reflux ratio is constantly adjusted to keep the distillate composition constant. Section 5.4.2 presents a detailed description of the control strategy involved in this operating mode.

Example 5.3
Rework the problems in Example 5.2 for the variable reflux mode. For the various iterations of R, use the following 10 reflux ratios: 0, 1.3343, 1.4057, 1.5091, 1.6234, 1.7498, 1.9283, 2.0902, 2.2718, and 2.5926.

Solution
Because the type of operation is variable reflux mode (Figure 5.4), the distillate composition is held constant at $x_D = x_{D,avg} = 0.9001$. The bottom composition can be obtained from the McCabe–Thiele graphical method, while the distillate can be obtained from the Rayleigh equation for the variable reflux condition (Equation 5.9):

$$D = F - B = F\left(1 - \frac{x_D - x_F}{x_D - x_B}\right).$$

Then the resulting x_B and D are:

R	0	1.3343	1.4057	1.5091	1.6234	1.7498	1.9283	2.0902	2.2718	2.5926
x_B	0.6	0.59694	0.58757	0.57463	0.56119	0.54728	0.52926	0.51440	0.49920	0.47550
D	0	1.3421	5.2893	10.3663	15.2282	19.8725	25.3683	29.5152	33.4386	38.969

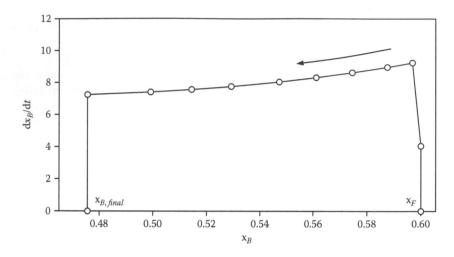

FIGURE 5.4 Graphical integration for batch time under the variable reflux mode.

The operation is stopped when the amount of distillate is greater than or equal to 38.969 mol, when the bottom composition at this condition is 0.4755. The time required for the distillation is given by Equation 5.10. By plotting the quantity $dx_B/dt = (R+1)/V \times \left\{ F\left(x_D - x_F\right) \right\}/\left(x_D - x_B\right)^2$ vs. x_B, the area under the curve between x_B equal to 0.6 and 0.4755 is the batch time. The time required is 0.994 hours.

5.2.2.4 Optimal Reflux Mode

The optimal reflux mode is a third mode of operation in which neither the distillate composition nor reflux is kept constant, as shown in Figure 5.5. This operating mode is a reflux profile that optimizes the given indices of column performance chosen as the objectives. The indices used in practice generally include the minimum batch time, maximum distillate, or maximum profit functions. This reflux mode is essentially a trade-off between the two operating modes and is based on being able to yield the most profitable operation from optimal performance. The calculation of this policy is a difficult task and relies on optimal control theory. The batch distillation literature is rich in papers on this policy; therefore, a separate section (Section 5.4.1) is dedicated to discussing the solution procedures for this operating mode. Although the first optimal reflux policy was discussed as early as 1963, practical implementation of this procedure has only been possible recently because of the advent of computers.

Example 5.4

Rework the problems in Example 5.2 for the optimal reflux mode. Consider the following reflux profiles for optimal batch operation.

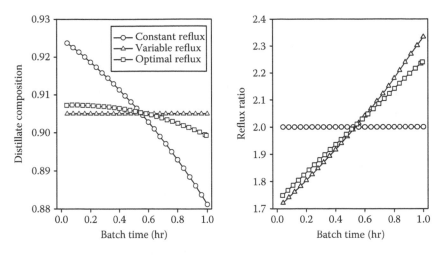

FIGURE 5.5 Three operating modes of batch rectification column.

x_D	0.97959	0.91127	0.90750	0.90228	0.89402	0.88141
R	0	1.5479	1.6699	1.7919	1.9343	2.0919

Solution

Because the operation is neither constant composition nor variable reflux, the bottom compositions are calculated at each given reflux profile using the McCabe–Thiele graphical method, resulting in:

x_B	0.6	0.59632	0.57311	0.54781	0.51533	0.47558

The distillate is calculated using the Rayleigh equation, in which the area under the curve of $1/(x_D - x_B)$ vs. x_B (Figure 5.6) is equivalent to

$$\ln\left(\frac{B}{F}\right) = \int_{x_F}^{x_B} \frac{dx_B}{x_D - x_B} = -\text{Area under the curve of } \frac{dx_B}{x_D - x_B} \text{ versus } x_B = -0.3467,$$

$$D = F\left(1 - \frac{B}{F}\right) = 38.969 \text{ moles.}$$

Thus, we can see that the distillates are equal under the three types of operation. The average distillate composition calculated using Equation 5.7 results in $x_{D,avg} = 0.9001$, which is exactly the same result as for the constant and variable reflux modes. Batch time T required when neither the distillate composition nor reflux is constant is found to be:

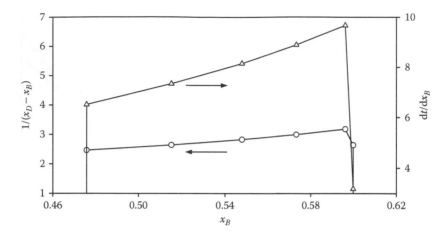

FIGURE 5.6 Graphical integration for Rayleigh equation (open circle) and batch time (open triangle) for the optimal reflux model.

$$\int_0^T dt = \int_B^F \frac{R+1}{V} dB \quad \text{and} \quad dB = B \frac{dx_B}{x_D - x_B}$$

$$\Rightarrow T = \int_{x_B}^{x_F} \frac{B}{V} \frac{R+1}{x_D - x_B} dx_B.$$

So, the time required is the area under the curve of $(B/V)((R+1)/(x_D - x_B))$ vs. x_B, as shown in Figure 5.6. The value of T is found to be 0.992 hours, smaller than for the constant and variable modes of operation.

5.3 HIERARCHY OF MODELS

As seen in Section 5.2, the earlier models of the batch rectifier were built on assumptions of negligible liquid holdup and ideal binary systems. Computers have played an important role in relaxing these assumptions, especially the negligible holdup assumption. Distefano[9] analyzed the numerical differential equations for multicomponent batch distillation in 1968 for the first time. The rigorous models of batch distillation in current state-of-the-art computer packages are based on his pioneering work; however, it is recognized that, due to the severe transients in batch distillation, a hierarchy of models is necessary to capture the dynamics of this flexible operation. This section presents the hierarchy of models ranging from the rigorous model similar to the one presented by Distefano to the simplest shortcut model.

5.3.1 RIGOROUS MODEL

A rigorous model in batch distillation involves consideration of column dynamics along with the reboiler and condenser dynamics. A detailed analysis of the

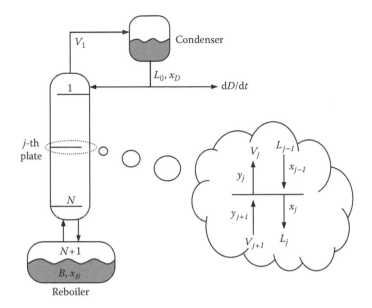

FIGURE 5.7 Schematic of a batch distillation column.

characteristics of differential mass and energy balances associated with the complete dynamics of a multicomponent batch distillation column was presented by Distefano.[9] He pointed out that the system of equations presented for batch distillation is much more difficult to solve than that for continuous distillation due to several factors. For example, in the case of batch distillation, plate holdup is generally much smaller than reboiler holdup, while in continuous distillation the ratio of reboiler holdup to plate holdup is not nearly as great. In addition, in batch distillation severe transients can occur, unlike continuous distillation, where variations are relatively small. Distefano's work forms the basis for almost all of the later work on rigorous modeling of batch distillation columns.

Figure 5.7 represents a schematic of a batch distillation column, where the holdup on each plate is responsible for the dynamics of each plate. For an arbitrary plate j, the total mass, component, and energy balances yield the governing equations, summarized in Table 5.1. This table lists all the equations involved in the dynamic analysis of the batch column and the assumptions behind these equations.

As the governing equations represent a generalized form of the batch rectifying column, treatment of an individual operating mode (i.e., constant reflux, variable reflux, or optimal reflux) exploits the same governing equations but with different specifications. Furthermore, the governing equations of the stripper, middle vessel column, and multivessel columns can be similarly derived.

From the system of differential equations in Table 5.1, we can easily see that the problem has no analytical solution, and we must resort to numerical solution techniques. The governing differential equations of batch distillation often fall

TABLE 5.1
Complete Column Dynamics for a Rigorous Model[2]

Assumptions
Negligible vapor holdup
Adiabatic operation
Theoretical plates
Constant molar holdup
Finite difference approximations for the enthalpy changes

Composition Calculations
Condenser and accumulator dynamics:

$$\frac{dx_D^{(i)}}{dt} = \frac{V_1}{H_D}\left(y_1^{(i)} - x_D^{(i)}\right), \, i = 1, 2, \dots, n$$

Plate dynamics:

$$\frac{dx_j^{(i)}}{dt} = \frac{1}{H_j}\left[V_{j+1}y_{j+1}^{(i)} + L_{j-1}x_{j-1}^{(i)} - V_j y_j^{(i)} - L_j x_j^{(i)}\right], \, i = 1, 2, \dots, n; j = 1, 2, \dots, N$$

Reboiler dynamics:

$$\frac{dx_B^{(i)}}{dt} = \frac{1}{B}\left[L_N(x_N^{(i)} - x_B^{(i)}) - V_B(y_B^{(i)} - x_B^{(i)})\right], \, i = 1, 2, \dots, n$$

Flow Rate Calculations
At the top of the column:

$$L_0 = R\frac{dD}{dt}; V_1 = (R+1)\frac{dD}{dt}$$

On the plates:

$$L_j = V_{j+1} + L_{j-1} - V_j; \quad j = 1, 2, \dots, N$$

$$V_{j+1} = \frac{1}{J_{j+1} - I_j}\left[V_j(J_j - I_j) + L_{j-1}(I_j - I_{j-1}) + H_j \delta I_j\right], \, j = 1, 2, \dots, N$$

At the bottom of the column:

$$\frac{dB}{dt} = L_N - V_B$$

TABLE 5.1 (CONTINUED)
Complete Column Dynamics for a Rigorous Model[2]

Heat-Duty Calculations
Condenser duty:

$$Q_D = V_1(J_1 - I_D) - H_D \delta_i I_D$$

Reboiler duty:

$$Q_B = V_B(J_B - I_B) - L_N(I_N - I_B) + B\delta_i I_B$$

Thermodynamics Models
Equilibrium relations:

$$y_j^{(i)} = f((x_j^{(k)}, k = 1, \ldots, n), TE_j, P_j)$$

Enthalpy calculations:

$$I_j = f((x_j^{(k)}, j = 1, \ldots, n), T_{E,j}, P_j)$$

$$J_j = f((x_j^{(k)}, j = 1, \ldots, n), T_{E,j}, P_j)$$

---

into the category of stiff differential equations. The solution of stiff differential equations contains a component that contributes very little to the solution but can cause errors that accumulate over time, resulting in an incorrect solution. Most recent batch distillation models[10,11] use stiff numerical methods based on a backward difference formula (BDF), and one of the well-known BDF techniques is the Livermore Solver for ordinary differential equations (LSODE)[12] method.

Because the computational intensity of the stiff algorithms is generally more severe than for non-stiff algorithms, it is better to switch to non-stiff algorithms. The quantifying measures, such as the stiffness ratio or computational stiffness, both based on eigenvalue calculations, can be used to decide whether or not to switch;[2] however, eigenvalue calculations are computationally expensive and are not normally used for large systems of differential equations. Further, it should be noted that, for highly stiff systems, it is difficult to apply any numerical integration method unless the system is transformed in some way to reduce the stiffness of the system. This can happen in batch distillation of wide boiling systems or for columns where the holdup inside the column is significantly smaller than that of the still. The semirigorous model can be used to circumvent this problem.

5.3.2 Low Holdup Semirigorous Model

For columns where the plate dynamics are significantly faster than the reboiler dynamics (due to very small plate holdups or wide boiling components), the stiff integrator often fails to find a solution (see Example 5.4 in this section). The solution to this problem is to split the system into two levels: (1) the reboiler, where the dynamics are slower, can be represented by differential equations; and (2) the rest of the column can be assumed to be in the *quasi-steady state*. Thus, the composition changes in the condenser and accumulator ($dx_D^{(i)}/dt$), the composition changes on plates ($dx_j^{(i)}/dt$), and the enthalpy changes in the condenser and on plates ($\delta_t I_D$ and $\delta_t I_j$) in Table 5.1 can be assumed to be zero. This results in a zero holdup model, so this approach can be used for simulating the semirigorous model of batch distillation. Bernot et al.[13,14] developed and compared semirigorous models of the batch rectifier and stripper for the behavior of multicomponent azeotropic distillation. Diwekar and coworkers[2,11] developed the software packages, BATCH-DIST and MultiBatchDS, in which a semirigorous model is available for cases when the rigorous model fails to obtain solutions, as can be seen in Example 5.5.

The holdup effects can be neglected in a number of cases where this model approximates the column behavior accurately. This model provides a close approximation of the Rayleigh equation, and for complex systems (e.g., azeotropic systems) the synthesis procedures can be easily derived based on the simple distillation residue curve maps (refer to Section 5.2 for details). Note, however, that this model involves an iterative solution of nonlinear plate-to-plate algebraic equations, which can be computationally less efficient than the rigorous model.

Example 5.5

An equimolar mixture containing 100 mol of a four-component mixture having relative volatilities of 2.0, 1.5, 1.0, and 0.5 is to be distilled in a batch distillation column. The column has 10 theoretical plates with a holdup of 0.001 mol per plate and a condenser holdup of 1 mol. The vapor boilup rate (V) of the reboiler is 100 mol/hr. The column is operating under a constant reflux mode with a reflux ratio equal to 5.0. Simulate a 1-hr operation of the column using the rigorous model presented in Table 5.1. Repeat the simulation using the semirigorous model.

Solution

Figure 5.8a shows the transient profiles obtained using the rigorous model, and Figure 5.8b was obtained using the semirigorous model. The rigorous model could not integrate the column because the step size became so small that rounding errors dominated the performance, thus switching to the semirigorous model is required in this case.

5.3.3 SHORTCUT MODEL AND FEASIBILITY CONSIDERATIONS

As seen in Section 5.1, the rigorous model of batch distillation operation involves a solution of several stiff differential equations. The computational intensity and memory requirement of the problem increase with an increase in the number of plates and components. The computational complexity associated with the rigorous model does not allow us to derive global properties such as feasible regions

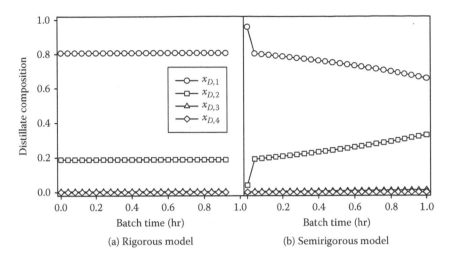

(a) Rigorous model   (b) Semirigorous model

FIGURE 5.8 Transient composition profiles for (a) rigorous model, and (b) semirigorous model.

of operation, which are critical for optimization, optimal control, and synthesis problems. Even if such information is available, the computational costs of optimization, optimal control, or synthesis using the rigorous model are prohibitive. One way to deal with these problems associated with the rigorous model is to develop simplified models such as the shortcut model and the collocation-based model. These simplified models are abstractions of the rigorous model, and their accuracy depends on the simplifying assumptions embedded within them. The process of abstraction can be viewed as a trade-off between simplicity and accuracy. The usefulness of abstracted models depends on the ease with which they can be analyzed for global behaviors without compromising accuracy. Moreover, the abstracted models are expected to be computationally simpler to analyze.

The shortcut model of batch distillation proposed by Diwekar[11] is based on the assumption that the batch distillation column can be considered equivalent to a continuous distillation column with the feed changing at any instant. Because continuous distillation theory is well developed and tested, the shortcut method of continuous distillation is modified for batch distillation, and the compositions are updated using a finite-difference approximation for the material balance (based on the Rayleigh equation). The other assumptions of the shortcut method include constant molar overflow and negligible plate holdups. As described earlier, the functional relationship between the distillate composition (x_D) and the bottom composition (x_B) is crucial for the simulation, and the Fenske–Underwood–Gilliland (FUG) method is used for estimating this relation.

Shortcut methods have also been modified to incorporate holdup issues using a compartmental modeling approach and extended to complex mixtures containing binary and ternary azeotropes.[15] Lotter and Diwekar[16] applied a similar shortcut approach to emerging batch columns, such as stripper and middle vessel columns.

The shortcut model is very useful in feasibility analysis. In order to maintain the feasibility of design, we must place certain constraints on the variables, especially for the design variables such as the number of plates (N) and reflux ratio (R). The shortcut model helps to identify these bounds on the design parameters. The bounds on the parameters depend on the operating modes. The feasible region of operation has been identified using the short-cut model and is summarized in Table 5.2. In this table, R_{min} is the Underwood minimum reflux ratio, which is different from R_{MIN}. R_{MIN} is defined as the value of R required to obtain the distillate composition of the key component equal to the specified average distillate composition at the initial conditions for the given N. Recently, Kim and Diwekar[17] defined new performance indices, such as the N-feasibility index and the R-feasibility index, for analyzing feasible regions of various column configurations. These new indices can identify distinctive feasibility regions for various configurations and provide useful guidelines for optimal column selection.

The shortcut model has been found to be extremely efficient and reasonably accurate for nearly ideal mixtures and for columns with negligible holdup effects. For further details, please refer to the book written by Diwekar.[2]

5.3.4 COLLOCATION-BASED MODELS

The next simplified model in the simulation hierarchy is the reduced-order model based on the orthogonal collocation approach. The collocation approach was first proposed in the context of continuous staged separation processes by Cho and

---

TABLE 5.2
Feasible Region for Multicomponent Batch Distillation Columns[2]

Variable Reflux	Constant Reflux	Optimal Reflux
Final still composition:		
$0 \leq x_{B,\infty}^{(1)} \leq x_D^{(1)}$	$0 \leq x_{B,\infty}^{(1)} \leq x_D^{(1)}$	$0 \leq x_{B,\infty}^{(1)} \leq x_D^{(1)}$
Distillate composition:		
$x_B^{(1)} \leq x_D^{(1)} \leq 1$	$x_B^{(1)} \leq x_D^{(1)} \leq 1$	$x_B^{(1)} \leq x_D^{(1)} \leq 1$
Reflux ratio:		
$R_{min} \leq R_{initial} \leq R_{max}$	$R_{MIN} \leq R \leq \infty$	$R_{MIN} \leq R \leq \infty$
Number of plates:		
$N_{min,f} \leq N$	$N_{min} \leq N$	$N_{min} \leq N$

Joseph.[18] The collocation approach to model reduction is based on approximating the column stage variables by using polynomials rather than discrete functions of stages; thus, it is widely used for packed batch column design. The orthogonal collocation technique can change partial differential equations to ordinary differential equations (ODEs) or algebraic equations, and ODEs to a set of algebraic equations. In the case of batch distillation, we encounter ordinary differential equations, and the orthogonal collocation technique can be used to reduce this system of ODEs into nonlinear algebraic equations.

Srivastava and Joseph[19] developed the orthogonal collocation method of a simplified packed batch column using the fourth-order polynomial. For a quasi-steady-state batch distillation with total reflux, Aly et al.[20] used the Galerkin method as the weighting function over the finite elements. Even though the Galerkin method is one of the best known approximation methods for weighted residuals, this method is difficult to implement.

Note that the orthogonal collocation model can also be used to reduce the order of optimization problems. It is not always advantageous to convert ordinary differential equations to nonlinear algebraic equations. The converted large systems of algebraic equations are computationally time consuming. Instead of using orthogonal collocation to reduce the ODEs to nonlinear algebraic equations, one can use it to reduce the order of ODEs.[21] This model is especially useful when other simplified models cannot be used to describe the column (e.g., for highly nonideal systems or systems for which constant molar flow assumptions cannot be used).

5.3.5 Model Selection Guidelines

So far several batch distillation simulation models with varying complexity have been presented. Figure 5.9 shows general guidelines to choosing the best one among the hierarchy of batch distillation models. With these models as the basis, numerous batch distillation tasks such as optimization and optimal control (Section 5.4), emerging batch column configurations (Section 5.5), and complex batch distillation systems (Section 5.5) have been developed.

5.4 OPTIMIZATION AND OPTIMAL CONTROL PROBLEMS

The previous sections concentrated on the design and simulation of batch distillation columns using a hierarchy of models. Optimal design and operation in a batch distillation process are challenging decision-making problems that involve several time-dependent and -independent decisions in the face of operating and thermodynamic constraints. Mathematical optimization theory makes the decision-making process easier and more systematic. With the advent of computers, it is possible to exploit these theories to the maximum extent, provided that the problem is properly formulated in terms of the objective functions and constraints and the suitable solution method from the optimization

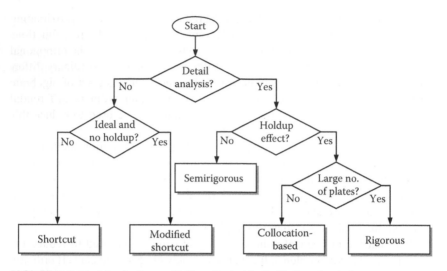

FIGURE 5.9 Model selection guidelines for batch distillation simulation.

theory is identified. Optimization methods are also used in solving and implementing control problems in batch distillation. This section presents design optimization, optimal control, and closed-loop control problems.

Literature on the optimization of the batch column is focused primarily on the solution of optimal control problems, including optimizing the indices of performance such as maximum distillate, minimum time, and maximum profit; however, literature on the optimal design of batch distillation for performing specified operations using the constant reflux or variable reflux modes is very limited. This section describes optimal control problems within the context of performance indices and optimization techniques. Some recent articles address the problem of design and optimal control policy together by combining optimal control theory and numerical optimization methods. This approach for simultaneous optimal design and operation is described later. Closed-loop control of columns is presented at the end of this section.

5.4.1 Optimal Control Problems

This subsection is devoted to optimal control problems in batch distillation, which have received considerable attention in the literature. In general, control refers to a closed-loop system where the desired operating point is compared to an actual operating point and a knowledge of the error is fed back to the system to drive the actual operating point toward the desired one; however, the optimal control problems we consider here do not fall under this definition of control. Because the decision variables that will result in optimal performance are time dependent, the control problems described here are referred to as optimal control problems; thus, use of the control function here provides an open-loop control. The dynamic nature of these decision variables makes these problems much more difficult to solve as compared to normal optimization, where the decision variables are scalar.

These problems are categorized by: (1) performance indices and (2) solution methods. The following subsection discusses the performance indices for optimal control problems (maximum distillate, minimum time, and maximum profit) and is followed by a subsection on mathematical techniques used to solve optimal control problems: calculus of variations, Pontryagin's maximum principle, dynamic programming, and nonlinear programming (NLP) techniques. The first three techniques treat the decision variables as vectors, while the NLP approach requires the variables to be transformed into scalars. For details about these methods, please refer to Diwekar.[22,23]

5.4.1.1   Performance Indices for Optimal Control Problems

Optimal control problems can be classified as:

- *Maximum distillate problem*, where the amount of distillate of a specified concentration for a specified time is maximized.[5,24–27] This problem can be represented as follows:

$$\max_{R_t}\ \ J = \int_0^T \frac{dD}{dt}\, dt = \int_0^T \frac{V}{R_t + 1}\, dt, \tag{5.11}$$

  subject to the material and energy balances. Converse and Gross[5] first reported the maximum distillate problem for binary batch distillation, which was solved using Pontryagin's maximum principle, the dynamic programming method, and the calculus of variations. Diwekar et al.[24] extended this optimization model to multicomponent systems and used the shortcut batch distillation model along with the maximum principle to calculate the optimal reflux policy. Logsdon et al.[25] used the orthogonal collocation approach on finite elements and NLP optimization techniques over the shortcut model, and they extended this method to the rigorous batch distillation model,[27] in which they considered the effect of column holdups on optimal control policy.

- *Minimum time problem*, where the batch time required to produce a prescribed amount of distillate of a specified concentration is minimized.[22,28] Although there are several different formulations for the minimum time problem, Diwekar[22] derived the following formulations to establish a unified theory for all the optimal control problems:

$$\min_{R_t}\ \ J = \int_0^T \frac{dt^*}{dt}\, dt. \tag{5.12}$$

where t^* is a dummy variable as a state variable.

- *Maximum profit problem*, where a profit function for a specified concentration of distillate is maximized.[25,29-31]

Much of the recent research on optimal control problems can be classified into this problem. Kerkhof and Vissers[29] were the first to use the profit function for maximization in batch distillation, and they solved the optimal control problem. They obtained the following simple objective function:

$$\max_{R_t,T} J = \frac{DP_r - FC_F}{T + t_s},$$ (5.13)

subject to purity constraints and column modeling equations.

Diwekar et al.[24] used a different objective function to solve the profit maximization problem under the constant and variable reflux conditions. Logsdon et al.[25] formulated a new profit function and solved the differential algebraic optimization problem for optimal design and operation. Li et al.[30] developed a detailed dynamic multifraction batch distillation model, discretized the model using the orthogonal collocation method on finite elements, and finally solved the maximum profit model using an NLP optimizer. Mujtaba and Macchietto[31] considered a rigorous reactive distillation system for the maximum conversion problem, which can also be classified as the maximum profit problem. The detailed dynamic system is then reduced by using polynomial curve-fitting techniques and solved by using an NLP optimizer.

A variant of this objective function is to minimize the mean rate of energy consumption when the market size for the product is fixed by the current demand. The objective function is given by Furlonge et al.:[32]

$$\min J = \frac{\int_0^T Q_R(t)dt}{T + t_s},$$ (5.14)

$$\text{s.t. } x_{D,avg} \geq x^*,$$

$$D \geq D^*,$$

where Q_R is the reboiler heat duty. They used this objective function for optimal control of multivessel columns for the first time. Hasebe et al.[33] also presented the optimal operation policy based on energy consumption for the multivessel column.

5.4.1.2 Solution Techniques

To solve the optimal control problems, the following four solution techniques have been used in the literature; of these, Pontryagin's maximum principle and nonlinear programming techniques are commonly used today:

- *Calculus of variations* — The theory of optimization began with the calculus of variations, which is based on the vanishing of the first variation of a functional ($dJ = 0$) according to the theorem of minimum potential energy, which involves the definition of stationary values for a function. This leads to the Euler equation and natural boundary conditions.[5]
- *Pontryagin's maximum principle* — The maximum principle was first proposed in 1956 by Pontryagin.[34] The objective function formulation is represented as a linear function in terms of the final values of a state vector and a vector of constants. Like the calculus of variations, this method is only applicable to optimal control problems for fixed scalar variables. The maximum principle necessitates repeated numerical solutions of two-point boundary value problems, thereby making it computationally expensive. Furthermore, it cannot handle bounds on the control variables.
- *Dynamic programming* — The method of dynamic programming is based on the principle of optimality, as stated by Bellaman.[35] In short, the principle of optimality states that the minimum value of a function is a function of the initial state and the initial time. This method is best suited for multistage processes; however, the application of dynamic programming to a continuously operating system leads to a set of nonlinear partial differential equations.
- *NLP optimization techniques* — NLP optimization techniques are the numerical tools used by models involving nonlinear algebraic equations. Obviously, applying NLP techniques to optimal control problems involves discretization of the control profile by applying either the orthogonal collocation on finite elements,[25,30,33] the control vector parameterization approach,[36,32] or the polynomial approximation.[26] These discretization approaches add nonlinearities to the system as the number of nonlinear equations increase; therefore, they require good initializations and may result in suboptimal solutions. On the other hand, the polynomial approximation methods depend on the crucial decision of choosing the right type and order of polynomials for approximating the control profile.

A new approach to optimal control problems in batch distillation, proposed in a paper by Diwekar,[22] combines the maximum principle and NLP techniques. This algorithm reduces the dimensionality of the problem (caused by NLP techniques) and avoids the solution of the two-point boundary value problems (caused by the maximum principle). Furthermore, it was shown that for batch distillation problems, bounds could be imposed on the control vector by virtue of the nature of the formulation.

5.4.2 CLOSED-LOOP CONTROL

The two traditional batch operation policies, constant reflux and variable reflux policies, involve different control strategies. For the constant reflux policy, where the distillate composition is continuously changing, the average distillate composition can only be known at the end of operation unless proper feedback from the operation is obtained. The control of the average distillate composition is, then, of an open-loop control nature; however, the variable reflux policy is inherently a feedback operation because the reflux ratio is constantly adjusted to keep the distillate composition constant. The purpose of designing a closed-loop control scheme is to reduce the sensitivity of the plant to external disturbances. Because batch distillation begins with total reflux to obtain a steady state and the distillate is withdrawn after that point, the reflux ratio and distillate composition may oscillate if a controller gain is not properly selected. This is the reason why the constant composition control proves to be very challenging. This subsection describes recent research efforts on closed-loop control problems.

Quintero-Marmol et al.[37] proposed and compared several methods for estimating the online distillate composition by feedback control under constant reflux operating mode in a batch rectifier. An extended Luenberger observer for tracking the distillate composition profile proved to provide the best result.

Bosley and Edgar[38] considered modeling, control, and optimization aspects of batch rectification using nonlinear model predictive control (NMPC) and implemented an optimal batch distillation policy that was determined *a priori* by the offline optimization. NMPC can determine the set of control moves that will yield the optimal trajectory and allow explicit constraints on inputs, outputs, and plant states. It is known that NMPC is one of the best approaches for distillate composition control; however, the control scheme is computationally intensive because optimization problems are solved inside this control loop. This work was further studied by Finefrock et al.,[39] who studied nonideal binary batch distillation under the variable reflux operating policy. Because the gain space can be changed significantly after a switch to the production phase, they suggested a gain-scheduled proportional and integral (PI) controller based on NMPC if the instantaneous distillate composition is known.

Besides NMPC, Barolo and Berto[40] provided a framework for obtaining composition control in batch distillation using a nonlinear internal model control (NIMC) approach. NIMC can exactly linearize the system input-output map and be easily tuned by using a single parameter for each component. The distillate composition is estimated by the selected temperature measurements. They also used an extended Luenberger observer for a composition estimator. Although this approach can be reliable and easily implemented, the authors pointed out the problem of selecting the best temperature measurement locations and the problems with using the extended Luenberger observer for a batch column with a large number of trays. For a tighter composition control, more research is necessary to develop a robust and fast closed-loop control scheme.

Closed-loop control schemes have also been applied to new column configurations and complex batch systems. For the control of the middle vessel column, Barolo et al.[41] first proposed and examined several control schemes with or without product recycling. They showed the experimental results of the proposed control structures for dual composition control with or without impurity. Farschman and Diwekar[42] proposed dual-composition control in which the two composition control loops can be decoupled if the instantaneous product compositions are known. The degree of interaction between the two composition control loops can be assessed using the relative gain array technique.

Hasebe et al.[43] proposed a single-loop cascade control system to control the composition of each vessel in the multivessel column. The vessel holdup under total reflux is the manipulated variable, and the reflux flow rate from each vessel is, then, controlled by a simple PI controller. Skogestad et al.[4] developed a simple feedback control strategy in which the temperature at the intermediate vessel is controlled by the reflux rates from the vessels, thereby adjusting the vessel holdups indirectly. Further, Furlonge et al.[32] compared different control schemes, including optimal control problems, in terms of energy consumption.

Future work in closed-loop control problems can involve identifying the proper temperature measurement locations, easy parameter tuning, and focusing on tracing the optimal profiles, as well as on-spec products.

5.5 EMERGING BATCH COLUMNS, COMPLEX SYSTEMS, AND SYNTHESIS

In the previous sections, we described various aspects of batch distillation, including the development of a hierarchy of models ranging from simplified to rigorous, optimization, and optimal control of the batch distillation (rectification) operation. This section presents discussion on alternative emerging column configurations and thermodynamically or kinetically complicated batch distillation systems such as azeotropic, extractive, and reactive distillations. In addition, this section describes how these complex batch column configurations and complex systems result in difficult batch distillation synthesis problems.

5.5.1 EMERGING BATCH COLUMNS

Figure 5.2 shows a batch stripper (Figure 5.2d), a middle vessel column (Figure 5.2e), and a multivessel column (Figure 5.2f) as emerging batch columns. These column configurations and their advantages are described here.

5.5.1.1 Batch Stripper

Although the batch stripper, often called an *inverted batch column* and originally proposed by Robinson and Gilliland,[44] is not a true emerging batch column, it has gained much attention in recent literature. In this column configuration, the feed mixture is charged into the top reflux drum, and the products are withdrawn at the bottom reboiler.

Bernot et al.[14] developed a semirigorous model of the batch stripper for multicomponent azeotropic distillation and showed that the batch stripper, compared to the rectifier, is essential to break a minimum boiling point azeotrope. Sørensen and Skogestad[45] compared the batch stripper with the batch rectifier in terms of batch time and proposed that the inverted column configuration is better than the regular column for separations where the light component in the feed is present in a small amount. They also reported that in some cases the stripper can separate feed mixtures while the rectifier design is infeasible for that separation. Kim and Diwekar,[17] based on this shortcut model, derived more generalized heuristics for column selection using various performance indices — namely, product purity and yield, feasibility and flexibility, and thermodynamic efficiency.

Example 5.6

From Example 5.2, the bottom product composition of a heavy component (B) of the batch rectifier is 0.5247. Repeat the simulation using a batch stripper when the product throughput is the same (i.e., $B = 38.969$ mol), and compare the bottom product compositions of the heavy component (B).

Solution

The reflux ratio of the batch rectifier in Example 5.2 should be converted to the reboil ratio (R_B) of the batch stripper. Because we assume a constant boilup rate (V), the relationship between the reflux and reboil ratios is:

$$R = \frac{L}{dD / dt} = \frac{V - dD / dt}{dD / dt} = \frac{V}{dD / dt} - 1 = \frac{V}{dB / dt} - 1 = R_B - 1$$

So, the reboil ratio becomes 2.82, and the other design and operating conditions remain the same. The bottom composition profiles are shown in Figure 5.10, in which the composition of a heavy component ($x_{B,B}$) is higher than that from the batch rectifier. The average bottom product composition is 0.6819. Thus, if a bottom product is a main concern, it would better to use a batch stripper to obtain highly pure bottom product. For detailed comparisons of the batch rectifier and stripper, please refer to the literature by Sørensen and Skogestad[45] and Kim and Diwekar.[17]

5.5.1.2 Middle Vessel Column

This column configuration consists of a middle vessel between two sections of the batch column. The feed is initially charged into the middle vessel, and the

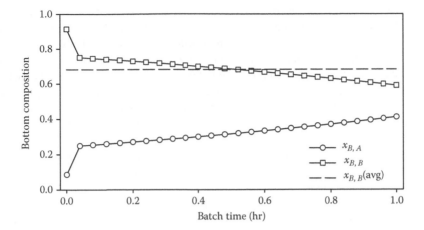

FIGURE 5.10 Bottom composition profiles of a batch stripper from Example 5.6.

products are simultaneously withdrawn from the top and the bottom of the column. The middle vessel column can be an ideal configuration for ternary batch systems. This column configuration has been known since the 1950s; however, only recently has an analysis of this column configuration been published. Davidyan et al.[3] analyzed the dynamic behavior of the middle vessel column for ideal binary and ternary and azeotropic ternary systems. They found additional steady states that are stable or unstable singular points of a dynamic system describing the column. They also introduced a new parameter (q'), which is the ratio of the vapor boilup rate in the rectifying section to the vapor boilup rate in the stripping section. Depending on the value of variable q', the column shows a qualitatively different behavior for a domain of the reflux and reboil ratio. Figure 5.11 shows the effect of q' on the top and bottom product purities. For $q' = 1$,

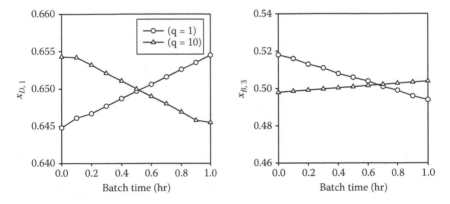

FIGURE 5.11 The effect of q on the top and bottom product purities in the middle vessel column.

the distillate composition of the more volatile component increases with time, and this is a favorable trend for the light key distillate; however, the bottom composition of the least volatile component is decreasing. These trends are opposite to those of batch rectification, for which the trends are similar to the case of $q' = 10$; therefore, the new degree of freedom (q') is an important parameter to be used in optimizing the operation. Meski and Morari[46] extended their previous work under the infinite separation and the minimum reflux conditions and suggested that the middle vessel column always outperforms the rectifier and stripper in terms of batch time. For a binary separation system, they also found that the steady-state operation corresponding to $q' = 1$ is the optimal control policy.

This column configuration is very flexible and effective; hence, one can, in theory, simultaneously obtain very pure components in the top, bottom, and middle vessel columns. For example, Safrit et al.[47] investigated extractive distillation in the middle vessel column and found that this column can recover all of the pure distillate product from an azeotropic feed with a relatively small size of reboiler, while a rectifier alone would require a still pot of infinite size.

5.5.1.3 Multivessel Column

Similar to a middle vessel column is the multivessel column. Hasebe et al.[43] presented a heat-integrated, multieffect batch distillation system (MEBDS) as an alternative to continuous distillation (Figure 5.12). The feed was initially distributed among all the middle vessels and operated in total reflux mode. They proposed a composition control system in which the vessel holdups are manipulated by level controllers. They concluded that this new emerging column configuration can have better separation performance than continuous distillation for systems having a larger number of products. Hasebe et al.[48] published an optimal operation policy for this column using variable holdup modes. They optimized

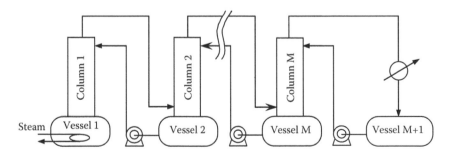

FIGURE 5.12 Multiple-effect batch distillation system. (From Hasebe, S. et al., Comparison of the Separation Performances of a Multieffect Batch Distillation System and a Continuous Distillation System, preprints of IFAC Symposium on Dynamics and Control of Chemical Reactors, Distillation Columns, and Batch Processes (DYCORD '95), Helsingor, Denmark, 1995, pp. 249–254. With permission.)

the liquid flow rates in order to minimize the batch time and concluded that the varying holdup mode resulted in up to 43% more distillate than that of the constant holdup mode. Recently, Hasebe et al.[33] optimized the holdup of each vessel as a function of time for the total reflux multivessel system. When they compared the optimal reflux mode with the constant reflux and variable reflux modes, they found that the performance index, defined as the amount of products per batch per total batch time, for a variable reflux ternary system was approximately 18 to 38% greater.

Skogestad et al.[4] reported a new column configuration they referred to as a *multivessel column*. This column is operated under total reflux conditions. They showed that the steady-state compositions in the intermediate vessels could be maintained regardless of the initial feed composition by controlling the liquid rate from the middle vessel so the temperature of the tray just below the middle vessel remained constant. This operation policy can be the ideal operation policy of batch distillation, especially for the middle vessel and multivessel columns. The total reflux mode is commonly used for the multivessel column[4] because multiple products can be accumulated in each vessel according to their relative volatilities. As a variant of this operating mode, the cyclic operation mode has also been studied. Some literature can be found on the cyclic operation policy, which is essentially a variant of the total reflux condition. Recently, Sørensen[49] presented a comprehensive study on optimal operation of the cyclic operating mode of the batch rectifier, stripper, and middle vessel columns. The computational results and experiments showed a significant savings in batch time for some separations.

Furlonge et al.[32] extended their previous study to optimal control problems and developed more detailed rigorous equations with dynamic energy balance equations, liquid and vapor holdups, and dry and wet head losses on each tray. They compared various operating modes in terms of mean energy consumption rate and found that the optimal initial feed distribution greatly improves the column performance, resulting in an energy consumption rate half that of the rectifier.

5.5.2 COMPLEX BATCH DISTILLATION SYSTEMS

Thermodynamically and kinetically complex systems such as azeotropic, extractive, and reactive systems pose additional bottlenecks in the design and operation of batch columns. The operational flexibility offered by batch distillation, along with new emerging designs, can provide promising alternatives for circumventing the bottlenecks. The following sections describe the methods for analyzing these complex systems. These methods also provide heuristics for the synthesis of these columns, especially in terms of the different cuts obtained in a single column or performance comparison of the complex columns.

5.5.2.1 Azeotropic Batch Distillation

Azeotropic distillation is an important and widely used separation technique as a large number of azeotropic mixtures are of great industrial importance. Despite their importance, azeotropic distillation techniques remain poorly understood from a design standpoint because of the complex thermodynamic behavior of the system. Theoretical studies on azeotropic distillation have mainly centered around methods for predicting the vapor–liquid equilibrium data from liquid solution models and their application to distillation design; however, only during the past two decades has there been a concerted effort to understand the nature of the composition region boundaries. Doherty and coworkers[13,14] in their pioneering works proposed several new concepts in azeotropic distillation. They established the use of ternary diagrams and residue curve maps in the design and synthesis of azeotropic continuous distillation columns. In batch distillation, they outlined a synthesis procedure based on the residue curve maps.

The residue curve map graphs the liquid composition paths that are solutions to the following set of ordinary differential equations:

$$\frac{dx_i}{d\xi} = x_i - y_i \qquad i = 1, 2, \ldots, n-1, \tag{5.15}$$

where n is the number of components in the system, and the independent variable, warped time (ξ), is a monotonically increasing quantity related to real time. One can see that Equation 5.15 is one form of the Rayleigh equation described earlier. The residue curve map occupies a significant place in the conceptual design stage of column sequencing in continuous distillation and fractions (cuts) sequencing in batch distillation.[13,14,50]

Despite the advances in the thermodynamics for predicting azeotropic mixture, feasible distillation boundaries, and sequence of cuts, the azeotropic batch distillation system is still incipient in terms of design, optimization, and optimal control. The design problems of these complex systems are described in Section 5.5.3.

Example 5.7

A residue curve map of the propylamine–acetonitrile (ACN)–water system[52] are given in Figure 5.13, in which the curves show liquid composition profiles from the lightest component to the heaviest component. Find the batch distillation regions and define the product cuts for each region.

Solution

Because the curve from the propylamine apex to the ACN–water azeotrope distinguishes two different product paths, this system has two distillation regions. For the left distillation region, the product sequence is propylamine, ACN–water azeotrope, and water. For the right region, the product sequence is propylamine, ACN–water azeotrope, and ACN. This example shows that conventional

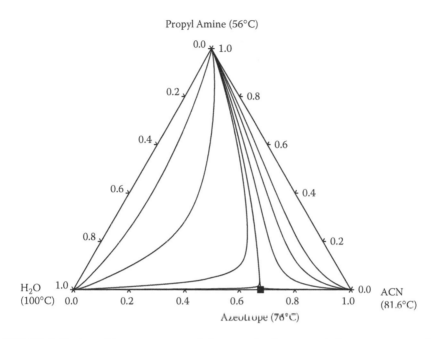

FIGURE 5.13 A residue curve map of the propylamine–acetonitrile–water system for Example 5.7.

distillation cannot obtain pure water and pure ACN cuts at the same time due to the distillation barrier; therefore, this system requires a mass separating agent to cross over this barrier, resulting in a novel batch distillation synthesis problem.

5.5.2.2 Extractive Batch Distillation

Extractive batch distillation can provide advantages of both batch distillation and extractive distillation; thus, this process can be very useful for separation and recovery of waste solvent streams that generally form multicomponent azeotropes. However, most of the recent research efforts on this kind of distillation have been limited to feasibility analysis. Safrit et al.[47] and Safrit and Westerberg[51] investigated batch extractive distillation in the middle vessel column. They showed that the extractive process is comprised of two steps (operations 2 and 3) and requires a much smaller still pot size. They also identified feasible and infeasible regions and showed that, by varying column conditions such as the product rate, reflux ratio, and reboil ratio, one can steer the middle vessel composition to avoid an infeasible region.

Figure 5.14 shows operational fractions of batch extractive distillation using a middle vessel column, recently developed by Kim,[52] in order to separate acetonitrile from an aqueous mixture. Fraction 1 is a total reflux and total reboil condition for startup without entrainer (E) feeding. The entrainer is fed to the bottom section of the middle vessel column in fraction 2, where the entrainer can

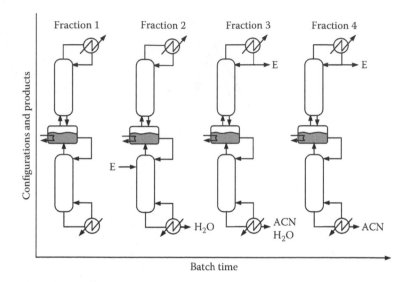

FIGURE 5.14 Operational fractions of batch extractive distillation in a middle vessel column.

increase relative volatility and separate highly pure water as a bottom product. In this fraction, the middle vessel column operates as a stripper. Fraction 3, where entrainer feeding stops, recovers the entrainer as a top product and a waste cut as a bottom product. Fraction 4 separates the entrainer and ACN as top and bottom products, respectively. This example shows the usefulness of batch extractive distillation using a middle vessel column. This process can provide flexibility in selecting a mass separating agent (entrainer) over batch azeotropic distillation and in obtaining proper product sequences and can exhibit seamless batch distillation operation between fractions. Detailed advantages of this process are explained later.

5.5.2.3 Reactive Batch Distillation

Although reactive distillation was acknowledged as a unit operation as early as in the 1920s, it has gained its research interest as an excellent alternative to both reaction and separation since the 1980s. For example, most of the new commercial processes of methyl-*tert*-butyl ether (MTBE, an anti-knocking agent) are based on continuous reactive distillation technologies. The analysis of a reactive batch distillation model in a staged column was first published by Cuille and Reklaitis.[53] Using a stiff integrator for the differential and algebraic equations, they presented a numerical solution technique for the esterification of 1-propanol and acetic acid. Wajge et al.[54] developed a new solution technique based on the orthogonal collocation method and the finite-element method for the reactive batch distillation of a packed column. The differential contactor model of a packed column, originally designed by Hitch and Rousseau,[55] was then reduced to low-order polynomials with the desired accuracy. They compared the results with those from the finite difference method and global collocation method for nonreactive packed-bed batch distillation systems and showed that their approach was more efficient. Wajge and Reklaitis[56] extended their previous work to the optimal campaign structure for reactive batch distillation, which can offer reasonably sharp separations between successive cuts and reduce the amount of waste off-cuts. To obtain the optimal reflux policies or profiles for the maximum distillate or minimum time problem, multiperiod reflux optimization can be applied. Macchietto and Mujtaba[57] showed that, for the same production rate, the waste generation can be significantly reduced under the optimal campaign structure.

An efficient optimization approach for reactive batch distillation using polynomial curve fitting techniques was presented by Mujtaba and Macchietto.[31] After finding the optimal solution of the maximum conversion problem, polynomial curve-fitting techniques were applied over these solutions, resulting in a nonlinear algebraic maximum profit problem that can be efficiently solved by a standard NLP technique. Four parameters in the profit function (maximum conversion, optimum distillate, optimum reflux ratio, and total reboiler heat load) were then represented by polynomials in terms of batch time. This algebraic representation of the optimal solution can be used for online optimization of batch distillation.

A dynamic rate-based model for packed-bed batch distillation was recently presented in which a solid catalyst was used first in the reactive batch distillation modeling.[58] The pilot-scale experiments were conducted with strong anion-exchange resins. The results were compared with the experimental data and with results from its counterpart, the equilibrium-based model. The rate-based model provides more accuracy, much higher physical significance, and more predictability of the experimental data even though the formulation of the rate-based model is complicated.

5.5.3 Batch Distillation Synthesis

The complexity of batch distillation design and operation is also reflected in the batch distillation synthesis problem. In continuous distillation, optimal column sequencing is the main focus of synthesis research. Several past reviews are available on this subject.[59,60] Unlike continuous distillation synthesis, the area of batch distillation synthesis is complicated by its transient nature. Decisions such as cut selection, operating mode, configuration type, and column sequencing enter into the synthesis problem. For complex systems such as azeotropic, extractive, and reactive distillation, identifying the distillation boundaries and steering toward feasible and optimal regions add further complications to the problem; however, as seen in the previous section, theoretical and geometric analyses can point toward optimal synthesis solutions in this area.[13,14,50]

Example 5.8

For the residue curve map of the propylamine–acetonitrile–water system given in Example 5.7, find the necessary batch distillation cuts to separate all three pure components if the initial feed (F_0) is (0.58, 0.20, 0.22). Discuss which column configuration is best for this separation.[52]

Solution

Because the initial feed is located in the left distillation region, draw a line from a stable node (H_2O apex) through the feed point to the distillation boundary. The top product (D_1) is on the distillation barrier. Draw a line from an unstable node (propylamine apex) through D_1 (i.e., now F_2) to the ACN–water boundary line, which will be B_2. Now there are only two components in the system, thus the top and bottom product cuts are ACN–water azeotrope and ACN, respectively. The product cuts are summarized in Table 5.3.

If a batch rectifier is used, shutdown and setup time is required before cut 2 because the top product from cut 1 is the feed to the bottom reboiler in the next cut. Similarly, if a batch stripper is used, shutdown and setup time is required before cut 3. Because a middle vessel column performs both rectifying and stripping operations, a middle vessel column configuration is the optimal column configuration. The operational fractions or cuts of this batch synthesis problem are shown in Figure 5.3.

State-of-the-art techniques used in the solution of synthesis problems include: (a) a heuristic approach, which relies on intuition and engineering knowledge; (b) a physical-insight approach, which is based on exploiting basic physical principles; and (c) an optimization approach. In this section, two common approaches, heuristics and optimization, are discussed.

The recent literature in batch distillation has been devoted to comparing emerging column configurations with the conventional one, thereby obtaining heuristics for optimal column configuration, optimal design, and optimal operating conditions.[17,43,45,46,61] In these studies, parameters such as product purity, batch time, or total cost were evaluated to compare the performance of column configurations. Chiotti and Iribarren[61] compared the rectifier with the stripper in terms of annual cost and product purity. They noted that the rectifier is better for the more volatile component products while it is more economical to obtain less volatile component products using the stripper. Meski and Morari[46] compared three column configurations in terms of the batch time under fixed product purity and infinite number of plates. They observed that the middle vessel column always

---

TABLE 5.3
Product Cuts for Example 5.8

Cut No.	Feed	Bottom Product Cut	Top Product Cut
1	$F_1 = F_0$	B_1 = pure water	D_1 = propylamine–ACN
2	$F_2 = D_1$	B_2 = ACN–water	D_2 = pure propylamine
3	$F_3 = B_2$	B_3 = pure ACN	D_3 = ACN–water azeotrope

had the shortest batch time, and the rectifier had the next shortest time. Sørensen and Skogestad[45] studied two competing column configurations, rectifier and stripper, in the context of minimum optimal operating time and also described the dynamic behavior of these columns. They concluded that the stripper is the preferred column configuration when a small amount of the more volatile component is in the feed and that the rectifier is better when the feed has a high amount of the more volatile component. Although several studies support the same heuristics, some studies present contradictions among the suggested heuristics. For example, the batch time studies of Meski and Morari[46] and Sørensen and Skogestad[45] give conflicting results with respect to feed composition. This is due to the limited ranges of parameters and systems considered, as well as the complexity and difficulty of the problem of column selection.

In order to elicit comprehensive heuristics, the analysis must cover a wider range of column configurations, operation policies, and design variables, and various performance indices must be included. Kim and Diwekar[17] extended the column configuration problem using four performance indices: product purity, yield, design feasibility and flexibility, and thermodynamic efficiency. It is generally observed that the rectifier is a promising column configuration for the more volatile component product and that the stripper is better in the opposite case. Feasibility studies based on the minimum number of plates and minimum reflux ratio addressed the flexibility of such a high-purity configuration for changing operating conditions. It was found that the rectifier and the stripper have distinctive feasibility regions in terms of the feed composition. Thermodynamic efficiency indicates how close a process or system is to its ultimate performance and also suggests whether or not the process or system can be improved. The rectifier can also be a promising column configuration in terms of thermodynamic efficiency, but in some conditions higher efficiencies of the stripper or the middle vessel column can be observed. Furthermore, for the middle vessel column, the thermodynamic efficiency is greatly affected by an added degree of freedom (q'). This systematic and parametric study by Kim and Diwekar[17] concluded that the trade-offs between performance indices should be considered within a multiobjective framework.

5.5.4 COMPUTER-AIDED DESIGN SOFTWARE

It is difficult to analyze batch distillation without using computers due to the two reasons stated before: (1) the process is time varying, so one has to resort to complex numerical integration techniques and different simulation models for obtaining the transients; and (b) this ever-changing process also provides flexibility in operating and configuring the column in numerous ways. Based on the current state of the art in batch distillation techniques and computer simulation technology, Table 5.4 identifies the required functionality and the rational behind it. Several commercial software packages are available for simulations, optimizations, and optimal controls of batch distillation (see Table 5.5). These include Bdist-SimOPT (Batch Process Technologies), BatchSim (Simulation Sciences),

TABLE 5.4
Batch Distillation Software Requirements[2]

Features		Why
Windows		User-friendly state-of-the-art input/output interface
Databank		Ability to generate data from structural information
Operations	Constant reflux	Yield improvement due to operational flexibility;
	Variable reflux	systematic optimization/optimal control methods
	Fixed equation optimal	
	Optimal reflux	
Models	Shortcut	Hierarchy of models for numerical stability, design
	Semirigorous	feasibility, and advanced system designs
	Design feasibility	
	Optimization	
Options	Reactive distillation	Advanced feature
	Three-phase distillation	
	Uncertainty analysis	
Configurations	Semibatch	Emerging designs provide promising directions for
	Recycle waste cut	effective designs to obtain purer products
	Rectifier	
	Stripper	
	Middle vessel column	

BatchFrac (Aspen Technology, based on Boston et al.[10]), and MultiBatchDS (Batch Process Research Company). Bdist-SimOPT and MultiBatchDS are derived from the academic package BATCHDIST.[11] Most of these packages, except MultiBatchDS, are usually limited to conventional systems as they were developed in early or late 1980s. The educational version of MultiBatchDS can be obtained from the AIChE CACHE website (http://www.che.utexas.edu/cache/).

5.6 SUMMARY

This chapter presented a complete review of the batch distillation literature, beginning from the first theoretical analysis to the current state of the art in computer-aided design and optimization methods. The new advances in batch distillation include novel column configurations, optimal designs, optimal operation policies, and new methods of analysis. These new advances can increase the possibility of using batch distillation profitably for a wide variety of separations, but they also present a bewildering array of problems regarding the selection of proper configurations, the correct operating mode, and optimal design parameters. Thus, we will certainly see future researchers working on various aspects of design, analysis, and synthesis of batch distillation, some of which are outlined below:

TABLE 5.5
Batch Distillation Software Comparison[2]

Features		BATCHSIM	BatchFrac	MultiBatchDS
Windows		Yes	No	Yes
Databank		SIMSCI	Aspen Plus	Cranium/Others
Operations	Constant reflux	Yes	Yes	Yes
	Variable reflux	Yes (limited)	Yes (limited)	Yes
	Fixed equation optimal	No	No	Yes
	Optimal reflux	No	No	Yes
	Shortcut	No	No	Yes
Models	Semirigorous	No	No	Yes
	Reduced order	No	No	Yes
	Rigorous	Yes	Yes	Yes
	Design feasibility	No	No	Yes
	Optimization	No	No	Yes
Options	Reactive distillation	No	Yes (limited)	Yes (limited)
	Three-phase distillation	No	Yes	Yes
	Uncertainty analysis	No	No	Yes
	Semibatch	No	No	Yes
	Recycle waste-cut	No	No	Yes
Configurations	Rectifier	Yes	Yes	Yes
	Stripper	No	No	Yes
	Middle vessel	No	No	Yes

- A more extensive analysis for each column configuration must be carried out. For example, the effect of q' on the performance of the middle vessel column has not been fully investigated. For the multi-vessel column, no general guidelines or heuristics exist for column holdups and operating modes because of the additional degrees of freedom.
- Azeotropic, extractive, and reactive distillations and off-cut recycling operations have been studied extensively in recent years, but they are in the developing stage. For instance, for continuous reactive distillation, solid catalysts are commonly used, but only a few applications of solid catalysts in batch reactive distillation exist in theory and practice.
- Comprehensive heuristics for optimal design and synthesis should be derived. Several heuristics and trade-offs between heuristics can be found, but they are still limited to the systems considered.
- Batch processes often encounter feed composition variations and other operational uncertainties. Consideration of uncertainties at various stages of design and operation can provide useful and cost-effective solutions to the batch processing industries.

- An important research area in this field is batch distillation synthesis. Based on future advances in batch distillation, batch distillation synthesis from a superstructure can lead to the most promising and flexible column configuration with the appropriate operation mode and conditions.

NOTATION

α	relative volatility
B	amount of bottom residue (mol)
dB/dt	bottom product flow rate or change of bottom product (mol/hr)
D	amount of distillate (mol)
dD/dt	distillate rate (mol/hr)
E	entrainer feed rate (mol/hr)
F	amount of feed (mol)
H_j	molar holdup on plate j (mol)
H_0, H_D	condenser holdup (mol)
I_D	enthalpy of the liquid in the condenser (J/mol)
I_j	enthalpy of the liquid stream leaving plate j (J/mol)
J_j	enthalpy of the vapor stream leaving plate j (J/mol)
L_j	liquid stream leaving plate j (mol/hr)
L_0	liquid reflux at the top of the column (mol/hr)
n	number of components
N	number of plates
q'	ratio of the top vapor flow rate to the bottom vapor flow rate
Q_R	reboiler heat duty
R	reflux ratio (L/D)
R_t	reflux ratio as a function of time
T	batch time (hr)
V_j	vapor stream leaving plate j (mol/hr)
x	liquid-phase mole fraction
x_D	liquid-phase mole fraction of the distillate
$x_{D,avg}$	average distillate mole fraction
x_F	liquid-phase mole fraction of the feed
y	vapor-phase mole fraction

REFERENCES

1. Rayleigh, L., On the distillation of binary mixtures, *Philos. Mag.*, (vi)4, 521, 1902.
2. Diwekar, U.M., *Batch Distillation: Simulation, Optimal Design and Control*, Taylor & Francis, Washington, DC, 1996.
3. Davidyan, A.G., Kiva, V.N., Meski, G.A., and Morari, M., Batch distillation column with a middle vessel, *Chem. Eng. Sci.*, 49(18), 3033–3051, 1994.
4. Skogestad, S., Wittgens, B., Litto, R., and Sørensen, E., Multivessel batch distillation, *AIChE J.*, 43(4), 971–978, 1997.

5. Converse, A.O. and Gross, G.D., Optimal distillate rate policy in batch distillation, *Ind. Eng. Chem. Fund.*, 2, 217–221, 1963.
6. McCabe, W.L. and Thiele, E.W., Graphical design of fractioning columns, *Ind. Eng. Chem.*, 17, 605–611, 1925.
7. Smoker, E.H. and Rose, A., Graphical determination of batch distillation curves for binary mixtures, *Trans. Am. Inst. Chem. Eng.*, 36, 285–293, 1940.
8. Bogart, M.J.P., The design of equipment for fractional batch distillation, *Trans. Am. Inst. Chem. Eng.*, 33, 139, 1937.
9. Distefano, G.P., Mathematical modeling and numerical integration of multicomponent batch distillation equations, *AIChE J.*, 14, 190–199, 1968.
10. Boston, J.F., Britt, H.I., Jirapongphan, S., and Shah, V.B., An advanced system for the simulation of batch distillation operations, in *Foundations of Computer-Aided Chemical Process Design*, Vol. 2, Engineering Foundation, New York, 1983, 203–237.
11. Diwekar, U.M. and Madhavan, K.P., Batch-Dist: A comprehensive package for simulation, design, optimization and optimal control of multicomponent, multifraction batch distillation columns, *Comp. Chem. Eng.*, 15(12), 833–842, 1991.
12. Hindmarsh, A.C., LSODE and LSODI: two new initial value ODE solvers, *ACM SIGNUM*, 15, 10, 1980.
13. Bernot, C., Doherty, M.F., and Malone, M.F., Patterns of composition change in multicomponent batch distillation, *Chem. Eng. Sci.*, 45, 1207–1221, 1990.
14. Bernot, C., Doherty, M.F., and Malone, M.F., Feasibility and separation sequence in multicomponent batch distillation, *Chem. Eng. Sci.*, 46, 1311–1326, 1991.
15. Diwekar, U.M., An efficient design method for binary azeotropic batch distillation, *AIChE J.*, 37, 1571–1578, 1991.
16. Lotter, S.P. and Diwekar, U.M., Shortcut models and feasibility considerations for emerging batch distillation columns, *Ind. Eng. Chem. Res.*, 36, 760–770, 1997.
17. Kim, K.J. and Diwekar, U.M., Comparing batch column configurations: a parametric study involving multiple objectives, *AIChE J.*, 46(12), 2475–2488, 2000.
18. Cho, Y.S. and Joseph, B., Reduced-order steady-state and dynamic models of separation processes. Part I. Development of the model reduction procedure, *AIChE J.*, 29, 261–269, 1983.
19. Srivastava, R.K. and Joseph, B., Simulation of packed bed separation processes using orthogonal collocation, *Comp. Chem. Eng.*, 8(1). 43–50, 1984.
20. Aly, S., Pibouleau, L., and Domenech, S., Treatment of batch, packed distillation by a finite-element method. Part I. Steady-state model with axial dispersion, *Int. Chem. Eng.*, 30(3), 452–465, 1990.
21. Diwekar, U.M., Simulation, Design, Optimization, and Optimal Control of Multicomponent Batch Distillation Columns, Ph.D. thesis, Indian Institute of Technology, Department of Chemical Engineering, Bombay, India, 1988.
22. Diwekar, U.M., Unified approach to solving optimal design-control problems in batch distillation, *AIChE J.*, 38(10), 1551–1563, 1992.
23. Diwekar, U.M., *Introduction to Applied Optimization*, Kluwer Academic, Boston, MA, 2002.
24. Diwekar, U.M., Malik, R., and Madhavan, K., Optimal reflux rate policy determination for multicomponent batch distillation columns, *Comp. Chem. Eng.*, 11, 629–637, 1987.

25. Logsdon, J.S., Diwekar, U.M., and Biegler, L.T., On the simultaneous optimal design and operation of batch distillation columns, *Trans. IChemE*, 68, 434–444, 1990.

26. Farhat, S., Czernicki, M., Pibouleau, L., and Domenech, S., Optimization of multiple-fraction batch distillation by nonlinear programming, *AIChE J.*, 36, 1349–1360, 1990.

27. Logsdon, J.S. and Biegler, L.T., Accurate determination of optimal reflux policies for the maximum distillate problem in batch distillation, *Ind. Eng. Chem. Res.*, 32, 692–700, 1993.

28. Coward, I., The time-optimal problems in binary batch distillation, *Chem. Eng. Sci.*, 22, 503–516, 1966.

29. Kerkhof, L.H. and Vissers, H.J.M., On the profit of optimal control in batch distillation, *Chem. Eng. Sci.*, 33, 961–970, 1978.

30. Li, P., Wozny, G., and Reuter, E., Optimization of multiple-fraction batch distillation with detailed dynamic process model, *Inst. Chem. Eng. Symp. Ser.*, 142, 289–300, 1997.

31. Mujtaba, I.M. and Macchietto, S., Efficient optimization of batch distillation with chemical reaction using polynomial curve fitting technique, *Ind. Eng. Chem. Proc. Design Devel.*, 36, 2287–2295, 1997.

32. Furlonge, H.I., Pantelides, C.C., and Sørensen, E., Optimal operation of multivessel batch distillation columns, *AIChE J.*, 45(4), 781–801, 1999.

33. Hasebe, S., Noda, M., and Hashimoto, I., Optimal operation policy for total reflux and multi-batch distillation systems, *Comp. Chem. Eng.*, 23, 523–532, 1999.

34. Pontryagin, L.S., Some Mathematical Problems Arising in Connection with the Theory of Automatic Control System (in Russian), paper presented at Session of the Academic Sciences of the USSR on Scientific Problems of Automatic Industry, Paris, France, 1956.

35. Bellman, R., *Dynamic Programming*, Princeton University Press, Princeton, NJ, 1957.

36. Charalambides, M.S., Shah, N., and Pantelides, C.C., Synthesis of batch reaction/distillation processes using detailed dynamic model, *Comp. Chem. Eng.*, 19, s167–s174, 1995.

37. Quintero-Marmol, E., Luyben, W.L., and Georgakis, C., Application of an extended Luenberger observer to the control of multicomponent batch distillation, *Ind. Eng. Chem. Res.*, 30(8), 1870–1880, 1991.

38. Bosley, J.R. and Edgar, T.F., Application of Nonlinear Model Predictive Control to Optimal Batch Distillation, paper presented at IFAC Dynamics and Control of Chemical Reactors (DYCORD+92), College Park, Maryland, 1992, 303–308.

39. Finefrock, Q.B., Bosley, J.R., and Edgar, T.F., Gain-scheduled PID control of batch distillation to overcome changing system dynamics, AIChE Annual Meeting, 1994.

40. Barolo, M. and Berto, F., Composition control in batch distillation: binary and multicomponent mixtures, *Ind. Eng. Chem. Res.*, 37, 4689–4698, 1998.

41. Barolo, M., Berto, F., Rienzi, S.A., Trotta, A., and Macchietto, S., Running batch distillation in a column with a middle vessel, *Ind. Eng. Chem. Res.*, 36, 4612–4618, 1996.

42. Farschman, C.A. and Diwekar, U.M., Dual composition control in a novel batch distillation column, *Ind. Eng. Chem. Res.*, 37, 89–96, 1998.

43. Hasebe, S., Kurooka, T., and Hashimoto, I., Comparison of the separation performances of a multieffect batch distillation system and a continuous distillation system, preprints of IFAC Symposium on Dynamics and Control of Chemical Reactors, Distillation Columns, and Batch Processes (DYCORD '95), Helsingor, Denmark, 1995, 249–254.

44. Robinson, C. and Gilliland, E., *Elements of Fractional Distillation*, 4th ed., McGraw-Hill, New York, 1950.

45. Sørensen, E. and Skogestad, S., Comparison of regular and inverted batch distillation, *Chem. Eng. Sci.*, 51(22), 4949–4962, 1996.

46. Meski, G.A. and Morari, M., Design and operation of a batch distillation column with a middle vessel, *Comp. Chem. Eng.*, 19, s597–s602, 1995.

47. Safrit, B.T., Westerberg, A.W., Diwekar, U.M., and Wahnschafft, O.M., Extending continuous conventional and extractive distillation feasibility insights to batch distillation, *Ind. Eng. Chem. Res.*, 34, 3257–3264, 1995.

48. Hasebe, S., Noda, M., and Hashimoto, I., Optimal operation policy for multi-effect batch distillation system, *Comp. Chem. Eng.*, 21(suppl.), s1221–s1226, 1997.

49. Sørensen, E., A cyclic operating policy for batch distillation: theory and practice, *Comp. Chem. Eng.*, 23, 533–542, 1999.

50. Cheong, W. and Barton, P.I., Azeotropic distillation in a middle vessel batch column. 1. Model formulation and linear separation boundaries, *Ind. Eng. Chem. Res.*, 38, 1504–1530, 1999.

51. Safrit, B.T. and Westerberg, A.W., Improved operational policies for batch extractive distillation columns, *Ind. Eng. Chem. Res.*, 36, 436–443, 1997.

52. Kim, K.J., Solvent Selection and Recycling: A Multiobjective Optimization Framework for Separation Processes, Ph.D. thesis, Carnegie Mellon University, Department of Environmental Engineering, Pittsburgh, PA, 2001.

53. Cuille, P.E. and Reklaitis, G.V., Dynamic simulation of multicomponent batch rectification with chemical reactions, *Comp. Chem. Eng.*, 10(4), 389–398, 1986.

54. Wajge, R.M., Wilson, J.M., Pekny, J.F., and Reklaitis, G.V., Investigation of numerical solution approaches to multicomponent batch distillation in packed beds, *Ind. Eng. Chem. Res.*, 36, 1738–1746, 1997.

55. Hitch, D.M. and Rousseau, R.W., Simulation of continuous contact separation processes multicomponent batch distillation, *Ind. Eng. Chem. Res.*, 27, 1466–1473, 1988.

56. Wajge, R.M. and Reklaitis, G.V., An optimal campaign structure for multicomponent batch distillation with reversible reaction, *Ind. Eng. Chem. Res.*, 37, 1910–1916, 1998.

57. Macchietto, S. and Mujtaba, I.M., *Design of Operation Policies for Batch Distillation*, NATO Advanced Study Institute Series F143, Springer-Verlag, Berlin, 1996, 174–215.

58. Kreul, L.U., Górak, A., and Barton, P.I., Dynamic rate-based model for multicomponent batch distillation, *AIChE J.*, 45(9), 1953–1961, 1999.

59. Nishida, N., Stephanopoulos, G., and Westerberg, A.W., Review of process synthesis, *AIChE J.*, 27, 321–351, 1981.

60. Grossmann, I.E., MINLP optimization strategies and algorithms for process synthesis, in *Proc. Foundations of Computer-Aided Process Design '89*, Elsevier, Amsterdam, 1990.

61. Chiotti, O.J. and Iribarren, O.A., Simplified models for binary batch distillation, *Comp. Chem. Eng.*, 15(1), 1–5, 1991.

6 Batch Crystallization

Priscilla J. Hill

CONTENTS

6.1 INTRODUCTION

Many chemical manufacturing processes involve solid particles, either as feeds or products or as internal process enhancers such as catalysts. These processes span a range of industries including pharmaceuticals, pigments, agricultural chemicals, and foods, as well as polymers. Solution crystallization, where a solvent (or a mixed solvent) with one or more solutes is involved, is a critical operation in solids processing because it can strongly affect the characteristics of a solid product. The crystallization step can perform several functions in a process, including separation or purification, as well as particle formation.

Crystallization is a key operation because it affects both product composition and morphology, which often determine product quality and process operability. The product specifications usually include the production rate of solids, required purity levels, crystal size distribution (CSD), and morphology. Process operability often depends on characteristics such as the flowability of a solid, which is largely controlled by the crystal shape. The challenge is to operate the crystallizer so crystals with the desired attributes will be produced.

The crystallizer unit can be operated as a batch, semibatch, or continuous process. One of the primary differences between batch and continuous crystallizers is that both temperature and supersaturation cannot be held constant simultaneously during batch operation, but this can be done with a continuous crystallizer when the unit reaches steady-state operation. Batch crystallization is often used in industry for low-volume, high-value-added specialty chemicals such as pharmaceuticals.

The purpose of this chapter is to discuss batch crystallization from solution. The first sections of the chapter provide a basis for the later sections. Fundamental concepts such as solubility and supersaturation are discussed in Section 6.2 and crystallization kinetics are discussed in Section 6.3. This review provides a background for the design models in Section 6.4 and the design and modeling decisions presented in Section 6.5. Section 6.6 introduces some of the practical concerns in instrumentation of a batch crystallizer. Section 6.7 introduces a general methodology for batch crystallizer design, and Section 6.8 gives examples of how to apply the principles presented in this chapter. Concluding remarks are given in Section 6.9.

6.2 FUNDAMENTAL CONCEPTS

A simplified sketch of a batch unit is given in Figure 6.1. It is usually a jacketed vessel with a stirrer, and it can contain internal structures such as baffles and a draft tube. The jacket controls the temperature, and the internal structures facilitate mixing. The initial feed is often a liquid, although it can contain solids for seeding. The initial liquid composition is known, as is the particle size distribution (PSD) and composition of the seeds. The operating temperature and pressure, as well as the stirring speed, are operating parameters that can be controlled during operation. In order to understand what is occurring in a crystallizer, it is first necessary to understand the basic phenomena that influence crystallization, such as solid–liquid equilibria; for example, material will not crystallize unless the

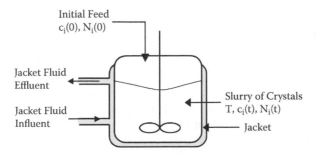

FIGURE 6.1 Simplified diagram of a batch crystallizer.

solution is supersaturated. Therefore, it is first necessary to know the process conditions under which crystals can form.

6.2.1 SOLUBILITY

The solubility is defined as the maximum amount of solute that will remain dissolved in a solution under specified equilibrium conditions. The solubility is often expressed as the quantity of solute per quantity of solution (e.g., g solute per 100 g solution, where the solution includes the solute as well as the solvent) or as the quantity of solute per quantity of solvent (e.g., g solute per 100 g solvent). Sometimes, the quantities are expressed in terms of moles or volume, but mass is more commonly used. In the case where the crystal is a hydrate, the solubility specifies whether the data are for the anhydrous material or for the hydrate. Because solubility is a function of temperature, it is often reported over a temperature range. If data are not available in the literature, the solubility can be estimated or measured.

Literature data can be found in a number of locations, including technical articles on specific systems, solubility handbooks (such as the one by Seidell[1]), and more general-purpose handbooks.[2,3] Some data are also available in books on crystallization.[4,5] The solubility can also be estimated from thermodynamics. These estimations are based on the phase equilibrium concept that the chemical potential of the solute in the solid form is equal to the chemical potential of the solute in solution. This may be written as:

$$f_{i,solid} = x_i \gamma_i f_{i,scl} \tag{6.1}$$

where $f_{i,solid}$ is the fugacity of the solute (component i) in the solid, and $f_{i,scl}$ is the fugacity of pure solute i in a subcooled liquid state.[6]

The simplest estimation method is to assume that the solute and solvent form an ideal solution. In this case, the solubility can be estimated from the van't Hoff equation:[4,6]

$$\ln x_i = \frac{\Delta H_{m,i}}{R} \left(\frac{1}{T_{m,i}} - \frac{1}{T} \right) \tag{6.2}$$

where x_i is the solubility given as a mole fraction, R is the ideal gas law constant, $\Delta H_{m,i}$ is the molar enthalpy of fusion for the solute, $T_{m,i}$ is the melting point of solute i, and T is the temperature of interest. For this equation, absolute temperatures must be used. Because the van't Hoff equation only uses data for the solute, it does not consider the nature of the solvent.

For nonideal liquid solutions, the activity coefficient is not equal to one. Walas[6] showed that for phase equilibria we have:

$$x_i = \frac{1}{\gamma_i} \exp\left[\frac{\Delta H_{m,i}}{R} \left(\frac{1}{T_{m,i}} - \frac{1}{T} \right) \right] \tag{6.3}$$

where γ_i is the activity coefficient of solute i. The activity coefficient can be calculated from existing methods for calculating the activity coefficient in a liquid solution. Because no one method is applicable to all systems, it is necessary to choose a method appropriate to the system under consideration. A review of these methods[7] is not limited to but includes the Scatchard–Hildebrand theory for nonpolar solutions and the UNIFAC group contribution method for nonelectrolyte solutions.

Both of the previous equations assume that the solid is a pure solid; however, this is not always the case. For a solid solution, the solubility calculations must be corrected to include the solid-phase activity coefficient $\gamma_{i,solid}$ and the composition in the solid phase, $x_{i,solid}$:

$$x_i = \frac{\gamma_{i,solid} x_{i,solid}}{\gamma_i} \exp\left[\frac{\Delta H_{m,i}}{R} \left(\frac{1}{T_{m,i}} - \frac{1}{T} \right) \right] \tag{6.4}$$

where the subscript "solid" indicates the solid phase. The solid-phase activity coefficient may be estimated for nonpolar solutes and solvents with a modified Scatchard–Hildebrand theory.[7] As with liquids, no single method is applicable to all systems. All of the solubility equations given account for the temperature dependence of the solubility; however, they are only valid over a temperature range where the heat of fusion is constant.

Solubility data have also been fit to various empirical expressions, including:

$$C_i^{sat} = C_{1,i}^{sat} + C_{2,i}^{sat} T + C_{3,i}^{sat} T^2 \tag{6.5}$$

where C_i^{sat} is the solubility of component i; $C_{1,i}^{sat}$, $C_{2,i}^{sat}$, and $C_{3,i}^{sat}$ are constants for component i; and T is the temperature.[8] These equations are convenient in that they may be easily coded in a computer program.

In order to get a phase diagram, we must have solubility information, which can come from literature data, calculations, or experiments. Although calculations can be used for the initial estimates, they should be checked against experimental data. Impurities in the solution can have a strong effect on the solubility. Experiments can be performed at an early stage to determine whether or not the impurities have a significant effect on the solubility.

6.2.2 Phase Diagram

A phase diagram is one method for showing the feasible region of operation of a crystallizer. With this diagram, one can immediately eliminate some process operating conditions that are not feasible. This is demonstrated for a binary system with a binary eutectic diagram and with a ternary phase diagram.

6.2.2.1 Eutectic Diagram

In the binary eutectic diagram shown in Figure 6.2, the solubility is plotted as temperature vs. composition for components A and B. This is for a simple eutectic system that has perfect immiscibility of solid A and solid B. This diagram may be constructed from experimental data or from the equations given for solubility. The general procedure for constructing such a diagram is to plot the pure component melting points, $T_{m,A}$ and $T_{m,B}$, to then plot the liquidus curves using one of the solubility equations given above, and then to plot the eutectic point where the two solubility curves cross. The eutectic point may also be calculated using the solubility equations given above and the concept that the mole fractions add to unity at the eutectic point. For the ideal system case, the eutectic point is determined by finding the temperature where the following is true:

$$
x_A + x_B = 1 = \exp\left[\frac{\Delta H_{m,A}}{R}\left(\frac{1}{T_{m,A}} - \frac{1}{T}\right)\right] + \exp\left[\frac{\Delta H_{m,B}}{R}\left(\frac{1}{T_{m,B}} - \frac{1}{T}\right)\right]
$$

$$(6.6)$$

For other cases, substitute the appropriate expressions for the mole fractions. The solidus line is the line drawn at the eutectic temperature.

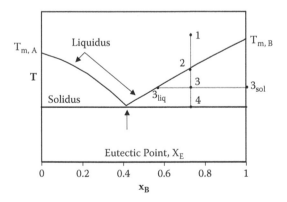

FIGURE 6.2 Eutectic diagram.

The eutectic diagram indicates what happens when a liquid of a given composition is cooled. At point 1, the mixture of components A and B is in the liquid phase with no solids. At point 2, solid component B starts coming out of solution. As the solution is cooled, more of component B solidifies. As B solidifies, it is removed from the liquid solution. This causes the composition of the liquid to change in such a way that it follows the composition of the liquidus. This is shown at point 3, where the overall composition is given by point 3, the solid composition is given by 3_{sol}, and the liquid composition is given by 3_{liq}. At the eutectic temperature, the remainder of the material crystallizes at the same composition as the liquid. From a design viewpoint, the eutectic diagram shows that to remove pure component B it is necessary for the initial solution concentration to be between that of the eutectic composition (x_E) and that of pure component B ($x_B = 1$). A similar argument can be made for component A where the starting composition must be between that for pure A and the eutectic composition. Other types of eutectic diagrams often exhibit more complex phase behavior; however, the simple eutectic diagram shown in Figure 6.2 represents many systems.

6.2.2.2 Ternary Diagram

Let us now consider the phase diagram in Figure 6.3, which shows the solubility of solutes A and B in solvent S at a fixed temperature. The point C_{SA} represents the solubility of A in S at the given temperature, while the solubility of B in S is represented by C_{SB}. The liquidus curves are the dark lines between the two solubilities. The intersection of the two liquidus curves is the double saturation point for A and B in solvent S. The region between the liquidus and 100% S will not contain any solids at the given temperature. Tie lines are shown as dashed lines.

The ternary phase diagram provides design information with regard to what separations are possible. If the goal is to take pure A out of solution, the crystallizer must be operated at the given temperature at a composition shown in the

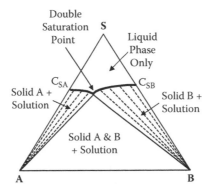

FIGURE 6.3 Isothermal ternary diagram for two solutes A and B and a solvent S.

region marked as "Solid A + Solution." This region is bounded by the liquidus, the line between S and A, and a line between pure A and the double saturation point. Similarly, to remove pure B the crystallizer must be operated in the region between the liquidus, the line between S and B, and a straight line between pure B and the double saturation point. Operating below the given temperature in the triangular region marked "Solid A + Solid B + Solution" will produce a mixture of solid A and B.

6.2.3 SUPERSATURATION

Supersaturation generally indicates that the solution concentration is higher than the solubility. Although it depends on the chemical potential of the solute, it is usually expressed in terms of concentration. One example is the driving force ΔC, which is defined as:

$$\Delta C = C - C^{sat} \tag{6.7}$$

where C is the concentration, and C^{sat} is the concentration in the liquid at saturation at a given temperature and pressure. A second equation expresses the supersaturation as a ratio:

$$\text{Supersaturation} = S = C/C^{sat} \tag{6.8}$$

A third expression combines the above two expressions as follows:

$$\text{Supersaturation} = \sigma = \frac{\Delta C}{C^{sat}} = \frac{C - C^{sat}}{C^{sat}} = S - 1 \tag{6.9}$$

This last expression is sometimes referred to as the *relative supersaturation*.[4,9] Because various expressions are used, it is crucial to know which expression was used in any reported literature data. For example, although the supersaturation is dimensionless in both Equations 6.8 and 6.9, it has different values. Mistaking the supersaturation calculated from Equation 6.8 as the one calculated from Equation 6.9 would give different conditions.

6.2.4 METASTABLE ZONE WIDTH

When cooling a solution, the maximum supersaturation reached before the material begins to solidify is the metastable zone limit, which is shown graphically in Figure 6.4, where the concentration is plotted as a function of temperature. Two curves are shown: the solubility curve and the metastable limit. The region in between these two curves is the metastable zone. To the right of the solubility curve, everything is in solution. If a solution of fixed composition is cooled, it is possible that it could begin to form solid particles once the solubility temperature

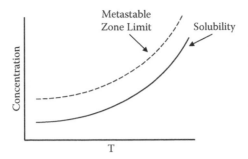

FIGURE 6.4 Metastable zone width.

is reached; however, the solution may be subcooled well into the metastable zone before particles form. Nuclei may start forming at any point in the metastable zone. When the solution has cooled to the metastable limit, the solute must come out of solution as a solid. The liquid cannot be cooled beyond the metastable limit without forming solids. The metastable zone width (MSZW) varies depending on the system being studied. It is usually quite narrow for small ionic crystals, such as NaCl, but can be much wider for organic molecules, such as citric acid. Also, the actual MSZW achieved depends on the cooling rate being used.[10] In general, the MSZW increases with the cooling rate.

6.2.5 CRYSTAL STRUCTURE AND CRYSTAL SHAPE

Crystals are characterized by their internal structure and external shape. Their internal structure is determined by an ordered arrangement of ions (e.g., NaCl), atoms (e.g., the carbon in diamond), or molecules (e.g., organic molecules such as adipic acid) referred to as the *crystal lattice*. Each of these lattices can be described by unit cells that are described by the angles between the major axes and the lengths of the major axes. These are shown schematically in Figure 6.5.

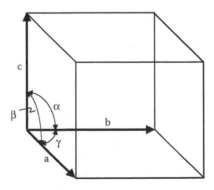

FIGURE 6.5 Elementary cell with lattice parameters.

TABLE 6.1
Crystal Lattices

Category	Bravais Lattice	Primary Lengths	Axis Angles
Cubic	Cubic	$a = b = c$	$\alpha = \beta = \gamma = 90°$
	Body-centered cubic		
	Face-centered cubic		
Tetragonal	Primary tetragonal	$a = b \neq c$	$\alpha = \beta = \gamma = 90°$
	Body-centered tetragonal		
Orthorhombic (rhombic)	Simple orthorhombic	$a \neq b \neq c$	$\alpha = \beta = \gamma = 90°$
	Body-centered orthorhombic		
	Base-centered orthorhombic		
	Face-centered orthorhombic		
Triclinic	Triclinic parallelepiped	$a \neq b \neq c$	$\alpha \neq \beta \neq \gamma$
Monoclinic	Monoclinic parallelepiped	$a \neq b \neq c$	$\alpha = \gamma = 90° \neq \beta$
	Base-centered monoclinic		
Hexagonal	Hexagonal prism	$a = b \neq c$	$\alpha = \beta = 90°$ $\gamma = 120°$
Rhombohedral (trigonal)	Trigonal rhombohedron	$a = b = c$	$\alpha = \beta = \gamma \neq 90°$

The seven main categories of unit cells are cubic (regular), tetragonal, orthorhombic, triclinic, monoclinic, hexagonal, and rhombohedral. Bravais classified crystals into 14 different lattice systems, which can be organized into these seven main categories and as shown in Table 6.1.[4]

While the external shape is influenced by the crystal lattice, it is also influenced by other factors such as the solvent properties, impurities, and operating conditions. If crystal shape is controlled by thermodynamics, the crystal must be grown under conditions very close to saturation conditions; that is, the supersaturation must be very low during crystal growth. This is rather unlikely during batch operation. It is much more common in industrial crystallization for crystal shape to be controlled by kinetics.[11]

Experiments have shown that crystals grown in different solvents have different morphologies. This happens because the faces of a crystal may have different growth rates. The relative facial growth rates of a crystal may vary under different operating conditions. Although two crystals may have the same basic lattice, they may have different proportions. An example of this is shown in Figure 6.6 for a cubic lattice where the first crystal is plate shaped, the second crystal is a cube, and the third crystal is needle shaped. The relative lengths of the sides are quite different for each of these three cases. This difference in morphology has been reported in various experimental studies.[12,13] The effect of kinetics on crystal shape in batch crystallizers has been specifically discussed in the literature.[14]

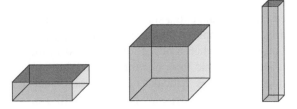

FIGURE 6.6 Regular crystals: plate, cube, and needle.

6.2.6 POLYMORPHISM

Polymorphs are formed when a material crystallizes into different forms; that is, the chemical formulas of the crystals are identical, but their internal structures are different. One of the best known examples is carbon, which can crystallize into graphite or diamond. Graphite has a hexagonal structure, and diamond has a cubic structure. The number of polymorphs is not limited to two and depends on the material. For example, carbon has other structural forms, such as buckminsterfullerene. In addition, many of the triacylglycerols (glycerin fatty acid esters) exhibit three polymorphs, denoted α, β', and β.[15] Controlling the polymorphic form is crucial as the polymorphs may have very different properties; for example, polymorphs may have different densities, different solubilities at the same temperature, or even different crystal habits.[9] One example is the hardness of carbon where graphite is a soft material that is usually opaque, while diamond is a very hard material that is often clear or translucent.

Polymorphs are not limited to carbon. In general, many long-chain molecules exhibit polymorphism. For example, the amino acids L-glutamic acid and L-histidine both exhibit two polymorphs.[16] For both acids, polymorphs of the same compound have different solubilities at a given temperature. Although both polymorphs of L-glutamic acid have an orthorhombic lattice, the crystals are quite different due to the lattice spacing. The metastable α crystal has a flat plate shape, but the stable β crystals are acicular or needle shaped. For L-histidine, the stable A form is orthorhombic while the metastable B form is monoclinic.

Other materials that exhibit polymorphs are fats and lipids.[15] These are important because they are common in food, cosmetics, and pharmaceuticals and because they determine the product quality of materials with fat crystals, such as chocolate and margarine. For example, cocoa butter, the main fat in chocolate, has six polymorphs denoted by the roman numerals I through VI.[17,18] These forms are numbered in order of their ascending melting points (17.3, 23.3, 25.5, 27.5, 33.8, and 36.3°C). Although form VI is the most stable form, form V is the one most desired for food products.

One concern with polymorphs is that some forms are not stable. The stable forms do not change with time, but the metastable forms may change to another polymorph. Sometimes this change is very slow. Frequently the metastable polymorph has a higher supersaturation at a given temperature than the stable polymorph. This causes the metastable polymorph to come out of solution first which is of great concern in the pharmaceutical industry, where a metastable form can be manufactured but can change to a stable form before it is used, thus changing the bioavailability. An example of this is ritonavir, a drug for AIDS. The drug was already in production when a new form appeared in the manufacturing of plants.[19] This new polymorphic form of ritonavir had different solubility properties. For pharmaceuticals, it is critical to control the crystal form because the bioavailability of the compound depends on the crystal form.[20]

Because the different polymorphs are different phases, they can be represented on phase diagrams. In general, the phase diagrams indicate which polymorph is the stable form at a given temperature and pressure. Other polymorphs may appear under the same conditions, but they are metastable under those conditions. The exception to this is when one stable polymorph form transitions to another stable polymorph form at a given temperature. At the transition temperature, two stable polymorphs may exist.

6.2.7 ENANTIOMORPHS

Enantiomorphs describe two unsymmetric molecules where one is a mirror image of the other one. Although many of these enantiomorphous crystals are optically active, not all of them are.[4] The ones that are optically active are referred to as being *chiral*. These crystals can rotate a plane of polarized light to the right or left and are denoted as either a D-form or an L-form. Alternatively, the D- and L-forms are sometimes denoted as R and S, or as + and −. This is very common with biologically produced materials such as amino acids. A racemic mixture contains both forms of the material in equal amounts, making the mixture optically inactive. Crystallization may sometimes be used to separate optical isomers.[20]

6.3 CRYSTALLIZATION KINETICS

While thermodynamics gives information on the maximum amount of material that will crystallize as a solid, it does not give the rate at which the solids are produced. Crystallization kinetics are needed to provide design information such as the production rate of crystals, the crystal size distribution, and the crystal shape. The evolution of particle size and shape is governed by the following mechanisms: growth, nucleation, agglomeration, and breakage. Each one of these mechanisms is discussed separately.

6.3.1 GROWTH

Particle growth refers to a particle increasing in size due to the addition of material from solution. This growth rate is often mass transfer controlled. A variety of models have been developed to model crystal growth. One of the first models was the McCabe ΔL law:[8,21]

$$G_L(L) \equiv \frac{dL}{dt} = G_0 \tag{6.10}$$

where $G_L(L)$ is the length-based growth rate defined as dL/dt, L is the representative particle length, and G_0 is constant with respect to particle size. This law assumes that the growth rate is independent of particle length. The parameter G_0 is often a function of temperature and supersaturation, as sometimes represented by

$$G_0 = k_G \left(C_i - C_i^{sat} \right)^n \tag{6.11}$$

$$k_G = A_G e^{(-E/RT)} \tag{6.12}$$

where C_i^{sat} is the solubility of i, C_i is the concentration of i, k_G is a constant at a given temperature, and A_G is a constant.

Later researchers[22–24] determined that the size-independent assumption was not always true. A list of some possible growth functions is given in Table 6.2, where L is the characteristic particle length and v is the particle volume. The size-dependent functions account for the fact that different diameter particles grow at different rates. The table also shows growth functions on a volume basis[25] as $G_V(v)$, where the growth rate is defined as dv/dt and v is the particle volume. These variables may also be volume independent or volume dependent.

Typical values of growth rates for selected materials are available from experimental studies. A review by Garside and Shah[26] give data obtained from continuous crystallizers. Because the basic growth kinetics are the same for both continuous and batch crystallizers, these data should be similar for batch processes. The data range from 0.18 to 13.8 μm/min depending on the system and the operating conditions. Given that the growth rates span a range of 2 orders of magnitude, it is clear that a single default value cannot be used for all cases.

TABLE 6.2
Growth Functions

Description	Function	Ref.
Size independent	$G_L(L) = G_0$	McCabe[21]
Size dependent	$G_L(L) = G_0 L^\alpha$	Bransom[22]
	$G_L(L) = G_0(1 + \gamma L)$	Canning and Randolph[23]
	$G_L(L) = G_0(1 + \gamma L)^\alpha$	Abegg et al.[24]
Volume independent	$G_V(v) = G_1$	Ramabhadran et al.[25]
Volume dependent	$G_V(v) = G_1 v^\alpha$	Ramabhadran et al.[25]

Mersmann et al.[27] provide a discussion of the fundamental mechanisms of crystal growth at the microscopic level. They note that, while a number of steps in the growth process can be rate limiting, most growth rates in industrial practice are limited by bulk diffusion and surface integration. A theoretical approach has been taken for estimating the mean crystal growth rate of a single solute in a solvent.[27,28] This approach considers whether heat transfer, surface integration, or bulk diffusion is the limiting step in particle growth. Mersmann et al.[28] determined that the theoretical growth rate (G_t) is based on:

$$G_t = \cfrac{1}{\cfrac{1}{G_h} + \cfrac{1}{G_{dif}} + \cfrac{1}{G_{surf}}} \tag{6.13}$$

where G_h is the heat-transfer-limited growth rate, G_{dif} is the bulk-diffusion-limited growth rate, and G_{surf} is the surface-integration-limited growth rate. Furthermore, because surface integration can occur by several mechanisms, the term G_{surf} is defined as:

$$G_{surf} = G_{BCF} + G_{B+S} + G_{PN} \tag{6.14}$$

where G_{BCF} accounts for growth due to the Burton, Cabrera, and Frank mechanism; G_{B+S} accounts for growth due to the birth and spread mechanism, and G_{PN} accounts for growth due to the polynuclear mechanism. Previous research[29] has shown that size-dependent growth rates can be explained by size-dependent surface integration kinetics.

6.3.2 NUCLEATION

Nucleation refers to the formation of solid particles from solution; more specifically, solute molecules can form small clusters. Below a critical size, it is possible for the clusters to grow or to decrease in size. Above a critical size, it is more likely that the cluster of molecules will grow. This critical size is the minimum size for a nucleus. Nucleation is characterized by both primary and secondary nucleation. The key difference is that primary nucleation occurs in the liquid spontaneously even if no solids are present, but secondary nucleation occurs due to the presence of a solid interface. The rates are generally combined as follows:

$$B^0 = B^0_{primary} + B^0_{secondary} \tag{6.15}$$

where B^0 is the nucleation rate. Secondary nucleation often depends on the magma density (M_T) and growth. A common empirical relation for this is:

$$B^0_{\sec ondary} = B_{k2} \, M_T^{B_i} \, G_L^{B_j} \tag{6.16}$$

where

$$B_{k2} = A_{N2} \, e^{(-E_{N2}/RT)} \tag{6.17}$$

For primary nucleation,

$$B^0_{primary} = B_{k1} \, S^{B_{j2}} \tag{6.18}$$

where

$$B_{k1} = A_{N1} \, e^{\left(-E_{N1}/RT\right)} \tag{6.19}$$

As is shown, nucleation is a function of supersaturation, temperature, and magma density.

The secondary nucleation rate (B^0) is sometimes represented by:

$$B^0 = K_R N_I^h M_T^j G_L^i \tag{6.20}$$

where K_R is a constant that is a function of temperature, hydrodynamics, and impurity concentration; N_I is the impeller speed in rpm; M_T is the magma density or the mass of solids per unit volume of slurry; G_L is the length-based growth rate; and h, i, and j are exponents. The constants depend on the system being studied. In empirical expressions, the exponents on some of the factors may be equal to zero for some cases.

Empirical data for nucleation have been summarized for various systems over a range of operating conditions using Equation 6.20.[26] Much like the growth rates, nucleation rates can span several orders of magnitude, with potassium chloride having nucleation rates of approximately 2.5×10^{24} no. $L^{-1}s^{-1}$ and urea having nucleation rates of approximately 3.5×10^{-10} no. $L^{-1}s^{-1}$. Unlike growth rates, nucleation rates span a much wider range of values. Clearly, data are needed to model nucleation effects in a crystallizer.

Mersmann et al.[28] provided an alternative approach, where coefficients for nucleation are estimated based on physical models. These equations are for activated primary nucleation, secondary nucleation excluding attrition, and secondary nucleation, including attrition. If experimental data are not available, these equations may be used to estimate nucleation rates; however, these equations require physical property data such as interfacial tension between the crystal and the mother liquor as well as the contact angle.

6.3.3 AGGLOMERATION

Particle agglomeration is often neglected, although it is significant for some systems. Agglomeration considers two particles colliding and sticking together to form a larger particle. It is typically modeled by using an agglomeration kernel that predicts the probability of two particles sticking together. The kernel can be a function of the sizes of the two particles colliding. A size-independent kernel is given by:

$$a(v,w) = a_0 \qquad (6.21)$$

where $a(v,w)$ is the agglomeration kernel, v and w are particle volumes, and a_0 is a constant.

Studies have been performed to test the effect of supersaturation and agitation rate on agglomeration. For calcium oxalate monohydrate (COM),[30] the agglomeration rate increased with increasing supersaturation, or, in this case, correlated more accurately with the free oxalate ion concentration. In these COM experiments, the agglomeration rate decreased with increasing agitation rate; therefore, it is necessary to take into account both supersaturation and the agitation rate when modeling agglomeration.

Although the size-independent kernel works well for some systems, many systems require a size-dependent kernel. One of the more commonly known forms is the one by Golovin:[31]

$$a(v,w) = a_0(v + w) \qquad (6.22)$$

where v and w are particle volumes, and a_0 is a constant for a given supersaturation and agitation rate. Other forms of agglomeration kernels are shown in Table 6.3, where u and v are particle volumes. Other listings of agglomeration kernels are available.[28,43,44] Many of these forms indicate that agglomeration is more likely for larger particles than for smaller particles. The parameter a_0 is a constant only with respect to particle size; it is frequently a function of the temperature, fluid viscosity, shear rate, average fluid velocity, and rate of energy dissipation per unit mass.

6.3.4 BREAKAGE AND ATTRITION

Although often neglected,[45] particle breakage can occur in stirred vessels such as crystallizers. Larger particles may produce smaller fragments due to breakage or attrition. In fact, attrition is frequently a mechanism for secondary nucleation.[46-48] Although many studies have been performed on continuous crystallizers, breakage and attrition may occur in any stirred crystallizer including batch units. In breakage and attrition, the particle that is broken is referred to as the parent particle and the particles formed from the broken parent particle are referred to as child particles. There are two primary factors for modeling breakage: the specific rate of breakage and the breakage distribution function.

TABLE 6.3
Agglomeration Kernels

Description	Kernel	Ref.
Turbulent diffusion	$a_0(u + v)$	Golovin[31]
Brownian motion	a_0	Ramabhadran et al.;[25] Scott[32]
	$a_0(u^{1/3} + v^{1/3})(u^{-1/3} + v^{-1/3})$	von Smoluchowski[33]
	$a_0(u^{1/3} + v^{1/3})^2 (u^{-1} + v^{-1})^{1/2}$	Friedlander[34]
Shear	$a_0(u^{1/3} + v^{1/3})^3$	von Smoluchowski[33]
	$a_0(u^{1/3} + v^{1/3})^{7/3}$	Shiloh et al.;[35] Tobin et al.[36]
Gravitational settling	$a_0(u^{1/3} + v^{1/3})^2 \mid u^{1/3} - v^{1/3} \mid$	Berry[37]
	$a_0 \dfrac{(u - v)^2}{(u + v)}$	Thompson[38]
	$a_0(u^{1/3} + v^{1/3})^2 \mid u^{2/3} - v^{2/3} \mid$	Schumann;[39] Drake[40]
Turbulent flow	$a_0(u^{1/3} + v^{1/3})^2$	Kruis and Kusters[41]
Kinetic theory	$a_0 \dfrac{(u^{1/3} + v^{1/3})(uv)^{1/2}}{(u + v)^{3/2}}$	Thompson[38]
Shear and diffusion	$a_0(u^{1/3} + v^{1/3}) + a_1$	Melis et al.[42]

The specific rate of breakage gives the rate at which the particles of a given size break. That is, it gives the average rate as the fraction of particles that break in a given time. It is typically expressed as a function of the particle size because larger particles usually have a higher breakage rate than smaller particles. The specific rate of breakage is commonly expressed as:

$$S(v) = S_c v^{\alpha} \tag{6.23}$$

where $S(v)$ is the specific rate of breakage for a particle of volume v, S_c is a constant, and α is the breakage rate exponent. Not all particles will exhibit breakage under a given set of conditions. Typically, attrition does not occur until a particle is larger than a minimum size.[45]

When a parent particle breaks, it forms a distribution of smaller particles. This distribution is referred to as the *breakage distribution function*. This distribution function can be expressed on either a mass or a number basis. Suppose v is the volume of the child particle and w is the volume of the parent particle. The mass-based breakage distribution function $b_M(v,w)$ indicates the mass fraction of particles of volume v formed by a particle of volume w breaking. The number-based breakage distribution function $b(v,w)$ indicates the number of particles of volume v formed by a particle of volume w breaking. The number-based breakage distribution function can take many possible functional forms. Selected number-based breakage distribution functions are shown in Table 6.4, where u and v are particle volumes, and L_{min} and L_{max} are minimum and maximum particle lengths.

TABLE 6.4
Breakage Distribution Functions

Description	Breakage Distribution Function	Ref.
Uniform	$2u/v^2$	Hill and Ng[49]
Parabola	$24\dfrac{u}{v^4}\left(1-\dfrac{hv}{2}\right)\left(u^2-uv+\dfrac{v^2}{4}\right)+\dfrac{hu}{v}$	Hill and Ng[49]
Milling	$\dfrac{\phi\gamma}{3v}\left(\dfrac{u}{v}\right)^{\frac{\gamma}{3}-1}+\dfrac{(1-\phi)\beta}{3v}\left(\dfrac{u}{v}\right)^{\frac{\beta}{3}-1}$	Austin et al.[50]
	$\dfrac{(u/v)^{-2/3}\exp(-u/v)}{3v(1-1/e)}$	Broadbent and Calcott[51]
	$\dfrac{1}{v}\left[ac\left(\dfrac{u}{v}\right)^{c-1}+(1-a)d\left(\dfrac{u}{v}\right)^{d-1}\right]$	Shoji et al.[52]
Theoretical	$1/3v(u/v)^{2/3}$	Reid[53]
	$n(u/v)^{n-1}/v$	Randolph and Ranjan[54]
	$\dfrac{\dfrac{2}{\sqrt{2\pi}\,\sigma}\left(\dfrac{u}{v}\right)\exp\left[-\dfrac{(u-v/2)^2}{2v^2}\right]}{erf\left(\dfrac{v}{\sqrt{2}\,\sigma}\right)}$	Pandya and Spielman[55]
	$\dfrac{2.25}{L_{min}^{-2.25}-L_{max}^{-2.25}}L^{-3.25}$	Gahn and Mersmann[56]
Fly Ash Formation	$(2u/v^2)(2n+1)(2u/v-1)^{2n}$	Peterson et al.[57]
	$\dfrac{K}{v}\exp\left[-\ln^2\left(\dfrac{u}{cv}\right)\right]\Big/2\ln^2\sigma$	Peterson et al.[57]

6.3.5 COMMENTS ON SIZE ENLARGEMENT

The two primary mechanisms for size enlargement are growth and agglomeration. In addition, two possible growth mechanisms are size-dependent growth (SDG) and growth rate dispersion (GRD). The size-dependent growth model is based on the assumption that the growth rate varies with the particle size as was shown with some of the models in Table 6.2. Growth rate dispersion, however, assumes that crystals of the same size have a distribution of growth rates.[58,59] The many studies that have been performed on GRD have been briefly reviewed by other researchers.[58,59]

Because both growth and agglomeration cause an increase in particle size, it can be difficult to determine which mechanism is causing the increase in particle size. A systematic study performed to discriminate between these mechanisms[58] provides an example for a potassium sulfate/water system. In this study, it was possible to distinguish between SDG, GRD, and agglomeration.

6.4 DESIGN MODELS

6.4.1 PARTICLE SIZE DISTRIBUTIONS AND POPULATION BALANCES

6.4.1.1 Particle Size Distributions

Particle size distributions (PSDs) can be represented in several forms. One common method is to use the cumulative distribution $N(L)$, where L is the particle diameter and $N(L)$ is the number of particles of size L or smaller that are in a unit volume of slurry. A typical distribution is shown in Figure 6.7, where the distribution is normalized. As shown, the cumulative distribution approaches a limiting value.

Another typical variable for describing the PSD is the number density, $n(L)$, which is also called the *population density*. It is defined as:

$$n(L) = \frac{\partial N(L)}{\partial L} \qquad (6.24)$$

and has units of number/volume/length, or no./(m^{3}μm). A typical plot of this is shown in Figure 6.8 for normalized values. Often used in calculations, the number density is a continuous function that changes with time, and it is sometimes written as $n(L,t)$. The number density can also be written on a volume basis where

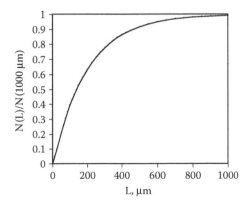

FIGURE 6.7 Normalized cumulative number distribution function.

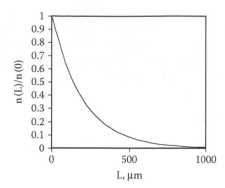

FIGURE 6.8 Normalized population density.

$N(v)$ is the cumulative distribution of the number of particles of volume v or smaller in a unit volume of slurry. The corresponding number density is:

$$n(v) = \frac{\partial N(v)}{\partial v} \tag{6.25}$$

The population density on a volume basis has units of number/volume/particle volume or no./(m³μm³).

6.4.1.2 Population Balance Equations

Population balance equations (PBEs) are a convenient method for following the changes in particle size due to the mechanisms of growth, nucleation, agglomeration, and breakage. These equations are commonly written in terms of the number density function. The advantage of PBEs is that they are flexible in that the user can choose which mechanisms to include in the model. Much of the early framework for PBEs was developed by Hulburt and Katz,[60] while Randolph and Larson[61] developed the method for particulate systems. On a volume basis, the PBE may be written as:

$$\frac{\partial n(v)}{\partial t} = \frac{1}{2} \int_0^v a(v-w,w)n(v-w)\,n(w)\mathrm{d}w$$

$$-n(v) \int_0^\infty a(v,w)n(w)\mathrm{d}w - \frac{\partial G_V(v)n(v)}{\partial v} \tag{6.26}$$

$$+\int_v^\infty b(v,w)S(w)n(w)\mathrm{d}w - S(v)n(v)$$

The factors in this equation are the coalescence kernel for collisions between particles of volume v and w, $a(v,w)$; the number-based breakage distribution function, $b(v,w)$; the volume-based growth rate, $G_v(v)$; the volume-based population density, $n(v)$; the number-based specific rate of breakage, $S(v)$; time t; and particle volumes v and w. These factors in the PBE are the same as those described in Section 6.3. Although the population density, $n(v)$, is a function of time as well as particle size, time is not explicitly listed.

The first two terms on the right-hand side of the PBE account for agglomeration. The first of these two terms accounts for a particle of volume w combining with a particle of volume $v - w$ to create a particle of volume v. The second agglomeration term accounts for a particle of volume v combining with a particle of volume w to form a particle of volume $v + w$. The first term accounts for the birth of particles into the size range v to $v + \delta v$, while the second term accounts for the death or disappearance of particles from the size range v to $v + \delta v$. The last two terms are birth and death terms accounting for particle breakage and attrition. The first term accounts for a particle of volume v produced when a particle of volume w is broken. The second term accounts for a particle of volume v disappearing from the size range when it is broken into smaller particles.

One might notice that Equation 6.26 does not have a separate term for nucleation. In fact, B^0 does not directly appear in the PBE. Nucleation is applied as a boundary condition for $n(0)$; that is, there is a boundary condition of $n(0,t) = B^0$ where t is the time. Although the nuclei are not of zero size, they are sufficiently small compared to the size of the crystals that this is a good approximation. One may also set initial conditions that give an initial size distribution.

The term with the growth rate accounts for both birth and death in a size interval due to particle growth. Particles are born into the size interval when smaller particles grow into the size range, while particle disappear from the interval when they grow larger than the size range.

Similarly, the PBE is often expressed on a length basis. For the length basis, the volumes v and w are replaced by the particle lengths L and λ. The other functions such as the coalescence kernel, number-based breakage function, and specific rate of breakage are the same. However, the growth rate is the length-based growth rate of dL/dt.

$$
\frac{\partial n(L)}{\partial t} = \frac{1}{2} \int_0^L a\left(\left(L^3 - \lambda^3 \right)^{1/3}, \lambda \right) n\left(\left(L^3 - \lambda^3 \right)^{1/3} \right) n(\lambda) d\lambda
$$

$$
- n(L) \int_0^\infty a(L,\lambda) n(\lambda) d\lambda - \frac{\partial G_L(L) n(L)}{\partial L} \tag{6.27}
$$

$$
+ \int_L^\infty b(L,\lambda) S(\lambda) n(\lambda) d\lambda - S(L) n(L)
$$

In the length-based birth term for agglomeration, the agglomeration kernel is a function of $(L^3 - \lambda^3)^{1/3}$ instead of being a function of $L - \lambda$. This accounts for the fact that the volumes of two particles will add up to the combined volume of the two particles, whereas this may or may not be true for the representative particle lengths.

6.4.1.3 Solution of Population Balance Equations (PBEs)

The main difficulty with PBEs is solving them. The degree of difficulty depends on which mechanisms are included and on the functional forms of the mechanisms. For example, if agglomeration and breakage can be neglected, the PBE becomes a partial differential equation; however, if either agglomeration or breakage is included, the PBE becomes a partial integro–differential equation. Many of these concerns are discussed by Ramkrishna.[62] Analytical solutions have been developed for some special cases, such as fragmentation, for a few functional forms.[63] While most cases do not have analytical solutions, a variety of techniques can be used to obtain solutions. These solution techniques include successive approximation, Laplace transforms, the method of moments, orthogonal collocation, similarity solutions,[57] Monte Carlo simulation methods, and discretization techniques.[62,64–71] More recently, other researchers have used wavelets to solve PBEs with growth, nucleation, and agglomeration.[72] Many of these techniques have been applied to crystallization or to mechanisms used in crystallization. Analytical solutions have been developed for special cases with specified initial distributions and limited functional forms for the mechanisms of growth, agglomeration, or a combination of the two;[32,73] however, the difficulty is that an analytical solution may not exist for the specific functional form that the real system exhibits. For this reason, it is often necessary to use numerical techniques such as the ones described above.

6.4.2 MASS BALANCE: DESUPERSATURATION AND POPULATION BALANCES

A batch crystallizer is always a dynamic process until crystallization has ended; therefore, if supersaturation is induced, the solution will stay supersaturated until saturation is reached. When the crystals begin to form, they remove the solute from the solution. This process is referred to as *desupersaturation*. Because desupersaturation changes with time, it must be modeled in a batch process. In order to model crystallization, it is necessary to consider the mass balance between the solute in the liquid and in the solid. To model this, several parameters such as liquid density and magma density must be considered.

6.4.2.1 Magma Density

Magma density is defined as the mass of solids per unit volume of slurry. It can be calculated from

$$M_T = \rho_s \phi_v \int_0^\infty L^3 n(L) \partial L \tag{6.28}$$

where ϕ_v is a shape factor relating L^3 to v. In practice, integration is between zero and the maximum expected particle size. The magma density accounts for the total production of solids in the crystallizer.

6.4.2.2 Liquid Density

For the case where the change in the liquid density is negligible, the volume of the slurry in the batch crystallizer can be modeled as a constant. However, in some cases the liquid density changes as the solute comes out of solution. In this case the change in liquid volume must be modeled. Solid density is input as a parameter. However liquid density can change due to composition or temperature changes. Many of the correlations are empirical. For a single solute in a solvent, the density of the saturated solution may be given by a formula of the form

$$\rho_{L,i} = \rho_{1,i} + \rho_{2,i} T \tag{6.29}$$

where $\rho_{1,i}$ and $\rho_{2,i}$ are density constants for solute i in a given solvent, and T is the temperature.[8] Other work has shown that a plot of density vs. concentration is continuous across the saturation point.[74,75] According to some research,[75] Equation 6.29 is valid even in supersaturated regions. A second option[75] is to express the density as a function of supersaturation:

$$\rho_L = \rho_3 + \rho_4 \sigma \tag{6.30}$$

where σ is the supersaturation, ρ_L is the density of the liquid (mass/unit volume), and ρ_3 and ρ_4 are constants for a given system at a given temperature.

6.4.2.3 Liquid Concentrations

These have the same units as density. Because one component is crystallizing out of solution, the solution composition changes. For noncrystallizing components, the concentration is given by one of the following, depending on the chosen definition of concentration:

$$C_i = 100 \, F_{in,i} / F_{in,solvent} \tag{6.31a}$$

$$C_i = 100 \, F_{in,i} / F_T \tag{6.31b}$$

$$C_i = F_{in,i} / V_L \tag{6.31c}$$

where $F_{in,i}$ is the mass of component i in the feed liquid, $F_{in,solvent}$ is the mass of solvent in the feed liquid, F_T is the total mass of the liquid, and V_L is the total volume of the liquid phase. The feed liquid is the liquid initially charged to the batch crystallizer. Equations 6.31a–c give the concentration of the noncrystallizing component in units of g i per 100 g solvent, g i per 100 g solution, and mass i per unit volume of solution, respectively. For the crystallizing component, the concentration is given by:

$$C_i = 100 \ (F_{in,i} - M_{sol})/F_{in,solvent} \qquad\qquad (6.32a)$$

$$C_i = 100 \ (F_{in,i} - M_{sol})/F_T \qquad\qquad (6.32b)$$

$$C_i = (F_{in,i} - M_{sol})/V_L \qquad\qquad (6.32c)$$

where M_{sol} is the total mass of the solid formed. Equations 6.32a–c give the concentration of the crystallizing component in units of g i per 100 g solvent, g i per 100 g solution, and mass i per unit volume of solution, respectively. Other definitions are possible such as mole fractions or molarity.

The mass of solids (M_{sol}) can be determined from the magma density:

$$M_{sol} = M_T V_T \qquad\qquad (6.33)$$

where V_T is the total volume of slurry in the crystallizer. Clearly, if more than one component is crystallizing, Equations 6.32a–c would have to be modified; however, for a simple crystallizer we can assume that only one component is crystallizing. These equations would also need to be modified if the crystallizer is operated in semibatch mode.

6.4.3 PARAMETER ESTIMATION

A primary problem in the modeling of crystallizers is obtaining the data, a problem referred to as the *inverse problem*. Instead of using the kinetic parameters to predict the results, the data from an experiment are used to determine the kinetic parameters. Because both agglomeration and growth increase the particle size and because both nucleation and attrition produce small particles, it can be difficult to distinguish between mechanisms. To separate the effects of different mechanisms, the problem is often divided into the various mechanisms, such as the inverse breakage problem, the inverse agglomeration problem, and the inverse growth and nucleation problem.[62] For each of the mechanisms, it is common to propose a functional form for the mechanism, to use the proposed form to model a system, and to compare the simulation results with experimental results. In many cases, optimization techniques are used to determine the parameters. Each mechanism is discussed in more detail below.

6.4.3.1 Growth and Nucleation

Because growth and nucleation are two of the most influential mechanisms in crystallization and because agglomeration and breakage are negligible in many cases, much of the work in crystallization has focused on growth and nucleation. Several techniques are used to obtain the kinetic parameters. Primarily, the experimental equipment is either a batch or a continuous crystallizer. The advantage of the continuous crystallizer is that a data point is obtained at steady state at a given temperature and supersaturation, and the analysis is relatively straightforward. The disadvantages are the time required due to the fact that a separate experiment is run for each data point and many tests are required to obtain the kinetics as a function of a variable, the difficulty of running the experiments, and the amount of material required to perform this large number of tests.[76] The batch crystallizer has the advantage that more data can be obtained from a single test, but it has the disadvantage that the calculations are more difficult as the supersaturation changes with time during the experiment.

Many experimenters have proposed methods for obtaining data from batch crystallizers.[76–82] Most of the work presented assumes a functional form for the particle growth and fits data to the form. Recent work[83] discusses an approach where the functional form is determined during the solution of the inverse problem. A unique method for measuring growth kinetics has been proposed[84] where a very small sample is analyzed using differential scanning calorimetry. For this case, growth kinetics are obtained from the desupersaturation curve. An advantage of this method is that growth kinetics can be determined over a wide range of temperatures and pressures using a very small sample.

6.4.3.2 Agglomeration

Other techniques must be used to determine agglomeration kinetics. Several studies were performed with calcium oxalate monohydrate (COM)[30,43,85] to determine agglomeration kinetics. In these studies, an agglomeration kernel was chosen for a simulation and the simulation results were compared against experimental results. A more rigorous method for determining agglomeration kinetics using the self-similar concept is given by Ramkrishna.[62]

6.4.3.3 Breakage and Attrition

The breakage equation assumes that only the mechanism of breakage is significant. Work has been done where simulations were performed with empirical expressions and compared against experimental results to determine which empirical breakage distribution function best fit the data.[86] Later work used a nonlinear parameter estimation technique to simultaneously estimate parameters for nucleation and attrition.[45] Ramkrishna[62] discusses the use of self-similarity concepts in solving for the breakage parameters. Mersmann[87] provides a method for estimating particle attrition based on first principles.

6.4.4 MIXING

One reason for so much discussion about mixing is that it is desirable to keep the particles suspended. If the particles are not suspended, they are not fully exposed to the solution. A second reason to be concerned about mixing is that both concentration and temperature gradients can exist in a batch vessel. These gradients allow crystallization to occur under different conditions in the vessel. Many models are simplified in that they assume uniform mixing. To determine if particles are adequately suspended, Zwietering[88] developed an empirical correlation to determine the condition where all of the particles are suspended. This correlation does not guarantee homogeneity of the suspension, but it does guarantee that none of the particles will be resting on the bottom of the vessel. The agitation rate to keep the particles just suspended (N_{js}) can be estimated from:

$$N_{js} = K_1 (D_T / D_I)^{K_2} \left(\frac{\rho_s - \rho_L}{\rho_L} \right)^{0.45} g^{0.45} v^{0.10} L^{0.20} M_s^{0.13} \qquad (6.34)$$

where D_T is the tank diameter, D_I is the stirrer diameter, ρ_s and ρ_L are the solid and liquid densities, K_1 and K_2 are constants depending on the impeller, g is the acceleration due to gravity, v is the kinematic viscosity, L is the particle diameter, and M_s is the mass percent of solids in the slurry. This correlation is often used to calculate a minimum mixing speed. In practice, the agitator is often run at a higher speed.

Another indicator of mixing in a stirred tank is the Reynold's number for mixing, which is often used to determine whether the flow in the stirred tank is laminar or turbulent. It is written as:

$$N_{Re} = \frac{N_I D_I^2 \rho_L}{\mu} \qquad (6.35)$$

where N_{Re} is the impeller Reynold's number, N_I is the rotational speed of the impeller in revolution per unit time, D_I is the impeller diameter, ρ_L is the fluid density, and μ is the fluid viscosity. Turbulent flow occurs when $N_{Re} > 10,000$, and $N_{Re} < 10$ for laminar flow.

Other correlations for mixing are reviewed by Mersmann.[89] These correlations depend on the vessel internals. For example, agitator speed depends on whether or not a draft tube is used, on the impeller type, and on the geometric ratios such as the ratio of the impeller diameter to the vessel diameter. The minimum time required for macromixing is usually only a few seconds for turbulent flow but can be much larger in the transition and laminar flow regions. Other correlations on the mixing of slurries as well as concerns specific to crystallization are given in the *Handbook of Industrial Mixing*.[90]

Perfect mixing rarely occurs in industrial practice,[90–91] and heuristics for scaling up crystallization processes rarely work well; therefore, another approach must be taken that includes the hydrodynamics as well as the thermodynamics and kinetics. Regarding modeling, a variety of approaches have been taken. Some researchers have modeled crystallizers by coupling models for crystallization in a uniformly mixed cell with computational fluid dynamics (CFD) codes.[92] CFD uses numerical techniques to solve fluid flow equations. In general, CFD models the geometry of a processing vessel by dividing the operating space into smaller cells. One difficulty has been determining how many cells are needed in modeling the crystallizer. Other researchers[91,93] have used compartmental models to separate the hydrodynamics from the kinetics. With compartmental models, the main idea is that certain sections of the crystallizer can each be adequately represented as a well-mixed crystallizer. In each of these regions or compartments, the temperature and supersaturation gradients are negligible. In general, the compartmental models consist of a small number of compartments on the order of three to five compartments, whereas CFD divides the crystallizer into many more cells. In both the CFD and compartmental modeling approaches, the real crystallizer is modeled as a network of ideally mixed crystallizers where the flow patterns between the ideal units may be determined by CFD or by other rules. While many of these models were used for continuous crystallizers, they are also applicable to batch crystallizers.

6.5 DESIGN AND MODELING DECISIONS

6.5.1 Methods for Inducing Supersaturation

For crystallization to occur, there must be a driving force. This is commonly represented as supersaturation, which can be induced by a variety of methods including cooling, evaporation, addition of a separating agent, or reaction. Each method has advantages and disadvantages.

6.5.1.1 Cooling

In the cooling method, the solution temperature is cooled below the saturation temperature for the solution. This procedure is easy to implement, but it is difficult to model due to the simultaneous changes in temperature and supersaturation. It is primarily used when the solubility increases significantly with temperature;[89] for example, because the effect of temperature on the solubility of sodium chloride in water is slight, cooling crystallization is usually not used for this system. Cooling crystallizers are often designed to run so the supersaturation remains almost constant. One operational problem with extensive cooling is that the crystallizer can have scaling at the cooling interface.

6.5.1.2 Evaporation

A second method is to concentrate the solution by evaporating the solvent. This can be done at a fixed temperature by heating the solution to its boiling point. This is often used for systems where the solubility is not a strong function of temperature (e.g., sodium chloride in water). The boiling point is sometimes lowered by operating the system under a vacuum. This method has the advantage that the temperature stays relatively constant, but it is more difficult to implement than cooling. One difficulty with evaporation is that the solution is usually more concentrated and hence more supersaturated at the evaporation surface. This can cause scaling across the liquid surface.

6.5.1.3 Separating Agent

A third method for inducing crystallization is adding a separating agent to change the solubility. Frequently an agent is added to reduce the solubility of the solute. This can be done by salting out, drowning out, or changing the pH of the solution. A salting-out agent is a solid that will dissolve when added to the system and move the composition of the solution into a region where the desired compound is less soluble and will crystallize out. Drowning out refers to the addition of a miscible liquid nonsolvent that reduces the solubility of the solute; for example, if a solute is soluble in water and sparingly soluble in an organic solvent, an organic solvent may be added to an aqueous solution to cause drowning out. Biological compounds such as amino acids often have a solubility that depends on the pH of the solution; therefore, adding an acid or a base changes the pH and the solubility. This separating agent method has the advantage that it can be operated at a constant temperature. Basically this method is operated in a semibatch mode.

6.5.1.4 Reactive Crystallization

Reactive crystallization can also be operated in semibatch mode at constant temperature. This is achieved by having one reactant in the vessel and adding a second reactant during operation. Depending on the solubility of the product formed by the reaction, the supersaturation can sometimes be controlled by the rate of addition of the second reactant. If the reaction product is insoluble, it will precipitate immediately. This may form amorphous materials as well as crystals. Clearly, this is only used in cases where there is a reaction. One example of reactive crystallization is the formation of calcium carbonate by mixing solutions of sodium carbonate and calcium chloride. Reactive crystallization has additional complexities due to the fact that the reaction kinetics must be considered in addition to the crystallization kinetics and due to the mixing of the reactants. In reactive systems, it is necessary to consider the relative time scales due to macromixing, mesomixing, micromixing, crystallization, and reaction. If reaction and crystallization are slow compared to mixing, the hydrodynamics may not be a major concern and the vessel may behave much like a well-mixed vessel; however, if reaction and crystallization are fast compared to mixing, then the hydrodynamics has a strong impact on crystallizer operation. The primary concern is that the reactions may primarily occur in the region where the second reactant is added and that this region will have high supersaturation. This would give very different results from a system where the reactant is uniformly mixed and the supersaturation is uniform in the vessel. The effect of mixing has been experimentally studied for the crystallization of benzoic acid where hydrochloric acid and sodium benzoate are the reactants.[94] This work also reviews the effects of the agitation rate and the feed rate on the mean crystal size in semibatch processes. Other work performed with calcium oxalate monohydrate showed the effects of micromixing on precipitation;[95] it was determined that the feed point location had a stronger influence than the stirring speed on micromixing.

6.5.2 Control of Supersaturation

Control of supersaturation is critical to controlling product quality in batch crystallizers (e.g., particle size distribution, shape, and purity). A problem with batch crystallizers is that the product often does not have the desired properties, largely due to control of the supersaturation. Much of crystallization is controlled by the interaction between the mechanisms. It is well known that a sudden increase in supersaturation will cause a very large number of nuclei to form. If the nucleation rate is very high compared to the growth rate, many nuclei will form but they will grow very little. For two cases, the same quantity of solids can be formed, but the one with the greater number of nuclei will have smaller particles. This will produce many fine particles, which is not desirable.[96]

Another concern is that for a batch crystallizer a sudden drop in temperature will cause a temperature gradient at the cooling interface. When this interface is much cooler than the bulk solution temperature, the solute will crystallize out primarily at the interface and cause scaling. This scaling can cause significant operating problems. A third difficulty is the purity of the final product. Sometimes pockets of solvent are trapped inside the crystal being formed. This pocket is referred to as an *inclusion*. The quantity of solvent and impurities entrapped inside the crystal depend on crystal growth history. It has been shown that inclusions become more likely as the growth rate increases;[4] therefore, it is necessary to control the supersaturation to lower the growth rate and reduce the crystal impurities. The crystal shape is also influenced by the growth rate. As previously discussed in Section 6.2.4, the facial growth rates are a function of the supersaturation, and changes in supersaturation may affect the relative facial growth rates; therefore, the final crystal shape depends on the characteristics of the system and the supersaturation.

6.5.2.1 Cooling Curves

One of the earlier methods suggested for controlling supersaturation in a batch cooling crystallizer is to control the cooling curve. Mullin and Nyvlt[96] predicted an optimal cooling curve and tested it experimentally against uncontrolled

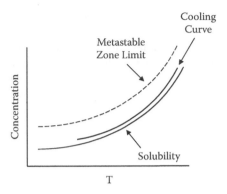

FIGURE 6.9 Cooling curve on a concentration–temperature plot.

cooling. Their results showed significant improvement in the crystal quality and an increase in the crystal size. The main goal of the cooling curve is to induce supersaturation to allow some nuclei to form and then to control cooling so that the original nuclei grow. Other researchers have recommended keeping the driving force (ΔC) constant. The concept is shown graphically in Figure 6.9. The cooling curve remains close to the solubility curve during operation and holds the driving force constant. Other researchers have studied the cooling curve problem from the view of optimization and process control. They compare natural cooling with linear cooling and optimal cooling. The key concept is how to define optimal cooling. Rawlings et al.[97] discuss the goals and results of cooling curve research. Such goals can include producing the largest particle size, producing a large mean particle size with a narrow CSD, or any other goal based on average particle size and CSD width. The general result is that optimal cooling can produce larger crystals than natural cooling or linear cooling.

6.5.2.2 Evaporation and Addition of Antisolvents

The concept of controlling evaporation and adding antisolvents is similar to the concept for controlled cooling. The heat supplied to the crystallizer can be adjusted to control the rate of evaporation. As with cooling, evaporation is controlled so a small number of nuclei are formed and allowed to grow. Evaporative crystallization is a semibatch operation as the solvent is being removed during operation. The goal of adding antisolvents is to control the supersaturation; therefore, not all of the antisolvent is added at once. Some antisolvent is added to induce nucleation, then the remainder is slowly added to control both growth and nucleation.

6.5.3 EFFECT OF SEEDING

Seeding is another method of controlling supersaturation. It is commonly combined with one of the three methods for inducing supersaturation. Seeds are added to batch crystallizers to maintain consistency from batch to batch. If they are added too early when the system is subsaturated, the seeds can dissolve. If they are added too late, nucleation may have started. It is generally desirable to add seeds when the system becomes saturated. Seeds are often added to start secondary nucleation at a desired time in the process. In systems that have large metastable zone widths, nucleation may not occur until the solution is highly supersaturated. In this case, once nucleation starts desupersaturation occurs very rapidly and allows for little crystal growth. By seeding, the seed crystals are allowed to grow at lower supersaturation levels. A second problem in systems with large metastable zone widths is that without seeding a significant amount of variation can occur from batch to batch with regard to when nucleation actually starts. This does not allow for the consistent production of crystals with the desired properties.

For reproducible batches, it is necessary to add seeds of a given composition with a narrow size distribution. Often, the seeds must be rinsed to remove the fine crystals. Another concern is the location of the addition. Dropping in dry crystals from the top of the vessel may cause some particles to form a layer at the slurry interface; therefore, it is often better to add the seed crystals in a slurry near the stirrer.

The major concerns regarding seeding are determining the quantity of seeds to be added and the size distribution of the seeds. Although studies have been performed with seeds and models have been developed for seeded batch crystallizers,[98] no general guidelines exist as to seed quantity and characteristics. Researchers have noted that, although much study has been performed on controlled cooling, work in the area of seeding is negligible.[99,100] Bohlin and Rasmuson[99] concluded that the product weight mean size could increase or decrease as a result of seeding; therefore, seeding should be done carefully. In a recent review of seeding and optimization of seeding,[100] the authors pointed out that seeding depends on the objective — for example, whether the objective is to produce the largest mean crystal size or largest number average crystal size.

6.5.4 OTHER CONSIDERATIONS

In designing a batch process, other considerations include the optimum operating temperature range. This is largely determined by the solubility. Operating at very low temperatures is often expensive due to the cost of refrigeration. Also, in some cases the stable polymorph is formed more easily at higher temperatures than at lower temperatures. An upper temperature limit can exist, as well; for example, some biological materials decompose at higher temperatures. These are just some of the concerns that should be addressed when designing a system.

6.6 INSTRUMENTATION

To control the process, it is necessary to measure solution properties such as temperature, concentration, and supersaturation, as well as solid properties such as the crystal size distribution. Some of the difficulty in controlling crystallizers can be attributed to limitations in sensor technology.[101–103] Another challenge is measuring the concentration (or supersaturation) and CSD.

6.6.1 CRYSTAL SIZE ANALYSIS

Traditionally, crystal size measurement has been performed offline. The earliest techniques included sieving; however, this is a very time-consuming process. Later techniques included Coulter counters, which often required dilution of the sample before measurement could take place. Newer techniques include Fraunhofer and laser diffraction methods. One limitation of the offline methods is that a sample must be taken and processed before it can be analyzed. The crystals cannot be left in the mother liquor for any significant period of time. If the temperature changes, the crystals could either dissolve or grow, producing a size distribution different from that in the crystallizer. Even if the sample was held at the desired temperature in the mother liquor, aging could change the CSD. Another difficulty is that samples have to be taken in such a way that they are representative and their withdrawal does not affect crystallizer operation, thus limiting the number of samples that can be taken during an experiment. For this reason, efforts are being made to develop online sensors. Some of these sensors have flow cells where the crystallizer contents are pumped through the measurement device (e.g., light-scattering techniques, image analysis under a microscope).[102]

Techniques that have sensors installed *in situ* include the Lasentec focused-beam reflectance measurement (FBRM) and particle vision measurement (PVM) systems,[103] as well as the ORM 2D system.[104] The FBRM system measures chord lengths as crystals flow in front of the probe window. With the FBRM system, it is necessary to account for the fact that the chord lengths do not necessarily represent the particle diameters but are chords measured across particles. The PVM system is an image analysis system that takes pictures of microscopic particles as they flow in front of the probe window. These techniques have advantages and disadvantages. The *in situ* techniques eliminate the need to send crystals outside of the crystallizer, which prevents the slurry from varying from the vessel temperature; however, microscopes can give better quality images than the PVM system.[102] One of the more novel and promising techniques is the use of ultrasonic attenuation spectroscopy.[105,106] One advantage of this technique is that it can measure dense slurries; in theory, it can measure particles in the size range of 0.01 to 1000 μm.

6.6.2 CONCENTRATION MEASUREMENT

A variety of techniques have been used to measure concentration, including refractometry,[107] densitometry,[108] conductivity,[109] calorimetry,[80,101,110] and attenuated total reflection–Fourier transform infrared (ATR–FTIR) spectroscopy.[103,111,112] In choosing a method, it is necessary to consider the types of systems to which the technique can be applied; for example, densitometry is primarily used for one solute in a single solvent. The goal of measuring concentration is to obtain supersaturation information. Usually the concentration is measured and compared with the solubility to obtain the supersaturation;[113] although a novel approach has been developed to measure the supersaturation directly;[113] although not yet tested in an industrial environment, the new sensor showed promising results in a laboratory batch vessel.

6.6.3 Polymorph Characterization

Traditionally, analytical tools for polymorph analysis were offline. This has changed in recent years due to the use of a Raman spectroscopy analyzer interfaced with an immersion probe. This probe can be used *in situ* in a suspension of solids and liquids. This approach has been used in both academic[114,115] and industrial research.[116] The advantage of this method is that it can be used to monitor polymorphic transformations online.

6.7 GENERAL METHODOLOGY

The challenge in designing a batch crystallizer is to design a system that will produce the desired quantity of crystals with the desired characteristics. Due to the many concerns that must be addressed, it is necessary to have a methodology for designing a batch crystallization process. One of the main concerns is determining the operating parameters that will produce the desired product. Crystallizer design can be done in a series of steps, as shown in Figure 6.10: gathering initial information, determining physical properties, determining crystallization kinetics, modeling, experiments, and final design. Some of these steps may occur in parallel. Although this is not an exhaustive list of all the possible steps, it does provide a general guideline for developing a design.

6.7.1 Step 1: Information

The first step is to gather any available information on the system and crystallization. This includes product specifications, such as production rate, purity requirements, polymorphic form (if applicable), CSD constraints, and shape requirements; physical properties of the system, such as solid–liquid equilibrium data, melting points, viscosity, metastable zone width, and decomposition temperature (if applicable); crystallization kinetic data, such as parameters and functional forms for nucleation, growth, agglomeration, and breakage; and any process constraints such as temperature operating ranges for the equipment.

6.7.2 Step 2: Physical Property and Crystallization Kinetics Data

The second step is to obtain any missing data, whether physical property data or crystallization kinetic data. For systems that may exhibit polymorphs, it is

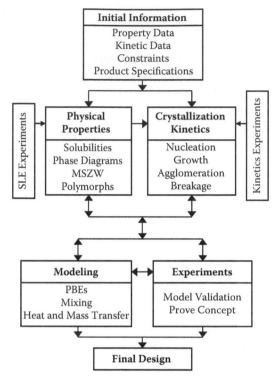

FIGURE 6.10 General methodology for batch crystallization design.

necessary to determine the number of possible polymorphs and the one that will be stable under the desired conditions. If solid–liquid equilibrium (SLE) data do not exist, they must be obtained as they determine the possible operating conditions of the crystallizer. It is necessary to have SLE data to determine the conditions under which crystallization kinetics data are required. This information is fed into the crystallization kinetics block. Crystallization kinetics data are often unavailable and must be determined experimentally. Part of the purpose of this step is to determine which mechanisms are important. For example, agglomeration may be significant for some systems and negligible for other systems. Typically, this information is required over a range of temperatures and supersaturation levels. Also, kinetics determines the size and operation time for the crystallizer.

6.7.3 STEP 3: MODELING AND VALIDATION EXPERIMENTS

The third major step is to use the data for modeling and validation experiments. Without modeling, many unnecessary data collection experiments might be performed, and some experiments that are necessary might not be done. Without these experiments, the modeling cannot be validated. As noted by Gerstlauer et al.,[117] the challenge is to integrate both microscopic and macroscopic behavior into the model. For modeling, it is necessary to divide the slurry into two phases: a continuous liquid phase and a dispersed solid phase. The dispersed solid phase is modeled using the PBEs described in Section 6.4.1 and the crystallization kinetics described in Section 6.3, and the liquid phase is modeled and connected to the solid-phase behavior as described in Section 6.4.2. Other details may be added to the model as necessary.

Modeling and simulation are used to determine the sensitivity of the system to different variables. Simulations can be performed to determine which parts of the underlying physics are critical. This, in turn, can determine how accurate the data must be. If necessary, more crystallization experiments can be performed to obtain better data. This transfer of information is indicated by the bidirectional arrows between the second and third steps, indicating that although modeling requires data from step 2 it can also determine whether or not more fundamental data are needed, thus sending the procedure back to step 2.

Another major task in this step is the model validation experiments. Although the models may indicate optimum operating conditions, it is necessary to test such findings in an actual vessel. If the experiments produce significantly different results from the model, it is necessary to correct the model. Without modeling, design would be done by a trial-and-error approach. The purpose of modeling is to give direction to the experiments and reduce the overall number of experiments.

6.7.4 STEP 4: FINAL DESIGN

The final design is based on the results of the previous steps. It specifies all of the operating conditions during operation including changes in temperature and the addition of seeds. In the final design additional items are specified including the instrumentation needed for process monitoring and control.

The advantage of this design methodology is that it combines theory with modeling and experiments. Therefore, if changes need to be made in the future, there is a basis for knowing how the crystallizer operating conditions should be adjusted.

6.8 CASE STUDIES

Actual cases vary widely in scale and span a range of industries. A few case studies are given here to illustrate concepts given earlier.

6.8.1 POPULATION BALANCE EQUATION SOLUTIONS

Section 6.4.1.3 briefly discusses the solution of PBEs. This is not a trivial task, and researchers should be careful when solving these partial integro–differential equations. One concern with these solutions is whether or not they can simultaneously predict changes in both number and mass. Typically, crystallization is concerned with the number density of particles; however, due to the requirement that there must be a mass balance for a model, changes in mass must be considered as well. For several mechanisms, either mass or number will be constant and the other one will change; for example, if growth is the only mechanism occurring, then the number of particles will remain the same and the mass will increase. If breakage is the only mechanism occurring, the total particle mass remains constant and the total number of particles increases. The total mass of particles also remains constant when agglomeration is the only active mechanism but agglomeration causes a reduction in the total number of particles. Frequently, the simpler solution techniques will accurately model both mass and number simultaneously.

To illustrate this, consider a simple system that only has agglomeration occurring at a fixed agglomeration rate. Suppose this system has size-dependent agglomeration of the form given in Equation 6.22. Several methods could be used to find a solution for this system. One of these is the conventional discretized equation, which is a discretized version of the continuous equation. Rather than having a continuous number distribution, particle size is divided into discretized size intervals where N_i indicates the number of particles in interval i. Similarly, there is a discretized agglomeration kernel, $a_{j,i-j}$. The conventional discretized agglomeration equation has the form:[68]

$$\frac{dN_i}{dt} = \frac{1}{2} \sum_{j=1}^{i-1} a_{j,i-j} N_j N_{i-j} - N_i \sum_{j=1}^{\infty} a_{i,j} N_j \qquad (6.36)$$

The equivalent discretized expression of Equation 6.22 is given as:

$$a_{i,j} = a_0 \left(\overline{v}_i + \overline{v}_j \right) \qquad (6.37)$$

where $\overline{v}_i$ is the average volume in interval i and is usually defined as $\overline{v}_i = (v_i + v_{i-1})/2$. These equations are expected to work best for equal volume intervals. For equal volume intervals, the first interval ranges from zero to a chosen value of v_1, and any subsequent volume interval i ranges in volume from $(i - 1)v_1$ to iv_1; therefore, each volume range has the same width. The concern with this approach is its accuracy.

An alternative discretized equation for equal intervals[68] is given as:

$$\frac{dN_i}{dt} = \frac{1}{4} \sum_{j=1}^{i-1} a_{i-j,j} N_j N_{i-j} + \frac{1}{4} \sum_{j=1}^{i} a_{i-j+1,j} N_j N_{i-j+1} - N_i \sum_{j=1}^{\infty} a_{i,j} N_j \qquad (6.38)$$

where Equation 6.37 is used for the discretized agglomeration kernel. To illustrate the difference between Equations 6.36 and 6.38, consider the case where particles in intervals $i - j$ and j agglomerate with each other. Whereas the conventional discretized equation assumes that all of the particles formed will fall into interval i, Equation 6.38 accounts for the reality that they may fall into intervals i and $i - 1$.

To test various solution methods it is necessary to use a case for which an analytical solution exists. One such analytical solution is given by Gelbard and Seinfeld[73] for the case where there is an initial distribution of:

$$n_0(v) = \frac{N_0}{\upsilon_0} \exp\left(-\frac{v}{\upsilon_0}\right) \tag{6.39}$$

where N_0 denotes the total particle count of the PSD and υ_0 denotes the PSD initial mean volume. The analytical solution for this initial distribution and the linear kernel given in Equation 6.22 is:

$$n(v,t) = \frac{N_0(1-T)}{vT^{1/2}} \exp\left[-(1+T)\,\tilde{v}\right] I_1\left(2\,\tilde{v}\,T^{1/2}\right) \tag{6.40}$$

where $\tilde{v} = v/\upsilon_0$, $\tau_1 = N_0 \upsilon_0 a_0 t$, $T = 1 - \exp(-\tau_1)$, and I_1 is the modified Bessel function of the first order.

In this set of simulations, equal size intervals are used. The discretized number N_i is in units of number per liter. The agglomeration kernel is of the form given in Equation 6.22 with the agglomeration constant a_0 set to 3×10^{-16} L/number/min. The initial size distribution is determined using Equation 6.39, where υ_0 is 4000 μm, the solid density is 3.217g/cm^3, and N_0 is set so the initial solid mass is 100 g/L of slurry. For comparison purposes, two sets of simulations are performed. Each set has a different size for the equal interval. The results are shown in Figure 6.11a and b, where the equation used and volume of each equal interval is denoted in the legend.

The analytical solution is used to test the accuracy of the two discretized equations. As expected, all methods predicted a decrease in the number of particles formed. This is shown in Figure 6.11a, where it is clear that all of the methods do well in predicting the decrease in the number of particles. For the mass, there should be mass conservation; that is, the total mass of solids does not change. This is true for the analytical solution and for Equation 6.38; however, the conventional discretized equation does not conserve mass, as shown in Figure 6.11b. The error for the conventional discretized equation depends on the size of the intervals. In general, there is less error with smaller intervals, as expected, but the total number of size intervals increases, which can significantly increase the computer processing time. For the case where the equal interval is 1000 μm^3, the number of size intervals used was 250, which yielded a maximum volume of 250,000 μm^3. If these were spherical particles, the maximum diameter would be 78.2 μm. Using this number of intervals for such a small size range is prohibitive. For the case where the interval size was 500 μm^3, 500 size intervals were used so the maximum particle size was the same.

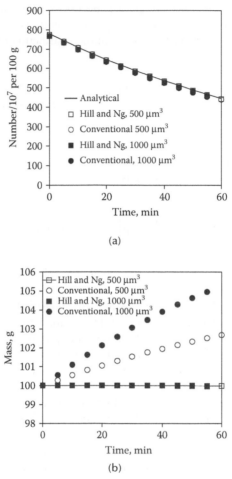

(a)

(b)

FIGURE 6.11 Comparison of solution techniques with analytical solution for a linear agglomeration kernel and equal volume intervals: (a) change in particle number with time; (b) change in total mass with time.

6.8.2 BAYER ALUMINA PROCESS

The Bayer alumina process is a large-scale operation. In this process, aluminum is leached out of the bauxite ore using a hot NaOH solution. A key step in the process is the recovery of alumina trihydroxide, $Al(OH)_3$, through crystallization. Because much of the sodium from the leaching step remains in solution, this system is primarily a ternary $Al_2O_3 \cdot Na_2O \cdot H_2O$ system. Although this is a commodity chemical, the crystallization step is usually run in the batch mode.

Step 1

The first step in the design of the process is to gather information. Before studying the crystallization kinetics, it is first necessary to determine the solubility.

Previous research[118] has shown that the solubility of aluminum trihydroxide in NaOH solutions can be represented by a semiempirical correlation:

$$\ln\left(\frac{g\,Al_2O_3}{g\,Na_2O}\right) = 6.2106 - \frac{2486.7}{T+273} + \frac{1.08753\,C_N}{T+273} \qquad (6.41)$$

where T is the temperature (in degrees C), and C_N is the NaOH concentration (in g Na_2O per L).

The crystallization kinetics of this ternary system has been studied by several researchers.[119–121] Both the nucleation and growth rates affect the design of the system. It has been noted[119–120] that this system is very stable in that the homogeneous nucleation occurring under industrial operating conditions is negligible; therefore, the crystallizer must be seeded for economical operation. A second concern is the growth rate. For cases with temperatures varying from 45 to 95°C and varying supersaturation, the growth rates range up to 4.5 μm/hr, or 0.075 μm/min.[119] Due to the low growth rate, high residence times of 35 hr or greater are needed. Because the high residence times require a prohibitively large vessel for a continuous process, a batch process is commonly used.

The second item in the information-gathering step is the goal of the process. The goals are to achieve a high rate of alumina recovery to improve process economics and to maintain a particle size distribution that does not cause operational problems; that is, fine particles must be avoided because they commonly cause operating problems in units downstream of the crystallizer. The designer needs to be aware that it may be difficult to meet both constraints simultaneously, particularly given the low growth rate of the particles. In this process, the crystallizer effluent is sent to classifiers. The classifiers remove a chosen size range of the smaller particles to use as crystallizer seed. It is not sufficient for avoiding fines; it is essential to produce crystals for both seed and product.

Step 2

Because some literature data are available for this project, experiments are not necessary to obtain basic data; however, experiments can be done at this stage to verify the literature data, if desired. This is particularly useful when the experiments have been performed with pure laboratory chemicals, because actual systems have impurities that can significantly affect parameters such as the solubility or crystal growth rate.

Step 3

The next step is to use the data in a model. For example, in this case, Misra and White[119] give the growth rate as:

$$G = \frac{dL}{dt} = 1.96 \times 10^6 \exp(-14.3 \times 10^3 / RT)(C_A - C_A^{sat})^2 \qquad (6.42)$$

where G is the growth rate (in μm/hr), $C_A(C_A^{sat})$ is the concentration (solubility) of Al_2O_3 (in g/L of solution), R is the ideal gas constant (in cal/mol·K), and T is the temperature (in K). The nucleation rate is also required. It should be noted that one should be selective here since Misra and White present the following four models:

$$B^0 = 2.13 \times 10^7 \tag{6.43}$$

$$B^0 = 1.0 \times 10^4 \, (C_A - C_A^{sat})^2 \tag{6.44}$$

$$B^0 = 2.6 \times 10^2 \, (C_A - C_A^{sat})^2 s \tag{6.45}$$

$$B^0 = 0 \text{ for } t < 2 \text{ hours} \tag{6.46}$$
$$= 3.1 \times 10^7 \text{ for } t > 2 \text{ hours}$$

where B^0 is the number of nuclei per hour per milliliter of solution, and s is the crystal surface area per unit volume of slurry (in cm^2/mL). Two models that do not work well are Equations 6.43 and 6.46. As noted by Misra and White,[119] it is unrealistic to have a nucleation rate that does not depend on the supersaturation. This is further illustrated by simulation. In the plant, the batch crystallizers have residence times ranging from 35 to 75 hours.[119] If constant nucleation rates are used in the modeling, a crystallizer reaches its equilibrium concentration well before 35 hr. At a typical operating temperature of 60°C, the slurry reaches equilibrium in less than 2 hr if Equation 6.43 is used with Equation 6.42 for a seed charge of 100 g seeds per L.

A simulation was performed for the case where the growth rate was as defined in Equation 6.42. For the nucleation rate, Equation 6.45 was used as it has a dependence on both supersaturation and the crystal surface area; however, because of the high initial supersaturation, Equation 6.45 was modified so the nucleation rate was zero at less than 2 hr and was modeled by Equation 6.45 after 2 hours of residence time. Other operating data are given in Table 6.5. The feed concentrations and the operating temperature are within the range for industrial crystallizers.[119] An impurity of $Fe_2O_3Na_2O$ is included because iron is often leached out of the ore by the hot NaOH and it is usually present in the crystallizer feed. The seed concentration was varied in these simulations from laboratory conditions of 10 g seed per L to industrial conditions of 100 g seed per L. Similar to previous work, the seeds are mainly made up of crystals smaller than 8 μm.

Results from these simulations are shown in Figure 6.12 for 10 and 100 g seed per L. For comparison purposes, the PSD of the original seeds is shown. In this plot, the original seeds are shown as a solid line, and the simulation results are plotted as symbols in the center of each particle diameter range. In this plot, the size distribution is shown as the number fraction of particles over a given size range. Four simulations were performed. Two of the simulations are for 10

TABLE 6.5
Simulation Operating Data for Alumina Crystallizer

Feed concentrations

$Al_2O_3Na_2O$	209 g/L
NaOH	26.4 g/L
H_2O	899 g/L
$Fe_2O_3Na_2O$	30 g/L
Operating temperature	60°C
Residence time	35 hours

Crystallization kinetics

Growth (μm/hr)

$$G = \frac{dL}{dt} = 1.96 \times 10^6 \exp(-14.3 \times 10^3 \, / \, RT)(C_A - C_A^{sat})^2$$

Nucleation
(number/hr/mL)

$$B^0 = 2.6 \times 10^2 \; (C_A - C_A^{sat})^2 \, s \,, t > 2 \text{ hr}$$

$$B^0 = 0, \, t < 2 \text{ hr}$$

Seed

Concentration	10 g seed per L
Seed size distribution (mass %)	80% < 8 μm
	8 μm < 20% <15 μm

g seeds per L (open symbols) and two simulations are for 100 g seeds per L (closed symbols). Two of the simulations (denoted by circles) use Equation 6.45 as written for nucleation; however, it was noticed that this equation predicts very high rates of nucleation initially. Therefore Equation 6.45 was modified so the nucleation rate was set to zero for times less than 2 hr (denoted by squares).

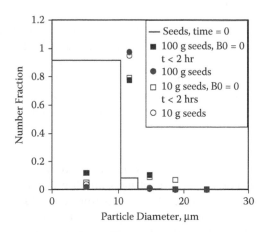

FIGURE 6.12 Comparison of seed size distribution with size distribution after 35 hours.

As expected, the number fraction of particles in the larger size ranges increased with time for all cases. More specifically, the number fraction of particles in the first size range decreased and the number fraction of particles in the second size range increased substantially. If an average growth rate of 1 μm/hr were assumed, one would expect to have crystals approximately 35 μm in diameter after 35 hours of operation; however, because the growth rate is a function of the supersaturation, it changes significantly during crystallizer operation. Therefore, although the process starts out with a higher growth rate, the growth rate decreases with desupersaturation and the actual average growth rate is lower. This is true for all cases. An inspection of the graph shows the effect of seeding on this operation. In general, adding more seeds produced a larger fraction of larger particles. In the third size interval, the number of crystals for the case of 100 g seeds per L is over 10 times more than for the case where 10 g seeds per L were used.

Step 4

The final design decisions are made in this step. Because nucleation is usually quite low initially in these crystallizers, the model that initially sets the nucleation rate to zero is probably more accurate. For this case, initial seeding with 100 g seeds per L produces about twice as many particles per liter as seeding with 10 g seeds per L. If more crystals are desired per unit volume, then 100 g seeds per L should be used. The PSDs for the two seeding cases are very similar; however, the lower amount of seeding yields slightly larger crystals. It should be noted that this decision is based on the models described earlier. As noted in Section 6.5.3., seeding must be done very carefully. The initial PSD, the quantity of seeds, and the crystallization kinetics for the system all affect the results. Every effort should be made to verify these models. If the models are incorrect, the predictions will also be incorrect.

6.8.3 L-GLUTAMIC ACID

L-Glutamic acid is of interest because it is an amino acid and because it has two polymorphic forms: the metastable α form and the stable β form. While both forms are orthorhombic, they have different lattice spacings and are easily distinguishable by their visual appearance. The α form is a flat plate, while the β form is acicular or needle shaped. L-Glutamic acid is produced by fermentation and purified downstream by crystallization. Crystallization is a key step since it determines which polymorph is produced.

Step 1

In the information gathering step, it is necessary to determine what is known about the two polymorphs. In general the solubility of the metastable form is higher than that of stable form. This is also true for L-glutamic acid over the temperature range shown in Figure 6.13.[16,122] Also, literature data exist regarding the processing conditions for both polymorphs. Experiments performed at various supersaturation rates at a fixed temperature indicate that supersaturation has a negligible effect on the percentage of α crystals formed; however, the temperature

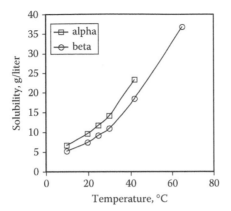

FIGURE 6.13 Solubility of α and β forms of L-glutamic acid as a function of temperature. (From Kitamura, M., *J. Crys. Growth*, 96, 541–546, 1989 and Kitamura, M., *J. Crys. Growth*, 237–239, 2205–2214, 2004. With permission.)

has a strong effect.[122] The α form is predominant at lower temperatures.[16,122,123] Specifically, over 95% of the material formed at or below 25°C is the α form, while at 50°C less than 40% of the material is of the α form. The other parameter that has a large effect on the selection of polymorph form is the cooling rate. Other experiments were conducted to determine the effect of the cooling rate by cooling a solution from 50 to 15°C. At cooling rates of 0.4°C/min and higher, the α form was prevalent, whereas the β form was prevalent at slow cooling rates of 0.1°C/min.[106]

The existing information sets limits on the operating supersaturation. At large initial supersaturations, the material starts nucleating before the solution reaches the desired operating temperature; however, at low initial supersaturation, the nucleation induction time is long after reaching the desired operating temperature.[16] The desired product for this example is the stable form (β) of L-glutamic acid. Although most companies prefer the α form because of its handling properties, the desired product for this particular application is the stable form. In fact, it is typically available commercially in the stable form.[106] The other specification is the length of the major axis. In this case, particles 550 μm in length are desired. Because the process will produce a PSD and not a single particle size, the goal is to find operating conditions where the largest number fraction of particles is near 550 μm for the major axis length.

Step 2

To obtain the stable form of L-glutamic acid, the literature indicates that it is better to operate at higher temperatures and lower cooling rates; therefore, it is necessary to perform experiments under the conditions under which the crystallizer is expected to operate. All of the existing data for crystal growth of the β crystal is at 25°C.[124–125] Although some literature data are available for the L-glutamic acid/water system, it is advisable to run some tests to verify the growth rates in this system and to verify the proportion of α to β crystals under various conditions. Preliminary tests were performed on L-glutamic acid by creating a saturated solution at 69°C, slowly cooling the solution at 0.1°C/min to 63°C, and then holding the temperature at 63°C.[125] No visible nuclei were noted before the solution cooled to 63°C, and after reaching the final temperature the induction time was about 13 minutes. Crystal samples taken at different residence times were analyzed under a microscope to determine the length of the major axis. In general, a close approximation is made with the expression:

$$G_{max}(\mu m/min) = 0.28(C - C^{sat})^{1.4} \qquad (6.47)$$

where G_{max} is the growth rate of the major axis and the concentrations are in g/L. This expression is similar to a form obtained at 25°C.[124] It is apparent that the fine particles disappear as the holding time is increased.[106,125] This may be due to the fact that, when crystals are present, the material coming out of solution tends to grow on existing particles rather than nucleate. It appears that the nucleation rate decreases rapidly with supersaturation. The experimental data at 63°C are approximated using the following nucleation model:

$$B^0(number/L/min) = 5.6 \times 10^{-18} (C - C^{sat})^{28} \qquad (6.48)$$

In these experiments, the aspect ratio averaged around 10 and was not influenced by supersaturation in this region.

Step 3

The next step is to decide how to operate the crystallizer. Because data are available at 63°C, this is the temperature used in the modeling. The model uses an induction time of 13 minutes with the nucleation rate given by Equation 6.48 after 13 minutes and the growth rate given in Equation 6.47. Agglomeration and breakage are assumed to be negligible. A second consideration is the particle volume. Because the crystal is known to have a needle shape, it may be approximated in modeling as a rectangular parallelepiped. The model uses the major axis in the population balance equations and the aspect ratio to determine the crystal volume. The model is solved using discretized PBEs and coupling them with a mass balance.

Step 4

To make the final decision, a graph is made of the average major axis length as a function of residence time. The crystallizer residence time in Figure 6.14 is the residence time in the crystallizer after cooling to 63°C. The representative major axis length is defined as the length that has the largest number fraction of particles. The graph illustrates that the representative major axis length is not a linear function with respect to residence time; therefore, it is necessary to have enough points to construct a curve. Based on the simulation results, the crystallizer should be operated for a residence time of 63 minutes.

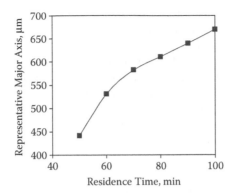

FIGURE 6.14 Representative major axis length as a function of crystallizer residence time.

6.9 CONCLUDING REMARKS

Batch crystallization plays an important role in the chemical industry. It is particularly important to the pharmaceutical industry where over 90% of the pharmaceutical products contain particles and many of these particles are in crystalline form.[126] Shekunov and York[126] discuss the difficulties in crystallizing a pharmaceutical product with the desired properties. This is not limited to the pharmaceutical industry, but also applies to other specialty chemicals such as food and agrichemicals.[103,112,127] Despite the requirements for crystal product control across a range of industries, there is still much that is not yet known about designing batch crystallizers to produce the desired product. For example, it would be useful to have a better understanding of seeding, nucleation, and the effects of impurities on solubility and kinetics; however, it is still possible to improve the design of batch crystallizers with what is currently known.

NOTATION

a_0	agglomeration constant
$a(L,\lambda)$	agglomeration kernel for collisions between particles of diameter L and λ (no min/m^3)$^{-1}$
$a(v,w)$	coalescence kernel for collisions between particles of volume v and w (no μm^3 min/μm^3/μm^3)$^{-1}$
A_G	constant for calculation of G_k
A_{n1}, A_{n2}	constants in nucleation rate calculations
$b(L,\lambda)$	number-based breakage function (number/number)
$b(v,w)$	number-based breakage function (number/number)
$b_M(v,w)$	mass-based breakage function (number/number)
B^0	nucleation rate (number/min)
B_{k1}, B_{k2}	nucleation rate factors

B_i, B_j, B_{j2}	nucleation rate constants
C_i	concentration of component i (e.g., g/100 g solvent or g/L solution)
$C_{1,i}^{sat}, C_{2,i}^{sat}, C_{3,i}^{sat}$	constants for calculation solubility
$C_i^{sat}(T)$	saturated concentration (solubility) at temperature T (e.g.. g/100 g solvent or g/L solution)
D_I	impeller diameter
E	constant for calculation of G_k
E_{n1}, E_{n2}	constants in nucleation rate calculations
E_{Si}	eutectic composition between components S and i (e.g., g/100 g solvent or g/L solution)
$F_{in,i}$	mass of component i in feed liquid (kg)
F_T	total mass of feed liquid (kg)
g	acceleration due to gravity
$G_L(L)$	length based growth rate (dL/dt, μm/min)
G_k, G_n	constants in growth equations
G_0	growth factor
$Gv(v)$	volume based growth rate (dv/dt, μm^3/min)
k_G	constant in growth rate equation
K_R	constant in nucleation rate equation
K_1, K_2	constants in Zwietering correlation
L	particle length (μm)
L_{min}, L_{max}	minimum and maximum particle sizes (μm)
$M_{i,L}$	mass of i in the liquid (kg)
M_L, M_{sol}	mass of liquid, mass of solid (kg)
M_s	mass percent solid of solids in the slurry
M_T	magma density (kg/m^3)
$n(L)$	number-based density function (number/m^{3}μm); this is actually a function of time: $n(L,t)$
$n(v)$	volume-based population density (number/m^3/μm^3)
$N(L)$	cumulative number-based distribution function per unit slurry volume, written as a function of particle length (number/m^3)
$N(v)$	cumulative number-based distribution function per unit slurry volume, written as a function of particle volume (number/m^3)
N_{Re}	impeller Reynold's number
N_I	rotational speed of the impeller (rpm)
N_{js}	impeller speed to keep solids just suspended (rpm)
R	ideal gas law constant
S	supersaturation expressed as a ratio
$S(L)$	number-based specific rate of breakage using particle length (min^{-1})
$S(v)$	number -based specific rate of breakage using particle volume (min^{-1})
t	time (min)
T	temperature (°C)

$T_{m,i}$	melting temperature of component i (°C)
u, v, w	particle volume (μm^3)
V_L	volume of liquid in slurry (m^3)
x_i	mole fraction of component i

Greek

γ_i	liquid phase activity coefficient
ΔC	supersaturation, driving force (e.g., g/100 g solvent or g/L solution)
$\Delta H_{m,i}$	heat of crystallization of component i
λ	particle length (μm)
μ	fluid viscosity
ν	kinematic viscosity
ρ_L, ρ_s	liquid and solid densities (kg/m^3)
ρ_i	density factor for component i (kg/m^3)
ϕ_v	shape factor

Subscripts

i	component i
L	liquid
primary	primary nucleation
secondary	secondary nucleation
solid	solid

Superscripts

B_i, B_j, B_{j2}	nucleation expression exponents
h, j, i	nucleation expression constants
n	exponent
sat	saturated

REFERENCES

1. Seidell, A., *Solubilities of Inorganic and Metal Organic Compounds*, 3rd ed., Vol. 1, Van Nostrand, New York, 1940.
2. Dean, J.A., Ed., *Lange's Handbook of Chemistry*, 13th ed., McGraw-Hill, New York, 1985, 10-7–10-21.
3. Liley, P.E., Reid, R.C., and Buck, E., Physical and chemical data, in *Perry's Chemical Engineers' Handbook*, 6th ed., Green, D.W., Ed., McGraw-Hill, New York, 1984, 3-96–3-103.
4. Mullin, J.W., *Crystallization*, 4th ed., Butterworth-Heinemann, Boston, 2001.
5. Mersmann, A., *Crystallization Technology Handbook*, 2nd ed., Marcel Dekker, New York, 2001, 728–737.
6. Walas, S.M., *Phase Equilibria in Chemical Engineering*, Butterworth, Boston, 1985, 397–421.
7. Prausnitz, J.M., Lichtenthaler, R.N., and de Azevedo, E.G., *Molecular Thermodynamics of Fluid-Phase Equilibria*, 3rd ed., Prentice Hall, Upper Saddle River, NJ, 1999.

8. McCabe, W.L., Crystal growth in aqueous solutions. II. Experimental, *Ind. Eng. Chem.*, 21, 112–119, 1929.

9. Mersmann, A., Physical and chemical properties of crystalline systems, in *Crystallization Technology Handbook*, 2nd ed., Mersmann, A., Ed., Marcel Dekker, New York, 2001, 1–44.

10. Nyvlt, J., Rychly, R., Gottfried, J., and Wurzelova, J., Metastable zone width of some aqueous solutions, *J. Crys. Growth*, 6, 151–162, 1970.

11. Dirksen, J.A. and Ring, T.A., Fundamentals of crystallization: kinetic effects on particle size distributions and morphology, *Chem. Eng. Sci.*, 46, 2389–2427, 1991.

12. Davey, R.J., Mullin, J.W., and Whiting, M.J.L., Habit modification of succinic acid crystals grown from different solvents, *J. Crys. Growth*, 58, 304–312, 1982.

13. Davey, R.J. and Garside, J., *From Molecules to Crystallizers*, Oxford University Press, New York, 2000, 36–43.

14. Ring, T.A., Kinetic effects on particle morphology and size distribution during batch precipitation, *Powd. Technol.*, 65, 195–206, 1991.

15. Sato, K., Crystallization behaviour of fats and lipids: a review, *Chem. Eng. Sci.*, 56, 2255–2265, 2001.

16. Kitamura, M., Controlling factor of polymorphism in crystallization process, *J. Crys. Growth*, 237–239, 2205–2214, 2002.

17. Fryer, P. and Pinschower, K., The materials science of chocolate, *MRS Bull.*, 25(12), 25–29, 2000.

18. Hartel, R.W., Phase transitions in chocolates and coatings, in *Phase/State Transitions in Foods: Chemical, Structural, and Rheological Changes*, Marcel Dekker, New York, 1998, 217–252.

19. Rouhi, A., The right stuff, *Chem. Eng. News*, 81(8), 32–35, 2003.

20. Paul, E.L. and Rosas, C.B., Challenges for chemical engineers in the pharmaceutical industry, *Chem. Eng. Prog.*, 86, 17–25, 1990.

21. McCabe, W.L., Crystal growth in aqueous solutions. I. Theory, *Ind. Eng. Chem.*, 21, 30–34, 1929.

22. Bransom, S.H., Factors in the design of continuous crystallizers, *Br. Chem. Eng.*, 5, 838–844, 1960.

23. Canning, T.F. and Randolph, A.D., Some aspects of crystallization theory: systems that violate McCabe's delta L law, *AIChE J.*, 13, 5–10, 1967.

24. Abegg, C.F., Stevens, J.D., and Larson, M.A., Crystal size distributions in continuous crystallizers when growth rate is size dependent, *AIChE J.*, 14, 118–122, 1968.

25. Ramabhadran, T.E., Peterson, T.W., and Seinfeld, J.H., Dynamics of aerosol coagulation and condensation, *AIChE J.*, 22, 840–851, 1976.

26. Garside, J. and Shah, M.B., Crystallization kinetics from MSMPR crystallizers, *Ind. Eng. Chem. Proc. Des. Dev.*, 19, 509–514, 1980.

27. Mersmann, A., Eble, A., and Heyer, C., Crystal growth, in *Crystallization Technology Handbook*, 2nd ed., Mersmann, A., Ed., Marcel Dekker, New York, 2001, 81–144.

28. Mersmann, A., Braun, B., and Loffelmann, M., Prediction of crystallization coefficients of the population balance, *Chem. Eng. Sci.*, 57, 4267–4275, 2002.

29. Garside, J., Phillips, V.R., and Shah, M.B., On size-dependent crystal growth, *Ind. Eng. Chem. Fund.*, 15, 230–233, 1976.

30. Bramley, A.S., Hounslow, M.J., and Ryall, R.L, Aggregation during precipitation from solution: kinetics for calcium oxalate monohydrate, *Chem. Eng. Sci.*, 52, 747–757, 1997.

31. Golovin, A.M., The solution of the coagulation equation for raindrops taking condensation into account, *Sov. Phys. Doklady*, 8, 191–193, 1963.

32. Scott, W.T., Analytic studies of cloud droplet coalescence, part I, *J. Atmos. Sci.*, 25, 54–65, 1968.

33. von Smoluchowski, M., Versuch einer mathematischen Theorie der Koagulationskinetik kolloider Losungen, *Zeitschr. Phys. Chem.*, 92, 129–168, 1917.

34. Friedlander, S.K., *Smoke, Dust and Haze*, Wiley, New York, 1977.

35. Shiloh, K., Sideman, S., and Resnick, W., Coalescence and break-up in dilute polydispersions, *Can. J. Chem. Eng.*, 51, 542–549, 1973.

36. Tobin, T., Muralidhar, R., Wright, H., and Ramkrishna, D., Determination of coalescence frequencies in liquid–liquid dispersions: effect of drop size dependence, *Chem. Eng. Sci.*, 45, 3491–3504, 1990.

37. Berry, E.X., Cloud droplet growth by collection, *J. Atmos. Sci.*, 24, 688–701, 1967.

38. Thompson, P.D., A transformation of the stochastic equation for droplet coalescence, *Proc. Int. Conf. Cloud Physics*, Toronto, 1968, 115–125.

39. Schumann, T.E.W., Theoretical aspects of the size distribution of fog particles, *Q. J. Roy. Meteor. Soc.*, 66, 195–207, 1940.

40. Drake, R.L., A general mathematical survey of the coagulation equation, in *Topics in Current Aerosol Research*, Part 2, Hidy, G.M. and Brock, J.R., Eds., Pergamon Press, New York, 1972, 201.

41. Kruis, F.E. and Kusters, K.A., The collision rate of particles in turbulent flow, *Chem. Eng. Comm.*, 158, 201–230, 1997.

42. Melis, S., Verduyn, M., Storti, G., Morbidelli, M., and Baldyga, J., Effect of fluid motion on the aggregation of small particles subject to interaction forces, *AIChE J.*, 45, 1383–1393, 1999.

43. Hartel, R.W. and Randolph, A.D., Mechanisms and kinetic modeling of calcium oxalate crystal aggregation in a urinelike liquor. II. Kinetic modeling, *AIChE J.*, 32, 1186–1195, 1986.

44. Smit, D.J., Hounslow, M.J., and Paterson, W.R., Aggregation and gelation. I. Analytical solutions for CST and batch operation, *Chem. Eng. Sci.*, 49, 1025–1035, 1994.

45. Jager, J., De Wolf, S., Kramer, H.J.M., and De Jong, E.J., Estimation of nucleation and attrition kinetics from CSD-transients in a continuous crystallizer, *Chem. Eng. Res. Des.*, 69, 53–62, 1991.

46. Shamlou, P.A., Jones, A.G., and Djamarani, K., Hydrodynamics of secondary nucleation in suspension crystallization, *Chem. Eng. Sci.*, 45, 1405–1416, 1990.

47. Van der Heijden, A.E.D.M., Van der Eerden, J.P., and Van Rosmalen, G.M., The secondary nucleation rate: a physical model, *Chem. Eng. Sci.*, 49, 3103–3113, 1994.

48. Meadhra, R.O., Kramer, H.J.M., and Van Rosmalen, G.M., Model for secondary nucleation in a suspension crystallizer, *AIChE J.*, 42, 973–982, 1996.

49. Hill, P.J. and Ng, K.M., New discretization procedure for the breakage equation, *AIChE J.*, 41, 1204–1216, 1995.

50. Austin, L., Shoji, K., Bhatia, V., Jindal, V., Savage, K., and Klimpel, R., Some results on the description of size reduction as a rate process in various mills, *Ind. Eng. Chem. Proc. Des. Dev.*, 15, 187–196, 1976.

51. Broadbent, S.R. and Calcott, T.G., Coal breakage processes, *J. Inst. Fuel*, 29, 528, 1956.
52. Shoji, K., Lohsrab, S., and Austin, L.G., The variation of breakage parameters with ball and powder filling in dry ball milling, *Powd. Technol.*, 25, 109–114, 1980.
53. Reid, K.J., A solution to the batch grinding equation, *Chem. Eng. Sci.*, 20, 952–963, 1965.
54. Randolph, A.D. and Ranjan, R., Effect of a material-flow model in prediction of particle size distributions in open- and closed-circuit mills, *Int. J. Min. Proc.*, 4, 99, 1977.
55. Pandya, J.D. and Spielman, L.A., Floc breakage in agitated suspensions: theory and data processing strategy, *J. Coll. Int. Sci.*, 90, 517–531, 1982
56. Gahn, C. and Mersmann, A., Brittle fracture in crystallization processes. Part A. Attrition and abrasion of brittle solids, *Chem. Eng. Sci.*, 54, 1273–1282, 1999.
57. Peterson, T.W., Scotto, M.V., and Sarofim, A.F., Comparison of comminution data with analytical solutions of the fragmentation equation, *Powd. Technol.*, 45, 87–93, 1985.
58. Van Peborgh Gooch, J.R., Hounslow, M.J., and Mydlarz, J., Discriminating between size-enlargement mechanisms, *Chem. Eng. Res. Des.*, 74, 803–811, 1996.
59. Jones, C.M. and Larson, M.A., Characterizing growth-rate dispersion of $NaNO_3$ secondary nuclei, *AIChE J.*, 45, 2128–2135, 1999.
60. Hulburt, H.M. and Katz, S., Some problems in particle technology: a statistical mechanical formulation, *Chem. Eng. Sci.*, 19, 555–574, 1964.
61. Randolph, A.D. and Larson, M.A., *Theory of Particulate Processes*, 2nd ed., Academic Press, New York, 1988, 50–79.
62. Ramkrishna, D., *Population Balances: Theory and Applications to Particulate Systems in Engineering*, Academic Press, New York, 2000.
63. Ziff, R.M., New solutions to the fragmentation equation, *J. Phys. A Math. Gen.*, 24, 2821–2828, 1991.
64. Gelbard, F., Tambour, Y., and Seinfeld, J.H., Sectional representations for simulating aerosol dynamics, *J. Coll. Int. Sci.*, 76, 541–556, 1980.
65. Sastry, K.V.S. and Gaschignard, P., Discretization procedure for the coalescence equation of particulate processes, *Ind. Eng. Chem. Fundam.*, 20, 355–361, 1981.
66. Hounslow, M.J., Ryall, R.L., and Marshall, V.R., A discretized population balance for nucleation growth and aggregation, *AIChE J.*, 34, 1821–1832, 1988.
67. Litster, J.D., Smit, D.J., and Hounslow, M.J., Adjustable discretized population balance for growth and aggregation, *AIChE J.*, 41, 591–603, 1995.
68. Hill, P.J. and Ng, K.M., New discretization procedure for the agglomeration equation, *AIChE J.*, 42, 727–741, 1996.
69. Kumar, S. and Ramkrishna, D., On the solution of population balance equations by discretization. I. A fixed pivot technique, *Chem. Eng. Sci.*, 51, 1311–1332, 1996.
70. Kumar, S. and Ramkrishna, D., On the solution of population balance equations by discretization. II. A moving pivot technique, *Chem. Eng. Sci.*, 51, 1333–1342, 1996.
71. Kumar, S. and Ramkrishna, D., On the solution of population balance equations by discretization. III. Nucleation, Growth and Aggregation of Particles, *Chem. Eng. Sci.*, 52, 4659–4679, 1997.
72. Liu, Y. and Cameron, I.T., A new wavelet-based method for the solution of the population balance equation, *Chem. Eng. Sci.*, 56, 5283–5294, 2001.

73. Gelbard, F. and Seinfeld, J.H., Numerical solution of the dynamic equation for particulate systems, *J. Comp. Phys.*, 28, 357–375, 1978.
74. Kubota, N., Shimizu, K., and Itagaki, H., Densities and viscosities of supersaturated potash alum aqueous solutions, *J. Crys. Growth*, 73, 359–363, 1985.
75. Frej, H., Balinska, A., and Jakubczyk, M., Density and viscosity of undersaturated, subsaturated, and supersaturated aqueous ammonium oxalate solutions from 287 K to 325 K, *J. Chem. Eng. Data*, 45, 415–418, 2000.
76. Tavare, N.S. and Garside, J., Simultaneous estimation of crystal nucleation and growth kinetics from batch experiments, *Chem. Eng. Res. Des.*, 64, 109–118, 1986.
77. Garside, J., Gibilaro, L.G., and Tavare, N.S., Evaluation of crystal growth kinetics from a desupersaturation curve using initial derivatives, *Chem. Eng. Sci.*, 37, 1625–1628, 1982.
78. Farrell, R.J. and Tsai, Y.C., Modeling, simulation and kinetic parameter estimation in batch crystallization processes, *AIChE J.*, 40, 586–593, 1994.
79. Qiu, Y. and Rasmuson, A.C., Estimation of crystallization kinetics from batch cooling experiments, *AIChE J.*, 40, 799–812, 1994.
80. Monnier, O., Fevotte, G., Hoff, C., and Klein, J.P., Model identification of batch cooling crystallizations through calorimetry and image analysis, *Chem. Eng. Sci.*, 52, 1125–1139, 1997.
81. Chung, S.H., Ma, D.L., and Braatz, R.D., Optimal model-based experimental design in batch crystallization, *Chemometrics Int. Lab. Sys.*, 50, 83–90, 2000.
82. Tadayon, A., Rohani, S., and Bennett, M.K., Estimation of nucleation and growth kinetics of ammonium sulfate from transients of a cooling batch seeded crystallizer, *Ind. Eng. Chem. Res.*, 41, 6181–6193, 2002.
83. Mahoney, A.W., Doyle, F.J., and Ramkrishna, D., Inverse problems in population balances: growth and nucleation from dynamic data, *AIChE J.*, 48, 981–990, 2002.
84. Mohan, R., Boateng, K.A., and Myerson, A.S., Measuring crystal growth kinetics using differential scanning calorimetry, *J. Crys. Growth*, 212, 489–499, 2000.
85. Collier, A.P. and Hounslow, M.J., Growth and aggregation rates for calcite and calcium oxalate monohydrate, *AIChE J.*, 45, 2298–2305, 1999.
86. Mazzarotta, B., Abrasion and breakage phenomena in agitated crystal suspensions, *Chem. Eng. Sci.*, 47, 3105–3111, 1992.
87. Mersmann, A., Attrition and attrition-controlled secondary nucleation, in *Crystallization Technology Handbook*, 2nd ed., Mersmann, A., Ed., Marcel Dekker, New York, 2001, 187–234.
88. Zwietering, T.N., Suspension of solid particles in liquid by agitators, *Chem. Eng. Sci.*, 8, 244–253, 1958.
89. Mersmann, A., Design of crystallizers, in *Crystallization Technology Handbook*, 2nd ed., Mersmann, A., Ed., Marcel Dekker, New York, 2001, 323–392.
90. Paul, E.L., Atiemo-Obeng, V.A., and Kresta, S.M., Eds., *Handbook of Industrial Mixing: Science and Practice*, Wiley Interscience, New York, 2004.
91. Bermingham, S.K., Kramer, H.J.M., and van Rosmalen, G.M., Towards on-scale crystalliser design using compartmental models, *Comp. Chem. Eng.*, 22, S355–S362, 1998.
92. ten Cate, A., Derksen, J.J., Kramer, H.J.M., Van Rosmalen, G.M., and Van den Akker, H.E.A., The microscopic modeling of hydrodynamics in industrial crystallizers, *Chem. Eng. Sci.*, 56, 2495–2509, 2001.
93. Ma, D.L., Braatz, R.D., and Tafti, D.K., Compartmental modeling of multidimensional crystallization, *Int. J. Mod. Phys.*, 16, 383–390, 2002.

94. Torbacke, M. and Rasmuson, Å.C., Influence of different scales of mixing in reaction crystallization, *Chem. Eng. Sci.*, 56, 2459–2473, 2001.

95. Marcant, B. and David, R., Experimental evidence for and prediction of micro-mixing effects in precipitation, *AIChE J.*, 37, 1698–1710, 1991.

96. Mullin, J.W. and Nyvlt, J., Programmed cooling of batch crystallizers, *Chem. Eng. Sci.*, 26, 369–377, 1971.

97. Rawlings, J.B., Miller, S.M., and Witkowski, W.R., Model identification and control of solution crystallization processes: a review, *Ind. Eng. Chem. Res.*, 32, 1275–1296, 1993.

98. Chianese, A., Di Berardino, F., and Jones, A.G., On the effect of secondary nucleation on the crystal size distribution from a seeded batch crystallizer, *Chem. Eng. Sci.*, 48, 551–560, 1993.

99. Bohlin, M. and Rasmuson, Å.C., Application of controlled cooling and seeding in batch crystallization, *Can. J. Chem. Eng.*, 70, 120–126, 1992.

100. Chung, S.H., Ma, D.L., and Braatz, R.D., Optimal seeding in batch crystallization, *Can. J. Chem. Eng.*, 77, 590–596, 1999.

101. Fevotte, G. and Klein, J.P., Calorimetric methods for the study of batch crystallization processes: some key results, *Chem. Eng. Res. Des.*, 74, 791–796, 1996.

102. Patience, D.B. and Rawlings, J.B., Particle-shape monitoring and control in crystallization processes, *AIChE J.*, 47, 2125–2130, 2001.

103. Braatz, R.D., Fujiwara, M., Ma, D.L., Togkalidou, T., and Tafti, D.K., Simulation and new sensor technologies for industrial crystallization: a review, *Int. J. Mod. Phys. B*, 16, 346–353, 2002.

104. Laube, N., Mohr, B., and Hesse, A., Laser-probe-based investigations of the evolution of particle size distributions of calcium oxalate particles formed in artificial urines, *J. Crys. Growth*, 233, 367–374, 2001.

105. Hipp, A.K., Walker, B., Mazzotti, M., and Morbidelli, M., *In situ* monitoring of batch crystallization by ultrasound spectroscopy, *Ind. Eng. Chem. Res.*, 39, 783–789, 2000.

106. Mougin, P., Wilkinson, D., Roberts, K.J., and Tweedie, R., Characterization of particle size and its distribution during the crystallization of organic fine chemical products as measured *in situ* using ultrasonic attenuation spectroscopy, *J. Acoust. Soc. Am.*, 109, 274–282, 2001.

107. Helt, J.E. and Larson, M.A., Effect of temperature on the crystallization of potassium nitrate by direct measurement of supersaturation, *AIChE J.*, 23, 822–830, 1977.

108. Garside, J. and Mullin, J.W., Methods apparatus: new product research, process development, and design-continuous measurement of solution concentration in a crystallizer, *Chem. Ind.*, Issue (48), 2007–2008, 1966.

109. David, R., Villermaux, J., Marchal, P., and Klein, J.P., Crystallization and precipitation engineering. IV. Kinetic model of adipic acid crystallization, *Chem. Eng. Sci.*, 46, 1129–1136, 1991.

110. Fevotte, G. and Klein, J.P., Application of on-line calorimetry to the advanced control of batch crystallizers, *Chem. Eng. Sci.*, 49, 1323–1336, 1994.

111. Wang, F. and Berglund, K.A., Monitoring pH swing crystallization of nicotinic acid by the use of attenuated total reflection Fourier transform infrared spectrometry, *Ind. Eng. Chem.*, 39, 2101–2104, 2000.

112. Lewiner, F., Klein, J.P., Puel, F., and Fevotte, G., On-line ATR FTIR measurement of supersaturation during solution crystallization processes: calibration and applications on three solute/solvent systems, *Chem. Eng. Sci.*, 56, 2069–2084, 2001.

113. Loffelmann, M. and Mersmann, A., How to measure supersaturation?, *Chem. Eng. Sci.*, 57, 4301–4310, 2002.

114. Wang, F., Wachter, J.A., Antosz, F.J., and Berglund, K.A., An investigation of solvent-mediated polymorphic transformation of progesterone using *in situ* Raman spectroscopy, *Org. Proc. Res. Dev.*, 4, 391–395, 2000.

115. Ono, T., ter Horst, J.H., and Jansens, P.J., Quantitative measurement of the polymorphic transformation of L-glutamic acid using *in situ* Raman spectroscopy, *Crys. Growth Des.*, 3, 465–469, 2004.

116. Starbuck, C., Spartalis, A., Wai, L., Wang, J., Fernandez, P., Lindemann, C.M., Zhou, G.X., and Ge, Z., Process optimization of a complex pharmaceutical polymorphic system via *in situ* Raman spectroscopy, *Crys. Growth Des.*, 2, 515–522, 2002.

117. Gerstlauer, A., Motz, S., Mitrovic, A., and Gilles, E.D., Development, analysis and validation of population models for continuous and batch crystallizers, *Chem. Eng. Sci.*, 57, 4311–4327, 2002.

118. Misra, C., Solubility of aluminum trihydroxide (hydrargillite) in sodium hydroxide solution, *Chem. Ind.*, Issue (19), 619–623, 1970.

119. Misra, C. and White, E.T., Kinetics of crystallization of aluminum trihydroxide from seeded caustic aluminate solutions, in Crystallization from Solution: Factors Influencing Size Distribution, *Chem. Eng. Prog. Symp. Ser.*, 67, 53–65, 1971.

120. Halfon, A. and Kaliaguine, S., Alumina trihydrate crystallization. Part 1. Secondary nucleation and growth rate kinetics, *Can. J. Chem. Eng.*, 54, 160–167, 1976.

121. Halfon, A. and Kaliaguine, S., Alumina trihydrate crystallization. Part 2. A model of agglomeration, *Can. J. Chem. Eng.*, 54, 168–172, 1976.

122. Kitamura, M., Polymorphism in the crystallization of L-glutamic acid, *J. Crys. Growth*, 96, 541–546, 1989.

123. Kitamura, M. and Ishizu, T., Growth kinetics and morphological change of polymorphs of L-glutamic acid, *J. Crys. Growth*, 209, 138–145, 2000.

124. Kitamura, M. and Ishizu, T., Kinetic effect of L-phenylalanine on growth process of L-glutamic acid polymorph, *J. Crys. Growth*, 192, 225–235, 1998.

125. Vedantham, K., Effect of Operating Parameters on the Growth Rate of Solution Grown Crystals, M.S. thesis, Mississippi State University, Mississippi State, MS, 2004.

126. Shekunov, B.Y. and York, P., Crystallization processes in pharmaceutical technology and drug delivery design, *J. Crys. Growth*, 211, 122–136, 2000.

127. Lewiner, F., Fevotte, G., Klein, J.P., and Puel, F., Improving batch cooling seeded crystallization of an organic weed-killer using on-line ATR FTIR measurement of supersaturation, *J. Crys. Growth*, 226, 348–362, 2001.

7 Crystallization and Associated Solid–Liquid Separations

Sean M. Dalziel and Thomas E. Friedmann

CONTENTS

7.1 INTRODUCTION

Industrial crystallization is widely used to form a crystalline product from a dissolved state. The two most common motivations to choose crystallization as a unit operation are:

- To recover a particulate form of a substance (e.g., so it can be packaged as a flowable powder)
- To purify the substance, commonly via multiple recrystallizations

Integral to a crystallization process is the solid–liquid separation step that follows the crystallizer in the process train. Figure 7.1 shows a typical process flow diagram for a chemical product identifying crystallization as well as upstream and downstream unit operations. Crystallization is an important unit operation in such a process chain as it is the primary stage of the product particle formation. Thus, the particle size distribution, purity, and yield are largely determined by the performance of the crystallizer. Although the crystallizer has a major

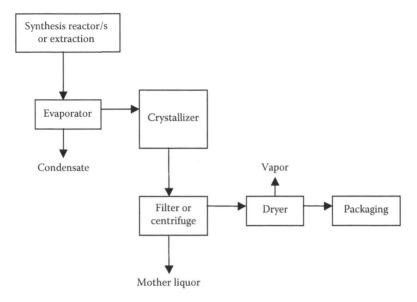

FIGURE 7.1 Process flow diagram identifying the typical configuration of a crystallizer for a chemical product.

impact on the product properties and quality, the slurry and crystal properties (size, density, habit) greatly influence the investment required in the downstream unit operations — particularly filtration/centrifugation and the dryer.

The interplay between crystallizer performance and solid–liquid separation devices often substantially affects the capital investment and operating costs of such a process. Unless holding tanks are available, batch crystallization processes require separation equipment that can process the entire crystallizer slurry at a higher rate than would be required for continuous crystallizers of the same production rate. This is because crystallizer tank emptying must be completed as quickly as possible to allow cleaning and turnaround for the next crystallizer batch; hence, solid–liquid separation capacity can present a bottleneck in turn-around time. Further, the filter or centrifuge usually will be sitting idle during the majority of the crystallization batch, so it is highly desirable to maximize the separability of a slurry product entering the solid–liquid separations device. It is important to adequately size the solid–liquid separation equipment, bearing in mind that the cost of these devices is typically not insignificant.

Given the integral nature of solid–liquid separation to crystallization and precipitation processes, this chapter addresses practical aspects of developing batch crystallization processes and supporting solid–liquid separations (batch or continuous). This chapter is intended to be a guide for laboratory and pilot testing, as well as for scale-up. Several other texts address crystallization in more detail and are recommended for further information, including Mullin,[1] Myerson,[2] Mersmann,[3] Tavare,[4] Davey and Garside,[5] and Perry et al.[6]

7.2 FUNDAMENTALS

7.2.1 CRYSTALLIZATION

The crystal lattice configuration enables a tight packing of molecules and rejects many foreign materials as impurities if they do not fit into the lattice. To fit into the lattice requires meeting chemical, steric, and free energy criteria; hence, the crystal lattice is a highly purified and concentrated solid phase. The crystalline state enables uniform physical and chemical properties of a solid product. Some key properties include melting point, solubility and rate of dissolution, surface chemistry, mechanical strength, optical properties, and stability at various humidities and temperatures.

The crystals of many chemical substances exhibit polymorphism, which is when the same chemical structure can pack into more than one lattice structure. Each polymorph may exhibit different physical and chemical properties. This can be an advantage (e.g., to enable a higher solubility or a higher melting point for the same chemical substance), but it can also be a major concern for scale-up and manufacturing as far as ensuring that a processes reliably produces only the desired polymorph.

7.2.1.1 Mass Transfer

Crystallization is a mass transfer unit operation and can be analyzed using these principles.[7] Material initially exists in a dissolved state and transfers to a solid phase via deposition onto the growing faces of a solid crystal. Chemical potential drives this mass transfer. In practice, however, supersaturation is the driving force used to describe crystallization mass transfer. Supersaturation is calculated from the solute concentration (c) in the solution of the crystallizing species and its solubility concentration (c^*) at the same conditions of temperature, pressure, solvent composition, and impurity levels. Supersaturation can be expressed as follows:

Absolute supersaturation	$c - c^*$
Relative supersaturation	$(c - c^*)/c^*$
Supersaturation ratio	c/c^*

Process engineering studies primarily use absolute or relative supersaturation, while thermodynamics studies tend to use the logarithm of the supersaturation ratio. For industrial purposes, the decision is arbitrary, but the absolute supersaturation form is marginally less affected by poorer solubility data. Because supersaturation can be calculated from measurements of solute and solubility concentrations, it is a practical way to determine the driving force of crystallization mass transfer.

Crystallization includes two different processes:

- Nucleation
- Crystal growth

Nucleation is the birth process for creating the solid-phase lattice, while crystal growth is the bulk of the mass transfer from solution to solid. Precipitation processes are quite similar to crystallization, although the solid phase lacks the ordered lattice configuration. Precipitation processes nucleate solids but they do not grow through molecules locking into a lattice; instead, they grow by random deposition and agglomeration. So, precipitation processes are typically nucleation dominated, while crystallization processes include both nucleation and growth processes in particle formation. The term *precipitation* is often used loosely to refer to the process of recovering solids from solution. The development of precipitation processes shares a number of aspects with crystallization development, particularly in terms of scale-up approach, vessel design, and solid–liquid separations. Scale-up of crystallization processes tends to have additional importance with regard to heat transfer and slurry mixing, while for precipitation processes the distribution of reacting feed streams tends to be a more dominating scale-up parameter.

Crystal growth kinetics is commonly correlated by a power law as:

$$G = k_G \cdot \sigma^n$$

where G is the growth rate (μm/hr), k_G is the growth rate constant, and σ is the supersaturation. The units of the growth constant depend on the units of solubility and whether absolute or relative supersaturation is used. The exponent (n) is the growth order, which is dimensionless. The growth order can be an indicator of the mechanism of crystal growth. Where $n = 1$, the rate of crystal growth is generally controlled by diffusion transport of the solute through the boundary layers to the growing face. Where the growth order is $n > 1$, the rate of crystal growth is usually controlled by surface integration kinetics, which is the orientating and docking of molecules into the correct configuration at the growing face of the lattice.

The overall crystallization rate is often mistakenly construed as the crystal growth rate. The overall crystallization rate is the rate of solute concentration decrease with time; that is,

$$R = -\frac{dc}{dt} \tag{7.1}$$

where R is the overall rate of crystallization, c is the solute concentration, and t is time. The crystal growth rate is a different quantity, being defined as the change in size of a crystal with time:

$$G = \frac{dL}{dt} \tag{7.2}$$

where L is the crystal size. The overall rate of crystallization can be predicted by using a population balance and a material balance together. This has been described in detail by Randolph and Larson.[8] The population balance is a means of accounting for the change in the number and size distribution of crystals in a population through the processes of nucleation (birth), dissolution (death), growth, agglomeration, and breakage (attrition). The overall rate of crystallization is a compilation of the effects of each aforementioned term; therefore, the overall rate of crystallization is not necessarily indicative of the crystal growth rate. Furthermore, the rate of mass transfer of solute from solution to the crystal phase is dependent on the interfacial area. So, the rate of crystallization can be manipulated through seeding, nucleation, and crystal growth.

7.2.1.2 Nucleation in Industrial Crystallizers

In an industrial or typical lab batch crystallization process, two nucleation mechanisms can occur:

- Primary heterogeneous nucleation
- Secondary nucleation

Typically primary heterogeneous nucleation is the mechanism that initiates crystallization for an unseeded batch crystallization process. This is where supersaturation is high enough for molecules to cluster into a critical nucleus and change phase from dissolved to a solid crystal. Practically speaking, homogeneous primary nucleation is not directly relevant to industrial crystallization and need not be investigated during process development.

Secondary nucleation is defined as the birth of new crystals in the presence of other crystals. After birth of the first crystal through primary heterogeneous nucleation, by definition secondary nucleation is the only type of nucleation for the remainder of the batch. Collisions of crystals with other crystals, the type of internal surface of the vessel or the impeller, or fluid shear forces acting on a crystal surface all can lead to small solid fragments or molecular clusters separating from the parent crystal. Where supersaturation is great enough, these fragments repair and grow as mature crystals, a process referred to as *contact nucleation*. Additionally, where the supersaturation in a slurry exceeds the metastable zone width, spontaneous nucleation will occur via the heterogeneous primary nucleation mechanism. Because nucleation in a batch crystallizer often includes simultaneous molecular clustering and collision mechanisms, the more effective correlations of nucleation in a batch crystallizer include terms for supersaturation, crystal content, and agitation intensity. Many correlation equations have been proposed. Perry et al.[6] provided a summary of the numerous nucleation equations that have been reported. Equation 7.3 is an example of a commonly used nucleation equation relevant to batch crystallization:

$$B = k_n \cdot \sigma^j \cdot M_T^k \cdot N^l \tag{7.3}$$

where B is nucleation rate (number per unit volume per unit time), k_N is the nucleation constant, σ is the supersaturation (absolute or relative), M_T is the crystal content of the slurry, and N is an agitation parameter (such as power input per volume or tip speed). The exponents j, k, and l indicate the relative dependence of each parameter on the overall nucleation rate. Systems that more readily nucleate in response to increasing supersaturation have a higher j exponent. Systems with more fragile crystals are typically more susceptible to collision nucleation mechanisms; these systems often exhibit both higher values of k and l.

For batch crystallizers that are not well controlled, an initially high rate of primary nucleation occurs which decreases with time as supersaturation is depleted. Simultaneously, the secondary nucleation rate increases over time due to the increasing crystal content and particle collision frequency. The potential for these two parameters to induce batch-to-batch variability is important to understand for scale-up and setting process parameters. By contrast, continuous crystallizers ideally operate at steady state and are generally self-stabilizing. In this case, the secondary nucleation mechanism predominates.

7.2.1.3 Phase Diagram and Nucleation Thresholds

When studying the crystallization behavior of a compound, it is useful to assemble a phase diagram showing the equilibrium solubility as a function of the relevant parameters, such as temperature, antisolvent concentration, pH, and salt concentration. On the same graph, nucleation thresholds can be drawn in to indicate nucleation kinetic behaviors that are time-dependent, nonequilibrium properties.

Many systems exhibit a useful metastable zone; this represents the supersaturated region, where no significant nucleation will occur within the time frame of the batch process. At supersaturations above the metastable zone, secondary nucleation will readily occur. This process is also referred to as *contact nucleation*, as it is where new crystals form only in the presence of other crystals. At supersaturations greater still, heterogeneous primary nucleation occurs within the time frame of the batch process. Figure 7.2 demonstrates how such a graph may appear. High measurement scatter is usually seen in such data due to the difficulty in making accurate assessments of when nucleation occurred and due to the stochastic nature of the nucleation processes. The data are commonly processed by fitting the data with a smooth curve having the shape of the solubility curve. Doing so establishes the nucleation threshold. The existence of metastability is due to the Gibbs–Thompson effect. In this instance, very fine fragments of crystals may have a higher solubility than larger crystals; hence, the fate of a crystal fragment has a probabilistic component. Depending on its size and the extent of supersaturation in the solution around it, the crystal fragment may dissolve completely or remain and grow.

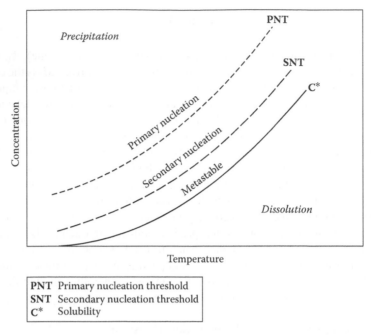

FIGURE 7.2 Solubility phase diagram with nucleation thresholds overlaid (secondary and heterogeneous primary).

The trajectory of a crystallization batch can be charted on a phase diagram with nucleation thresholds to indicate the path taken as a function of batch process time. Figure 7.3 gives three examples of such a diagram for different modes of batch crystallization.

Accurate measurement of nucleation rates is not a trivial task in batch crystallization. Further, batch-to-batch variability in nucleation behavior often occurs due to one parameter being highly sensitive and difficult to control precisely between batches. Nucleation thresholds provide a very useful compromise, as they bound the regions where the differing nucleation regimes occur. Using solubility data and nucleation thresholds as boundaries is a useful way to rapidly establish the set points for a batch crystallization process and improve batch-to-batch consistency.

7.2.1.4 Self-Nucleating or Seeding Batch Crystallization

Nucleation is the most difficult phenomenon to control during batch crystallization. It is known to be the most common cause of batch-to-batch variability in crystal size distribution and the performance of solid–liquid separation devices. The preferred method of avoiding the complications of poorly controlled nucleation is to perform seeding. To be effective, seed crystals must be added when the supersaturated solution is in the metastable zone. The crystallizer must

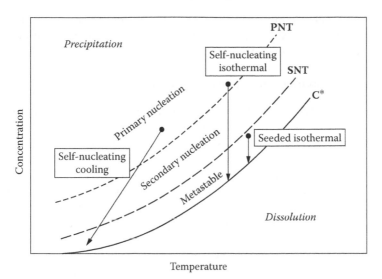

FIGURE 7.3 Trajectory of three example batch crystallizations across the phase diagram: (i) self-nucleating, cooling; (ii) self-nucleating, isothermal; and (iii) seeded isothermal.

be controlled so it remains inside the metastable zone to avoid nucleation or dissolution.

Where seeding is carried out correctly, the accurately known number, mass, and surface area of seed crystals allows control of the final crystal size distribution. For the desirable case of no significant nucleation, the number of crystals in the product slurry is equal to the number of crystals in the seed charge. The surface area of the seed initially defines the available area for mass transfer (which controls the batch time), and the mass of the seed crystals is simply a function of number and the mass mean size of the seed. Controlling a seeded batch crystallizer properly is the best way to minimize batch-to-batch variability and ensure reliable performance of downstream operations (filtration/centrifugation, drying, milling, and packaging).

7.2.1.5 Measurement of Solubility Data

Solubility data are measured by equilibrating a crystallization liquor at a given temperature with an excess of the crystal phase for which data are being generated. The equilibration time depends on the kinetics, the surface area, and whether equilibration is being performed by dissolution or by crystallization. Most studies perform equilibration by dissolution over 12- to 24-hour time frames, but this is not always sufficient time. If precise data are being generated, equilibration by crystallization should be carried out over a period of weeks to months. For each case, samples of equilibrated slurry are filtered and the dissolved content of the filtrate is measured by an appropriate method, such as high-performance liquid chromatography (HPLC), ultraviolet (UV)/vis spectrophotometry, gravimetric or conductivity analyses, measurement of the refractive index, or other means. More information on the measurement of solubility from crystallizing and dissolving systems can be found in Dalziel et al.[9]

7.2.1.6 Measurement of Nucleation Thresholds

Several techniques have been reported to measure the nucleation thresholds and thus the metastable zone width. Automated lab reactors, with online turbidity or particle sizing devices can be used to rapidly develop data. Less sophisticated techniques can be just as effective. Cherdrungsi et al.[10] reported a simple technique where supersaturated liquor is sealed in a sample jar and agitated at constant temperature in a shaker bath. The time period for visual observation of nucleation (observation of crystals or cloud) is plotted for various supersaturations. This procedure is repeated over a range of temperatures and concentrations. These experiments are performed in the absence of seed for the heterogeneous primary nucleation threshold and in the presence of a few large seed crystals for the secondary nucleation threshold. The metastable zone is determined by the extent of supersaturation that can exist without nucleation over the time period of the batch crystallization process.

7.2.1 SOLID/LIQUID SEPARATION

Batch processing is commonly encountered in solid/liquid separation. In the batchwise operation mode, the individual solid/liquid separation steps are not synchronized with each other and are independently controllable. A semicontinuous operation mode can be achieved by having several machines operating at staged batch times. For filtration, these individual steps are cake formation, cake washing, and cake dewatering. For sedimentation, the steps would involve the sedimentation and sludge discharge steps. Different driving forces and mechanisms are used to achieve solid/liquid separations. The main mechanisms used are sedimentation (by gravity or centrifugal field) and filtration (using a pressure difference generated by pumps, gas pressure, vacuum, centrifugal field, or combinations thereof). The various mechanisms are described in the following sections.

7.2.2 SEDIMENTATION

Sedimentation processes are based on the gravitational forces acting on the particles to be separated; therefore, a density difference between the solid particles and the fluid is required. The sedimentation velocity of a sphere is described by Stokes' law:

$$v = \frac{x^2}{18 \cdot \eta} \cdot \left(\rho_s - \rho_f \right) \cdot g \qquad (7.4)$$

The settling time can be accelerated by reducing the viscosity (η) or increasing the density difference (ρ_s, solids density; ρ_{fl}, fluid density) or particle size (x). The gravitational force (g) can be increased by using centrifuges. To describe the increase of gravitational force in the centrifuge equipment, the term *g-force* or *relative centrifugal force* (RCF) is commonly used. In European literature, the term *C-value* or *Z-value* is often encountered for the same relationship described in Equation (7.5), where r is the centrifuge radius and ω is the angular velocity:

$$C = \frac{r \cdot \omega^2}{g} \tag{7.5}$$

Stokes' law is valid for single spheres sedimenting in a Newtonian fluid. Settling behavior of suspensions containing higher solids concentrations have been investigated in the past and are documented in the literature. An equation used primarily to describe the sedimentation velocities as a function of particle concentration was derived by Richardson and Zaki.[11] The exponent is a function of the Reynolds number, and for laminar conditions the value of 4.65 (as shown in Equation 7.6) can be used:

$$\frac{v}{v_{Stokes}} = (1 - C_v)^{4.65} \tag{7.6}$$

In this empirical relationship, v_{Stokes} is the unhindered settling velocity according to Stokes' law, and C_v is the particle volume concentration.

More recent research in this area has been directed toward a better understanding of settling and consolidation of fine particle suspensions and of polydisperse particle systems.[12] It is important to understand the fundamentals of sedimentation and their implication for proper testing and evaluation of sedimentation behavior, as described later in this chapter. Especially when dealing with fine particle suspensions, the physicochemical interactions become more important and effects such as agglomeration and flocculation can significantly influence the outcome of sedimentation. Good knowledge and control of the process conditions boundaries (pH, temperature, shear, ionic strength, upstream impurities or additives) are crucial for the successful design of a sedimentation approach.

7.2.2.1 Filtration

Filtration is the separation of a fluid/solids mixture involving the passage of most of the fluid through a porous barrier (filter medium), which retains most of the solid particulates contained in the mixture. This definition given by Perry[6] distinguishes the term *filtration* from the second unit operation in solid/liquid separation processes: *sedimentation*. Whereas in sedimentation a density difference between solids and fluid is mandatory, filtration is driven by hydrostatic head (gravity), by pressure applied upstream or reduced pressure (vacuum) applied downstream of the filter medium, or by centrifugal force across the filter medium. Filtration can be classified into two major categories:

- *Clarifying or depth filtration*, when the solids are withheld or trapped completely within or on the filter medium; retention can be accomplished by various mechanisms (mechanically, electrostatically)
- *Cake filtration*, when the layer of particles deposited on the filter medium acts as a filtering medium

7.2.2.1.1   Filter Media

Filter media function in several capacities to provide a clear filtrate. They can act as absolute filter media, where the pore size is designed to prevent any particles of a given size to pass the filter media. Such filter media are commonly used as guard filters or police filters. They can be located prior to the equipment to prevent coarse particles from plugging up nozzles, fouling filters, or having other negative impacts on the process. Guard filters can also be used downstream of filtration equipment to catch any particles that might have passed the main solid/liquid separation step. Filter media for deep filtration purposes are, in general, designed with a larger pore size than the particles to be retained. The filter medium in this case serves as a starting point for the cake formation (explained later). Various materials for filter media are used, the most common being nylon, polyethylene, and polypropylene. Metal wire meshes are also used in certain filter media or support media. The filter media material, the type of weave, and the surface finish are of importance for the such filter media properties as initial flow rate, retention efficiency, filter cake release, blinding resistance, and strength. Useful reviews on filter media types and the associated retention mechanisms can be found in Rushton and Griffiths.[13]

7.2.2.1.2   Depth Filtration

For depth filtration, the particles are retained within the filter media. The main application areas include water treatment and purification, in the beverage industry, and removal of particulates from solvents and polymer solutions. Depending on application and required volumes to be filtered, deep filters are built in a variety of configurations from large, slow-filtering sand filters (filtration velocities around 1 m/day) to small, rapid-filtering (1 m/hr) cartridge filters intended for single use. The particle retention efficiency in deep filters is a function of the transport mechanism of the particles to the filter media and the adhesion probability when the particle reach the surface. The transport mechanisms encountered in depth filtration are (1) direct interception, (2) sedimentation, and (3) diffusion. For a particle to be retained in the filter matrix it has to adhere permanently to the surface. The exact calculation of this adhesion probability is not straightforward for real filtering systems. Adhesion probability is highly dependent on particle charge and size as well as on the surface properties of the filter media.

The measurement of these properties is necessary for the proper selection and design of deep filters.

7.2.2.1.3 Surface Filtration

The term *surface filtration* is used when the particles are preferably retained at the surface of the filter medium. The different modes of surface filtration include sieve filtration, cross-flow filtration, and cake filtration. Sieve filtration refers to the removal of particles from a suspension with a low solids concentration, without the real formation of a filter cake. At a preset pressure loss, the filter media are exchanged or regenerated by backflushing or other cleaning mechanisms. In cross-flow filtration, the suspension flows parallel to the filter medium to avoid deposition of particles on the filter medium. The clear permeate is forced through the filter medium by the applied transmembrane pressure difference. The concentrate is discharged from the cross-flow apparatus in a highly concentrated but still pumpable form. For cross-flow operations, membranes are typically used as the filter media. Depending on the desired particle (or molecule) cut size, the membrane filtration technologies are classified as microfiltration, ultrafiltration, and nanofiltration/reverse osmosis. Because these technologies are commonly used as continuous processes they will not be described further here. A good reference for membrane filtration can be found in Ho and Sirkar.[14]

Cake Filtration

In cake filtration, the layer of particles on the filter medium acts as a filter medium. The main parameters in cake filtration are shown schematically in Figure 7.4. In cake filtration, a first layer of particles is formed by bridging mechanisms, where multiple particles arch over the filter medium opening. The pore size of the filter medium (filter cloth) can therefore be significantly larger than the smallest particle

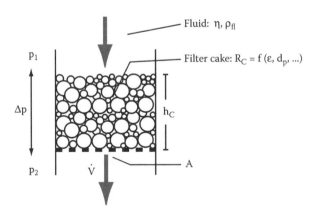

FIGURE 7.4 Parameters in cake filtration: pressure difference, $\Delta p = p_1 - p_2$; volumetric flow rate, $\dot{V}$; dynamic viscosity, η; fluid density, ρ_{fl}; filter cake height, h_C; filter cake resistance, R_C; filter area, A. (From Friedmann, T., in *Institute of Food Science: Laboratory of Food Process Engineering*, ETH Zürich, Zurich, 1999. With permission.)

size in the suspension to be separated. Filter cloth selection and the solids concentration of the suspension are determining parameters for the initiation of filter cake buildup. Selection of the appropriate filter media is discussed in Section 7.3.7.2. Due to the initial bridging mechanism, the first filtrate may still contain an increased amount of solids. After a first layer has built up, the filter cake itself acts as a filter medium, and the filtrate runs clear. During filtration, the cake height (h_C) increases due to deposition of solids at the filter cake surface. An increase of h_C consequently changes the filtrate flow rate and pressure difference as the filtration proceeds. If all particles are deposited and pure liquid flows through the filter cake, the volumetric flow rate can be assumed to be constant for a given pressure difference. These basic relationships for flow through porous media were already described by Darcy in the 19th century.[15] He also found that the volumetric flow rate (Q) was inversely proportional to the height of the packed bed. Darcy's law is given in Equation 7.7, where A is the filter area, Δp is the pressure drop across the filter cake, η is the fluid viscosity, and k is the specific cake permeability:

$$Q \propto \frac{A \cdot \Delta p \cdot k}{\eta \cdot h_C} \qquad (7.7)$$

The permeability (k) can also be expressed as the reciprocal of the filter cake resistance (r_C) as introduced in Equation 7.8:

$$k = \frac{1}{r_C} \qquad (7.8)$$

The overall resistance consists of the cake resistance (r_C) and the filter medium resistance (R_M):

$$R = r_C \cdot h_C + R_M \qquad (7.9)$$

The Darcy equation holds for the following assumptions:

- Laminar flow conditions
- Homogeneous suspension, no segregation
- Newtonian fluid properties
- Filter cake properties are constant (incompressible filter cake)

Whereas laminar flow conditions are practically always met in filtration processes, fluid behavior is often of a non-Newtonian nature. A certain degree of compressibility is found in almost every filter cake.

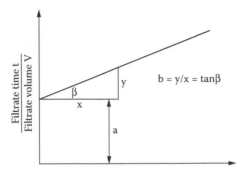

FIGURE 7.5 Schematic diagram of t/V over V diagram for determination of filter cake resistance and filter medium resistance.

Assuming that filter cake resistance increases linearly with cake height (incompressible filter cake) and that the suspension concentration remains constant during filtration time, the cake height (h_C) is proportional to the filtrate volume (V):

$$h_C = \kappa \cdot \frac{V}{A} \tag{7.10}$$

where κ is a proportionality constant that can be determined from a mass balance. For the assumptions stated before, the cake buildup rate can then be expressed as:

$$\frac{dh}{dt} = \frac{\kappa \cdot \Delta p}{\eta \cdot \left(r_C \cdot h_C + R_M \right)} \tag{7.11}$$

A useful method to determine the specific cake resistance (r_C) and medium resistance (R_M) from filtration experiments is the $t/V = f(V)$ method. The filtrate flow is measured during cake formation, and t/V is plotted over the filtrate volume (V) as showed schematically in Figure 7.5. The evaluation of cake and filter medium resistance is done by using the cake formation equation with a special arrangement of the parameters, as shown in Equation 7.12. The offset (a) and slope (b) can be evaluated from the data graph and used to determine r_C and R_M:

$$\frac{t}{V} = \left(\frac{r_C \cdot \kappa \cdot \eta}{2 \cdot A^2 \cdot \Delta p} \right) \cdot V + \frac{R_M \cdot \eta}{A \cdot \Delta p} = b \cdot V + a \tag{7.12}$$

Determining permeability and its relation to the physical characteristics of the porous medium has been the aim of many researchers. The main parameters

TABLE 7.1
Permeability Functions for Flow Through a Packed Bed of Particles

Ref.	Permeability Model	
Rumpf and Gupte (1971)	$k = \dfrac{d_p^2 \cdot \varepsilon_v^{5.5}}{5.6}$, where d_p is the particle diameter, and ε_v is the porosity	7.1
Carman and Kozeny (1938)	$k = \dfrac{\varepsilon_v^3}{K \cdot S_v^2 \cdot (1 - \varepsilon_v)^2}$, where K is the Kozeny constant (values from 4 to 5 are reported in the literature); ε_v is the porosity, and S_v is the specific surface area	7.2

found in empirical permeability models are the void volume fraction (ε_v; porosity) and particle or pore size. Some often-cited permeability functions are given in Table 7.1. The equations given in Table 7.1 hold for packed beds of rigid particles; however, as mentioned before, most filter cakes exhibit more or less compressible behavior.

Filter Cake Structure and Compressibility
Pressure has several effects on cake filtration. The desired effect of a proportional increase of filtrate flow rate with an increase in pressure drop across the cake is found only for rigid particles forming an incompressible filter cake. From experimental data, it is known that flow rate increases only slightly for compressible filter cakes, such as flocculent or other deformable precipitates. Some materials even have a critical pressure above which a further pressure increase results in an actual decrease of flow rate.

The deformation mechanisms of packed beds are shown schematically in Figure 7.6. Under a compressive load, an initial packed bed structure can be deformed irreversibly by rearrangement or disrupture of the particles. An elastic deformation of the particles leads to a reversible deformation of the packed bed. After release of the compressive load, the packed bed regains its initial structure. Particles of fibrous shape can bend under a load leading to a reversible (elastic) deformation as well.

Ruth[16] and Grace[17] developed a compression–permeability cell (CP cell) for studying the properties of compressible filter cakes. They found that the average cake resistance ($\bar{\alpha}$) can be described empirically by a power-law function, where the exponent n is a measure of compressibility:

$$\bar{\alpha} = \alpha_0 + \alpha' \cdot p_c^n \tag{7.15}$$

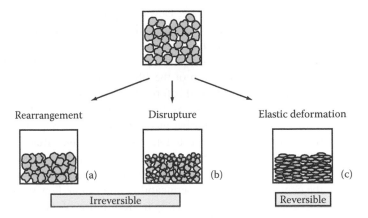

FIGURE 7.6 Deformation mechanisms of packed beds; from an initial state the packed bed can deform irreversibly (plastic) by rearrangement (a) or disrupture (b), or reversibly (elastic) (c). (From Friedmann, T., in *Institute of Food Science: Laboratory of Food Process Engineering*, ETH Zürich, Zurich, 1999. With permission.)

Here, p_c is the compressive pressure, and α' and n are empirical parameters determined from flow experiments; α_0 can be regarded as an initial (unstressed) cake resistance. The CP cell is described in the lab equipment section.

Considerable work on filtration with compressible filter cakes has been done by Tiller and coworkers. Tiller and Huang[18] proposed the following relationship for the local filter cake resistance (α_x) across the filter cake:

$$\alpha_x = \alpha_0 \cdot p_c^n, \qquad p_c > p_i \qquad (7.16)$$

$$\alpha_x = \alpha_i, \qquad p_c < p_i \qquad (7.17)$$

where n is again the cake compressibility, varying from 0 for rigid, incompressible cakes to 1.0 for highly compressible cakes. They assumed that α_x approaches a limiting value α_i at some low pressure p_i.

Many authors have followed these general approaches with various modifications. Absolute values of the compressibility constants may vary depending on the equipment used to measure compressibility. This makes it difficult to compare data from different experiments.[19]

7.2.2.2 Filtration in Centrifugal Field

Filter centrifuges are used for separation of solid–liquid systems and the ensuing washing and dewatering of the filter cake. Centrifugal acceleration is used as the driving force to separate solids from liquids. In contrast to pressure filtration, cake buildup is accomplished by sedimentation of the suspended particles. The remaining fluid passes through this filter cake, where separation of the finer particles still in suspension takes place. Dewatering can be more economic in filter centrifuges than in pressure filters due to the rapidly increasing amount of compressed air necessary (in pressure filtration) to reduce residual moisture after gas breaks through the filter cake. Unfortunately, the same centrifugal acceleration also leads to an enhanced compaction of the filter cake. Consequences of filter cake compression have been mentioned before. An alternative is the combination of pressure and centrifugal filtration, as found in the concept of hyperbaric centrifugation (explained later). Contributions to the literature regarding centrifugal filtration have been made by, for example, Stahl,[20] Leung,[21] Rushton et al.,[22] Zeitsch,[23] and Mayer.[24] The summary below is based primarily on these reviews. In centrifugal filtration, the area for flow and driving force increase with radial distance form the centrifuge axis. The permeability and filtrate flow rates through a filter cake can therefore deviate significantly from filtration of the same material under normal conditions (no centrifugation).

7.2.2.2.1 Dewatering

During the cake filtration and washing process, the entire void volume is filled with fluid; it is completely saturated. Desaturation of a filter cake takes place in the last process step: dewatering. Final residual moisture depends on the material and process parameters. To understand desaturation mechanisms, the different fluid (moisture) components have to be defined. Figure 7.7 shows the various fluid components of a filter cake according to Batel.[25]

Saturation (S) of a filter cake is defined as the ratio of fluid filled volume (V_{fl}) to the total void volume (V_v) of the filter cake:

$$S = \frac{V_{fl}}{V_v} \tag{7.18}$$

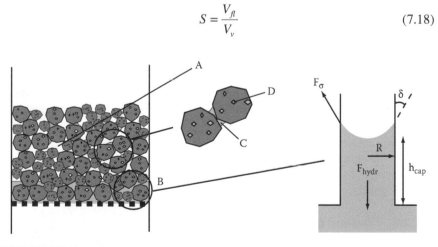

FIGURE 7.7 Fluid sources in a filter cake: (A) free moisture (bulk and surface), (B) capillary moisture, (C) pendular moisture, and (D) inherent/bound moisture. (Adapted from Batel, W., *Chemie Ingenieur Technik*, 28(5), 343–349, 1956.)

The equilibrium saturation that can be attained by centrifugal desaturation is given by the force balance of capillary and centrifugal forces. Mersmann proposed a desaturation model characterized by the dimensionless Bond number, which is the ratio of centrifugal force to capillary force.[26] With hydraulic diameter d_H, surface tension σ_i, and contact angle δ, the capillary force (F_{cap}) becomes:

$$F_{cap} \propto \sigma_i \cdot \cos \delta \cdot \pi \cdot d_H \qquad (7.19)$$

Considering the dimensionless relative centrifugal force (C) defined previously and the height of the fluid capillary (h_{cap}) in the filter cake, the centrifugal force F_Z becomes:

$$F_Z \propto \rho_{fl} \cdot C \cdot g \cdot h_{cap} \cdot \pi \cdot d_H^2 \qquad (7.20)$$

Hence, the Bond number (Bo_1) for the bulk (free) fluid component is defined as:

$$Bo_1 = \frac{\rho_{fl} \cdot C \cdot g \cdot h_c \cdot d_H}{\sigma_i \cdot \cos \delta} \qquad (7.21)$$

After desaturation of the bulk fluid, the pendular fluid remains at the particle contact points. The centrifugal force on such a pendular fluid element is assumed to be proportional to the hydraulic diameter (d_H) to the third power, as given in Equation 7.21:

$$F_Z \propto \rho_{fl} \cdot C \cdot g \cdot d_H^3 \qquad (17.22)$$

A second Bond number (Bo_2) for the pendular fluid component is therefore defined as:

$$Bo_2 = \frac{\rho_{fl} \cdot C \cdot g \cdot d_H^2}{\sigma_i \cdot \cos \delta} \qquad (17.23)$$

Desaturation of filter cakes is usually represented in so-called Bond diagrams, where equilibrium saturation S is plotted as a function of Bo_1 or Bo_2. A typical Bond diagram according to Mersmann is shown in Figure 7.8. The Bond diagram shows four stages of desaturation. At very low Bond numbers, the filter cake is completely saturated ($S = 1$). At a particular Bond number (increased C value), the capillary entrance pressure of the filter cake is overcome, and desaturation of the packed bed begins. Bulk or free fluid is dewatered at this stage. Because centrifugal force is proportional to capillary height in this domain, desaturation

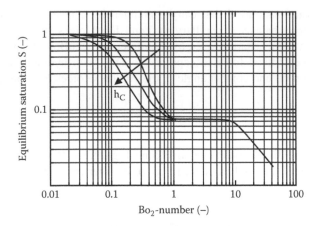

FIGURE 7.8 Bond-diagram. Equilibrium saturation S is plotted as a function of Bo_2 (pendular fluid component); with increasing filter cake height (h_C) desaturation of the bulk (free) fluid occurs at lower Bo_2 values. Desaturation of pendular moisture is independent of h_C. (Adapted from Mersmann, A., *Verfahrenstechnik*, 6(6), 203–206, 1972.)

depends on the filter cake height. Capillary rise (h_{cap}) is further reduced with increasing Bond number until the so-called Bond plateau is reached (Bo_2 1). The bulk fluid is drained at this point, but the centrifugal force is not sufficient to dewater pendular moisture. Extensive studies have shown that an increase in C value of nearly a decade is necessary to dewater the pendular fluid component in the filter cake.[24,26]

7.2.2.3 Washing

Washing the separated solids can be performed at various stages of the process, and the washing mechanisms can be differentiated as displacement and diffusion washing. During the displacement washing phase, the original mother liquor is displaced from the filter cake pores by the wash liquid. Ideally, all the mother liquor would be displaced by the wash liquid; however, this situation is almost never encountered in real systems. By increasing the viscosity of the wash liquid above the viscosity of the mother liquor, the displacement efficiency can be improved. After breakthrough of the wash liquid through the filter cake, diffusion primarily governs the additional washing. High wash ratios (wash liquid/mother liquor) are necessary to reduce the remaining mother liquor in the cake. It is not recommended to run the washing much longer than after breakthrough; if additional purity is required, a reslurrying step of the cake with wash liquid is commonly applied. Reslurrying can easily be accomplished in Nutsche-type (or Rosenmund) filters, but new developments also allow reslurry washing in filter centrifuges (e.g., Heinkel inverting filter centrifuges). It is important to consider the washing procedure during the selection and design phase of the solid–liquid separation equipment, as the flowrates and dewatering performance can change significantly after removal of the mother liquor due to new surface interactions and possible rearrangement of the filter cake.

7.2.2.4 Influence of Particle Size and Shape on Solid–Liquid Separation

It is well known that particle size and particle shape strongly influence the filtering and sedimenting behavior in the following separation step. Especially with biological products, the compressibility of the materials can further deteriorate the separation performance. Recent advances in correlating filter cake properties with particle collective characteristics (particle size distribution, particle shape, specific surface) are given by Sorrentino.[27] It is crucial to have a good understanding of the particulate system and to consider particle size and shape in the development of solid–liquid separation processes. Particle morphology changes are not always obvious and can occur by many means (pH, temperature, shear intensity, residence time in mixing tank). For industrial processes, variations in particle size, shape, or propensity for agglomeration can lead to major upsets in the downstream processes after crystallization or precipitation. Drying and bulk handling processes can also be affected. Laboratory-scale experiments are important and currently cannot be replaced by computational modeling alone. The laboratory-scale tests described in the following section are excellent tools for evaluating critical parameters for the selection, design, and optimization of batch or continuous solid–liquid separation equipment.

7.3 LAB TESTING

The selection of crystallization as the unit operation of choice must be tested at laboratory scale to avoid wasted capital investment. That is, the performance, product quality, yield, and likely processing costs must be evaluated with respect to alternate separation technologies. Further, basic crystallization process data are required to guide the scale-up, design, and operation of the industrial process.

7.3.1 Crystallization Solvent

A suitable solvent must be chosen for the crystallization. Solvent selection is determined either by convenience (e.g., the solvent that the crude material is supplied in) or by necessity after choosing from a range of possible solvents. Even from the earliest stage of development, it is recommended to choose solvents for scale-up that offer the greatest:

* Safety (e.g., flammability, toxicity, reactivity)
* Ease of regeneration or disposal
* Availability at the intended site of production
* Acceptability of traces in the final product

In the food industry, water is the primary solvent (e.g., for crystallization of sucrose, dextrose, lactose, sodium chloride). Some alcohols are occasionally used (e.g., for fructose crystallization in aqueous ethanol). Hydrophobic compounds such as flavor extracts, waxes, and oils are often processed with hydrocarbon or other organic solvents (e.g., cyclohexane). In the fine chemical, pharmaceutical, and agrochemical industries, a broader range of organic solvents is used for batch crystallization. Examples of widely used solvent classes include alcohols, hydro-carbons, esters, and chlorinated organics, among others.

The solubility in the chosen solvent determines much of the operational constraints. For example, temperature-dependent solubility suggests that cooling may be an appropriate mode of crystallization. By contrast, a weak temperature dependence on solubility suggests that an evaporative process may be superior. A highly temperature-dependent solubility can become a constraint, as it may lead to the need for additional jacketing, insulation, and temperature control of the slurry exiting the crystallizer and entering the solid–liquid separations device. This additional cost and operational sensitivity are necessary to avoid encrustation in the vessel or pipes and any change in the product properties due to fouling, bursts of nucleation, or rapid uncontrolled crystal growth.

7.3.2 SALT FORMS

Active pharmaceutical ingredients (APIs) are usually crystallized as organic salts, rather than in the free base (or free acid) form. The decision to scale-up this type of crystal is a judgment made during development. Salt forms of APIs generally exhibit greater aqueous solubility than their free base or acid equivalent forms; therefore, increased bioavailability is an advantage of drug products containing APIs as salts. The counter-ion is selected empirically from those that crystallize and which provide a stable form of the APIs, with minimized toxicity potential introduced by the counter-ion. The hydrochloride salt is the most common, although many others exist.[28]

7.3.3 FEASIBILITY TESTING CRYSTALLIZATION

When a solvent (or several candidates) has been selected, small-scale, single-batch crystallization should be conducted using a small quantity of the material. This is best conducted in a vial (10 to 100 mL). The vessel should be agitated (by shaker tray or overhead mixer, if available). Although magnetic stirring bars are commonly used at this point, doing so is discouraged because it leads to grinding of the crystals, which can produce misleading early data on crystal size distribution. If the feasibility test is completed within minutes, magnetic agitation is an acceptable compromise, but for longer periods it should be avoided.

Feasibility testing involves a simplified crystallization test, during which crystallization should be rapidly induced by the simplest and most convenient driving force (evaporation, cooling, or addition of an antisolvent, such as water or an alcohol, or by rapid titration with a reactant such as acid or alkali). The feasibility test should not be taken too literally. It is simply a first test to allow approximate purification levels to be considered; likewise, the mean crystal size of the product can be indicated within several hundred percent and the likely shape of the crystals can be noted. Both purity and particle size can be substantially improved with process development data and good engineering. In many cases, crystal shape can be manipulated, yet doing so requires more sophistication in the experimental techniques and approach.

If a system grossly fails the feasibility testing in terms of expected purity or particle size, then resources should be deployed early to screen alternative processes. For example, purification by chromatography, extraction, membrane processing, or other means could be screened. Similarly, particle formation by spray drying, granulation, extrusion, micronization, or other process could be examined as stand-alone or additional unit operations.

To reveal possible solid–liquid separation complications, the slurry from the direct strike can be subjected to a filter leaf test to determine the specific cake resistance. At this point, it is too early to establish design specifications for filtration or centrifugation equipment; however, a particularly fine product (i.e., where the product contains more than a few percent of particles around or less than 1 μm) may already suggest potential difficulties in separating the crystals from the mother liquor, such as particles breaking through the filter media into the filtrate or blinding of the filter media and a subsequent low filtrate flux.

7.3.4 VAPOR DIFFUSION CRYSTALLIZATION

Feasibility testing usually can be conducted for fine chemicals, food ingredients, and consumer products; however, for pharmaceuticals, agrochemicals, and recombinant bioproducts, often only milligram to gram quantities of material are available. For these situations, microtechniques must be used. At the smallest scale, vapor diffusion techniques can be used to crystallize microliter volumes of material. This approach is widely used for the crystallization of proteins and nucleic acids, although generally this is done to determine structure via x-ray crystallography. Figure 7.9 provides a schematic of a vapor diffusion crystallization system according to the hanging drop technique. Several microliters of the substance to be crystallized are suspended below a coverslip and mixed with several microliters of a formulation solution containing the solvents, salts, and buffers being screened for the crystallization chemistry. The drop hanging from the coverslip concentrates via vapor phase equilibrium with a bulk of the same formulation solution in a multiwell plate. Various kits are available that provide arrays of formulation solutions as well as the plates and coverslips.

After storage for days or weeks at the desired temperature, the microarray plate is observed under a microscope. Coverslips that have desirable crystals (rather than oils, amorphous solids, or other) are readily identified, which allows rapid identification of solvents, salts, and buffer ingredients that are conducive to crystallization of the target molecule. This technique was originally developed for protein crystallization to generate high-quality, single crystals for x-ray

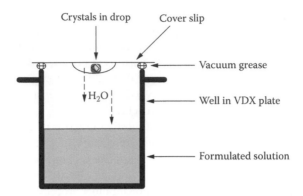

FIGURE 7.9 Vapor diffusion crystallization in the hanging drop mode. The diagram is a cross section of one well in a multiwell vapor diffusion screening plate.

diffraction studies of molecular structure; however, this same approach can be applied to proteins, nucleic acids, and small molecules that are intended to be crystallized at a larger scale when the solution chemistry has been identified. Other configurations of vapor diffusion crystallization can also be used, such as sitting drop, drop in oil, and vial in vial. Hampton Research (Aliso Viejo, CA) is one supplier of equipment and literature for vapor diffusion crystallization techniques.

7.3.5 MINIATURE FEASIBILITY CRYSTALLIZATION TESTS

For small-molecule crystallization screening (e.g., for pharmaceutical and agrochemical active ingredients), vapor diffusion techniques are used but far less widely. The broader use of organic solvents makes the disposable, plastic hardware of the hanging-drop techniques less convenient to use. In cases where only milligrams or grams of material are available, microscale feasibility tests are often conducted. Glass HPLC vials are convenient vessels for this purpose. Nitrogen or dry air blown into these vials with fine tubing allows concentration and subsequent crystallization through solvent evaporation. Antisolvent addition and cooling techniques are also used. Semi-automated, high-throughput screening techniques are emerging that allow crystallization to be assessed using smaller quantities of material and more broadly than testing by manual methods. A good review of this subject can be found in Morissette et al.[29]

7.3.6 LITER-SCALE STIRRED VESSEL CRYSTALLIZATION

Batch crystallization tests at a scale of approximately 0.25 to 5.0 L are very useful and important when conducted properly. They provide more rapid data generation than pilot testing and are less demanding on bulk material and operators. Figure 7.10 provides a diagram of a basic liter-scale batch crystallizer. Features of liter-scale crystallizers should include:

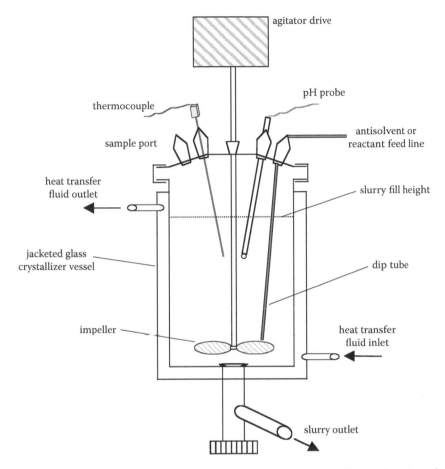

FIGURE 7.10 Diagram of a batch crystallizer apparatus suitable for liter-scale test work.

- Jacketed glass vessel (height/diameter ratio of approximately 1 to 2)
- Programmable recirculating water bath
- Glass or other suitable lid to minimize evaporation or thermal gradients
- Reflux condenser, if appropriate
- Immersed thermocouple or thermometer
- Immersed pH probe (if pH monitoring is appropriate)
- Multibladed axial flow impeller located 1/4 to 1/2 a diameter above the vessel base
- An impeller with a diameter equal to 1/3 to 1/2 that of the vessel diameter
- A dip tube located close to the impeller high shear zone, for introducing reactants or antisolvents
- A sampling port for withdrawing representative samples of slurry without interrupting the agitator, dip tube, or thermocouple

Some additional features that are desirable, if available, include:

- Bottom draw-off to dump the slurry product rapidly without disman-
  tling the crystallizer apparatus
- Boroscope to image crystals *in situ*
- Lasentec focused-beam reflectance measurement (FBRM) probe to
  measure chord length distributions *in situ*
- Attenuated total reflection–Fourier transform infrared (ATR-FTIR)
  probe to measure solution concentration *in situ* to determine supersaturation
- Raman probe to monitor the phase of the crystallizing solids *in situ*

At the liter scale, baffling is not essential. While several baffles are convenient, the intrusive dip tube, thermometer, and pH probe all perform a baffling function at this scale. Agitation should be set at the minimum revolutions per minute (rpm) required to suspend the crystals so they do not accumulate at the base of the vessel. Over-agitation is to be avoided. Small changes in rpm at the liter scale can easily lead to over-agitation, which gives rise to greater secondary nucleation.

A number of automated lab reactors are available that are very useful for crystallization tests at these scales. The LabMax® (Mettler-Toledo; Columbus, OH), Advantage Series™ (Argonaut Technologies; Redwood City, CA), and Auto-Lab™ (HEL; Hertfordshire, U.K.) reactors are three examples of such devices. Resin kettle reactor vessels without the automated peripherals (agitator, heat and cooling, peristaltic pumping, data logging, etc.) can adequately be used to achieve similar results with more manual programming and operator interven-tion. The integration of online particle sizing with automated crystallizer or reactor devices provides a useful function that improves data generation and productivity. Software can be set up to allow smarter operation, such as resetting automatic cooling, termination, and reheating for subsequent tests.

The desupersaturation profile of a liter-scale crystallization is typically of the form shown in Figure 7.11. Initially, the supersaturation is high and fairly flat, while the total surface area is still low. As crystal nucleation and growth progress, the total crystal surface area builds up, allowing the overall crystallization mass transfer to occur more rapidly. As the supersaturation is depleted, the rate of growth and overall mass transfer decreases. This continues as the batch approaches the equilibrium solubility concentration.[9]

The growth and nucleation rates from batch crystallizations can be interpreted using a variety of techniques. Misra and White[30] reported a technique that is still widely used. Their approach considers the shift of the cumulative number form of the size distribution on the size axis (for growth rate) and number axis (for nucleation rate). Thus, correlations of growth and nucleation rates with super-saturation over time allow determination of the kinetic parameters.

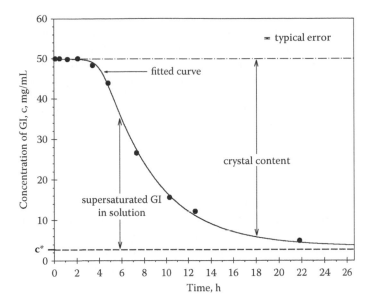

FIGURE 7.11 Example of the desupersaturation profile during crystallization of glucose isomerase enzyme.[38]

7.3.6.1 Experimental Objectives of Liter-Scale Crystallization

Much of the basic data generation for scale-up can be carried out at a scale of 1 to 5 L. These experiments should use prototypic feed liquor with an impurity profile similar to that of the intended larger scale process. Highly purified feed stocks should be used just to establish the effect of impurities (by comparison) or if difficulty is initially encountered in determining the experimental conditions (concentration, temperature, etc.). Each experiment should be designed to maximize the information gathered on the critical variable while minimizing the time required. This is a most common industrial practice, as timelines for process development often do not accommodate as thorough an investigation of each parameter as may be desired. Careful analysis of experimental data and observations is important in judging which variables should be investigated. The following section outlines areas of investigation that are most important in the absence of all data required for population-balance-based scale-up and design.

7.3.6.1.1 Process Set Points and Product Yield

Experiments seeking to maximize yield by controlling the initial concentration, final temperature, batch hold time, and solvent composition will allow determination of whether or not an adequate yield can be recovered from a single batch crystallization. Such a determination can also be inferred from solubility data by comparing the initial concentration with the solubility concentration at the final temperature. If a sufficient yield cannot be achieved through manipulation of temperature and solvent composition, the liter-scale crystallization apparatus can be used to generate multiple batches of prototypic mother liquor quantities to combine and use for testing and development of a secondary crystallization step. Because the impurities are usually of a greater concentration in a secondary crystallization, this unit process should be developed separately, as the impurities can have a substantial effect on solubility, crystallization, and product quality. The yield information from these experiments is important input for the material balance of the process flowsheet.

7.3.6.1.2 Seeding

If a sufficiently wide metastable zone width exists, the effectiveness of seeding can be tested to gain greater process control and minimize batch-to-batch variability. Seed can be added dry or as a slurry. Slurried seed is preferred, as it lacks adsorbed amorphous material from dried mother liquor and can be injected readily as a slurry, rather than dosed as a powder. Varying the seed mass charged to the crystallizer can provide control over the rate of crystallization as well as the final crystal product size distribution. This is the case when the batch is maintained within the metastable zone, so spontaneous nucleation is avoided. It is the total *surface area* of seed crystals added that primarily affects the mass transfer rate (more surface area leads to faster desupersaturation), and it is the *number* of these crystals that can provide control over the final product size (fewer seed crystals lead to a larger product). Despite this, the easiest means of adding seed crystals is on a dry *mass* or slurry *volume* basis; therefore, it is very useful for subsequent process data analysis and optimization that the seed crystals used for process development experiments be taken from a single stock supply and that this stock be well characterized in terms of the total surface area per unit mass and crystal number per unit mass. Particle size analyzers can provide this information in a crude fashion, while additional data from specific surface area analysis strengthens the confidence in the data. For initial purposes, the seed mass should be in the range of 1 to 5% of the anticipated crystal product mass. This mass can be increased or decreased substantially depending on the seed surface area, kinetics of the crystallization process, desired batch time, and desired product crystal size distribution.

7.3.6.1.3 Nucleation Onset

If seeding is not practical due to a narrow metastable zone, the sensitivity of the nucleation onset point can be determined with respect to deviations in operational parameters (e.g., temperature, concentration, reactant addition quantity). Manipulation of the early part of the cooling profile leading up to initial nucleation can be quite useful here. To gain better process control and lessen batch-to-batch variability, the objective of these experiments should be to establish a repeatable onset temperature of nucleation and minimize the nucleation rate. Gentle decreases in temperature (or slow addition of antisolvents) during the transition period from homogeneous solution to a nucleated slurry of fine crystals is the best strategy. Note that the cooling profile from natural heat loss from a vessel will provide an undesirably high rate of cooling during this transition period; therefore, management of vessel temperature through controlled cooling along with adequate mixing of the solution contents to avoid hot or cold spots is the best approach.

7.3.6.1.4 Particle Size and Filtration Performance

The effect of various parameters on the final particle size distribution should be investigated as much as time and resources permit; for example: (1) initial concentration, (2) final concentration, (3) temperature profiles (cooling rate), (4) antisolvent or reactant addition rate, (5) evaporation rate, (6) filterability of the slurries exhibiting significantly different size distributions or rheologies, (7) sensitivity of the filtrate to further crystallize with slight temperature changes, and (8) sensitivity of the crystals to breakage resulting from over-agitation.

7.3.6.1.5 Feed Liquor Impurities

Impurities present potential issues in processing by crystallization when they are not well understood. Careful experimentation is worthwhile to determine the sensitivity to various impurities. The influence of impurities can be investigated by comparison with equivalent tests using high-purity feed stocks. Impurities have the potential to affect all key aspects of the crystallization: growth and nucleation kinetics, metastability, product size distribution, yield, overall rate, crystal habit, and product quality. The solid–liquid separation properties of these slurries should also be examined closely. Variation in feed liquor impurities at the manufacturing scale is common. A filter leaf test is a good way to test for any significant effects of impurities on the potential performance of the solid–liquid separation device. Impurities commonly affect the crystal habit and can have a large impact on filterability and purity. Needle or plate-like crystals tend to stack in a way that traps impure mother liquor. These are shapes that are more conducive to forming a compressible filter cake, thus making it particularly difficult to draw the filtrate through. In attempting to improve the impurity profile of a product, it is useful to seek to understand the location of each impurity species. Washing tests with various solvents in which the impurity is soluble but the product is insoluble will allow determination of impurities adsorbed to the surface of the crystal product and those that are trapped internally within the crystal.

7.3.6.1.6 Effect of Washing

The effect of washing on purity and yield can be tested at the liter scale using various candidate wash solvents and proportions. Such testing will provide a more accurate estimate of the purity than is possible in the initial crystallization. Where greater purity is desired, the effect of multiple washes or the use of a different wash solvent is preferred over secondary crystallization processing. Purity analyses must be carried out on both the washed crystals and the separated liquor after washing. Because some product may be lost through dissolution, a total material balance approach allows logical tracking of impurities for each washing condition tested.

7.3.6.1.7 Effect of Recrystallization on Specific Impurities

Trials for one or several recrystallizations are often performed, even on a small scale. It is possible for the concentration of impurities to increase or decrease in the crystal product, so impurities must be analyzed at this point on a species basis (e.g., by HPLC, gas chromatography [GC], inductively coupled plasma [ICP] analysis), rather than at the more simplistic total impurity level. Impurities can be incorporated into the crystal product by several mechanisms: (1) being adsorbed to the exterior of the crystal through dried mother liquor (majority of impurity mass); (2) being entrained within the crystal through inclusions (trapped pockets of mother liquor); (3) by poisoning of one or more crystals faces.

7.3.6.1.8 Slurry Rheology

Observation and measurements of the rheological properties of the slurry are a useful way to ensure that mixing, suspension, and slurry pumping will be possible and not lead to a potential catastrophic event (such as not being able to resuspend a dense slurry after an agitator shutdown or a pump failure).

7.3.6.2 Purity and Mean Crystal Size

The purity of the final product can be affected by the performance of a crystallization process. As the majority of impurities are found on the outer surface of crystals, a larger mean crystal size can give a higher purity, all else being equal. Likewise, if the mean crystal size distribution is smaller, the higher surface area per unit mass gives a lower purity product. The impurities located on the exterior of the crystals are those that are reduced somewhat by washing during solid–liquid separation. Since the wash solvent may partially dissolve the product crystals, it is preferable to have larger crystals so the extent of yield loss is reduced. Since the separability of the slurry increases with larger particle sizes, a larger crystal size distribution is desirable for three reasons.

Control of the crystal size distribution is possible in batch crystallization. For bulk products, process manipulation efforts to increase the mean crystal size are more usual. In the best case of a seeded process that operates within the metastable zone, the mean crystal size can be varied substantially via the number of seed crystals charged to the vessel. Since no significant nucleation occurs, the mean crystal size of the product can be calculated directly from the batch product mass, initial seed number, and initial seed size (assuming size-independent growth and no growth rate dispersion). Fewer seed crystals (on a number basis) lead to a larger product mean crystal size. In the worst case, where the metastable zone width is too small for practical use, the mean crystal size can be effectively increased through fines destruction. In this mode of operation, the smallest crystals are classified out mid-batch (e.g., in a settling zone or cyclone), redissolved (e.g., by heating), and then brought back to the crystallizer. This is a very effective way to ensure that the larger crystals continue growing and the smaller crystals are removed. Longer batch times are required due to the recycling; however, fines destruction has proven itself a very effective approach to increasing the mean particle size as well as purity.

For intermediary cases, where seed control or fines destruction is not appropriate, manipulation of the cooling profile or antisolvent addition rate is a means of increasing the crystal size distribution. The goal should be to reduce the supersaturation, when fines are generated early in the batch, and additionally to reduce the agitation intensity. A cube law cooling profile can be tested. In this case, the rate of decrease in the batch temperature is nonlinear and not natural (via heat loss). Minimal decreases occur for most of the batch while the surface area is being generated, then a much faster decrease occurs toward the end of the batch. This allows more surface area to be generated for rapid mass transfer without large deviations in supersaturation that otherwise would induce massive nucleation.

Inclusion of mother liquor within crystals decreases the purity independent of crystal size and is a problem that is not improved through washing. Higher purity can be achieved by recrystallization, which dilutes out the impurities. Inclusions can be observed microscopically by placing the crystals into a non-solvent of the same refractive index as that of the pure crystal. The inclusions appear as dark specks, often in zones with shapes similar to those of the crystals. In some cases, inclusions may be reduced by reducing the growth rate; however, this does not simply mean increasing the hold time at the end of a batch. It requires manipulation of the cooling rate, antisolvent, or reactant addition rate so supersaturation is lower when the inclusions would otherwise be trapped into the crystal. Density measurement of crystalline products (such as sugar) is sometimes used as a batch quality control test for the level of inclusions. Generally, the greater the inclusion level, the less the density deviates from that of the pure crystalline material.

7.3.7 Selection of Appropriate Mode of Solid–Liquid Separation

If after the particle formation step (crystallization/precipitation) a dry product is desired, the appropriate mode of solid–liquid separation has to be selected. Basic filtration and sedimentation data are required to make the correct decisions regarding equipment selection and design. The following sections provide guidance on the available tools and procedures to facilitate this screening phase.

7.3.7.1 Pretreatment of Slurry

Special attention should be paid to the sample preparation prior to filtration experiments. This is an often neglected area that can easily lead to misinterpretation of lab results and false conclusions for the design and optimization of the solid–liquid separation process. The following discussion addresses what should be observed during testing.

In general, particulate systems are not stable over time. The particles can settle over time or agglomeration can change the properties significantly with regard to solid–liquid separation. Also, Ostwald ripening and dissolution of fines and the growth of larger crystals caused by temperature fluctuations during shipping have to be taken into account. If possible, the sampling and testing should occur as close to the process as possible. If shipment of samples from plant sites cannot be avoided, the consequences of shipping the slurry (shaking, changes in temperature, ongoing reactions or degradation of the components) must be evaluated carefully. If the properties change significantly, the data generated will not be valid and cannot be used to derive any useful design criteria, so every effort should be made to perform the testing onsite or to generate the material closer to the testing equipment.

In many cases filter aids or flocculants are added to the slurry to improve the solid–liquid separation performance. Testing and screening of different flocculent additions can be done by following the settling characteristics (see below) or by use of the capillary suction test. The capillary suction timer (CST) automatically measures the time for the filtrate to advance between radially separated electrodes when a fixed area of special filter paper is exposed to the suspension.

7.3.7.2 Filter Media Selection

Filter media should be tested and selected with the separation objective in mind. A clarifying filtration (to obtain clear filtrate) will have different criteria for the filter medium than a filtration application aimed at recovery of the solids, where a certain bleeding of particles through the filter medium can be tolerated. Usually, a compromise must be made between tight filter media that will give clear filtrate from the start or more open media that will allow for some bleeding of particles during the start-up phase. If the media are selected properly, the bleeding should stop after the filter cake builds up on the filter media. The performance of the filter media can be tested using lab-scale equipment as described below. The following list summarizes the requirements for a good-performing filter media:[6]

- Ability to retain particles and quickly bridge solids across the pores
- Minimum propensity to entrap solids in the pores and blind the media
- Minimum resistance to filtrate flow
- Chemical resistance
- Sufficient strength and resistance to mechanical wear
- Ability to cleanly discharge filter cake

A good selection guide is also given by Perry et al.[6] to optimize the requirements given in the list above.

Porosimetry involves the testing of filter media by measurement of the pore size distribution in the clean and used media by displacement of a wetting liquid from the pores through a steadily increasing pressure differential. This method can be very helpful to quantify the successive blinding of filter media after multiple filtration batches.

7.3.7.3   Lab-Scale Filtration Tests

7.3.7.3.1   Pressure Filters

Lab-scale filter rigs are available from various suppliers in a large variety of materials and degree of instrumentation and accessories. The basic setup for pressure filtration experiments is shown in Figure 7.12. Most commonly used systems include those offered by Cuno, Millipore, Sartorius, and Bokela. They all accommodate a variety of standard filter media as well as customized media. Also, most filter manufacturers will offer some sort of lab-scale device for screening and testing. Again, the measurements should simulate the process conditions as closely as possible, including temperature, agitation conditions, and timing. After addition of the slurry into the filter, the system is pressurized. The filtrate is collected and the filtrate mass is recorded over time. If necessary, a washing step can be simulated. The recorded data can be used to calculate specific cake resistance, permeability, and cake forming time. The final cake solids content gives an indication of the residual moisture to expect. By measuring the same material at various pressure levels, the compressibility of the material can be evaluated. The clarity of the filtrate and its solid content are characteristic of the filter medium performance. During the testing, it should be observed if the filtrate runs clear after an initial turbid phase. The bleeding of fine particles through the filter medium is typical for cake filtration and can be handled in the large-scale process by recirculation, if necessary. If the filtrate continues to contain unacceptable levels of particles, other filter media should be considered.

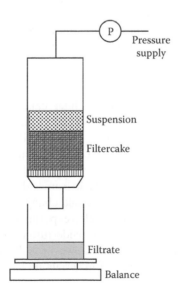

FIGURE 7.12 Lab-scale pressure filter.

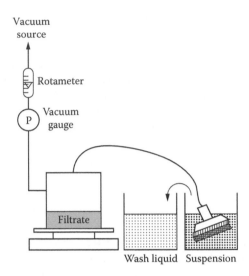

FIGURE 7.13 Lab-scale leaf filter test.

7.3.7.3.2 Leaf Filter

A very simple test is the laboratory filter leaf test. It consists of the filter media on the metal support and appropriate sealings connected to a vacuum source (Figure 7.13). The leaf filter is dipped into the agitated slurry tank to start the filtration. A cake forms on the filter leaf, and the filtrate is collected in the filtrate receiver. Data analysis is analogous to the pressure filter test described previously. Due to sedimentation effects during the filtration process, it is important for the filter leaf orientation in the slurry tank to match conditions in the large-scale process.

7.3.7.3.3 CP Cell

A common tool for the investigation of filter cake resistance and cake solids content (or porosity) with the applied pressure is the compression–permeability cell (CP cell), first proposed by Ruth.[16] The CP cell does have some disadvantages, as pointed out by Wakeman.[31] CP cell data do not show any effect of slurry concentration, which affects α and ε_v values in the actual filtration process. Inaccuracies in CP cell testing can result from sidewall friction, the time lag required to reach an equilibrium porosity, and changes in the cake characteristics over time. It is therefore desirable to develop experimental techniques and theories that obviate the need for CP cells and provide further insight into the formation and structure of filter cakes.[31] CP cells, however, still find application in classifying the behavior of various kinds of materials under stress.

7.3.7.4 Sedimentation Tests

Sedimentation performance is obviously influenced by the particle characteristics (density, size, shape) and the fluid characteristics (density, viscosity). In addition, the slurry properties will dictate the settling behavior. At very low concentrations and particle sizes larger than approximately 20 μm, settling of discrete particles will occur, obeying Stokes' law. With increasing solids concentration and smaller particle size (or broader size distribution), the settling will be in the zone settling regime, where all the particles in suspension settle at approximately the same velocity. Sedimentation behavior can be tested in various ways, the simplest being a transparent tube and observation of the settling front over time. Sampling of the slurry over time at various heights and characterizing the total solids and particle sizes provide a good indication of the settling properties and whether or not classification is occurring. It must be noted that for slurries that tend to flocculate over time the test-tube length should be adequate to simulate process-like conditions. A more detailed description of the long- and short-tube test is given in Perry et al.[6] Other means for characterizing sedimentation behavior include the measurement of light transmission or backscattering in a turbidity meter.

7.3.7.5   Lab-Scale Centrifugation Tests

7.3.7.5.1   Beaker Centrifuge
Beaker centrifuges are available in various designs for lab-scale separations or for testing and design of filter centrifuges. More sophisticated designs allow for online measurement of the dewatering kinetics. For kinetic measurements, the dewatered liquid is collected in a secondary beaker that contains a pressure sensor. The pressure values are transmitted wirelessly, and the dewatered liquid from the filter cake is calculated. The optical observation of the filter cake during process-ing (filter cake formation and dewatering) can also be helpful in understanding the compressibility of the filter cake under centrifugal acceleration. A special case of beaker centrifuge is the long-arm centrifuge, in which the beakers are located far enough from the rotational axis to neglect the influence of the varying centrifugal acceleration in the layers from the axis to outer diameter.

7.3.7.5.2   Lab Basket Centrifuge
Lab-scale basket centrifuges have proven to be useful to study filling procedures, cake buildup, and washing procedures in filter centrifuges.

7.3.7.5.3   Bottle Spin Test
A bottle spin test can be used to determine sedimentation in solid-bowl centri-fuges. In these tests, the suspension is poured into graduated centrifuge bottles, which are centrifuged at a given time and g force. The supernatant is transferred into a second graduated centrifuge bottle and centrifuged at a higher relative centrifugal force. This second centrifugation yields the volume percentage of solids remaining in the supernatant.

7.3.7.5.4   Analytical Centrifuge
Analytical centrifuges with integrated measurement of light transmission allow for rapid classification of stability and separation. This type of equipment is very

useful for small sample sizes and for screening and quality control purposes. It records the kinetics of light transmission changes for multiple samples simultaneously and can be temperature controlled.

7.3.8 Particle Size Measurement

Numerous devices are available to measure the size distributions of particles. Examples include sieving, Coulter counters, laser diffraction, time of flight, and image analysis. Particle sizing methods determine a distribution based on volume or number. Conversion from one form to the other is mathematically simple but can generate large artifacts if not performed with a good understanding of the sensitivity to analysis errors.

For spherical particles, methods of size analysis are in closer agreement; however, crystals are not spherical, and discrepancies arise because each method produces results biased by its interpretation of edges and shape. Laser diffraction devices are commonly used in crystallization studies. They rapidly measure a volume-based size distribution, which can be related through the crystal density to the solute desupersaturation to determine the overall crystallization rate. Microscopic measurement with image analysis is a useful technique because it requires only small quantities of crystals and suspending media to obtain a size distribution. Furthermore, the crystals can be readily recovered from the microscope slide after analysis; however, image analysis can be tedious and imprecise due to low number statistics. Regardless of the device chosen, it is important that the user understand the biases of the device and be critical in the interpretation of distributions generated by the instrument. Truncation at the coarse end of a volume-based distribution or at the fines end of a number-based distribution is commonly performed to avoid being misled by imprecise distribution tails.

Particle sizing sample preparation and stability are critically important in crystallization and solid–liquid separation. If a slurry is being used, the sensitivity of the slurry to changes in temperature must be considered. Samples withdrawn from a vessel at above ambient temperature are usually difficult to the sample after cooling to ambient conditions (due to further crystal growth and nucleation upon cooling). Dilution of a slurry sample into a more temperature-stable nonsolvent is one way to handle this issue. Another way is to rapidly filter samples at the process temperature, then resuspend immediately before sizing in a sizing medium that is equilibrated with the same crystal phase at the temperature of size measurement. A good reference for particle size analysis and devices is given by Allen.[32]

7.3.9 Suspension and Filtrate Rheology

Darcy's law holds for fluids with Newtonian flow behavior, which means viscosity is shear independent; however, many filtrates show non-Newtonian fluid properties. Viscosity (η) is then no longer a material constant but depends on actual flow conditions through the porous medium. A further complication is that laminar flow through porous media consists of shear and extensional flow components. It is, therefore, difficult to determine the actual stress experienced by the fluid. Changing pore structure (compressible filter cake) additionally alters flow behavior. The fluid rheology is also needed to determine the sedimentation behavior in the separation equipment and for mixing operations. Comprehensive overviews on rheology and rheometrical methods are given, for example, by Macosko[33] and Pahl et al.[34]

7.4 PILOT TESTING

Developmental testing of crystallization processes at a pilot scale is good practice for establishing the robustness of the process, determining the principal scale-up parameters, and generating prototypic product samples for evaluation. The size of equipment used for pilot testing varies substantially depending on the available equipment, the available feed stocks, and the eventual full-scale crystallizer volume. Reasonable volume-based scale-up factors from the lab to pilot level can range from 20 to 500. At high scale-up factors, greater risks and costs are associated with a failed test batch; further, it becomes more difficult to interpret unexpected results. When working with crystallizer or reactor vendors, often their equipment and facilities can be used for these tests.

For small-quantity products typical of pharmaceutical and some agrochemical activities (from hundreds of kilograms to several tonnes per batch at full scale), pilot studies address the simpler aspects of scale-up. Where time and resources permit, these should include statistical experimental designs (such as partial and full factorials designs) for understanding the impact, sensitivity, and interactions of key process variables on the reproduction of favorable lab-scale results with regard to:

- Yield
- Size distribution
- Bulk density
- Slurry product filterability
- Purity

The process variables are generally set to those of the lab-scale process, with controlled deviations in one or two driving forces for crystallization (e.g., cooling rate, antisolvent addition rate, evaporation rate). Similarly, scale-up variances in the influence of agitation rate, seed charge, and initial concentration are important to test. Where larger production batches are intended at full scale (i.e., more than several tonnes per batch), such as those common in the fine chemical and food industries, the pilot study parameters to be tested should include greater attention to mixing and heat transfer. As the diameter of the vessel increases, the agitator rpm does not scale up geometrically, neither does the relative available heat transfer surface.

Many authors have published differing bases of agitation scale-up. Some have included constant impeller tip speed, while others use a constant power input to the slurry. Power input per swept volume generally appears to be a reliable basis for agitator scale-up. More information on industrial mixing can be found in Paul et al.[35]

Reconfiguring the internals of the pilot crystallizer is an option that is useful for systems that prove to be quite sensitive to mixing. A draft tube is an excellent way to suspend a slurry for least agitator power input. Mechanically sensitive crystals (such as needle-like habits) can benefit from this feature. Further, a draft tube ensures vertical mixing of the slurry which minimizes the opportunity for solids settling. The downside of draft tubes for batch processes is that a constant fill height is required. If a draft tube is used for a batch crystallizer, windows are needed in the draft tube to avoid the settling of solids and not being able to resuspend them during filling and emptying. Hence, draft tubes are rarely used for batch crystallizers. Although draft tubes are not so appropriate for batch crystallizers, the use of baffles is a very effective way to achieve good mixing within the process vessel. Baffles disrupt the solid-body rotation and vortexing of a slurry. They lead to higher turbulence and improved suspension of crystals for a given agitator speed. The mixing requirements in batch crystallization vary according to:

- Suspension of crystals in the bulk of the vessel
- Dilution of antisolvent or reactant streams
- Dispersion of highly supersaturated regions of molecules or reactants

Insufficient macromixing leads to settling of solids, which can cause a catastrophic failure in a full-scale crystallizer, particularly if the vessel is not designed to allow this situation to be ameliorated. Although settling usually is not catastrophic in lab- and pilot-scale crystallizers, the point at which it occurs must be known so a safety factor can be built into the full-scale mixing design and operational procedures. Over-mixing at the macro scale is not wise, as it leads to breakage and greater secondary nucleation. This can shorten the batch time, but at the cost of loss of control, a lower mean crystal size, and a slurry that is more difficult to separate.

As the diameter of a jacketed vessel increases, the surface area increases proportionally to the diameter squared, while the volume to be cooled increases proportionally to the diameter cubed. Hence, at larger scales, relatively less heat transfer area is available for cooling if proportionally more heat transfer surface is not introduced. For this reason, macromixing must provide an adequate sweep of slurry past the cooling surface. If stagnant regions exist, the local slurry will be over-cooled, leading to uncontrolled nucleation, while other regions of slurry will be insufficiently cooled. This leads to a lack of crystallizer control, batch-to-batch variability, and greater dispersion of crystal size and possibly shape. So, the macromixing requirements must be balanced between maintaining reliable "just suspended" conditions and sufficient mixing to sweep the slurry over the heat transfer surface.

At shorter length scales, fast dilution of antisolvent streams (or reactants) with the bulk of the slurry must be achieved to avoid uncontrolled nucleation in the highly supersaturated zones. In the worst case (which is all too common), antisolvent or reactants are decanted or dumped above the surface of the slurry, while the impeller is at the bottom of the vessel. In the best case, the antisolvent or reactant stream is added as a high-velocity jet through one or more spargers as close as possible to the highest shear zone in the vessel (usually the impeller tip). Pilot studies can be designed to test such extremes or intermediate situations with other feed-point locations to determine the effect of this parameter on overall batch performance.

For the case of antisolvent-driven crystallization or reactive crystallization, the chemical reaction rate or dilution rate affects the kinetics sometimes more so than the crystal growth and nucleation kinetics. Therefore, poor mixing at the submicron eddy scale can cause local gradients in the driving force for crystal nucleation and growth which can lead to greater dispersion of the crystal size distribution and shape, especially the occurrence of higher aspect ratio crystal habits.

Impeller selection is important at the pilot and full scale. Mixing texts and available mixing options must be carefully considered, as geometric scale-up of impellers is not recommended. For higher aspect ratio or large tanks, multiple impellers on the same shaft can be used. Axial flow impellers are primarily used for batch crystallization. Retreating curved impellers are to be avoided for crystallization. They lack axial lift, hence over-agitation is required to suspend the crystals. Radial flow impellers are rare for batch crystallization, except for tanks with multiple agitators, in which a radial impeller is sometimes used at the base of the tank and an axial impeller higher up the shaft.

The thermal gradient (ΔT) at the heat transfer surface is a parameter that is useful to test on a pilot scale. To test the sensitivity of the crystallizing system to thermal gradients, bursts of significantly colder heat transfer fluid can be used to achieve the same cooling profile but with deliberately induced higher ΔT. If the equipment permits, a better way is to disconnect the cooling fluid from one or several zones of the heat transfer surface, which provides an acceptable ΔT and a heat transfer surface area that can be measured empirically. Higher thermal gradients tend to lead to regions of high supersaturation, uncontrolled nucleation, encrustation, and consequent variability from batch to batch.

At the pilot scale, as with lab scale, the use of a microscope (light microscope or scanning electron microscope [SEM]) to add qualitative observations is most strongly encouraged. These data can be quantified by the use of image analysis software, if warranted. Changes in crystal habit (shape) and multimodalities of the size distribution can quite strongly affect the separation as well as the final product characteristics (e.g., flowability, caking propensity, bulk density, and segregation). These subtle changes are usually not noticed in regular measurements of the size distributions.

Seeding should be tested at pilot scale if the lab study has recommended its use. The seed can be added as dry powder or preferably as a slurry. In either case, the size distribution, number, and surface area of crystals per unit mass of seed must be measured carefully and precisely prior to use of the seed. This will allow improved data interpretation and detection of significant nucleation, if it occurs. If seed is prepared as a slurry, it should be made up in a formula that minimizes the sensitivity of the seed slurry to thermal gradients and maximizes the dispersion stability of the seed (a surfactant may be useful here). This might require the use of a dispersion of seed in a solvent that is actually a diluted antisolvent for the compound under development. This approach enables one master seed tank to be used for an entire campaign or season of production. Accurate dispensing of the seed slurry is important (volumetrically or gravimetrically). The seed must be added to the vessel when the liquor becomes supersaturated but before it exceeds the secondary nucleation threshold. This should be timed according to the readout from a thermocouple immersed in the vessel near the site of where the seed should be charged. Adding seed too early leads to dissolution of the seed and no significant effect on the crystallization. Adding the seed too late (after self-nucleation) leads to the seed having little influence over control and reproducibility of the batch. Seeding, therefore, must be tested with care and precision.

Many authors describe aspects of crystallization process scale-up and design according to population balance and computational fluid dynamics modeling techniques; however, most industrial situations have a lack of these essential data to use these approaches effectively. Further, the time and costs required to measure these data for process design and establishing setpoints are often prohibitive, so an empirical approach is more commonly used. The insights of modeling and population balancing are very powerful for process optimization and post-installation improvements; yet, for initial design purposes, the empirical techniques still predominate, so no good substitute exists for well-conducted pilot studies in the development of industrial crystallization and solid–liquid separation processes.

For solid–liquid separation equipment, pilot-scale testing is also common practice and should be conducted parallel to the crystallization pilot testing whenever possible. Most vendors of solid–liquid separation devices offer pilot-scale equipment for testing. The selection of appropriate equipment will be based on previously performed lab-scale tests. The importance of good-quality data from lab-scale tests is again emphasized here. Just as an industrial-scale operation can only be designed based on valid pilot-scale data, pilot-scale tests will only be meaningful and successful if the proper equipment was selected for the pilot testing.

Whereas lab-scale tests provide guidance on the selection of appropriate modes of solid–liquid separation (filtration, sedimentation, centrifugation) and the general types of equipment necessary (e.g., basket filter centrifuge vs. inverting filter centrifuge), pilot testing confirms the appropriate selection of equipment and, more importantly, the correct design and specifications of the apparatus. Due to the complexity of solid–liquid systems and the difficulty in appropriately describing them, the use of simple tables and decision trees or mathematical modeling alone in most cases is not sufficient to select the necessary equipment.

Scaling of filtration equipment is calculated based on the mass of dry solids or filtrate volume flow rate per unit area and cycle. A safety factor of 25% on top of the calculated filter area is commonly applied.[6] Expert systems and software packages based on empirical models can be very helpful for selection and design if the required parameters can be measured reliably. A recent development in this direction was described by Nicolaou.[36] The system allows various inputs of lab-scale filtration data (cake formation, cake expression, cake washing, and cake dewatering) and yields design and scale-up data for the specified equipment.

7.5 INDUSTRIAL EQUIPMENT

7.5.1 CRYSTALLIZATION EQUIPMENT

7.5.1.1 General

A variety of commercial crystallizers is available as turn-key processes or with the flexibility to be customized. An endorsement is not given here for particular vendors, but a good starting point is the list of suppliers found at http://www.askache.com/AskaChEj.htm#crystallizers. Contracting with project engineering firms for custom design, fabrication, and installation of crystallizers is quite common and allows unique features to be readily built in (e.g., tank, agitator, drive motor, gearbox, pumps, instrumentation, controls, fines destruction systems). The materials of construction, cleaning and sanitation requirements, valves, and fittings are important aspects that must be addressed to suit particular applications. The pharmaceutical and agrochemical industries primarily use glass-lined tanks, which give a high level of chemical resistance to process solvents, reactants, and cleaning agents. By contrast, the food industry mostly avoids the use of glass in process areas and equipment. Crystallization vessels for food ingredients are typically made from a suitably acid- or corrosion-resistant grade of stainless steel (316 being the most common).

7.5.1.2 Modular Crystallization Equipment

Figure 7.14 provides a diagram of a Pfaudler glass-lined tank crystallizer. The agitator, baffle, and instrumentation are attached to the head of the vessel. These glass-lined tanks have a wide range of process temperatures and pressures which gives them the flexibility to be used in a modular fashion for batch processing. Of particular advantage to crystallization is their flexibility to be used in cooling, vacuum evaporative, and antisolvent modes. Reactions, mixing, distillation, and solvent exchanges are all commonly performed with the same type of vessels.

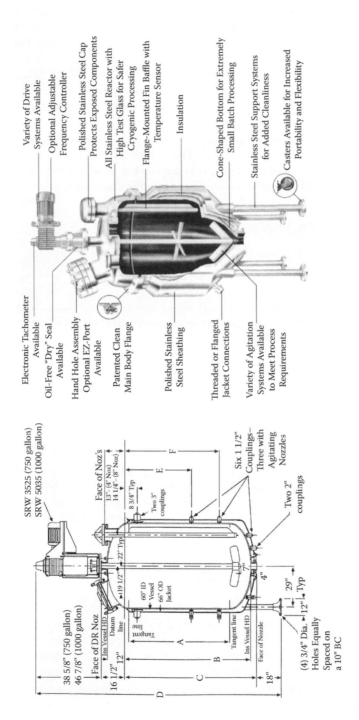

FIGURE 7.14 Examples of glass-lined tank crystallizers. (Courtesy of Pfaudler, Inc., Rochester, NY.)

7.5.1.3 Stainless Steel Agitated Tank Crystallizers

The draft tube baffled (DTB) crystallizer by Swenson Technology (Monee, IL) has been widely used for agitated-tank crystallization of inorganic and organic salts, minerals, fertilizers, and fine chemicals. This style of crystallizer has been applied to cooling, evaporative, and reactive crystallizations. Numerous other designs and suppliers are available, and stainless steel crystallizers are commonly custom built for an application rather than using "off-the-shelf" designs.

7.5.1.4 Scraped-Surface Crystallization

Scraped-surface heat exchange crystallizers are often used in the processing of waxes, some fine chemicals, ice cream, frozen concentrations (e.g., orange juice), and margarine. These devices are a tube-and-shell type of heat exchanger, with a low-speed internal agitator that runs the length of the pipe and removes nuclei from the chilled inner surface. Cooling is controlled as a profile along the length of the crystallizer using a counter-current cooling fluid in the outer region of the tube-in-tube heat exchanger. The diameter and throughput determine the residence time. Due to the small diameter and powerful agitation drives, scraped-surface crystallizers can operate at higher slurry viscosities than agitated-tank crystallizers. This lends their application to lower temperature processing (e.g., freeze crystallization of water in ice cream) and more concentrated slurries. Scraped-surface crystallizers can be configured to operate in batch or continuous mode. Cleaning and sanitation requirements (e.g., for food products) as well as throughput generally dictate the choice of process mode. Armstrong Engineering Associates (West Chester, PA) is one example of a scraped-surface crystallizer supplier.

7.5.1.5 Melt Crystallization

Sulzer Chemtech Ltd. (Houston, TX) is one supplier of melt crystallization equipment, which performs crystallization purification from the melted state of a crude organic material. Development of these processes is not covered in this chapter, but, briefly, after filling the melt crystallizer chamber with the molten feed, the contact surface is chilled to below the freezing point of the compound. The melt freezes onto the cold surface, after which the temperature is cycled above and below the melting point. This leads to sweating and drainage of lower melting impurities away from the crystalline mass. Finally, the surface temperature is raised above the melting point, and the purified product is recovered as a liquid. Naphthalene, waxes, and benzoic acid are examples of chemicals that are purified by industrial melt crystallization. A key advantage of melt crystallization is that crystallization purification can be achieved without investment in solid–liquid separation devices; however, this process delivers a liquid-phase product that would require further processing (e.g., prilling, spray chilling) to achieve a particulate form.

7.5.2 SOLID–LIQUID SEPARATION EQUIPMENT

This section compiles a list of the most common batch solid–liquid separation process equipment groups. This list has been limited to batch processing equipment, although sometimes it may be necessary to use a continuously operating separation device in combination with a batch operating crystallizer. Advantages of batch solid–liquid separation equipment include its flexibility with regard to adjustment of feed and washing and dewatering conditions, which allows their adaptation to a variety of products and process conditions. For centrifuges, discharge at a reduced speed ensures a gentle discharge of shear-sensitive products, such as fragile crystals.

7.5.2.1 Bag Filter

Bag filters are discontinuously operating filters, in which a filter bag sits in a supporting perforated basket. The filtrate is collected in the surrounding pressure tank. Bag filters are used primarily to clean up large volumes of liquid with low contamination. The filter bag is exchanged when the upper pressure loss across the bag filter is reached.

Lab test for design: Lab-scale pressure filter (with ability to filter larger amounts of feed, due to low solids concentrations usually found in bag filter applications).

7.5.2.2 Candle Filter

Candle filters (or external-cake tubular filters) can be designed for cake filtration or depth filtration. The tubes can be made of wire cloth, plastic, or metal and can be used with additional filter media or in combination with a precoat of filter aid (e.g., diatomaceous earth). Various solids discharge mechanisms, including backpulse or backflushing techniques, are available. Common designs have multiple candle elements combined in a filter housing to increase the available filter area per filter housing volume unit.

Lab test for design: Lab-scale pressure filter.

7.5.2.3 Dead-End Filtration

A discontinuous version of membrane processes is dead-end filtration. The filtration occurs through a micro- or ultrafiltration membrane until an upper pressure loss across the membrane is reached. At this point, the membrane cartridge has to be exchanged or regenerated.

Lab test for design: Lab-scale pressure filter (same comment as for bag filter might apply).

7.5.2.4 Filter Press

Characteristic of the filter press processes is the squeeze or pressing cycle after the filtration step which leads to a reduction of the filter cake pore volume and consequently a further reduction in moisture content in the pressed cake. Many designs are available, the most common being the plate and frame type. Filter presses have the advantage of relative low cost and small footprint per filter area. The systems can be operated at high pressures to yield dense, dry filter cakes with good solids-handling properties. In addition, the filter capacity can be adjusted easily by adding additional frames. Some disadvantages of filter presses include potential leaking from the filter frames and exposure of the operators to the filter contents. It is of particular importance to test for the cake release properties of the filter medium, as the discharge operation relies on the cake dropping off the vertical plates by gravity when they are opened. Some presses can apply backpulses to assist the release. Washing of the filter cake is also possible in filter presses, but variable cake density can lead to uneven wash results. Complete systems can be obtained from equipment vendors such as Netzsch (Selb, Germany).

Lab test for design: CP cell.

7.5.2.5 Monoplate Pressure Filter/Nutsche Filter

Monoplate or Nutsche pressure or vacuum filters are used for high-solids-content suspensions in cake filtration mode. This very popular filter is usually designed to have a horizontal filter plate with top-down flow direction. Sizes vary over a wide range as well as the sometimes quite sophisticated discharge mechanisms for cake discharge. Nutsche filters are very versatile for washing operations as in most designs both displacement and slurry washing can be performed in the same equipment. The term *Rosenmund filter* is sometimes used and refers to the traditional manufacturer of this type of batch filters. Nutsche filters are available as completely closed systems that allow handling of hazardous materials.

Lab test for design: Lab-scale pressure filter.

7.5.2.6 Tubular Centrifuge

Tubular centrifuges are sedimenting centrifuges with a solid bowl. They usually operate at high *g* forces ($>10,000$ *g*) compared to their continuous counterparts for sedimentation centrifugation (decanter, disc stack centrifuge). The discharge mechanism can be time consuming with tubular centrifuges and can involve manual handling. Tubular centrifuges are often used for clarification of low solids concentration streams containing fine particles.

Lab test for design: Bottle spin test.

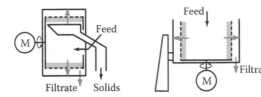

FIGURE 7.15 (Left) Peeler centrifuge and (right) vertical basket centrifuge. (From Friedmann, T., in *Institute of Food Science: Laboratory of Food Process Engineering*, ETH Zürich, Zürich, 1999. With permission.)

7.5.2.7 Peeler Centrifuge

The peeler centrifuge is a filtering centrifuge with a horizontal basket. Characteristic of this equipment is the discharge mechanism that allows it to plow through the filter cake with a knife at relatively high basket speeds (see also schematic drawing in Figure 7.15). This very efficient discharge mechanism can reduce cycle times significantly, although the high mechanical shear on the crystals has to be taken into account, and in some fragile systems the peeler centrifuge can generate too much attrition. The peeler centrifuge can also be operated with a siphon mechanism that provides an additional vacuum to enhance dewatering of the filter cake.

Lab test for design: Laboratory bucket centrifuge, lab-scale basket centrifuge.

7.5.2.8 Vertical Basket Centrifuge

The vertical basket centrifuge is usually the lowest-cost option for a filtering centrifuge. Discharge occurs through a valve in the bottom part of the screen bowl or by a peeling mechanism (Figure 7.15). Due to the vertical setup of the basket, uneven cake buildup is a potential problem. Basket centrifuges are often used for small-scale operations and frequently changing products. Also, for shear sensitive products, this centrifuge offers a gentler discharge of solids.

Lab test for design: Laboratory bucket centrifuge, lab-scale basket centrifuge.

7.5.2.9 Inverting Filter Centrifuge

In the inverting filter centrifuge, the filter cake is formed on a flexible filter cloth supported on a screen bowl. For the discharge of solids, the filter cloth is inverted by axial movement of the filter chamber (Figure 7.16 and Figure 7.17). A special feature of this centrifuge type is that the filtration bowl can be pressurized, offering an additional driving force for dewatering. This concept of hyperbaric centrifugal filtration offers many benefits for materials difficult to filter and dewater.[37]

Lab test for design: Laboratory bucket centrifuge, lab-scale basket centrifuge.

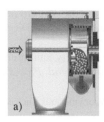 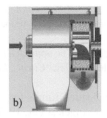  

FIGURE 7.16 Inverting filter centrifuge: (a) during filling/filtration, (b) during washing, (c) during dewatering, and (d) with inverted filter cloth during discharging of the filter cake. (Courtesy of HEINKEL AG.)

FIGURE 7.17 Open bowl of HF inverting filter centrifuge. (Image courtesy of HEINKEL AG.)

REFERENCES

1. Mullin, J.W., *Crystallization*, 3rd ed., Redwood Press, Wiltshire, 1993.
2. Myerson, A.S., *The Handbook of Industrial Crystallization*, 2nd ed., Butterworth-Heinemann, Stoneham, MA, 2001.
3. Mersmann, A., *Crystallization Technology Handbook*, 2nd ed., Marcel Dekker, New York, 2001.
4. Tavare, N.S., *Industrial Crystallization: Process Simulation, Analysis and Design*, Plenum Press, New York, 1995.
5. Davey, R.J. and Garside, J., *From Molecules to Crystallizers: An Introduction to Crystallization*, Oxford University Press, London, 2000.
6. Perry, R.H., Green, D.W., and Maloney, J.O., Eds., *Perry's Chemical Engineer's Handbook*, 7th ed., McGraw-Hill, New York, 1997.
7. McCabe, W.L., Smith, J.C., and Harriott, P., *Unit Operations of Chemical Engineering*, 6th ed., McGraw-Hill, New York, 2000.

8. Randolph, A.D. and Larson, M.A., *Theory of Particulate Processes: Analysis and Techniques of Continuous Crystallization*, 2nd ed., Academic Press, San Diego, CA, 1988.

9. Dalziel, S.M., White, E.T., and Johns, M.R., The approach to solubility equilibrium in crystallizing and dissolving systems, *Dev. Chem. Eng. Mineral Process.*, 10(5/6), 521–537, 2002.

10. Cherdrungsi, K. et al., Secondary Nucleation Limits for Lysozyme and Lactose, AIChE Annual Meeting, 295–300, Los Angeles, CA, 1997.

11. Richardson, J.F. and Zaki, W.N., The sedimentation of a suspension of uniform spheres under conditions of viscous flow, *Chem. Eng. Sci.*, 3(2), 65–73, 1954.

12. Bickert, G., Sedimentation feinster suspendierter Partikeln im Zentrifugalfeld, in *Fakultät für Chemieingenieurwesen*, Karlsruhe, Technische Hochschule Karlsruhe, 1997.

13. Rushton, A. and Griffiths, P.V.R., Filter media, in *Filtration*, Matteson, M.J., Ed., Marcel Dekker, New York, 1986, 163–200.

14. Ho, W.S.W. and Sirkar, K.K., *Membrane Handbook*, Kluwer Academic, Dordrecht, 1992.

15. Darcy, H.P.G., *Les Fontaines publiques de la ville de Dijon*, Victor Dalmont, Paris, 1856.

16. Ruth, B.F., Correlating filtration theory with industrial practice, *Ind. Eng. Chem.*, 38(6), 564–571, 1946.

17. Grace, H.P., Resistance and compressibility of filter cakes, part I, *Chem. Eng. Prog.*, 49(6), 303–319, 1953.

18. Tiller, F.M. and Huang, C.J., Theory on filtration equipment, *Ind. Eng. Chem.*, 53(7), 529–537, 1961.

19. Holdich, R.G., Prediction of solid concentration and height in a compressible filter cake, *Int. J. Mineral Process.*, 39(3-4), 157–171, 1993.

20. Stahl, W.H., *Industrie-Zentrifugen*, DrM Press, Maennedorf, 2004.

21. Leung, W.W.F., *Industrial Centrifugation Technology*, McGraw-Hill, New York, 1998.

22. Rushton, A., Ward, A.S., and Holdich, R.G., *Solid–Liquid Filtration and Separation Technology*, VCH, Weinheim, 1996.

23. Zeitsch, K., Centrifugal filtration, in *Solid–Liquid Separation*, Svarovsky, L., Ed., Butterworth & Co., London, 1990, 476–532.

24. Mayer, G., Die Beschreibung des Entfeuchtungsverhaltens von körnigen Haufwerken im Fliehkraftfeld, in *Fakultät Chemieingenieurwesen*, TU Karlsruhe, Karlsruhe, 1986.

25. Batel, W., Aufnahmevermögen körniger Stoffe für Flüssigkeiten im Hinblick auf verfahrenstechnische Prozesse, *Chemie Ingenieur Technik*, 28(5), 343–349, 1956.

26. Mersmann, A., Restflüssigkeit in Schüttungen, *Verfahrenstechnik*, 6(6), 203–206, 1972.

27. Sorrentino, J.A., *Advances in Correlating Filter Cake Properties with Particle Collective Characteristics*, Universitaet Karlsruhe (TU), Karlsruhe, 2002.

28. Stahl, P.H. and Wermuth, C.G., *Handbook of Pharmaceutical Salts: Properties, Selection and Use*, Wiley–VCH, Weinheim, Germany, 2002.

29. Morissette, S.L. et al., High-throughput crystallization: polymorphs, salts, co-crystals and solvates of pharmaceutical solids, *Adv. Drug Del. Rev.*, 56, 275–300, 2004.

30. Misra, C. and White, E.T., Kinetics of crystallization of aluminum trihydroxide from seeded caustic aluminate solutions, *Chem. Eng. Prog. Symp. Ser.*, 67, 53–65, 1971.
31. Wakeman, R.J., A numerical integration of the differential equations describing the formation of and flow in compressible filter cakes, *Trans. Inst. Chem. Eng.*, 56, 258–265, 1978.
32. Allen, T., *Particle Size Measurement*, 5th ed., Chapman & Hall, London, 1996.
33. Macosko, C.W., *Rheology: Principles, Measurements and Applications*, VCH, New York, 1993.
34. Pahl, M., Gleissle, W., and Laun, H.-M., *Praktische Rheologie der Kunststoffe und Elastomere*, VDI-Verlag, Düsseldorf, 1991.
35. Paul, E., Atiemo-Obeng, V., and Kresta, S., *Handbook of Industrial Mixing: Science and Practice*, Wiley-Interscience, Hoboken, NJ, 2003.
36. Nicolaou, J., An Innovative Computer Programme for Calculation and Design of Filter Presses, World Filtration Congress 9, New Orleans, LA, 2004.
37. Friedmann, T., Flow of non-Newtonian fluids through compressible porous media in centrifugal filtration processing, in *Institute of Food Science: Laboratory of Food Process Engineering*, ETH Zürich, Zürich, 1999.
38. Dalziel, S.M., *Crystallization Studies on Glucose Isomerase*, University of Queensland, St. Lucia, Australia, 2000.

8 Pollution Prevention for Batch Pharmaceutical and Specialty Chemical Processes

Andreas A. Linninger and Andrés Malcolm

CONTENTS

8.1 INTRODUCTION

The chemical process industries face continually increasing expectations from society and regulators to reduce releases to the environment. Until recently, end-of-pipe treatment methods have provided sufficient control to maintain regulatory limits. As environmental concerns tighten, waste treatment alone is insufficient to ensure low impact to air, water, and land. Good housekeeping and elimination of fugitive emissions have demonstrated impressive environmental impact reduction;[30,54,87] however, further emissions reductions may only be achieved through process design modification. The identification of process modifications to avoid waste generation is termed *pollution prevention* (PP). Many terms are used to describe PP activities: waste minimization, waste reduction, source reduction, waste diversion, pollution prevention, recycling, and reuse. Figure 8.1 introduces the hierarchy of pollution prevention efforts.[74] According to this pyramid, minimum waste generation at the source is superior to safe disposal. Stricter definitions of pollution prevention[96] recognize only the upper two tiers in the hierarchy: minimize generation and minimize introduction.

Due to the complex reaction chemistry, batch pharmaceutical manufacturing differs significantly from continuous chemical synthesis in the petroleum and

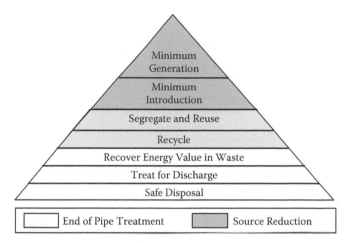

FIGURE 8.1 Pollution prevention hierarchy. (Adapted from Mulholland, K.L. and Dyer, J.A., *Pollution Prevention: Methodology, Technologies and Practices*, American Institute of Chemical Engineers, New York, 1999. With permission.)

bulk chemical industries (e.g., USEPA, 1976).[94a] In batch pharmaceutical industries, the problem of waste minimization and pollution prevention is much more challenging due to the required regulatory supervision by the Food and Drug Administration (FDA). The FDA drug approval process commands that operating conditions of each step in a pharmaceutical recipe be specified within tight limits to ensure consistent product quality. An approved manufacturing recipe cannot be altered. This approval eliminates options for ongoing process improvements. The specific challenges facing pharmaceutical manufacturers at each level of pollution prevention are discussed next:

1. *At-source waste reduction* — Drugs are manufactured in numerous stages of complex organic reaction steps transforming large organic precursor molecules into drug intermediates. In the majority of pharmaceutical processes, the final product is obtained in multiple stages, each one producing a single highly purified intermediate. The product purification and equipment cleaning necessary in each stage require additional solvents for extraction, crystallization, filter cake washing, etc. The extensive use of solvents often cannot be avoided in long and complex organic syntheses requiring sophisticated chemical reaction and separation pathways. Optimization of an existing batch recipe to reduce solvents is not permissible without expensive FDA reapproval. Because at-source waste reduction is only possible in the conceptual design phase before drug approval, it is often impractical for existing manufacturing recipes.
2. *Internal recycles opportunities* — Figure 8.2 compares the input–output structure of a multistage batch pharmaceutical process with a

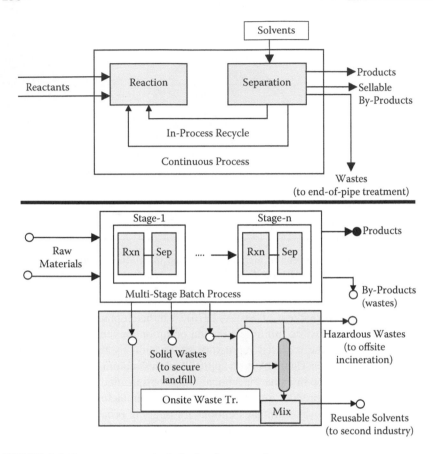

FIGURE 8.2 Input–output analysis for batch vs. continuous processes.

conventional continuous process. In batch processes, the recovered solvents cannot be recycled back into the process due to strict concerns of cross-contamination. Due to the attention to product quality, drug manufacturing generates huge amounts of waste loads per unit mass of finished product; however, recycling to other tasks of inferior demands may be possible. As an example consider organic solvents such as acetonitrile, methanol, and ethyl acetate, used for the extraction of polymer-based hydrophobic molecules during the manufacture of phenols and pesticides.[97] These used solvents can be recovered and reused for equipment washing in another drug-manufacturing campaign within the same plant. Solvents can also be recovered offsite by separation specialists using complex separations (e.g., azeotropic, extractive, reactive distillation).[33,49,51,56,91]

3. *External reuse and recovery* — If solvent-recovery is not economically viable, solvent-rich waste streams are incinerated either onsite or in an

offsite facility. An interesting alternative to waste treatment is converting effluents into marketable products. In pharmaceutical manufacturing, this approach is often quite attractive, as the byproduct streams contain expensive compounds. The complementary products are inferior in the quality spectrum (e.g., paint additives, line flush, wash solvents). Avoidance of wastes by conditioning all effluents into secondary products can be accomplished by chemical transformation and conditioning of byproducts, additional recovery steps, and blending. Details of this method, known as product-only manufacturing, can be found elsewhere.[24]

4. *End-of-pipe treatment: destructive treatment and offsite disposal* — Unavoidable residues and wastes after solvent recovery are destroyed by thermal incineration or disposed of offsite. End-of-pipe treatment of pharmaceutical and specialty chemical compounds is expensive due to the loss of material value and relatively high operating costs associated with destructive treatment. Moreover, oxidation of valuable organic precursors is not ecologically sound. Consequently, existing environmental legislation, such as the Resource Consumption and Recovery Act (RCRA), impose stringent regulations on end-of-pipe treatments of pharmaceutical wastes. The RCRA legislation also imposes a cradle-to-grave responsibility on manufacturers who are held liable even if they contract a third party to dispose their hazardous wastes. The current regulatory framework also dictates the use of specific treatment technology to reduce emissions (e.g., use of particulate scrubbers for treating flue gases from incinerators).

This chapter provides an overview of waste management options for batch manufacturing processes. Special attention is given to pharmaceuticals, although, except for FDA regulations, the discussion applies also to specialty chemical manufacturing. The relation between different regulatory models and manufacturing practices is also discussed. In Section 8.2, we briefly review standard batch process operations, sources of pollution, and typical pollution control technologies used in pharmaceutical and specialty chemical industry. Section 8.3 discusses regulatory incentives for pollution prevention. Section 8.4 demonstrates systematic approaches for identifying pollution prevention measures for batch industries. Industrial case studies illustrate the potential of computer tools for the systematic pollution prevention and waste reduction efforts. The chapter closes with conclusions and a summary.

8.2 POLLUTION SOURCES AND CONTROLS IN BATCH INDUSTRIES

This subsection provides a brief overview of batch operations in pharmaceutical and specialty chemical plants, introduces the sources of pollution in batch manufacturing, and reviews existing environmental regulations for handling these pollutants.

8.2.1 BATCH OPERATIONS IN PHARMACEUTICAL AND SPECIALTY CHEMICAL INDUSTRIES

8.2.1.1 Multipurpose Plant Operation

The pharmaceutical and specialty chemical industries utilize a fixed set of batch-type unit operations realized in standardized equipments at multipurpose plants. The standardization of operational tasks is necessary to allow entirely different synthesis routes to be manufactured in the same multifunctional equipment. Each pharmaceutical product is usually manufactured in a "campaign," during which one or more production lines are used for a few weeks or months to produce the amount necessary to satisfy the projected demand. After equipment cleaning, the same standard equipment can be used to manufacture a completely different product using other raw materials according to a different batch recipe. Most products can be synthesized in any multipurpose site. Campaigns are usually tightly scheduled, with detailed coordination extending from the procurement of raw materials to packaging and labeling of the product. The three main phases involved in the manufacturing of pharmaceutical and specialty chemicals are (1) product synthesis, (2) purification, and (3) formulation/dosage (see Figure 8.3). Table 8.1 lists the typical operations used in batch manufacturing processes in each of the stages:

- *Product synthesis* — Synthesis involves the chemical formation of a molecule with desired pharmacological properties or product qualities. New molecules can be produced through (organic) chemical reactions or extracted from natural sources (e.g., plants, minerals, or animals). Drugs and specialty chemicals are manufactured by three principal routes: (1) fermentation, (2) chemical (organic) synthesis, and (3) biological extraction steps. Sometimes fermentation is combined with organic chemical synthesis steps.
- *Fermentation/bioreaction* — Fermentation is a large-scale batch process used commonly for producing antibiotics and steroids.[90] In fermentation, the main ingredient is synthesized in mass cultures of microorganisms (biomass) growing in bioreactors. The biomass consumes the carbon substrate to synthesize the desired drug. Numerous organisms (e.g., yeast, bacteria) are used for different drugs.[11,12,55] Most bioreactions take place in aqueous phase in excess of 90% water. After completion of the fermentation step, the product mix contains the desired product suspended in water, byproducts (wax, defoamers, etc.), and biomass. All but the product are potential waste sources.
- *Chemical synthesis* — Chemical synthesis employs organic reactions without using microorganisms. Most drugs today are produced by

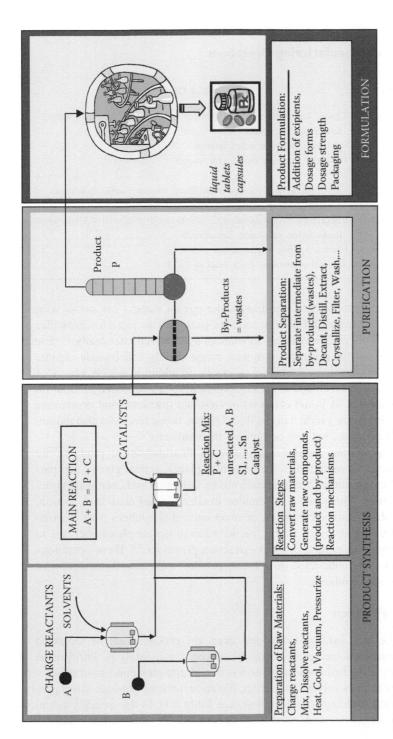

FIGURE 8.3 Production steps in batch pharmaceutical manufacturing.

TABLE 8.1
Typical Batch Manufacturing Operations

Group of Operation	Name of Operation
Material transfer	Charge, charge from recycle, transfer, transfer-through-heat exchanger, transfer intermediate
Heat transfer	Cool, heat, heat and reflux, quench
Operations on gases	Pressurize, vacuum, purge, vent, sweep
Operations on solids	Centrifuge, drying, ceramic microfilters, reverse osmosis, continuous filtration, filter in place, wash cake, crystallization
Liquid separations	Concentration (continuous/semicontinuous), distillation (batch/continuous), fractionation, extraction, decantation
Column operations	Elusion, loading, regeneration
Reactions	Age, pH adjustment, react, react in CSTR

chemical synthesis.[8,57] Cardiovascular agents, central nervous system agents, vitamins, antibiotics, and antihistamines are just a few examples of the bulk pharmaceutical substances synthesized chemically.[100] The precursors for chemical synthesis range widely and include organic and less commonly inorganic reactants. In addition, a wide variety of organic solvents listed as priority pollutants[100] are deployed for dilution of reactants, as liquid catalysts to facilitate reactions, and as solvents to stabilize the product in the liquid phase. Some reactions also require solid catalysts, but less commonly than solvents.

- *Biological and natural extraction* — Pharmaceutical products are extracted from natural sources such as plants, animal glands, and parasitic fungi through a series of volume reduction and chemical extraction steps. This process is common in allergy relief medicines, insulin, morphine, anticancer drugs, or other natural substances with pharmacological properties. Blood fractionation to isolate plasma belongs to the group of natural product extraction processes.[100] These operations are usually conducted on a much smaller scale than fermentation or chemical synthesis.

8.2.1.2 Purification

After the chemical synthesis of a new drug, the product must be separated to eliminate unreacted species, byproducts, solvents, and catalysts. Purification of drugs or specialty chemicals involves a series of complex condensed phase separation steps such as solvent extraction, filtration, reverse-osmosis, direct precipitation, and ion exchange or adsorption (see Table 8.1). In the specialty chemical industry, high purity and ultra-high specifications may require additional unit operations such as crystallization and recrystallization of the product, centrifugation for collecting and washing the product, forced-air drying to dry the intermediate product, classification for isolating various particle sizes, and finally milling to achieve the desired particle sizes and shapes. Some of the separation steps common in batch processes are described next:[68,73]

- *Solvent extraction* — Solvent extraction is often used to transfer the intermediate product from the aqueous broth into a more concentrated solution.[46,66,85] If contaminants are separated in multiple stages, the process is termed *extraction*. Splitting a liquid stream in a single stage into an aqueous and an organic phase is also known as *decantation*. After an extraction step, solvent evaporation, precipitation, batch distillation, or additional extraction steps may be necessary to further concentrate the product.

- *Ion exchange, adsorption, or chromatography* — These are separation processes transferring the product from the broth to a solid surface such as ion exchange resin, adsorptive resin, activated carbon, or the pores of a granular support. In chromatography, the solutes to be separated move through a chromatographic separator with an inert eluting fluid at different rates.[85] These sorption processes are often used for removal of trace contaminants.

- *Batch distillation* — Batch distillation is becoming increasingly important for solvent recovery in high-value, small-volume specialty and pharmaceutical industries. Batch distillation is the preferred unit operation for small-scale solvent recovery due to its flexibility. The optimal design of batch distillation dynamics as well as product-cut sequencing has been discussed extensively in the literature.[2,29,31,32,110] A single-stage distillation with recondensation in a condenser is a very common batch operation and is known as *concentration*. Vacuum distillation is used to lower the bubble points for the separation of thermally unstable products.

- *Crystallization* — Often the final drug purification step, crystallization is induced by lowering the temperature below the solubility line of the mixture (cooling). Alternatively, the addition of another solvent (cosolvent or antisolvents) may alter the liquid–solid phase equilibrium, causing crystals to form. Crystallization proceeds through three consecutive stages: (1) nucleation (i.e., initial aggregation of crystallization nuclei), (2) crystal growth (i.e., successive incorporation of molecules onto the crystal surface), and (3) Ostwald ripening (i.e., the aggregation of larger crystals at the expense of smaller crystals due to solubility differences). Control of crystallizers is an important topic for ensuring product quality.[45,69,108] In recent years, supercritical fluids have been introduced as crystallization media.[36] Crystallization from supercritical fluids may be achieved by adding gas antisolvent (GAS), by rapid expansion of supercritical solutions, or by precipitation with compressed antisolvents using either the supercritical antisolvent (SAS) process or an aerosol-spray extraction (ASE) system.

- *Filtration and drying* — The crystallizer sludge containing the product may be concentrated by boiling off the remaining solvent (e.g., vacuum concentration). The product sludge can be purified by filtration. The remaining solvent in the wet cake is removed by drying. Drying is a unit operation often employed during product purification and in formulation and dosage operations. The drying process may impact the quality of the final drug in several aspects: the polymorphic form of the active ingredient, degradation products formed due to the drying conditions, and the presence of residual solvents undesirable or above the permitted limits after drying.[48] Consequently, it is important to control the influence of all factors consistent with the desired product specifications. After the product purification (cleaning and drying), the crude drug product is shipped for formulation and dosage.

8.2.1.3 Formulation/Dosage

Formulation plants receive crude drug ingredients as raw materials with the purpose of turning them into a form and strength suitable for human use as tablets, liquids, capsules, ointments, etc. This last processing stage is often carried out at a different site than the crude drug manufacturing.[58,84] A pharmaceutical product can take a number of dosage forms (e.g., liquid, tablets, capsules, ointments, sprays, patches) and dosage strengths (e.g., 50, 100, 250, 500 mg). The final formulation incorporates substances other than the active ingredient (*excipients*) to improve the taste, stabilize active ingredients in tablet form, delay absorption of the drug into the body, or prevent bacterial growth in liquid or cream preparations. The unit operations involved in dosage and formulation include mixing and compounding.

8.2.2 POLLUTION FROM BATCH OPERATIONS

A batch pharmaceutical plant may produce hundreds of waste streams from multiple campaigns. A pilot study of a pilot pharmaceutical process indicated that, for every unit of finished drug product, 500 units of wastes were produced (see Figure 8.4).[60] Another case study led to a ratio of 3600 kg of waste to 150 kg of final product.[40] Figure 8.5 shows the discharge of different waste types in a typical pharmaceutical manufacturing plant. One can see that the maximum loads are wastewaters followed by organic solvents and solid wastes. The total volatile organic compound (VOC) air emissions from pharmaceutical manufacturing in the United States amounted to 37 kton in the year 1997.[100] Even though air emissions are smaller than the liquid and solid wastes, they have much higher environmental impact. Effluents emanating from a batch-manufacturing site can be categorized into five main groups: (1) wastewater, (2) sludges, (3) inorganic loads, (4) volatile air emissions, and (5) organic solvents. Table 8.2 lists the

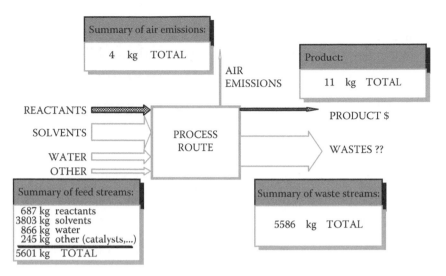

FIGURE 8.4 Overview of material streams in pharmaceutical pilot plant. (From Linninger, A.A. et al., *AIChE Symp. Ser.*, 90(303), 46–58, 1994. With permission.)

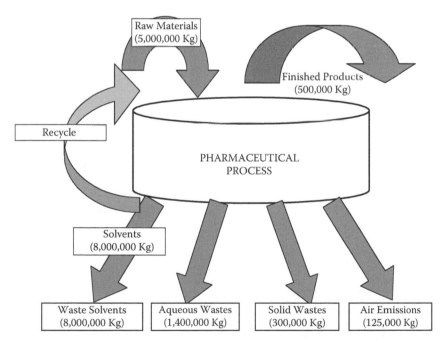

FIGURE 8.5 Typical waste loads and distribution in a synthetic organic medicine plant (USEPA, 1976).

TABLE 8.2
Pollution Sources in Pharmaceutical Manufacturing and Their Control

Waste Categories	Source	Pollution Control Options
Waste water	Fermentation, chemical synthesis and cleaning	Wastewater treatment plant
Sludges	Residual cells and waxes from fermentation; filter press from purification steps	Secured landfill
Organic solvent	Product purification operations: extraction, crystallization, and vacuum distillation	Recovery, recycle, and reuse; incineration; offsite disposal
Inorganic loads	Chemical synthesis operations	Neutralization, chemical oxidation, offsite disposal
Air emissions: VOCs	Volatile solvents used for synthesis and product purification	Scrubber, condensers, flares, heat, charge, vacuum distill, purge

sources for each of these waste categories and the common pollution prevention controls associated with these waste types. Table 8.3 correlates batch unit operations with the effluents they may produce.

8.2.2.1 Wastewater

Almost 70% of all wastewater stems from fermentation and chemical synthesis operations. Wastewater from chemical synthesis processes typically have high biological oxygen demand (BOD), chemical oxygen demand (COD), some suspended solids (traces of organic crystals), and pHs ranging from 1 to 11.[100] Major portions of the remaining 30% wastewater originate in equipment cleaning between operational steps or batches. Aqueous waste streams also result from filtrates, concentrates, wet scrubbers, and spills. Because of high organic concentration or toxicity, pretreatment may be required prior to sewer discharge. Wastewater loads also arise in purification steps (e.g., extraction, stripping, vacuum distillation). Normally, solvents used for final product purification are recovered and reused externally; however, small portions left in the aqueous phase can appear in the wastewater stream of the plant.

8.2.2.2 Organic Solvents

Organic solvents often originate in extraction, crystallization, and filter cake washing. Typical solvents used in the pharmaceutical industry include acetone, methanol, isopropanol, ethanol, tetrahydrofuran (THF), amyl alcohol, and methyl isobutyl ketone (MIBK).[101]

TABLE 8.3
Operations and Effluents in the Pharmaceutical Industry

Process	Inputs	Air Emissions	Wastewater	Residual Wastes
Chemical synthesis reactions	Solvents, catalysts, reactants	VOC emissions from reactor vents, manways, material loading and unloading, and acid gases; fugitive emissions from pumps, sample collections, valves, and tanks	Process wastewaters with spent solvents, catalysts, and reactants; pump seal waters and wet scrubber wastewater; equipment cleaning wastewater; wastewater may be high in BOD, COD, and TSS with pH of 1 to 11	Reaction residues and reactor bottom wastes
Separation	Separation and extraction solvents	VOC emissions from filtering systems that are not contained and fugitive emissions from valves, tanks, and centrifuges	Equipment cleaning wash waters, spills, leaks, and spent separation solvents	
Purification	Purification solvents	Solvent vapors from purification tanks; fugitive emissions	Equipment cleaning wash waters, spills, leaks, and spent purification solvents	
Drying	Finished active drugs or intermediates	VOC emissions from manual loading and unloading of dryers	Equipment cleaning wash waters, spills, leaks	
Natural product extraction	Plants, roots, animal tissues, extraction solvents	Solvent vapors and VOCs from extraction chemicals	Equipment cleaning wash waters and spent solvents; natural product extraction wastewaters have low BOD, COD, and TSS and pH of 6 to 8	Spent raw materials
Fermentation	Inoculum, sugars, starches, nutrients, phosphates, and fermentation solvents	Odoriferous gases, extraction solvent vapors, and particulates	Spent fermentor broth, fermentation wastewater containing sugars, starches, nutrients, etc.; wastewater tends to have high BOD, COD, and TSS and pH of 4 to 8	Waste filter cake and fermentation residues
Formulation	Active drug, binders, sugar, syrups, etc.	Tablet dusts and other particulates	Equipment cleaning wash waters (spent solvents), spills, and leaks; wash waters typically contain low levels of BOD, COD, TSS and have pH of 6 to 8	Particulates, waste packaging, rejected tablets, capsules etc.

Source: USEPA, *Profile of the Pharmaceutical Manufacturing Industry,* EPA/310-R-97-005, EPA Office of Enforcement and Compliance Assurance, Washington, DC, 1997.

8.2.2.3 Sludges

Unconverted raw materials and additives from the fermentation process make up the bulk of solid wastes (i.e., cell masses, waxes, and defoamers). The filter press materials may also generate additional solid wastes (e.g., dicolite used for ceramic microfilters). Solid effluents are generally sent to secured landfills.

8.2.2.4 Volatile Air Emissions

Most batch operations, such as charge, heat, or drying, involving VOCs result in air emissions. The free volume above the liquid level of batch equipment is saturated with vapors that include VOCs. When that saturated gas is displaced or liberated to the atmosphere, it carries with it the VOCs. The type and amount of emissions generated are dependent on the operating temperature and pressure, as well as on how the product is manufactured or formulated. Dryers belong to the largest sources of VOC emissions in batch manufacturing.[100] In addition to the loss of solvent during drying, manual loading and unloading of dryers can release solvent vapors into ambient air, especially when tray dryers are used. VOCs are also generated from reaction and separation steps via reactor vents and manways. Table 8.4 lists VOC-generating operations such as heat, charge, vacuum distillation, and purging. This table is based on the U.S. Environmental Protection Agency (USEPA) guidelines for reporting VOCs emissions from batch operations.[95] Typical controls for these emission sources include cryogenic condensers, scrubbers, carbon absorbers, and incinerators.[28]

8.2.3 Pollution Control Regulations

This section provides a brief overview of current environmental regulations in effect for pharmaceutical wastes and pollution control. Environmental regulations define what types of chemicals constitute a hazard, give guidelines on how to treat them, and in some cases prescribe specific treatment steps for emission control (see Table 8.5).

TABLE 8.4
Operations That Create Volatile Organic Carbons (VOCs)

Batch Operation	Cause of Pollution
Charging	Displacement of air saturated with VOCs
Evacuation (depressurizing)	Saturation of freeboard gas due to pressure change
Nitrogen or air sweep	Saturation of purge gas with VOCs
Heating	Increase in VOC vapor pressure due to temperature
Gas evolution	Displacement of air saturated with VOCs due to chemical reaction
Vacuum distillation	VOC generation due to compositional and pressure change

TABLE 8.5
Pollution Source and their Regulations

Wastes	Regulation	Discharge Type
Air	Maximum achievable control technology (MACT)	Hazardous air pollutants (HAPS); 188 listed
	National Ambient Air Quality Standards (NAAQs)	Criteria pollutants: NO_2, SO_2, CO, O_3, Pb, particulates
Water	National Pollution Discharge Elimination System (NPDES)	Direct discharge to sewer
	Site-specific regulations	Discharge to sanitary district
	Organic chemical plastic synthetic fibers (OCPSFs)	Industry specific discharge
Solids	Resource Consumption and Recovery Act (RCRA)	Solids, spilled liquids, and containerized liquids

8.2.3.1 Air Emissions Regulations

Both gaseous organic and inorganic compounds as well as particulates may be emitted during batch manufacturing. Some of the volatile organic compounds and inorganic gases are classified as hazardous air pollutants (HAPs) under the Clean Air Act (CAA). The CAA, originally passed in 1970 and amended in 1977 and 1990, maintains the following standards: (1) the National Ambient Air Quality Standards (NAAQs) for priority pollutants, and (2) maximum achievable control technology (MACT) standards for hazardous air pollutants. The NAAQs established six priority pollutants: ozone, lead, carbon monoxide, sulfur dioxide, nitrogen dioxide, and respirable particulate matter, as listed in Table 8.6. Primary and secondary standards were established to protect public health against adverse effects from direct exposure. The secondary standards for any adverse environmental effects are usually less stringent. MACT standards are technology-based air emission standards authorized by the CAA and designed to drastically reduce HAP emissions. The CAA amendments regulate 188 HAPs from different industrial sources; therefore, the MACT regulations are both industry specific and technology specific. The nearly 100 MACT standards are found in the Code of Federal Regulations (40 CFR Part 63). Each standard deals with a specific source category such as dry cleaners, petroleum refineries, or vegetable oil production. The pharmaceutical MACT program can be found in 40 CFR 63.1250 through 63.1261.[104]

8.2.3.2 Wastewater Regulations

The three types of discharges applicable to the pharmaceutical and specialty chemical manufacturing are (1) direct discharge, (2) indirect discharge, and (3) zero discharge. Direct discharge refers to the discharge of pollutants directly into lakes, streams, wetlands, and other surface waters. National Pollutant Discharge

TABLE 8.6
National Ambient Air Quality Standards

Pollutant	Primary Standards (Protective of Health)
Ozone	0.120 ppm (1-hour average)
Carbon monoxide	9 ppm (8-hour average)
	35 ppm (1-hour average)
Particulate matter (<10 μm)	150 mg/m^3 (24-hour average)
	50 mg/m^3 (annual arithmetic mean)
Sulfur dioxide	0.140 ppm (24-hour average)
	0.03 ppm (annual arithmetic mean)
Nitrogen dioxide	0.053 ppm (annual arithmetic mean)
Lead	1.5 mg/m^3 (arithmetic mean averaged quarterly)

Elimination System (NPDES) permits regulate all direct discharges. Indirect discharge concerns the discharge of pollutants indirectly through publicly owned treatment works (POTWs). Zero discharge prohibits the discharge of any pollutant to surface waters of the United States or to a POTW. Table 8.7 shows the final pollutant concentrations for direct and indirect discharges based on average information from several fermentation and chemical synthesis facilities.[101] The typical thresholds for indirect discharge to a local POTW are indicated in Table 8.8.

8.2.3.3   Regulations for Solid Wastes

Solid wastes (e.g., sludges from pharmaceutical and specialty chemical manufacturing processes) are treated according to the Resource Consumption and Recovery Act (RCRA) of 1976. According to RCRA regulations, spilled liquids from process pipelines and containerized liquids are considered to be solid wastes. This concept establishes a cradle-to-grave responsibility, thus preventing hazardous waste generators from delegating liability for waste disposal by a contractor. Even if the wastes are mishandled through the actions of a third party, the original

TABLE 8.7
Pollutant Concentrations in Final Effluents for Direct and Indirect Discharge

Pollutant	Final Effluent Concentration (mg/L)	
	Direct Discharge	Indirect Discharge
BOD$_5$	90	885
COD	530	2200
TSS	122	444

TABLE 8.8
Indirect Discharge: Typical Wastewater
Discharge Limits for a Local POTW
(North Shore Sanitary District, IL)

Pollutants	Daily Maximum (mg/L)	Monthly Average (mg/L)
COD	1800	1200
BOD_5	600	400
TSS	500	350
Ammonia	50	—
Nitrate	45	—
Phosphorus	20	—
Cyanides	0.3	—
Zinc	4	—
Iron	50	—
Arsenic	0.5	—
Selenium	14	—
pH	5–9	—
Sulfide (water)	0.5	—

generator is liable for improper disposal. According to the RCRA, wastes are classified either as hazardous or nonhazardous. An example of the logic for RCRA hazard qualification is depicted in Figure 8.6. RCRA hazardous wastes have very stringent permit programs for waste handling and disposal. These can be found in the Code of Federal Regulations (40 CFR 260–272).[4]

8.2.4 REGULATORY INCENTIVES FOR POLLUTION PREVENTION

The majority of existing environmental regulations follow a command-and-control type of approach. The regulators command hard thresholds on emissions and specify which treatment is considered state of the art. If a manufacturing site crosses thresholds, specific control measures are dictated (e.g., MACT rules that enforce the use of maximum available control technology). Table 8.9 provides typical command-and-control regulations.

While manufacturers operating within a command-and-control environment are usually in compliance, they have few incentives for further process improvement.[93] This type of regulation does not encourage process innovations or enhance ecological process performance beyond regulatory limits. In recent years, the USEPA has moved toward market-based regulations in order to encourage sustained pollution reduction efforts.[102] Because new market-based approaches already affect pharmaceutical and chemical manufacturers (e.g., VOC emission trading), the basic principles of emission trading and its impact on batch manufacturing are introduced next.

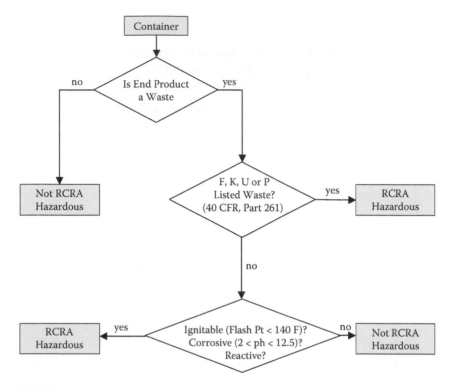

FIGURE 8.6 Logic for classification of RCRA hazardous wastes.

TABLE 8.9
Example of Command-and-Control Regulatory Framework

Command   (1) The Clean Air Act of 1990 regulates a maximum threshold of 10 ton/yr of a single hazardous air pollutant (HAP) or 25 tons per year of a combination of HAPs for any manufacturing facility.

                  (2) Threshold for SO_2 is <0.3 ppm (NAAQS, annual arithmetic mean standards).

Control   (1) Wastewater discharges covered by the National Pollution Discharge Elimination System (NPDES) must meet effluent limitations based on available technology; for toxic and nonconventional pollutants, the best available technology (BAT) must be used.

                  (2) Particulate scrubbers must be used for flue gases emanating from incinerators.

Economists have proposed a cap-and-trade regulatory model that exploits the forces of a competitive market to stimulate pollution prevention and accelerate technological improvement.[71,76,93] In a cap-and-trade market, a regulator collaborates with a group of manufacturers for the common goal of achieving acceptable levels of total emissions usually capping or reducing their total emissions. The regulator stipulates a tolerable level of emissions. This cap limits the total emissions of a region (e.g., state of Illinois) or the total emissions from a group of polluters (e.g., power generators). The regulated industries receive titles for the right to emit a certain amount of pollutant, a so-called *emission permit*. The volume of permits issued corresponds to the cap of permissible emissions. Each polluter must render permits equal to the amount of pollution they cause at the end of each period. A company can sell surplus permits on a secondary market for profit. A polluter can purchase additional permits to cover emissions in excess of its allowance if unused permits are available. This system encourages pollution reduction measures in exchange for revenues from selling surplus pollution rights. Despite the free trading, the total emissions from the regulated industry can never exceed the cap.

A limited emission trading market has existed since 1975. The first major environmental success was the sulfur dioxide emissions trading program, which cut SO_2 emissions from power generators to reduce acid rain and proved the effectiveness of the concept on a large scale. As of 2001, SO_2 trading encompassed nearly 2300 units at 1000 plants and had reduced emissions by more than 6.5 million tons compared to 1980 levels. By 2010, the program will achieve a cap of 8.95 million tons, which is 50% of the 1980 SO_2 emission levels.[103]

Market-based mechanisms for reducing greenhouse gases have achieved widespread intellectual and political support. The broad acceptance of emission trading was reflected in the Kyoto Protocol,[78] which has not yet been signed by the United States. Industrialized countries abiding by the treaty adopt legally binding commitments to cut back emissions to levels below those of 1990.

The Chicago Board of Trade administered the first SO_2 permit auction in 1993. Currently, pharmaceutical and specialty chemicals industries are regulated through the trading of VOC air emissions in the following states: Connecticut, Illinois, Florida, Maine, Michigan, New Jersey, and Virginia.[15]

8.2.5 CAP-AND-TRADE REGULATORY MODEL

The cap-and-trade model has the basic elements listed in Table 8.10: a maximum volume of emissions permits (cap), the distribution of allowances (or permits), and emission trading opportunities. The cap puts a ceiling on total tolerable emissions to be generated by all polluters. Each year the regulator allocates a maximum number of emission permit credits equal to a cap amount. This cap is firm and cannot be violated; hence, the total emission credits are designed such that the total emissions remain below acceptable limit. A critical question relates to a fair distribution of the rights to pollute. Three different permit allocations systems exist:[14,27,39]

TABLE 8.10
Elements of Emission Trading

Term	Definition
Emissions cap	A limit on the total amount of pollution that can be emitted (released) from all regulated sources (e.g., power plants); the cap is set lower than historical emissions to cause reductions in emissions
Allowance	Authorization to emit a fixed amount of a pollutant
Measurement	Accurate tracking of all emissions
Flexibility	Being able to choose how to reduce emissions, including whether to buy additional allowances from other sources that reduce emissions
Allowance trading	Buying or selling allowances on the open market
Compliance	At the end of each compliance period, each source owning at least as many allowances as its emissions

Source: USEPA, 2003 (Ref. 102).

1. The *one-time allocation* system provides for *gratis* allocations based on the history of previous emissions (e.g., SO_2 trading program).[103] Under one-time allocation or grandfathering, permits are extended to sources perpetually at the beginning of the program. A drawback of this system relates to the impossibility of permit allocations for new companies.

2. The *auctioning* approach knows no free distribution of permits; even original rights to pollute need to be acquired in an open auction (e.g., Chicago Board of Trade). Under a pure auction, all generators must pay up front for all allowances. All polluters in the program have an equal position in the auction, usually held once a year.

3. *Output-based allocation* permits are distributed according to the production outputs (e.g., power generation) from each unit in prior years. Allowance allocations are periodically redistributed, typically every 1 to 3 years. Output-based allocation also provides permits to entering companies. This aspect is an important factor in encouraging investments in newer plants and technologies.

Example

Figure 8.7 illustrates the basic concept of cap-and-trade pollution control under the assumption of only three polluters in a region (companies A, B, and C). The initial emissions of the entire region are assumed to amount to 60 tons/yr of a certain pollutant (e.g., SO_2: plant A, 20 tons/yr; Plant B, 18 tons/yr; Plant C, 22 tons/yr). The regulatory agency determines an emission reduction for the region of 25%; consequently, the regulator allocates new emission credits in accordance with the emissions history of each plant and by factoring the desired 25% reduction (e.g., plant A, 15 tons/yr; Plant B, 13 tons/yr; Plant C, 16 tons/yr). After installing the new abatement technology, company B emits 12 tons/yr, thus

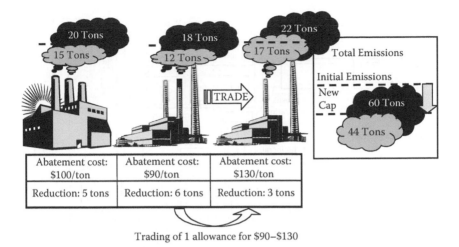

FIGURE 8.7 Emission trading between three companies (A, B, and C).

leaving 1 ton/yr of unused permits, which it can sell for profit. Company C emits 17 tons/yr, 1 ton more than the allocated amount. Company B can sell its excess permits to Company C. Although Company C emitted more than its original permit allocation, the total cap for the entire region was not surpassed (15 + 13 + 16 = 44 tons). This simplified example demonstrates the flexibility given plant managers to decide when to invest in new abatement technology while keeping emissions capped. A more detailed case study is available in Section 8.4.

8.3 IMPLEMENTATION OF POLLUTION PREVENTION

This section introduces technological options available for pollution prevention in batch industries. First, guidelines for choosing different waste management strategies are presented, and a brief survey of software tools to support pollution prevention efforts is given at the end of this section.

8.3.1 AVAILABLE POLLUTION PREVENTION TECHNOLOGIES

Figure 8.8 depicts a typical waste management facility at a pharmaceutical manufacturing site composed of a centralized incineration facility, dedicated solvent recovery unit, and wastewater treatment plant. A manufacturing site may also have its own hazardous waste landfill and tank farms for temporary storage of effluents before off-site treatment.

8.3.1.1 Solvent Recovery Plant

Spent solvents from different campaigns can be recovered in dedicated solvent recovery plants by batch or continuous distillation. Complex separations such as

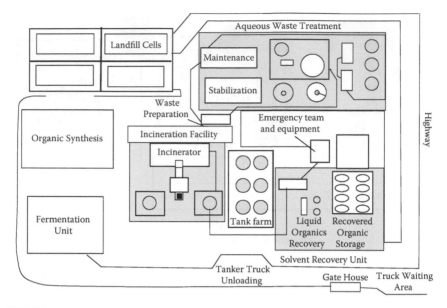

FIGURE 8.8 Illustration of typical waste management facilities at a batch manufacturing site.

extractive or azeotropic distillation are not commonly undertaken in multipurpose sites. Table 8.11 gives examples of standard industrial column sizes used for batch and continuous distillation. Recovered solvents are mostly directed to inferior uses (e.g., washing) or are sold for profit in a secondary industry (e.g., paint industry).

TABLE 8.11
Standard Sizes for Batch and Continuous Distillation Columns

Batch Column Capacity (L/batch)	Continuous Column Height (ft) × Diameter (in)
150	22 × 12
250	35 × 18
400	82 × 18
600	82 × 24
1000	200 × 36

Source: Coldberg, R., *Oral Communication*, Chemical Intermediates and Catalysis Research Lab, Kingsport, TN, 2002. With permission.

8.3.1.2 Wastewater Treatment Plant

Figure 8.9 shows a schematic of a wastewater treatment facility. Industrial wastewater treatment facilities for an entire manufacturing site can handle more than 5 million gallons of wastewater per day.[25] Organics are usually destroyed in biological reactors (anaerobic and aerobic digesters). The microorganisms operate under slightly basic conditions (pH range of 6 to 9). High pH variations in effluents entering the wastewater treatment facility are adjusted in an equalization basin. The neutralized wastewater passes through a carbonaceous aeration basin where the aerobic digestion takes place. After aerobic digestion, the biomass is typically settled in a clarifier, and part of the activated sludge is removed by filtration to maintain a stable population of microorganisms. The clean water from the clarifier is sent to a monitoring center to ensure that environmental health and safety standards are being met before discharging the water to a municipal sanitary district or sewer. Wastewater containing high concentrations of organics (up to 10 to 20%) must be treated in a high-strength equalization basin followed by anaerobic digestion. The pretreated effluents can be directed to the aerobic digester only after suitable reduction of organics. Wastewater containing nitrates may be subject to special denitrification steps.

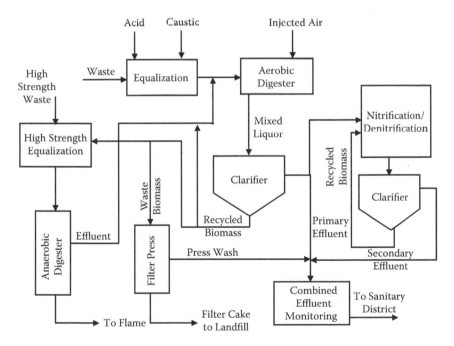

FIGURE 8.9 Typical wastewater treatment facility.

8.3.1.3 Incinerator Waste Plant

Thermal destruction of organic wastes is carried out in centralized incinerators. The thermal rating of hazardous waste incinerators varies from 30 to 120 BTU/hr (7500 to 30,000 lb/hr capacity).[26] Reduction in waste volume is attained by converting organics into flue gases (e.g., CO_2, water vapor, NO_x, SO_x, and HAPs) and solid ashes. According to federal regulations, an incinerator must destroy at least 99.99% of hazardous organics. Examples of federal performance standards for flue gases emanating from hazardous waste incinerators in the United States are listed below:[54]

- Particulates — 0.08 grains/dry standard cubic foot (180 mg/disc) corrected to 7% O_2 in the flue gas.
- Emission of HAPs — 4 lb/hr or 99% control; RCRA regulations will probably change these standards to risk-based limits for HAPs and chlorine.
- Carbon monoxide — 100 parts per million by volume as a 60-minute rolling average corrected to 7% oxygen measured on a dry basis.
- Metal emissions — Antimony, arsenic, beryllium, cadmium, chromium, lead, manganese, nickel, selenium, and mercury are listed as HAPs under Title III of the 1990 Clean Air Act amendments.
- Gaseous emission control — Incinerator off-gas may be subject to air-emissions control by scrubbing (gas washing). Wet scrubbers or gas absorbers are used to remove one or more constituents from a gas stream by treatment with a liquid. For deciding the applicability of a scrubbing pollution control step, the solubility of the contaminant in the absorbing liquid must be high. Scrubbing can treat acidic or basic compounds and VOCs with removal rates of up to a 90% of the contaminant.[100] Particulate scrubbers or electrostatic precipitators eliminate dust and heavy metal particulates. This equipment is used to capture solid particles resulting from drying intermediates or from formulation steps.

8.3.1.4 Off-Site Disposal and Landfill

Overloads from a manufacturing site can be also stored in tank farms and disposed of offsite. Off-site disposal costs depend greatly on whether the waste is solid or liquid and hazardous or not, the nature of the hazardous constituents, and even geographical location (e.g., supply and demand for offsite waste treatment capacity). Landfill is the least sustainable treatment option. Table 8.12 summarizes the typical offsite disposal costs for various waste types.

8.3.2 GUIDELINES FOR POLLUTION PREVENTION

Many companies have adopted a standardized protocol for pollution prevention. Figure 8.10 outlines a generalized three-phase pollution prevention guideline

TABLE 8.12
Typical Offsite Disposal Costs for Different Waste Types

Type of Waste	Specific Cost ($/lb)	Disposal Method
Bulk organic liquids	0.35	Incineration
Sludges (with organics)	1.15	Incineration
Sludges (with inorganics)	0.43	Stabilization and secure landfill
Solids (with organics)	0.85	Incineration
Solids (with trace organics)	0.13	Secure landfill

Source: Mulholland, K.L. and Dyer, J.A., *Pollution Prevention: Methodologies, Technologies and Practices*, AIChE, New York, 1999. With permission.

proposed by Mulholland and Dyer:[74] (1) the chartering phase, (2) the assessment phase, and (3) the implementation phase. In the chartering phase, the flowsheet of a manufacturing process is fixed. The operational steps and stream table are determined. Initial decisions on waste-reduction steps vs. end-of-pipe treatment decisions are made. The assessment phase entails the characterization of process and waste streams (hazardous, toxic, other adverse properties), the definition of pollution prevention goals (e.g., removal of SO_2 from a VOC stream, COD

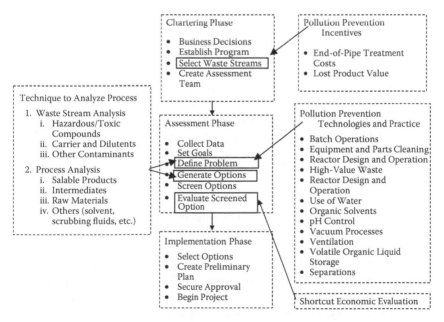

FIGURE 8.10 Pollution prevention guidelines. (Adapted from Mulholland, K.L. and Dyer, J.A., *Pollution Prevention: Methodology, Technologies and Practices*, American Institute of Chemical Engineers, New York, 1999. With permission.)

reduction in wastewater streams), and the screening of various pollution prevention options (e.g., onsite thermal incineration vs. condensation of VOC streams). In the implementation phase, the most suitable pollution prevention options are implemented. Approval for the successful pollution prevention project is secured from the company's management, and the technical implementation finally commences (e.g., equipment installation, treatment). All phases of pollution prevention require a lot of time and effort by experienced engineers. Time constraints, lack of information, and limited in-house expertise may lead to suboptimal decisions; therefore, several automatic software tools have been developed in recent years to assist plant managers in identifying PP opportunities consistently.[13] The next section discusses some of the features offered by software tools in support of pollution prevention.

8.3.3 SOFTWARE TOOLS FOR POLLUTION PREVENTION

Computer-aided pollution prevention tools enable designers to quantify environmental implications and generate suggestions for process modifications. Software tools are becoming more popular with plant managers and environmental health and safety specialists.[79] A number of commercial pollution prevention tools, primarily originating from government agencies and university research program, serve this relatively small market. Table 8.13 lists the available software and their features. A short description follows:

- Online manuals, such as the Pollution Prevention Electronic Design Guide (P2EDGE)[47] or Solutions Facilities P2 Plan Software,[88] offer guidelines for facility planning. P2EDGE is an electronic "idea notebook" of pollution prevention strategies. These tools can also be used in conjunction with other pollution prevention assessment protocols for minimizing the impact of hazardous material spills and inventories, stormwater pollution, and incorporate recycled materials into the construction of buildings and landscapes. Several EPA guidelines[99] offer step-by-step protocols and case studies for pollution prevention. OpsEnvironmental, from Environmental Software Providers, brings together data from permits and manufacturing and support operations into a single framework for calculations, reporting, and action-item notification. A collection of successful pollution prevention projects in the German and Swiss chemical and pharmaceutical industries can be found in DECHEMA.[30]
- The Waste Reduction Resource Center (WRRC)[109] database of pollution prevention articles and pamphlets belongs to the largest pollution prevention clearinghouses in the United States. The Waste Reduction Advisory System (WRAS) is a database of pollution prevention and waste minimization abstracts organized by both keyword and Standard Industrial Code (SIC); it was developed for the state of Illinois in 1987. In addition, several commercial Material Safety Data Sheets (MSDS)

TABLE 8.13
Pollution Prevention Software Summary

Application Type	Uses	Software Name
Online environmental manuals	Impact assessment	P2EDGE
		P2 Plan Software
		EPA webpage
		OpsEnvironmental
Databases of chemical impacts	Impact assessment	WRRC
		WRAS
		MSDS
Expert systems	Guide to cleaning options	SAGE
		IdeaFisher
		Expert Choice
Process simulation software	Design and development	BDK®
	of batch processes	Batchplus®
		EnviroPro
		Designer®
		Batch XL
		Batches
		Emission Master®
Decision analysis tools	Informed decision making	WAR
	for pollution prevention	EFRAT
	alternatives	DORT
		P2/FINANCE
		CPS

databases, hazardous material tracking systems, and waste tracking software products are also available for assessing pollution prevention opportunities and measuring program effectiveness.[75]

- Expert systems such as the Solvent Alternatives Guide (SAGE)[80,98] is a comprehensive guide designed to provide pollution prevention information on solvent and process alternatives for parts cleaning and degreasing. Detailed technology descriptions for each applicable option along with the rationale for their selection are presented. Other examples include creative thinking tools such as Idea Fisher,[42] which uses associative thinking for the expansion of design ideas. The main asset of Idea Fisher is that it is an "idea base" that contains thousands of questionnaires. Expert Choice[41] applies the analytical hierarchy process, a decision methodology that can be very useful in making complex decisions with incomplete information. Venkataramani et al.[107] developed the EASY expert system for the assessment of treatment options for liquid and vapor waste streams from batch pharmaceutical plants.

- Process analysis tools are state of the art for the design and development of batch processes. These simulation tools facilitate the creation of process flow diagrams and provide estimates of the expected waste quantities and compositions. Specialized tools are available as a constituent part of commercial packages:
  - Batch Plus® by Aspentech is a recipe-oriented batch process modeling environment that spans the pharmaceutical process development value chain from candidate drug selection through manufacturing.[5,6] EnviroProDesigner® from Intelligen performs material and energy balances for batch operations used in pharmaceutical and specialty chemicals and calculates the amount as well as type of waste generated.[79,83] Wastestreams are automatically classified as liquid, solid and emissions (vapor).
  - The Batch Design Kit (BDK®), available through Hyprotech/Aspentech, offers a virtual laboratory that enables process designers to experiment with conceptual batch recipes and assess the environmental impact caused by their design ideas.[60–63]
  - Batch XL, from ABB Eutech,[1] focuses on improving business operation performance by optimizing resource allocation, batch sizes and schedule.
  - Batches, from Batch Processing Technology,[7] is a versatile simulation system designed specifically for multiproduct pharmaceutical, specialty chemical and food industries. It contains accurate and easy-to-use batch operation modules.
  - Emission Master®, from Mitchell Scientific,[70] is useful for estimating HAP vent emissions for batch and continuous processes using computerized EPA models. Its process modeling environment offers computational models for filling, purging, heating, depressurization, vacuum, gas evolution, solids drying, and storage tanks considered in maximum achievable control technologies (MACTs).
- Decision-analysis tools help users evaluate the environmental and cost consequences of various pollution prevention options; they aim at achieving a better understanding of the hidden costs associated with environmental compliance. This category of software tools also assists users in evaluating the full life-cycle impacts of manufacturing processes.
- The Waste Reduction (WAR) algorithm[67] discriminates between process alternatives by performing a potential environmental impact (PEI) balance on alternative process flowsheets. The resulting global pollution index provides a measure of the environmental performance of a chemical process flowsheet. The Environmental Fate and Risk Assessment Tool (EFRAT)[86] assesses flowsheets in terms of human health and environmental damage based on an environmental risk index calculator (ERIC).

- The Design Options Ranking Tool (DORT)[94] provides the means to compare alternative process design options based on economics, environment, health, and safety as well as other user-defined criteria. The tool supports standard economic analysis such as discounted cash flow and net present value.
- The P2/FINANCE tool[92] features environmental cost-accounting principles, particularly for the evaluation of capital investment options such as process upgrades or modifications. The software estimates the environmental costs associated with a process by taking into account factors such as waste management and liability costs.

The existing methods to identify design alternatives still require expert knowledge and a significant number of skilled man-hours. The applicability of these methods, particularly during conceptual process design, can be limited due to a lack of data.[44] The next section presents a brief overview of various computer-aided methodologies for the synthesis of pollution prevention options with very scarce data.

8.3.4 COMPUTER-AIDED METHODS FOR THE SYNTHESIS OF WASTE REDUCTION OPTIONS

Computer-aided process synthesis aims at generating process alternatives automatically with little user intervention. Without claiming completeness, four categories of systematic synthesis approaches are discussed:

- Mass exchanger networks
- Batch process design with ecological considerations
- Waste reduction algorithms
- Combinatorial process synthesis (CPS)

8.3.4.1 Mass Exchanger Networks

El-Halwagi and coworkers[34,37,38,72] proposed the concept of a mass exchanger network (MEN) for the synthesis of optimal waste treatment options. A mass exchanger is any direct-contact, mass-transfer unit that employs a mass separating agent (MSA) to selectively remove pollutants from a waste stream. Successful industrial applications involve cleaning dilute systems such as wastewater and flue gases from incinerators. Pistikopoulos[50] discussed the simultaneous synthesis of reaction and separation steps using MEN modules. The IDEA framework[111] guarantees consideration of all alternative network designs and global optimality of the resulting process flowsheets. Analogous to pinch analysis for heat exchangers networks, this research has provided optimal criteria for the allocation of separation networks. Results have been obtained for the synthesis of energy-efficient distillation networks, mass exchange networks with single and multicomponent targets, optimal membrane networks, reactor networks, and power generation cycles.

8.3.4.2 Batch Process Design with Ecological Considerations

Few methods in computer-assisted process design have focused on batch process design,[3,35,52,81] batch operations,[9,10,59] or online optimization.[89] A pioneering recipe-oriented simulator was PROVAL, developed by Merck.[53] Linninger and Stephanopoulos[61,62] introduced a methodology for generation and assessment of batch processes with ecological considerations based on the concept of zero achievable pollution (ZAP) and minimum avoidable pollution (MAP). Venkatasubramanian[105,106] developed a computer-aided methodology for the automatic synthesis of batch operating procedures. Friedler's combinatorial algorithms[82,94] led to a maximum superstructure of state-task-networks based on graph theoretical considerations.

8.3.4.3 Waste Reduction Methodologies for Batch Processes

A practical software tool for ecological and economic assessment of waste treatment options has been developed by the ETH Safety and Environmental Technology Group and applied successfully to industrial problems in Switzerland.[16] The method was extended to model waste treatment selection and costing in the presence of uncertainty.[17] Simplified decision trees were used for determining possible treatment paths for each waste stream. Various uncertainty amounts and compositions gave rise to the least expensive path and most advantageous cost distribution under uncertainty. The ETH researchers have also developed a metaheuristic algorithm using stochastic optimization techniques for multi-objective design of multipurpose batch plants.[18] The ETH approach can generate the least expensive treatment and recycle options for a given waste stream. The design methodology focuses on steady-state analysis of single processes. The technology selections rules do not take into account existing site infrastructure or capacities.

8.3.4.4 Combinatorial Process Synthesis

Combinatorial process synthesis is a new flowsheet synthesis paradigm. Figure 8.11 outlines the architecture of the combinatorial process synthesis software, which generates a superstructure of all feasible recovery and treatment steps.[64,65] Superstructures for industrial problems typically encompass thousands of structurally different design policies. Optimal waste management strategies are identified via rigorous mathematical programming techniques to arrive at operating procedures offering optimal trade-offs among economic and ecological targets while still satisfying site-specific emission, logistic, and plant-specific capacity

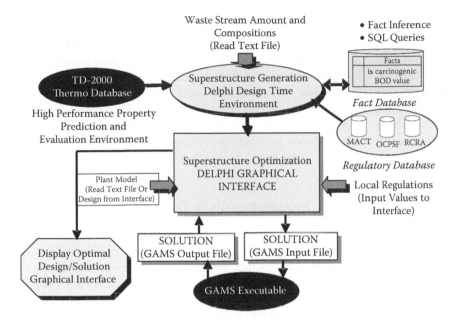

FIGURE 8.11 Architecture of the combinatorial process synthesis software.

constraints. The reasoning and superstructure optimization programming offers code generation features that automatically synthesize problem files for commercial optimization tools (e.g., GAMS, MATLAB). Comprehensive reporting functions including plant operation reports, investment reports, and facility allocation reports help plant managers analyze different solutions and their impact on the entire plant.[20] The combinatorial process synthesis was applied successfully in the following types of waste minimization synthesis problems:

- *Multi-objective synthesis and design* — Design of plant-wide waste management policies with the best trade-off between process economics and environmental impact[19]
- *Design under uncertainty* — Multi-objective design of plant-wide operations under uncertainty in the process streams[21]
- *Long-term operational and investment planning* — A predictive closed-loop control algorithm for optimal plant operation and investment decisions for the entire manufacturing site over a planning horizon of 5 to 10 years[22]
- *Source reduction/waste minimization* — Pollution prevention by optimally converting and conditioning all process streams into a changing portfolio of secondary products (e.g., product-only manufacturing paradigm)[24]

8.4 CASE STUDIES

This section discusses two hypothetical industrial case studies exemplifying computer-aided decision making for pollution prevention in the batch industries. The first example demonstrated how to arrive at operating procedures with optimal trade-offs among economic and ecological targets while satisfying site-specific emission, logistic, and plant-specific capacity constraints. The second case study quantifies the impact of various regulatory models on business decisions to illustrate that a market-based approach can lead to emission reductions at lower cost to the industry than would be achievable with existing command-and-control regulations.

8.4.1 CASE STUDY A. MULTIPERIOD WASTE TREATMENT SYNTHESIS

This design problem identifies the best recycle and treatment options for a 5-year planning horizon as well as the optimal investment schedule while increasing the production capacity of a multipurpose manufacturing site by 10%. It is expected that a command-and-control regulatory scenario currently imposing a limit of 65 ktons of CO_2 emissions per year will further reduce the threshold to 55 ktons in 5 years (see Figure 8.12).

Initial Plant Infrastructure

The inventory of our hypothetical batch-manufacturing site is depicted in Figure 8.8 and specified in Table 8.14: (1) a solvent recovery facility with 12 distillation columns ranging from 35 to 200 feet in height and 12 to 36 inches in diameter, (2) a wastewater treatment plant with a daily capacity of 5 million gallons, and (3) a centralized incineration facility with a capacity of 50 mBtu/hr. (See Reference 23 for more details)

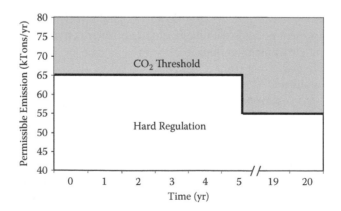

FIGURE 8.12 Command-and-control regulatory scenario for case study A.

TABLE 8.14
Initial Waste Management Facilities Infrastructure for Case Study

Site A	Equipment Type	Number of Units
Incinerator facility	Incinerator (50 mBtu/hr)	1
	Wet scrubber (10,000 cfm)	
Solvent recovery unit	200' $\leftrightarrow$ 36"	1
	82' $\leftrightarrow$ 24"	2
	82' $\leftrightarrow$ 18"	3
	82' $\leftrightarrow$ 12"	3
	35' $\leftrightarrow$ 18"	2
	35' $\leftrightarrow$ 12"	1
Wastewater plant	Clarifier, ion exchanger, equalizer (5 mgd)	1

Waste Forecast Scenarios

Market and business forecasts lead to expected plant production data for the entire planning period. From the projected production figures, one can infer the expected waste loads and compositions — the so-called waste forecast described in Table 8.15, Figure 8.14 illustrates the relationships among the business, market, and waste forecasts and displays the expected waste loads of the eight waste categories emitted at the site. The organic waste streams, W_1, W_2, and W_8, are associated with high-demand products; consequently, a sharp increment in these categories will occur (approximately 15 to 20% per year). The other streams are expected to grow slowly. The objective of the computed-aided analysis is to solve the

TABLE 8.15
Waste Composition for Case Study

Compounds	Composition (ton/yr)							
	W_1	W_2	W_3	W_4	W_5	W_6	W_7	W_8
1-Propanol	0	0	100	0	300	300	0	100
Methanol	0	40	0	16	0	0	2	0
Acetone	100	70	60		100	0	0	100
Acetonitrile	0	0	300	0	60	250	0	300
Water	0	10	0	50	0	30	22	0
Benzene	50	0	0	0	0	0	0	0
Ehtylene-dichloride	60	0	0	0	0	0	0	0
Toluene	59	0	0	0	0	0	0	0
Sodium chloride	0	0	0	25	0	0	53	0

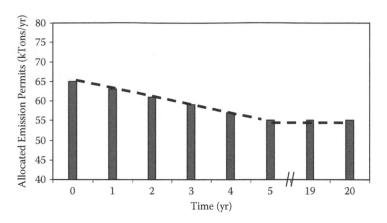

FIGURE 8.13 Permit allocation in cap-and trade regulatory scenario for case study B.

following multiperiod planning problem to find the optimal operating procedures and itemized capital investments:

> *Given* plant superstructure and the expected waste loads...
> *Find* the yearly operating policies (recycle and treatment options) and necessary investments (new recycle units or treatment equipment) at...
> *Minimum* total annualized cost...
> *Subject to* the plant capacity and regulatory limits (thresholds).

The exact formulation of this complex mathematical program is discussed in the literature (Linninger,[23] 2004b). Figure 8.15 displays a portion of the superstructure of feasible recovery and treatment steps for four of the eight waste streams (W_1, W_2, W_3, and W_4) automatically generated with combinatorial process synthesis methodology. The superstructure embeds 69 distinct treatment and recovery options with two to eight alternative treatment paths; therefore, this partial superstructure implicitly embeds a total of $8 \times 4 \times 5 \times 2 = 320$ different waste treatment alternatives. A treatment policy can only be implemented at a particular manufacturing site if the plant offers free capacity of the required equipment or plants (e.g., incinerators, distillation columns). The existing plant capacity can also be augmented by purchasing additional units (investments).

Due to a hard limit on CO_2 emissions, the optimal strategy approach (π_1) combines solvent recycle steps and incineration. The operational steps chosen in policy π_1 are depicted in gray in Figure 8.15. For example, for waste stream W_4 a liquid mixture of water, methanol, and sodium chloride is directed to a biological wastewater treatment step, T70. The final emissions are (1) cleaned wastewater (S55) and (2) off-gas (S54). S55 is discharged into the sewer (T72). S54 containing CO_2 and H_2O from the digestion of organics in the biomass is liberated into the atmosphere. In year three, the optimal treatment strategy swaps to a recycle policy (π_3), which deploys even more energy-intensive solvent recovery steps. This

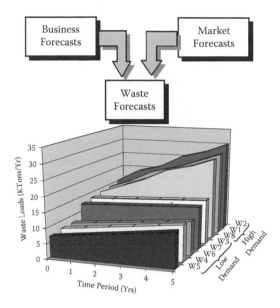

FIGURE 8.14 Mean waste loads from business and market forecasts and expected waste forecasts.

transition to more difficult separation is necessary to avoid incineration steps, thus satisfying the upcoming CO_2 threshold of 55 ktons/yr. The new, more expensive solvent recycle tasks are made possible by capital investments to upgrade the solvent recovery plant, as indicated in Table 8.16.

TABLE 8.16
Optimal Operating and Capital Costs under Different Regulatory Scenarios

	Command-and-Control Scenario		Cap-and-Trade Scenario		
Period	Operating Policy[a]	Capital Cost (M$)	Operating Policy[a]	Capital Cost (M$)	Emission Trading Cost (M$)
Year 0	π_0	—	π_0	—	—
Year 1	π_1	2.267	π_1	2.267	−1.54
Year 2	π_1	3.596	π_1	3.596	−0.35
Year 3	π_3	4.550	π_2	0.400	0.48
Year 4	π_3	0.400	π_2	—	1.38
Year 5	π_3	—	π_5	4.882	−3.16
Net present cost		88.897		81.821	

[a] Each policy (p_1 to p_5) indicates different flowsheets.

8.4.2   Case Study B. The Impact of Regulations on Manufacturing Practices

In earlier sections, we pointed out the lack of incentives for technological inno-vations in the command-and-control regulatory model. We also presented the benefits of the more innovative emission trading model. We now demonstrate the ability of different regulatory systems to induce pollution prevention and waste reduction efforts in the industry. Moreover, the case study also outlines a system-atic approach to assess the costs associated with improved environmental perfor-mance under any regulatory framework. Under a cap-and-trade regulatory sce-nario, the same company presented in case study A is allowed to participate in CO_2 trading with other firms. It is assumed that the site will receive 65 ktons of CO_2 emission credits initially (e.g., year 0). In order to achieve a reduction in output, permit allocation is assumed to decrease by 4% per year for the next 5 years, is depicted in Figure 8.12.

The optimal plant management strategy under the emission trading scenario shows that it is advisable to change the operating policy from π_0 to π_1 in year one. This policy shift is driven by the benefits associated with selling unused permits from year 0. From year three onward, a less expensive policy (π_2) avoids difficult separations in favor of inexpensive self-sustained incineration. The use of incineration, although inexpensive, requires more than the allocated CO_2 per-mits; hence, additional permits must be purchased on the market, as shown in Figure 8.17. In this simple example, the CO_2 permit price is assumed to be unaffected by the demand. The total cost for purchasing the necessary CO_2 credits is \$1.8 million. In year five, additional investments make possible a switch to high-volume solvent recovery, leading to long-term operations with low CO_2 output. This new low-polluting policy, π_5, profits from the benefits of selling the unused CO_2 permits on the market, as shown in Table 8.17.

Comparison of Regulatory Models

The optimal strategies under both regulatory models have similar investment schedules for the first two years; however, the cap-and-trade strategy allows the plant manager to decide when to invest in state-of-the-art pollution abatement equipment in contrast to the command-and-control strategy that commands invest-ment in a specific technology at a particular time (see Table 8.17). Flexibility with regard to the timing and size of investment decisions in accordance with ecological considerations as well as production plans is missing in a command-and-control regulatory scenario. Figure 8.16 plots the annual CO_2 emission from the command-and-control strategy, which always operates below its hard upper bound. Figure 8.17 depicts the CO_2 emissions for the cap-and-trade strategy which temporarily surpass the permit allocation in years three and four before adoption of a very clean strategy in year five. Despite the additional permits, under the cap-and-trade scenario the total cumulative CO_2 emissions over the 20-year pro-jected horizon are only 580 ktons, which is 63% lower than before induction of

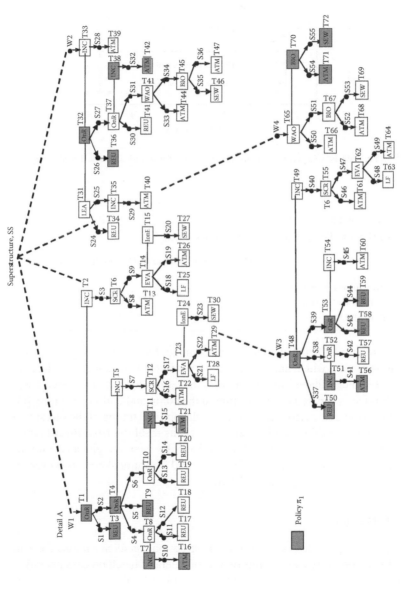

FIGURE 8.15 Superstructure of four waste streams. OnR, onsite recycle; INC, incineration; REU, reuse; EVA, evaporation; IonE, ion exchange; SCR, scrubber; WAO, wet air oxidation; LEA, leaching; LF, landfill; ATM, atmosphere; SEW, sewer.

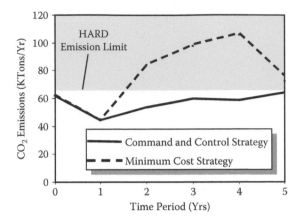

FIGURE 8.16 Yearly CO_2 emissions in the optimal command-and-control strategy.

TABLE 8.17
Total CO_2 Emissions for Strategies S_0 to S_2

Strategy	CO_2 Emissions (Ktons)	CO_2 Reduction (%)
Minimum-cost strategy (S_0)	1549	—
Command-and-control (S_1)	738	52
Cap-and-trade (S_2)	580	63

the new regulations. On the other hand, the command-and-control regulation leads to 735 ktons of CO_2, which amounts to only a 52% emission reduction. Moreover, the optimal strategy under emission trading has a lower total annualized cost of $81.8 million as compared to $89 million in the command-and-control (see Table 8.16). This better cost performance is due to the optimal timing of the investments and the revenue of selling surplus emission credits resulting from investments in technology. The data of Figure 8.17 and Table 8.16 show the globally minimum cost strategy, with the lowest possible capital and operational cost; however, that strategy causes very high pollution levels.

8.5 SUMMARY

This chapter has provided an overview of waste management options for batch manufacturing processes and a brief review of standard batch process operations, sources of pollution, and typical pollution control technologies used in pharmaceutical and specialty chemical industries. The presentation included a discussion of commonly used batch operations and the types of wastes associated with them. Existing types of environmental regulation were summarized, and a brief introduction to the novel cap-and-trade regulatory framework was provided. Available

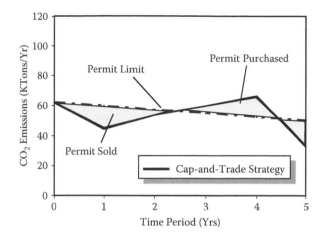

FIGURE 8.17 Yearly CO_2 emissions in the optimal cap-and-trade strategy.

pollution prevention techniques were explained, and systematic approaches for identifying pollution prevention measures for the batch industries were presented. Computer-aided design methodologies can aid plant managers in choosing optimal operating procedures and scheduling necessary investments for optimal pollution prevention strategies in anticipation of regulatory changes. Under the simplified assumptions of the case studies (e.g., arbitrary permit costs, no price flexibility), a cap-and-trade regulatory system achieved a higher level of emission reduction at lower cost. The quantitative analysis revealed that the flexibility of the emission trading model benefited plant managers in two aspects: (1) they can optimally time investments in accordance with their production plans, and (2) they can receive ongoing benefits from technological improvements by selling surplus permits. In conjunction, these two advantages may encourage the introduction of pollution prevention efforts in industry which are missing in the current command-and-control environmental regulations.

8.6 SUMMARY OF POLLUTION PREVENTION SOFTWARE

Online Manuals

- Pollution Prevention Electronic Design Guide (P2EDGE), available at http://www.ornl.gov/adm/ornlp2/p2edge.htm, offers a tool set to help engineers and designers incorporate pollution prevention strategies during the design of new products, processes, and facilities in order to reduce life-cycle costs and increase materials and energy efficiency.
- Solutions Facilities P2 Plan Software, available at http://www.env-sol.com/, specializes in providing public-domain documents (regulatory lists, manuals, etc.) on CD-ROM.

- OpsEnvironmental, from ESP (http://www.esp-net.com/ops/index. htm), brings together data from permits, manufacturing, and support operations into a single framework for calculations, reporting, and action-item notification.

Databases

- The WRRC (Waste Reduction Resource Center; http://wrrc.p2pays. org/) database of pollution prevention articles, pamphlets, and other documents is one of the largest pollution prevention clearinghouses in the U.S available at United States.
- WARS (Waste Reduction Advisory System) is a database of pollution prevention and waste minimization article abstracts, organized by both keyword and Standard Industrial Code (SIC); it was developed for the state of Illinois in 1987.
- MSDS databases, hazardous material tracking systems, and waste tracking software products are also available for assessing pollution prevention opportunities and measuring program effectiveness at http://www.ilpi.com/msds/.

Expert Systems

- The Solvent Alternatives Guide (SAGE; http://clean.rti.org/) provides an interactive interface that leads users through a series of questions that help the system narrow down the cleaning options based on a wide variety of part- and process-specific issues.
- IdeaFisher™ (http://www.ideacenter.com/) uses associative thinking for the expansion of ideas. The main feature of the software is an idea base that contains thousands of questionnaires.
- ExpertChoice[41] (http://www.expertchoice.com) applies the analytical hierarchy process (HAP), a decision methodology that can be very useful in making complex decisions with incomplete information.

Process Simulation Software Tools

- The Batch Design Kit (BDK), available through Hyprotech/Aspentech (http://www.hyprotech.com/bdk/default.asp) offers a virtual laboratory that enables process designers to implement their ideas and assess the environmental impact caused by their designs.
- Batch Plus by Aspentech (http://www.aspentech.com) is a recipe-oriented batch process modeling environment that spans the pharmaceutical process development value chain from candidate drug selection through manufacturing.
- EnviroProDesigner from Intelligen (http://www.intelligen.com/environmental.shtml) performs material and energy balances of integrated manufacturing facilities.

- Batch XL from ABB Eutech (http://www.abb.com/) focuses on improving business operation performance by optimizing resource allocation, batch sizes, and schedule.
- Batches, from Batch Processing Technology (http://www.bptech.com/), is a versatile simulation system designed specifically for the multiproduct pharmaceutical, specialty chemical, and food industries that contains rigorous, accurate, and easy-to-use operation modules.
- Emission Master, from Mitchell Scientific (http://www.mitchellscientific.com/EmissionMaster.htm), is useful for estimating HAP vent emissions for batch and continuous processes using computerized EPA . The process modeling environment offers such computational models as filling, purging, heating, depressurization, vacuum, gas evolution, solids drying, and storage tanks considered in the MACT standard.

Decision-Analysis Tools

- The waste reduction (WAR) algorithm (http://www.epa.gov/ord/NRMRL/std/sab/sim_war.htm) discriminates between process alternatives by performing a potential environmental impact (PEI) balance on alternative process flowsheets. The resulting global pollution index provides a measure of the environmental performance of each process flowsheet.
- The Environmental Fate and Risk Assessment Tool (EFRAT) (http://es.epa.gov/ncer_abstracts/centers/cencitt/year3/process/shonn2.html) discriminates between flowsheets based on an environmental risk index calculator (ERIC), which is indicative of human health and environmental damage, and on fate and transport characteristics.
- The Design Options Ranking Tool (DORT) (http://cpas.mtu.edu/tools/d0014/index.htm) provides a means to compare design options based on economics, environment, health, and safety as well as other user-defined criteria. Standard economic analyses such as discounted cash flow and net present value can be performed using DORT.
- P2/FINANCE tool (http://www.tellus.org/b&s/software/p2.html) illustrates environmental cost-accounting principles, particularly for the evaluation of capital investment options such as process upgrades or modifications. The software estimates the environmental costs associated with a process by taking into account such factors as waste management and liability costs.
- Combinatorial Process Synthesis, developed by the University of Illinois at Chicago (UIC) (http://vienna.che.uic.edu/) is a computer-aided decision-making tool for synthesizing plantwide waste management strategies; it provides options for generating a tree of feasible recycle and treatment options for effluents from batch manufacturing.

ACKNOWLEDGMENTS

Financial support from NSF Grant DMI-0328134 and the Environmental Manufacturing Management (EvMM) fellowship from the UIC Institute for Environmental Science and Policy is gratefully acknowledged.

REFERENCES

1. ABB, http://www.abb.com/, accessed 2004.
2. Ahmad, B.S., Zhang, Y., and Barton, P.I., Product sequences in azeotropic batch distillation, *AIChE J.*, 44(5), 1051–1070, 1998.
3. Allgor, R.J., Barrera, M.D., Barton, P.I., and Evans, L.B., Optimal batch process development, *Comput. Chem. Eng.*, 20(6/7), 885–896, 1996.
4. Aspen Law and Business, *RCRA Regulations and Keyword Index*, Aspen Publishers, New York, 2001.
5. Aspen Technology, *AspenPlus: Unit Operation Models*, Version 10.2, Aspen Technology, Cambridge, MA, 2000.
6. Batchplus, http://www.aspentech.com/includes/product.cfm?IndustryID=4&ProductID=91, accessed 2004.
7. Batch Processing Technology, http://www.bptech.com/, accessed 2004.
8. Bazin, H.G. and Linhardt, R.J., Properties of carbohydrates, in *Glycoscience: Chemistry and Chemical Biology*, Fraser-Reid, B., Tatsuta, K., and Thiem, J., Eds., Springer-Verlag, Heidelberg, 1999.
9. Bernot, C., Doherty, M.F., and Malone, M.F., Feasibility and separation sequencing in multicomponent batch distillation, *Chem. Eng. Sci.*, 46, 1311–1326, 1991.
10. Bhatia, T. and Biegler, L.T., Dynamic optimization in the design and scheduling of multiproduct batch plants, *Ind. Eng. Chem. Res.*, 35(7), 2234–2246, 1996.
11. Bizukojc, M. and Ledakowicz, S., Morphologically structured model for growth and citric acid accumulation by *Aspergillus niger*, *Enzyme Microbiol. Technol.*, 32, 268–281, 2003.
12. Boonme, M., Bridge, W., Leksawasdi, N., and Rogers, P.L., Batch and continuous culture of *Lactococcus lactis* NZ133: experimental data and model development, *Biochem. Eng. J.*, 14, 127–135, 2003.
13. Brennan, M. and Schwartz, E., A new approach to evaluating natural resource investments, in *The New Corporate Finance: Where Theory Meets Practice*, Chew, D., Ed., McGraw-Hill, New York, 1993, 98–107.
14. Burtraw, D., Carbon emission trading costs and allowance allocations: evaluating the options, *Resources for the Future*, 13–16, 2001.
15. Cantor, http://www.emissionstrading.com/index.htm, accessed 2004.
16. Cavin, L., Jankowitsch, O., Fischer, U., and Hungerbühler, K., Software tool for waste treatment selection using economic and ecological assessments, *European Symposium on Computer-Aided Process Engineering*, 10, Elsevier, Amsterdam, 2000, 889–894.
17. Cavin, L., Dimmer, P., Fischer, U., and Hungerbühler, K., A model for waste treatment selection under uncertainty, *Ind. Eng. Chem. Res.*, 40, 2252–2259, 2001.
18. Cavin L., Fischer, U., Glover, F., and Hungerbühler, K., Multi-objective process design in multi-purpose batch plants using a Tabu Search optimization algorithm, *Comp. Chem. Eng.*, 28(4), 459–478, 2004.

19. Chakraborty A. and Linninger A.A., Plant-wide waste management. 1. Synthesis and multi-objective design, *Ind. Eng. Chem. Res.*, 41(18), 4591–4604, 2002.

20. Chakraborty A., Plant-Wide Synthesis of Optimal Waste Treatment Policies Under Uncertainty, Dr. Eng. thesis, Department of Chemical Engineering, University of Illinois, Chicago, 2002.

21. Chakraborty, A. and Linninger, A.A., Plant-wide waste management. 2. Decision making under uncertainty, *Ind. Eng. Chem. Res.*, 42, 357–369, 2003.

22. Chakraborty, A. and Linninger, A.A., Plant–wide waste management. 3. Long term operation and investment planning under uncertainty, *Ind. Eng. Chem. Res.*, 42, 4772–4788, 2003.

23. Chakraborty, A., Malcolm, A., Colberg, R.D., and Linninger, A.A., Optimal waste reduction and investment planning under uncertainty, *Comput. Chem. Eng.*, 28, 1145–1156, 2004.

24. Chakraborty, A., Malcolm, A., and Linninger, A.A., Pharmaceuticals product-only design, FOCAPD 2004 Conference, July 11–16, Princeton University, Princeton, NJ, 2004.

25. Churn, C.C., Johnson, D.W., and Severin, B.F., Facility and process design for large scale activated sludge industrial wastewater plant, paper 51G, *Proc. of the 44th Industrial Waste Conference*, 1989.

26. Coldberg, R., *Oral Communication*, Chemical Intermediates and Catalysis Research Lab, Kingsport, TN, 2002.

27. Cramton, P. and Kerr, S., *Tradable Carbon Permit Auctions: How and Why to Auction Not Grandfather*, Discussion Paper 98–34, Resources for the Future, Washington, DC, 1998.

28. Crume, R. and Portzer, J., Pharmaceutical industry, in *Air Pollution Engineering Manual*, Buonicore, A.J. and Davis, W. T., Eds., Air and Waste Management Association/Van Nostrand Reinhold, New York, 1992.

29. Davidyan, A.G., KivaGeorge, V.N., Meski, A., and Morari, M., Batch distillation in a column with a middle vessel, *Chem. Eng. Sci.*, 49(18), 3033–3051, 1994.

30. Broschüre, *Produktionsintegrierter Umweltschutz in der Chemischen Industrie*, DECHEMA, Frankfurt, 1990.

31. Diwekar, U.M., How simple can it be? A look at the models for batch distillation, *Comput. Chem. Eng.*, 18, S451–S457, 1994.

32. Diwekar, U.M., *Batch Distillation*, Taylor & Francis, New York, 1995.

33. Doherty, M. and Malone, M., *Conceptual Design of Distillation Systems*, McGraw-Hill, New York, 2001.

34. Dunn, R.F., Zhu, M., Srinivas, B.K., and El-Halwagi, M.M., Optimal design of energy-induced separation networks for VOC recovery, *AIChE Symp. Ser.*, 90(303), 74–81, 1994.

35. Dyer, J.A., Mulholland, K.L., and Keller, R.A., Prevent pollution in batch processes, *Chem. Eng. Prog.*, 95(5), 24–29, 1999.

36. Edwards, A.D., Yu Shekunov, B., Kordikowski, A., Forbes, R.T., and York, P., Crystallization of pure anhydrous polymorphs of carbamazepine by solution enhanced dispersion with supercritical fluids (SEDS), *J. Pharmac. Sci.*, 90, 1115–1124, 2001.

37. El-Halwagi, M. and Manousiouthakis, V., Simultaneous synthesis of mass-exchange and regeneration networks, *AIChE J.*, 36(8), 1209, 1990.

38. El-Halwagi, M. and Manousiouthakis, V., Synthesis of mass exchange networks, *AIChE J.*, 35(8), 1233, 1989.

39. Energy and Environmental Analysis, *Analysis of Output-Based Allocation of Emission Trading Allowances*, Report for U.S. Combined Heat and Power Association, 2003.
40. Environmental Information Centre, http://www.cleantechindia.com/eicnew/successstories/bulk.html, accessed 2004.
41. Fernandez, A.A., Expert choice: ProVersion 9.0 for Windows confirms product as an outstanding choice for addressing complex, multicriteria problems, *ORMS*, 23(4), 1996.
42. Fisher, M., *The IdeaFisher: How to Land That Big Idea and Other Secrets of Creativity in Business*, Petersons, Boston, 1996.
43. Friedler, F., Varga, J.B., and Fan L.T., Algorithmic approach to integration of total flowsheet synthesis and waste minimization, in pollution prevention via process and product modifications, *AIChE Symp. Ser.*, 90(303), 86, 1994.
44. Fromm, C. H., Pollution prevention in process design, *Pollut. Prev. Rev.*, 389–401, 1992.
45. Fujiwara, M., Chow, P.S., Ma, D.L., and Braatz, R.D., Paracetamol crystallization using laser backscattering and ATR–FTIR spectroscopy: metastability, agglomeration, and control, *Crystal Growth Des.*, 2, 363–370, 2002.
46. Geankoplis, C.J., *Transport Processes and Separation Processes Principles*, 4th ed., Prentice Hall, Upper Saddle River, NJ, 2003.
47. Greitzer, F.L., Brown, B.W., Dorsey, J.A., and Raney, E.A., Pollution Prevention Electronic Design Guideline: A Tool for Identifying Pollution Prevention in Facility Design, 4th Annual Air Force Worldwide Pollution Prevention Conference and Exhibition, August 14–17, 1995, San Antonio, TX.
48. Guerrero, M., Albet, C., Palomer, A., and Guglietta, A., Drying in pharmaceutical and biotechnological industries, *Food Sci. Technol. Int.*, 9(3), 237–243, 2003.
49. Hilmen, E.K., Separation of Azeotropic Mixtures: Tools for Analysis and Studies on Batch Distillation Operation, Dr. Eng. thesis, Department of Chemical Engineering, Norwegian University of Science and Technology, Trondheim, 2000.
50. Ismail, S.R., Pistikopoulos, E.N., and Papalexandri, K.P., Synthesis of reactive and combined reactor/separation systems utilizing a mass/heat exchange transfer module, *Comp. Chem. Eng.*, 54, 2721–2729, 1999.
51. Kim, K.J. and Diwekar, U.M., Comparing batch column configurations: parametric study involving multiple objectives, *AIChE J.*, 46(12), 2475–2488, 2000.
52. Kondili, E., Pantelides, C.C., and Sargent, R.W.H., A general algorithm for short-term scheduling of batch operations, *Comp. Chem. Eng.*, 17, 21, 1993.
53. Kull, B. and Hsu, E., *PROVAL User's Manual*, Merck, Whitehouse Station, NJ, 1991.
54. LaGrega, M.D., Buckingham, P.L., and Evans, J.C., *Hazardous Waste Management*, McGraw-Hill, New York, 1994.
55. Lee, W. and Huang, C., Modeling of ethanol fermentation using *Zymomonas mobilis* ATCC 10988 grown on media containing glucose and fructose, *Biochem. Eng. J.*, 4, 217–227, 2000.
56. Lee, J.W., Hauan S., and Westerberg, A.W., Feasibility of a reactive distillation column with ternary mixtures, *Ind. Eng. Chem. Res.*, 40, 2714–2728, 2001.
57. Li, L.-H. and Tius, M.A., Stereospecific synthesis of cryptophycin, 1, *Org. Lett.*, 4, 1637–1640, 2002.
58. Lieberman, H.A., Rieger, M.M., and Banker, G.S., *Pharmaceutical Dosage Forms: Disperse Systems*, Marcel Dekker, New York, 1996.

59. Lin, X. and Floudas, C.A, Design, synthesis and scheduling of multipurpose batch plants via an effective continuous-time formulation, *Comput. Chem. Eng.*, 25(4–6), 665–674, 2001.

60. Linninger, A.A., Ali, S.A., Stephanopoulos, E., Han, C., and Stephanopoulos, G., Synthesis and assessment of batch processes for pollution prevention, in Pollution Prevention via Process and Product Modifications, *AIChE Symp. Ser.*, 90(303), 46–58. 1994.

61. Linninger, A.A., Ali, S.A., Stephanopoulos, E., Han, C., and Stephanopoulos, G., Generation and assessment of batch processes with ecological considerations, *Comp. Chem. Eng.*, 19, S7–S13, 1995.

62. Linninger, A.A., Ali, S.A., and Stephanopoulos, G., Knowledge-based validation and waste management of batch pharmaceutical process designs, *Comp. Chem. Eng.*, 20, S1431–1436, 1996.

63. Linninger, A.A., Salomone, E., Ali, S.A., Stephanopoulos, E., and Stephanopoulos, G., Pollution prevention for production systems of energetic materials, *Waste Manage. J.*, 17(2/3), 165–173, 1998.

64. Linninger A.A. and Chakraborty, A., Synthesis and optimization of waste treatment flowsheets, *Comp. Chem. Eng.*, 23, 1415–1425, 1999.

65. Linninger, A.A. and Chakraborty, A., Pharmaceutical waste management under uncertainty, *Comp. Chem. Eng.*, 25, 675–681, 2001.

66. Lo, T.C., Baird, M.H.l., and Hanson, C., Eds., *Handbook of Solvent Extraction*, Wiley-Interscience, New York, 1983.

67. Mallick, S.K., Cabezas, H., Bare, J.C., and Sikdar, S.K., A pollution reduction methodology for chemical processes simulators, *Ind. Eng. Chem. Res.*, 35, 4128, 1996.

68. McKetta, J.J., Ed., *Encyclopedia of Chemical Processing and Design*, Marcel Dekker, New York, 1992.

69. Mersmann, A., *Crystallization Technology Handbook*, Marcel Dekker, New York, 1995.

70. Mitchell Scientific, http://www.mitchellscientific.com/EmissionMaster.htm, accessed 2004.

71. Milliman, S.R. and Prince, R., Firm incentives to promote technological change in pollution control, *J. Environ. Econ. Manage.*, 17, 247–265, 1989.

72. Moureldin, M.B. and El-Halwagi, M.M., Pollution prevention targets through integrated design and operation, *Comput. Chem. Eng.*, 24(2–7), 1445–1453, 2000.

73. Muhrer, G., Gas Anti-Solvent Recrystallization of Specialty Chemicals, Ph.D. thesis, Swiss Federal Institute of Technology, Zurich, Switzerland, 2002.

74. Mulholland, K.L. and Dyer, J.A., *Pollution Prevention: Methodologies, Technologies and Practice*, American Institute of Chemical Engineers, New York, 1999.

75. Material Safety Data Sheets (MSDSs), http://www.msdsonline.com/, accessed 2004.

76. Nichols, A.L., *Targeting Economic Incentives for Environmental Protection*, MIT Press, Cambridge, MA, 1984.

77. Narayan, V., Diwekar, U., and Hoza, M., Synthesizing optimal waste blends, *Comp. Chem. Eng.*, 20(S2), S1443–S1448, 1996.

78. Oberthür, S. and Ott, H.E., *The Kyoto Protocol: International Climate Policy for the 21st Century*, Springer-Verlag, Berlin, 1999.

79. Petrides, D., Calandranis, J., and Flora, J., Clean water begins with a mouse: a comprehensive computer simulation can simplify the task of designing a cost-effective wastewater treatment plant, *Industrial Wastewater*, 33–40, 1997.

80. Research Triangle Institute, http://clean.rti.org/, accessed 2004

81. Salomone, H.E., Montagna, J.M., and Iribarren, O.A., A simulation approach to the design and operation of multiproduct batch plants, *Chem. Eng. Res. Des.*, 75(A4), 427–437, 1997.

82. Romero, J., Espuna, A., Friedler, F., and Puigjaner, L., A new framework for batch process optimization using the flexible recipe, *Ind. Eng. Chem. Res.*, 42(2), 370–379, 2003.

83. Santamarina, V.E., Modeling and optimization of a municipal treatment plant using EnviroProDesigner, *Environ. Prog.*, 16(4), 268–273, 1997.

84. Sarantopoulos, P.D., Altiok, T., and Elsayed, E.A., Manufacturing in the pharmaceutical industry, *J. Manuf. Syst.*, 14(6), 452–467, 1995.

85. Seader, J.D. and Henley, E.J., *Separation Process Principles*, John Wiley & Sons, New York, 1998.

86. Shonnard, D. and Hiew, D.S., Comparative environmental assessments of VOC recovery and recycle design alternatives for a gaseous waste stream, *Environ. Sci. Technol.*, 34, 5222, 2000.

87. Sikdar, S.K. and El-Halwagi, M., *Process Design Tools for the Environment*, Taylor & Francis, New York, 2000.

88. Solutions Software Corporation, http://www.env–sol.com/gov.html, accessed 2004.

89. Srinivasan, B., Palanki, S., and Bonvin, D., Dynamic optimization of batch processes. I. Characterization of the nominal solution, *Comput. Chem. Eng.*, 27(1), 1–26, 2003.

90. Stanbury, P.F. and Whitaker, A., *Principles of Fermentation Technology*, Pergamon Press, Elmsford, NY, 1984.

91. Tao, L. and Malone, M.F., Synthesis of azeotropic distillation systems with recycles, *Ind. Eng. Chem. Res.*, 42, 1783–1794, 2003.

92. Tellus Institute, http://www.tellus.org/b&s/software/p2.html#download, accessed 2004.

93. Tietenberg, T., *Environmental and Natural Resource Economics*, 4th ed., Harper Collins College Publishers, New York, 1996.

94. Toth, T.J. and Barna, B.A., Process Optimization with Environmental Constraints: The Design Option Ranking Tool, AIChE Spring National Meeting, New Orleans, LA, February, 1996.

94a. USEPA, *Pharmaceutical Industry: Hazardous Waste Generation, Treatment, and Disposal.* Solid Waste Management Series, SW–508, U.S. Environmental Protection Agency, Washington, DC, 1976.

95. USEPA, *Control of Volatile Organic Compound Emissions from Batch Processes*, Guideline Series, EPA-453/R-93-017, U.S. Environmental Protection Agency, Research Triangle Park, NC, 1993.

96. USEPA, *Beyond VOC RACT CTG Requirement*, EPA-453/R-95-010, Control Technology Center, U.S. Environmental Protection Agency, Research Triangle Park, NC, 1995.

97. USEPA, *Profile of the Organic Chemical Industry*, EPA-310/R-95-012, Office of Enforcement and Compliance Assurance, U.S. Environmental Protection Agency, Research Triangle Park, NC, 1995.

98. USEPA, SAGE 2.1, *Solvent Alternatives Guide: User's Guide*, EPA-600/SR-95/049, Air and Energy Engineering Research Laboratory, U.S. Environmental Protection Agency, Research Triangle Park, NC, 1995.

99. USEPA, *Incorporating Environmental Costs and Considerations into Decision Making: Review of Available Tools and Software*, EPA-742/R-95-006, U.S. Environmental Protection Agency, Research Triangle Park, NC, 1996.

100. USEPA, *Profile of the Pharmaceutical Manufacturing Industry*, EPA-310/R-97-005, Office of Enforcement and Compliance Assurance, U.S. Environmental Protection Agency, Research Triangle Park, NC, 1997.

101. USEPA, *Development Document for Final Effluent Limitations Guidelines and Standards for the Pharmaceutical Manufacturing Point Source Category*, EPA-821/R-98-005, U.S. Environmental Protection Agency, Research Triangle Park, NC, 1998.

102. USEPA, *Tools of the Trade: A Guide to Designing and Operating a Cap and Trade Program for Pollution Prevention*, EPA-430/B-03-002, U.S. Environmental Protection Agency, Research Triangle Park, NC, 2003.

103. USEPA, *EPA Acid Rain Program: 2002 Progress Report*, EPA-430/R-03-011, Clean Air Markets Division, U.S. Environmental Protection Agency, Research Triangle Park, NC, 2003.

104. USEPA, CFR 40 Part 63: Maximum Achievable Control Technology (MACT) standard, *http://yosemite.epa.gov/R10/AIRPAGE.NSF/0/7b53b04584f8da9a88256 d8f007a08e8?OpenDocument*, accessed 2004.

105. Viswanathan, S., Johnsson, C., Srinivasan, R., Venkatasubramanian, V., and Arzen, K.E., Automating operating procedure synthesis for batch processes. Part I. Knowledge representation and planning framework, *Comput. Chem. Eng.*, 22(11), 1673–1685, 1998.

106. Viswanathan, S., Johnsson, C., Srinivasan, R., Venkatasubramanian, V., and Arzen, K.E., Automating operating procedure synthesis for batch processes. Part II. Implementation and application, *Comput. Chem. Eng.*, 22(11), 1687–1698, 1998.

107. Venkataramani, E.S., House, M.J., and Bacher, S., *An Expert System Based Environmental Assessment System (EASY)*, Merck, Whitehouse Station, NJ, 1990.

108. Wankat, P., *Equilibrium Staged Separations*, Prentice Hall, Englewood Cliffs, NJ, 1988.

109. Waste Reduction Resource Center, http://wrrc.p2pays.org/, accessed 2004.

110. Westerberg, A.W., Woo, L.J., and Hauan, S., Synthesis of distillation-based processes for non-ideal mixtures, *Comput. Chem. Eng.*, 24(9–10), 2043–2054, 2000.

111. Wilson, S. and Manousiouthakis, V.I., IDEAS approach to process network synthesis: application to multi-component MEN, *AIChE J.*, 46(12), 2408–2416, 2000.

Part III

Batch Processing Management

9 Batch Process Modeling and Optimization

Andreas Cruse, Wolfgang Marquardt,
Jan Oldenburg, and Martin Schlegel

CONTENTS

9.1 INTRODUCTION

The interest in batch process modeling and optimization in the chemical industry has increased during the last years in response to the requirements of increasingly competitive markets. Typically, batch processes are considered as a process alternative:

- If a high-value product is produced at a low volume
- If there are large market-driven fluctuations in the demand or short product life-cycles
- If flexibility with respect to product grades, volume, and quality is required
- If technical difficulties such as long residence times, multiphase systems, or fouling are significant

Batch processes are characterized by some distinct features. Various products are often produced in the same plant. The process units and their connections to a plant may change with time. Often, the intermediate products are stored in buffer tanks and are processed (sometimes after blending) in the same but possibly reconfigured plant or in a different plant. The operation of batch processes is defined by some recipe that allocates processing tasks to process equipment and defines a sequence of time-varying controls, which are continuous and discrete in nature. The high flexibility of batch processes can only be fully exploited if design and operational decision making are supported by advanced modeling and optimization techniques.

As an illustration, we consider a reactive distillation process for the production of methyl acetate (MeAc) produced by a reversible reaction between acetic acid (HAc) and methanol (MeOH):

$$HAc + MeOH \rightleftharpoons MeAc + H_2O$$

We would like to produce methyl acetate from methanol and acetic acid in reactive and nonreactive distillation columns. The raw materials are pure components. The desired product streams are methyl acetate and water with some given purity specifications. The most favorable process consists of a semibatch reactive distillation column that produces methyl acetate of a desired purity as the top product. A schematic of this process is shown in Figure 9.1. The use of reactive distillation is especially advantageous for equilibrium reactions, as the continuous removal of the product increases conversion; however, it is not the only possible option for designing a process for this purpose, as we will see later.

9.1.1 WHY MODEL?

There are several reasons why we are interested in a model of a batch process. The most important engineering tasks, which should be carried out by means of a model, are process and recipe design, as well as planning and scheduling, in order to match the production demands of the customers. These tasks assume nominal operation and a reasonably valid model. Design of a process unit or a process plant given some production targets (comprised of type of product, the desired quantity, and the requested quality specifications) typically involves three steps. The first step takes place in the laboratory, where a new product is discovered or alternative ways for the production of an existing product are identified.

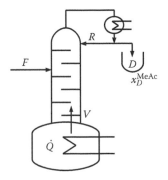

FIGURE 9.1 Semibatch reactive distillation column for the production of methyl acetate.

Then, the unit operations and their interconnections to a *batch plant* as well as feasible and efficient operational strategies for all process units, the so-called *recipes*, are determined in a process design step. Finally, the third step deals with *equipment allocation* and *batch process scheduling*, accounting for customer demand and employing the recipes obtained from the previous design step. In the following text, we focus first on the process design step, which comes in various degrees of complexity with respect to the structure of the process plant and the units to be chosen to implement the functionality.

Let us illustrate the different degrees of complexity by means of the methyl acetate example. Here, the reaction and a first guess of the process structure are given. Further, we have developed a rough idea on alternatives to the initial process structure. First, we consider the process alternative depicted in Figure 9.1, the semibatch reactive distillation column. If an existing column has to be employed, the remaining degrees of freedom to be fixed are:

- The feed rates of both raw materials (F in Figure 9.1)
- The heat supplied to the evaporator (or the boil-up rate, V)
- The distillate flow rate (D) (or the reflux rate, R)

as functions of time, as well as the amount of the raw materials put in the still before the process is started. Hence, we have decision variables that are time varying as well as time invariant. If, in addition, the column itself is not given, the additional degrees of freedom lie in the equipment configuration itself. They include the size and number of trays, location of the feed tray, and alternatives to the regular processing from a still at the bottom, such as a middle-vessel operation (see Figure 9.2), where the main reaction volume is placed somewhere in the middle of the column. In this case, some structural degrees of freedom have to be fixed once and are not reconsidered during process operation.

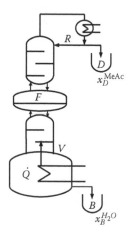

FIGURE 9.2 Process design option with middle vessel.

The design problem to be solved here belongs to the class of dynamic optimization (or optimal control) problems with differential–algebraic constraints and continuous degrees of freedom (time-dependent or time-varying) if the equipment structure is fixed or with continuous as well as discrete degrees of freedom if the equipment structure is also a subject to be decided on.

Up to this point, only the case of a single and given production target has been considered. In practice, batch plants are operated in campaigns and a multitude of products are manufactured. For example, a variety of esterifications may be carried out in the same semibatch column shown in Figure 9.1. Hence, the production target comprised of type of product, volume, and quality for a certain campaign has to be determined on a higher decision level involving planning and scheduling activities.

So far, we have assumed complete knowledge of the process and its operation. In particular, we have implicitly assumed a perfect model with known (or vanishing) disturbances as well as a perfect forecast of customer demand for the products. All these assumptions are not valid in a realistic setting. Various kinds of uncertainty have to be considered. Unknown time-invariant parameters (e.g., coefficients in the reaction kinetics or heat transfer relation) are a source of uncertainty. Furthermore, uncertainty is caused by time-varying disturbances such as drifting model parameters (e.g., a decreasing heat-transfer coefficient due to fouling) or changing environmental conditions. Additionally, structural model uncertainties may exist, for example, unknown side reactions in the esterification reaction mechanism in the example above. Finally, the process itself may be subject to such uncertainty as a sudden loss of cooling or heating due to a pump failure in the condenser or evaporator of the semibatch distillation column.

The uncertainties can be dealt with explicitly during the design phase. The general strategy is to design the process in such a way that it fulfills the requirements even in the presence of uncertainty. For example, we can define a possible range for the uncertain parameters and then develop a design of the process that ensures a feasible operation within this entire range. Typically, such approaches lead to conservative solutions.

Alternatively, we can deal with uncertainty in real time during the operation of a particular process. This is advantageous, because, in addition to the *a priori* knowledge used in the design phase, we also have measurement information available, which in conjunction with the model can be used to reduce the uncertainty. Not only do the measurements include a part of the process state but they also refer to some of the disturbances acting on the process as well as to changes in the production target due to varying customer demand. Hence, in this way we are able to adjust the process to the actual situation in real time to maximize profit at any instant in time to the extent possible. In this case, a combined estimation and optimization problem has to be solved in order to reconcile the model as well as to fix continuous and probably also discrete decision variables online.

9.1.2 Why Optimize?

We have learned about a number of design and operational problems that are supposed to be solved by the aid of a model. It is common practice to solve the design and operation problems in an iterative manner. The process (and control) structure is selected and fixed for a given production schedule. The model is formulated and implemented by means of a simulator. Then, parameter variations are carried out in order to match the design requirements for some configuration in an extensive search process. This procedure is not likely to succeed, given the immense complexity of batch design, control, and operation problems. In contrast to continuous processes, we have to decide on time-varying quantities and hence on a virtually infinite number of parameters that are required to fix the decision variables as a function of time.

Instead of simulation we advocate the use of optimization algorithms to improve the quality of the resulting design at reduced engineering effort. Even in the case of optimization, we will not be able to avoid the trial-and-error search during problem solving that is a typical feature of simulation-based design. However, this search is carried out on a higher level of abstraction. Whereas we search over the space of process structures, design parameters, and recipes for simulation-based problem solving, the search space is restricted to alternative objective functions and constraint sets for an optimization-based approach. Obviously, in the latter case, the designers' objectives can be formulated more directly and the search for appropriate process structures, design parameters, and recipes is carried out exhaustively by means of a numerical algorithm. Consequently, the solution quality can truly be expected to improve at reduced engineering effort if optimization is employed.

9.1.3 What to Gain from Modeling and Optimization?

Industrial batch and semibatch processes are often still operated using recipes based on heuristics and experience. In contrast to continuous processes, for which many rigorous optimization studies of practical relevance have been published, these techniques are not yet commonly applied for batch-type processes; however, supported by maturing dynamic optimization techniques, recently published studies show the benefits for batch process applications.

For example, Li et al.[86] applied modeling and optimization techniques to an industrial semibatch distillation process with a chemical reaction taking place in the reboiler. A detailed dynamic model was set up to describe the process. The model was validated with measured data from experiments conducted on the industrial site. With the help of model-based optimization of reflux ratio and feed rate policies, significant savings of 30% in terms of operational time when compared to the conventional operation have been achieved.

Abel et al.[1] described optimization of the operation of an industrial semibatch reactor. As in the publication discussed before, the objective of this case study was to minimize the batch time. Various operational, quality, and safety-related constraints were considered during the batch and at its final time. The optimization was based on a detailed process model derived from first principles. With the help of a reduced model optimization, computations were carried out that resulted in optimal trajectories for the operational variables of feed flow rate and reactor temperature. A reduction of the conventional operating time by 28% can be expected and has been verified experimentally in the laboratory and at the production scale.

9.1.4 CHAPTER OVERVIEW

The scope of this chapter includes modeling and optimization techniques for batch process design and operation assuming a fixed and given production target. In this context, we also address the unavoidable uncertainties that may be encountered during process design or during process operations. We will not deal with scheduling and planning, because dynamic process models using detailed physical knowledge are not used at that level. Scheduling and planning are based on coarser models that only represent nonlinearity and dynamics in a very approximate manner. Such problems are treated in Chapter 10. In Section 9.2, we discuss aspects of batch process modeling with a focus on model application in dynamic optimization. Section 9.3 gives an overview of various mathematical techniques for design optimization for processes with a fixed structure and those for which the structure itself is subject to optimization. Section 9.4 discusses concepts and algorithms for dynamic real-time optimization and control. Examples are used throughout to illustrate the theoretical considerations. We conclude with a summary and briefly discuss open research issues.

9.2 MODELING FOR OPTIMIZATION

In contrast to continuous processes, the discrete–continuous nature of batch processes has to be accounted for in the process model. Even in the case of a continuous representation of those physical phenomena that give rise to sharp transitions of the process variables (for example, in the case of a phase change or rupture of a safety disc), the recipes will unavoidably introduce discontinuities into the model. Furthermore, optimal control profiles are typically discontinuous by nature. The discrete–continuous behavior of batch processes can be divided into phases with continuous quantities. If all the sharp transitions of states caused by fast physicochemical phenomena are modeled in a continuous way, the discontinuities are only due to the discrete events imposed by the discrete stages of the recipe implemented in the supervisory control system.

Appropriate consideration of the various types of discontinuities, the transition conditions from one phase to the next, and the possible arrangement of the phases over time leads to a number of hybrid model structures. These different structures are discussed in detail in Section 9.2.3. As we will see, all these models consist of the dynamics necessary to describe the behavior in a certain phase, switching conditions that trigger a switch from one phase to the following, and mapping conditions that transfer the states at the end of one phase into the states at the beginning of the following phase.[17] These models form the core of the optimization models that are presented in Section 9.2.4.

Within each of these phases, the models governing the process behavior have to be derived from the process knowledge available. The systematic development of a mathematical model for a single phase of a batch process is not any different from the development of a continuous process model; hence, all the experience built up in recent years can also be employed here.

The next two sections cover the issue of how to deal with the unavoidable uncertainty and the lack of mechanistic understanding in batch process modeling. This is an important aspect due to the complex and poorly understood physico-chemical phenomena of the special unit operations and complex multiphase and multicomponent mixtures that often occur in batch processes.

9.2.1 FUNDAMENTAL–EMPIRICAL MODELING

Let us consider a batch process in a particular phase between two discontinuities such as the semibatch reactive distillation column for methyl acetate production (see Figure 9.1). For a systematic development of the process model, we follow the procedure suggested by Marquardt[39] and Foss et al.[59] We start with an abstraction of the process plant leading to a set of balance envelopes and their connections. In case of the reactive distillation column, the major balance envelopes are the trays, the still, and the reflux drum, or, more precisely, the liquid and vapor phases therein. In the next step, the balance equations are formulated as a sum of fluxes for the extensive quantities of interest. The balance equations can be interpreted to have the inherent hierarchy shown in Figure 9.3.

The fluxes are modeled by constitutive equations to specify, for example, the relation of a reaction rate with other process quantities. Recursively, these constitutive equations contain process quantities (e.g., the rate constant of a chemical reaction) that may result from other constitutive equations such as an Arrhenius law or a physical property correlation. We continue until we have a fully determined set of equations, which allows computation of the state variables from specified parameters and inputs, which can be of a constant or time-varying nature depending on the type of input.

The formulation of all the model equations is largely dependent on the level of process knowledge available. For every process quantity occurring in an equation on a certain level of the recursion, we can decide whether it is treated as a constant or a time-varying quantity to be specified as a degree of freedom in the simulation model or will be refined by another equation. In the latter case, these constitutive equations may have a mechanistic basis or, alternatively, may be of a physically motivated semiempirical or even completely empirical nature. In most cases, we are not able to incorporate or (for complexity reasons) are not interested in incorporating truly mechanistic knowledge (on the molecular level) to determine the constitutive equations. Instead, we correlate unknown process quantities by means of a (semi-)empirical equation.

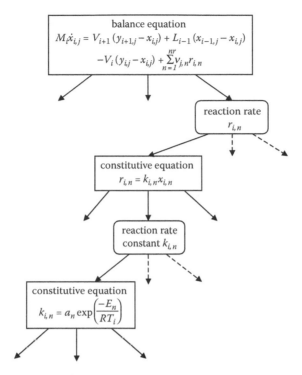

balance equation

$$M_i \dot{x}_{i,j} = V_{i+1}(y_{i+1,j} - x_{i,j}) + L_{i-1}(x_{i-1,j} - x_{i,j})$$
$$- V_i(y_{i,j} - x_{i,j}) + \sum_{n=1}^{nr} v_{j,n} r_{i,n}$$

reaction rate

$$r_{i,n}$$

constitutive equation

$$r_{i,n} = k_{i,n} x_{i,n}$$

reaction rate constant $k_{i,n}$

constitutive equation

$$k_{i,n} = a_n \exp\left(\frac{-E_n}{RT_i}\right)$$

FIGURE 9.3 Model hierarchy of balance equations and constitutive equations.

Hence, all process models are by definition *mixed fundamental–empirical models*, as they are comprised of fundamental as well as empirical model constituents. The fundamental model constituents typically represent the balances of mass, energy, and momentum and at least part of the constitutive equations that are required to fix the fluxes and thermodynamic state functions in terms of state variables. Empirical model constituents, on the other hand, have to be incorporated in the overall process model if no physically grounded constitutive equation is available to model some physical property or kinetic phenomenon. The two different types of parameterizations to implement these relations are *empirical regression models* and *empirical trend models*.

9.2.1.1 Empirical Regression Models

If we limit ourselves to lumped parameter systems, a set of differential–algebraic equations (DAE):

$$0 = f(\dot{x}(t), x(t), z(t), \xi(\cdot), \vartheta_1, u(t), p) \tag{9.1}$$

will show up at some level of refinement. Here, the variables $x \in \mathbb{R}^{n_x}$ denote the dynamic state variables, whereas $z \in \mathbb{R}^{n_z}$ are the algebraic state variables;

$\vartheta_1 \in \mathbb{R}^{n\vartheta_1}$ and $u \in \mathbb{R}^{nu}$ denote model parameters and time-dependent inputs, respectively; and $p \in \mathbb{R}^{np}$ are time-invariant design parameters. In the case of the reactive distillation column, the states x could be concentrations, temperatures, and holdups on the trays, in the still, and in the reflux drum. Molar fractions and reaction rates are typical examples for the algebraic states y. The time-dependent decision variables are the feed, reflux, and boil-up rates. The time-invariant degrees of freedom comprise the initial charge of materials in the still which is part of the initial conditions of the DAE (Equation 9.1).

The functions f are assumed to be known. They typically include the balances of mass and energy, which can always be formulated on the basis of the available process understanding. The n_ξ unknown functions $\xi(\cdot)$ represent physical quantities that are difficult to model mechanistically. Examples are a flux, kinetic constant, or physical property. In the methyl acetate example, a rate constant $k_{i,n}$ (see Figure 9.3) in the expression for the reaction kinetics could be unknown. Obviously, instead of postulating some model structure:

$$\xi = \xi(x(t), u(t), \vartheta_2) \tag{9.2}$$

based on physical hypotheses, as in fundamental modeling, any other purely mathematically motivated (empirical) model structure can be chosen to implement the constitutive equation for the computation of the unknown ξ. The probably unknown parameters in ϑ_1 of the fundamental model and the structure of the function $\xi(\cdot)$ and its parameters $\vartheta_2 \in \mathbb{R}^{n\vartheta_2}$ have to be estimated from plant data. Neural networks, linear multivariate regression models, and fuzzy and neuro-fuzzy models have been employed in the past. Mixed fundamental–empirical models with serial structure (see Figure 9.4b) are getting a lot of attention[98] and have found numerous applications in batch process applications. Satisfactory prediction quality can be obtained if sufficient data are available for training.[123]

9.2.1.2 Empirical Trend Models

The algebraic model (Equation 9.2) suggested in the last section as part of a serial hybrid model structure is most appropriate to represent the static relation between fluxes or physical properties and state variables. However, in some situations, the dynamic model:

$$\xi = \omega_1(x, u, \xi, \pi, \vartheta_3) \tag{9.3}$$

$$\pi = \omega_2(x, u, \xi, \pi, \vartheta_4) \tag{9.4}$$

may be more suitable to complement a fundamental model. Here, the quantities $\xi \in \mathbb{R}^{n\xi}$ and $\pi \in \mathbb{R}^{n\pi}$ are interpreted as part of the (extended) state vector rather than as a nonlinear state function. Often, due to a lack of better knowledge, the

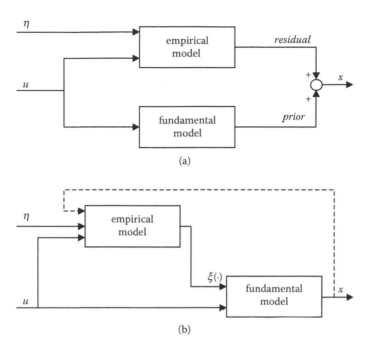

FIGURE 9.4 Mixed models: (a) parallel structure, and (b) serial structure; u and η denote the inputs (manipulated and measured variables), and x represents the states of the process to be modeled.

dynamic models for ξ and π are chosen simply as constant or linear trends.[108] Because the predictive quality of the model is very limited, its use is confined to real-time applications where predictions are only required on a short time interval.[67] An example is given later in Section 9.4.4.

9.2.2 MODEL AND PROCESS UNCERTAINTY

A structured combination of mechanistic and empirical model constituents as suggested in the previous section can be interpreted as an attempt to deal with *model uncertainty*; however, sufficient model structure information is often not available. For example, the choice of the balance envelopes, the spatial and chemical resolution of the physicochemical phenomena chosen, and the definition of the system boundary unavoidably introduce uncertainty, which is often unstructured.

In those ill-defined situations, the fundamental model is pragmatically upgraded by a corrective quantity. For example, Kramer and Thompson[80] suggest a parallel structure of a fundamental and an empirical model (Figure 9.4a), where the outputs of both constituents are summed up to form the output of the mixed model. Usually, both of these models are dynamic. The empirical model is often implemented as some type of neural network that correlates known process inputs.

It acts as an error model and compensates for any unstructured model uncertainty in the fundamental model.

In addition to the model uncertainty discussed so far, we often have to deal with *process uncertainty* or *disturbances*. Typical examples are the fouling of heat transfer areas, the clogging of distillation tower trays, fluctuating feed streams from an upstream process, or even the complete failure of some piece of equipment, such as a cooling water pump.

Often, process uncertainty and disturbances can be modeled the same way as the uncertainties stemming from inadequate understanding of the physico-chemical phenomena in nominal operation. For example, the parallel structure in the last subsection may also be used if the correction can be adequately correlated with measurable process quantities.

This assumption is not valid in those cases where the uncertainty is due to the dynamics of the environment or where internal process uncertainty cannot be directly correlated to the system state. While the model structure (Equation 9.1) is still valid in principle, the uncertain quantities are often assumed to enter the model in an additive manner. If we replace $\xi(\cdot)$ by the variables $d(t)$, which only depend on time, we obtain the model structure:*

$$0 = f(\dot{x}(t), x(t), z(t), \vartheta_1, u(t), p) + d(t) \tag{9.5}$$

The unknown quantities $d(t) \in \mathbb{R}^{n_d}$ are interpreted as disturbances. These unknown inputs can be represented by a trend model (Equations 9.3 and 9.4), where the right-hand sides ω_1 and ω_2 do not depend on x and u. Often, simple integrating models are used. Alternatively, the components of $d(t)$ are parameterized by some expansion:

$$d_i(t) = \sum_{j \in J} \vartheta_{5,i,j} \varphi_j(t) \tag{9.6}$$

the parameters of which have to be determined either by parameter estimation or by approximate model inversion (see, for example, Binder et al.[26]).

After parameterization of model or process uncertainty by means of an empirical regression, a trend model, or a time-varying function, as presented above, most of the model parameters in the vectors $\vartheta_i, i = 1,\ldots,4$, concatenated in $\vartheta \in \mathbb{R}^{n_\vartheta}$, are themselves uncertain. This remaining *parametric uncertainty* can be modeled in two different ways.

In a deterministic setting, the parameters are assumed to vary in a bounded set T. In particular, every single parameter ϑ may be bounded from above and from below. Hence, there is an infinite number of models for every realization $\vartheta \in T$. In order to approximately solve these infinite dimensional problems, a

---

* Obviously, the functions f in this expression are not the same as in (Equation 9.1). For the sake of simple notation, we do not distinguish different functions f here and subsequently.

representative sampling of ϑ has to be applied. For a particular realization of ϑ, this constitutes a single model.

Alternatively, the process parameters ϑ and time-varying disturbances $d(t)$ can be considered as random variables with an associated probability density function. Consequently, all the process states $x(t)$ and $z(t)$ become random variables also resulting in considerable model complexity.

For convenience, we assume an explicit parameterization of all unknown quantities by a finite number of uncertain parameters $\zeta \in \mathbb{R}^{n_\zeta}$, which are of a deterministic or stochastic nature. In the deterministic case, the parameters are assumed to lie in a compact set (e.g., $\zeta \in C$), whereas a joint probability density function, $F(\zeta)$, has to be defined in the stochastic case.

We want to briefly mention here a completely different approach to represent uncertainty in dynamic systems.[31,124] Rather than representing the time-varying uncertain quantities $d(t)$ by a series expansion with random variables weights (see Equation 9.6), the disturbance inputs are discretized in time. At any time instant, the discrete disturbance can assume a random value according to its probability density. Discrete control variables are introduced accordingly to counteract the influence of the disturbance in discrete time. This formulation leads to a scenario tree where the complexity of the tree is determined by the number of discretization intervals and the number of decisions possible in each control period. This modeling approach will not be considered in the sequel.

9.2.3 Discrete–Continuous Model Structures

Regardless of the modeling approach taken and the uncertainties present in the model, different types of *discrete–continuous model structures* arise due to the large variety of problems to be addressed in conjunction with batch process design and operation. Though it would be possible to generalize and come up with a model framework that covers all the various special cases, such a model would be difficult to comprehend. Further, due to the inherent complexity of all of these models, the model framework that best matches the requirements of an application is always chosen in order to limit complexity to the extent possible.

We will now introduce three different problem classes with increasing orders of complexity. In all cases, we assume that the structural (discrete) degrees of freedom have been fixed previously. This assumption will be relaxed later. The classification relies on the type of discontinuities present in the model of a batch process. *Implicit discontinuities* arise when a discrete change in the process is triggered by its physical behavior. Such a discontinuity occurs in general when a certain relation between process parameters and states is met. This is, for example, the case when the feed valve of the semibatch reactive distillation column is opened or closed depending on the amount of material or concentrations of reactants in the still, or if a thermodynamic phase change takes place in some piece of equipment. On the other hand, the opening of the feed valve at a time set in advance is considered an *explicit discontinuity*. This type of discontinuity depends only on the process time and has no relation to the process states and parameters.

The three model classes we want to distinguish are as follows: (1) they do not show any discontinuity, (2) they show discrete switching at given times, or (3) they may switch if some state- (and time-) dependent logical condition is true. These classes are denoted as *single-stage, multistage*, and *general discrete–continuous hybrid models*, respectively. In this section, we are not explicitly dealing with uncertainty. Obviously, most of the considerations in Section 2.1 are valid and can be taken into account in setting up the following models.

9.2.3.1 Single-Stage Models

Let us look again at the methyl acetate example and consider only the semibatch reactive distillation with fixed equipment structure. We have one unit and only one manufacturing task. The DAE model:

$$f(\dot{x}(t), x(t), z(t), u(t), p, \zeta, t) = 0, t \in [t_0, t_f] \tag{9.7}$$

$$l(\dot{x}(t_0), x(t_0), z(t_0), p, \zeta) = 0 \tag{9.8}$$

is, for convenience, written here in implicit form. We assume that the equation system can be transformed into semi-explicit form by simple algebraic manipulations. Further, the differential index is restricted to one. The final time t_f may also be considered as a design degree of freedom. The initial conditions l are assumed to be consistent with the model equations at time t_0. The model f is continuous and is comprised of a single set of dynamic and algebraic states x and z, controls u, design parameters p, and model parameters ζ; therefore, the model describes a process in a single mode or single discrete state and is referred to as a *single-stage model*.

9.2.3.2 Multistage Models

In contrast to the models introduced in the previous section, multistage models are used to describe situations where the process is subject to a sequence of modes or discrete states and where switching from one mode to the next occurs at some time t_k. The terms *stages* and *phases* are often used instead of *mode* or *discrete state* to reflect this strict sequence.

We define a set $K = \{1, \ldots, n_s\}$, comprised of the indices of all model stages. In every stage, the process dynamics can be modeled by a set of DAEs, as above. Then, the multistage model reads as:

$$f_k(\dot{x}_k(t), x_k(t), z_k(t), u_k(t), p, \zeta, t) = 0, t \in [t_{k-1}, t_k], \forall\, k \in K \tag{9.9}$$

$$l(\dot{x}_1(t_0), x_1(t_0), z_1(t_0), p, \zeta) = 0 \tag{9.10}$$

$$x_{k+1}(t_k) - x_k(t_k) = 0, \forall k \in K_m \tag{9.11}$$

Index k denotes quantities belonging to stage k. Multistage models contain additional stage transition conditions (Equation 9.11), which map the differential state variable values $x_k \in R^{n_{x_k}}$ across the stage boundaries. The mapping condition (Equation 9.11) occurs quite frequently but is a simple form of a more general formulation that is required for problems where the stage transition conditions depend on process parameters and algebraic states, in addition to the differential states. According to Leineweber et al.,[84] a generalized mapping condition can be formulated as:

$$x_{k+1}(t_k) - m_k(x_k(t_k), z_k(t_k), p, t_k) = 0, \forall k \in K_m \tag{9.12}$$

The indices of the mapping conditions are collected in the set $K_m = \{1, \ldots, n_s - 1\}$. With a fixed index $k = n_s = 1$, we obtain a single-stage model as introduced previously. Note that the final time t_f of the batch process corresponds to t_{n_s} in the multistage formalism. The mapping condition (Equation 9.12) might also be defined in a more general form where m_k is also a function of the control variables $u_k(t)$ and the parameters ζ. Moreover, Equation 9.12 might only be available in implicit form.[17]

In the example of methyl acetate production, a batch reactor and a nonreactive batch distillation column operated in sequence could be employed as an alternative to the semibatch reactive distillation column shown in Figure 9.1. In this case, the final product of the reactor is fed to the still of the batch column; hence, we have $n_s = 2$ stages, where the reaction is stage 1 and the separation is stage 2, as depicted in Figure 9.5.

Note that this process and therefore the corresponding process model are discontinuous. For the modeling and illustration of discrete event systems, Petri nets are used quite frequently.[102] Such a representation is also shown in Figure 9.5. A Petri net consists of so-called places (circles) representing the discrete system states, and transitions (rectangles), which describe the switching from one discrete state to the next. The currently active state is marked by a token. In our example, the first discrete state corresponds to the operation of the reactor. When the reaction is finished, the transition to the second state (separation in the distillation column) is carried out.

9.2.3.3 General Discrete–Continuous Hybrid Models

In fact, the multistage process model discussed in the previous section could also be referred to as discrete–continuous, but it is a special model of this type due to the explicit nature of the switching condition and the fixed number and

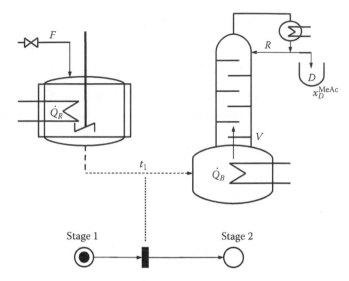

FIGURE 9.5 Two-stage batch process with reactor and distillation column: equipment (top) and event-oriented (Petri net) (bottom) representations.

sequence of modes. To understand this, we introduce the notion of a switching or triggering function that is used to locate the point in time where a new process stage is activated; that is, the model index k is increased by one. Following a definition given by Carver,[37] we can state the switching function:

$$\varphi_k(x_k(t), z_k(t), p, t) \tag{9.13}$$

on each model stage $k \in K$ except for the first ($k = 1$), for which initial conditions are specified in Equation 9.10. For simplicity, we assume that the switching function is independent of the control variables $u_k(t)$ and the uncertain parameters ζ. A transition takes place at time t_k when the sign of the switching function changes; that is, $\varphi_k(t_k) = 0$. For multistage models, φ_k takes a very simple form (φ_k^{ms}), which is independent of the state and parameter values:

$$\varphi_k^{ms} = t - t_k, \forall k \in K_m \tag{9.14}$$

If the final time t_k of a batch process stage itself is a degree of freedom, the reformulation $t = t_k + \tau_k(t_k - t_{k-1})$ is introduced where $\tau_k \in [0,1]$ is a nondimensional time coordinate and the final time t_k is a free parameter in stage k. In this case, the switching function results in $\varphi_k^{ms} = \tau_k - 1$. Note that this function is again independent of the states and parameters of the process.

The discrete–continuous models from the last subsection can be generalized if the switching conditions are not restricted to depend on explicitly known points in time. Instead of switching at a given point in time t_k from one discrete mode to the next, the switching may be triggered implicitly by some states or parameters in the process at some unknown time instant t_-^*.* In this case, the switching condition that applies for arbitrary model stages reads as:

$$\varphi_k(x_k(t_-^*), z_k(t_-^*), p, t_-^*) = 0 \tag{9.15}$$

The presence of implicit discontinuities increases the problem complexity tremendously because the number of discrete modes is not known *a priori* as is the case for a multistage model with a given number of stages n_s. Moreover, the numerical solution of these models is in general more difficult, as the location of the switching point has to be determined in parallel to the solution of the model equations (see, for example, Park and Barton[95]). An appropriate transition condition for example, continuity of the differential state variables (Equation 9.11) similar to the one introduced in the previous section for multistage models is required for each discrete phase change. For discrete–continuous hybrid models this leads to:

$$x_k(t_+^*) - x_k(t_-^*) = 0 \tag{9.16}$$

where the t^* is again employed instead of t_k. For more general mapping conditions, which also include algebraic state variables and parameters, refer to, for example, Barton and Pantelides.[17] In fact, an interesting link between multistage and general discrete–continuous models can be seen when comparing Equation 9.16 with the mapping condition shown in Equation 9.11.

In many cases, a discrete–continuous hybrid model can be easily represented by a so-called *state graph*.[17] If, for example, weir hydraulics are considered for the stages of a reactive batch distillation column,[51] we have a discrete–continuous hybrid model. The state graph representation of a single tray is depicted in Figure 9.6 (left).

FIGURE 9.6 State graph (left) and Petri net (right) representation of tray hydraulics.

---

* The notation t_-^* indicates that the triggering of the switching takes place immediately before the actual switching time t^*.

Here, the switching is triggered by a function that changes its sign when the actual level h on a tray reaches the height of the weir h_{weir}; that is, $\varphi_{weir} = h - h_{weir}$. For $\varphi_{weir} < 0$, we have no liquid flow leaving the tray (i.e., $L_{out} = 0$); otherwise, if $\varphi_{weir} > 0$, L_{out} is related to tray holdup M by a Francis weir equation. Note that for a batch distillation column, this formalism leads to a rather complex model because the discrete–continuous hybrid elements have to be employed for every tray.

This type of discontinuity is termed *reversible* because switching in both directions, back and forth, is possible. This is not the case for *irreversible* discontinuities, such as switching due to a bursting rupture disc or simply the multistage process shown in Figure 9.5. In more complex cases (i.e., when a large number of potential discrete phases including concurrent or alternative discrete modes exist), an event-oriented representation such as a Petri net has advantages over state graphs. Figure 9.6 (right) shows the Petri net for the weir hydraulics example.

Most discrete–continuous hybrid models can be represented by a *direct* sequence (such as the weir hydraulics mentioned above), a sequence containing *concurrent* events, or a sequence of *alternative* events.[93] If two heated tanks with reactant are used to feed the reactive column with methanol at a predefined temperature, we would have two concurrent discrete phases with two switching functions that trigger the filling phase of the column as soon as the desired temperature is reached. Alternative sequences are treated in the next section, where structural design decisions are considered.

9.2.3.4 Accommodating Structural Design Decisions

In all previous cases we have neglected parameterization of the models with respect to structural degrees of freedom. For example, we have fixed the number of trays and the location of the feed tray in the reactive distillation column. Structural degrees of freedom can show up in a variety of different ways in all the three model classes. Raman and Grossmann[99] proposed employing logic-based *superstructure models* involving disjunctions expressed by Boolean variables: $Y \in \{\text{True, False}\}^{n_Y}$. In the methyl acetate example, the Boolean variables could be used to model the decision whether to connect a side feed stream to the column instead of employing a classical batch operation or whether a middle-vessel column should be used ($Y_{middle-vessel} = \text{True}$) or not ($Y_{middle-vessel} = \text{False}$). These structural design options would be modeled by different sets of differential equations, of which only one is used depending on the Boolean variable value. In mathematical terms, this leads to a set of *disjunctive equations*, each of which is related to a Boolean variable. Such a problem formulation is a rather natural choice due to the relation between process synthesis problems and their disjunctive representation.

If the location of the side feed stream of the batch reactive distillation column and the number of reactive and nonreactive stages are additional degrees of freedom, this leads to a problem of combinatorial complexity. In order to avoid undesired design configurations and to reduce the complexity, we usually relate possible combinations to each other by propositional logic expressions. For example, we may want to state that only nonreactive stages are implemented above the side feed stream if a semibatch mode of operation is chosen. Such logical relationships between parts of the superstructure model, termed *implications*, are frequently employed in design problems. Another typical implication would be:

$$Y_{reactive-column} \Rightarrow \neg Y_{reactor} \tag{9.17}$$

which expresses that no reactor will be chosen when a reactive semibatch distillation column is selected. In this case, we would combine the structural design alternatives depicted in Figure 9.1 and Figure 9.5 into one disjunctive model. Logic expressions, such as the implication in Equation 9.17, are usually transformed into the representation $\Omega(Y)$, which involves logical operations only. Such a representation of Equation 9.17 could be stated as:

$$\Omega(Y) = \neg Y_{reactive-column} \vee \neg Y_{reactor} = \text{True} \tag{9.18}$$

We generally aim at propositional logic expressions due to the fact that they can be easily transformed into an algebraic representation comprised of linear equality and inequality constraints, as we will see later.

Summarizing, the following problem formulation is obtained by incorporating Boolean variables and disjunctive sets of equations into the multistage model stated in Equations 9.9 to 9.11:

$$f_k(\dot{x}_k(t), x_k(t), z_k(t), u_k(t), p, \zeta, t) = 0, t \in [t_{k-1}, t_k], k \in K \tag{9.19}$$

$$l(\dot{x}_1(t_0), x_1(t_0), z_1(t_0), p, \zeta) = 0 \tag{9.20}$$

$$x_{k+1}(t_k) - x_k(t_k) = 0, k \in K_m \tag{9.21}$$

$$
\begin{bmatrix}
Y_i \\
q_{k,i}(\dot{x}_k(t), x_k(t), z_k(t), \\
u_k(t), p, \zeta, t) = 0 \\
t \in [t_{k-1}, t_k], k \in K \\
s_i(\dot{x}_1(t_0), x_1(t_0), z_1(t_0), p, \zeta) = 0 \\
x_{k+1}(t_k) - x_k(t_k) = 0 \\
k \in K_m
\end{bmatrix}
\vee
\begin{bmatrix}
\neg Y_i \\
B_{k,i}[x_k(t)^T, z_k(t)^T, \\
u_k^T(t), p^T, x_k(t_{k-1})^T]^T = 0 \\
t \in [t_{k-1}, t_k] \\
k \in K
\end{bmatrix}
\tag{9.22}
$$

$$i = 1, \ldots, n_Y$$

$$\Omega(Y) = \text{True} \tag{9.23}$$

Because $f_k(\cdot) = 0$, $l(\cdot) = 0$ and the stage transition conditions (Equation 9.21) hold globally, further equations $q_{k,i}(\cdot) = 0$, initial conditions $s_i(\cdot) = 0$, and stage transition conditions are included in Equation 9.22 that are only enforced if a corresponding Boolean variable (Y_i) is true. Otherwise, if Y_i is false, a subset of the state and control variables, time-invariant parameters, initial values of differential states for nonexisting units are set to zero. Thus, $B_{k,i}$ is a square diagonal matrix with constant 1- or 0-valued matrix elements.

Note that, with a fixed choice of the Boolean variables Y and given control variables u_k, design parameters p, initial values $x_k(t_{k-1})$, and known uncertain parameters ζ, then the combined set of DAEs $f_k(\cdot) = 0$ and $q_{k,i}(\cdot) = 0$, the corresponding initial conditions $l(\cdot) = 0$ and $s_i(\cdot) = 0$, and the stage transition conditions (Equations 9.19 to 9.22) are assumed to uniquely determine the state variable vector, $x_k(t)$, $z_k(t)$.

The multistage model with structural design decisions can be regarded as a generalization of the multistage model presented in Section 9.2.3.2. Here, the switching conditions are also known explicitly and are, thus, independent of the process state and of the model parameters; however, multiple alternatives exist as to which model will be active during model stage $k \in K$, a fact that underlines the combinatorial character of the formulations shown in Equations 9.19 to 9.23.

Differences between the two models show up when we look at the different Petri net representations corresponding to the two model classes. In multistage models, as discussed in Section 9.2, the Petri net takes the form of a direct sequence (see Figure 9.5), whereas the model defined by Equations 9.19 to 9.23 contains alternative sequences that are determined by the Boolean variables Y. For example, two alternatives to separate the intermediate product of the batch reactor for the multistage process shown in Figure 9.5 could be formulated as a disjunctive multistage model containing two alternative options (distillation column and membrane unit) to perform the separation. The corresponding Petri net presentation with two switching functions of the type shown in Equation 9.13 is shown in Figure 9.7.

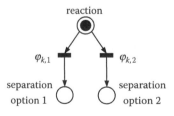

FIGURE 9.7 Alternative sequence in a Petri net.

9.2.3.5 The Treatment of Implicit Discontinuities

So far, structural design decisions have been considered only in conjunction with explicit discontinuities where the switching times t_k are set explicitly (see Equation 9.14); however, it is possible to generalize the disjunctive multistage model (Equations 9.19 to 9.23) to allow for implicit discontinuities in each of the stages where the switching time is determined by solving the nonlinear switching condition (Equation 9.15). Then, discrete–continuous hybrid models comprising implicit discontinuities, discussed in Section 9.3, would enter Equations 9.19 and 9.22). Such a generalization would increase the complexity of the model tremendously because a large number of explicit transitions between model stages as well as between discrete modes triggered by implicit discontinuities (at *a priori* unknown points in time t^*) might occur. For example, Avraam et al.[9] show how disjunctions can be used to model implicit discontinuities in discrete–continuous models. Using their formalism, the discrete–continuous multistage model (Equations 9.19 to 9.23) can be generalized to two layers of disjunctions. The first layer is comprised of all design decisions, including all the explicit discontinuities modeled by a switching function of the type shown in Equation 9.14. Implicit discontinuities stemming, for example, from physical discrete transitions (such as phase changes) lead to a disjunctive form of the DAE systems $f_k(\cdot) = 0$ and $q_{k,i}(\cdot) = 0$ themselves, as different sets of model equations apply for different discrete modes of the process within a stage k. Thus, for a fixed stage index k the global DAE system $f_k(\cdot) = 0$ is represented by different sets of differential equations. Each set is used to determine the state variable vector in the corresponding discrete mode of the process (see Avraam et al.[9] for more information). Because the equations $q_{k,i}(\cdot) = 0$ are dependent on external design decisions, they are already expressed in a disjunctive form (see Equation 9.22). Thus, a second layer of disjunctions is required to model implicit discontinuities that can potentially occur in the equations $q_{k,i}(\cdot) = 0$ in stage k if the corresponding Boolean variable Y_i is true.

9.2.4 OPTIMIZATION MODELS

All models described in the previous section define simulation problems on a finite time interval when Y (the process structure and sequence of the stages) and the duration (t_k) of the stages are fixed, when the continuous degrees of freedom u_k and p have been specified, and when the nominal values of the uncertain parameters ζ (representing model parameters ϑ and discretized disturbances $d(t)$) are available in all the stages. Such specifications would be largely arbitrary and would typically not account for production constraints and process economics. Hence, in order to specify requirements on the design, control, and operation of a batch process, an objective function and various constraints have to be formulated in addition to the model equations to finally result in a dynamic optimization instead of a simulation problem. These model extensions are formulated in this section.

9.2.4.1 Constraints

Constraints are formulated to enforce design and operational specifications, as well as to guarantee safety. A process operation is *safe* if all process variables considered to be safety relevant stay within certain predefined bounds during the entire operation cycle. Typical safety constraints are limits on the temperature or pressure in a reactor. An example of an operational constraint is a prespecified product concentration that is to be met during or at the end of the batch operation. An optimization problem (and the process described thereby) is *feasible* if all design, operational, and safety constraints are enforced during the entire operation.

The constraints can be cast into relations between the variables in a discrete mode of the discrete–continuous models. These are typically formulated as additional inequality (or equality) constraints that are valid at a particular time instant, such as a switching point in the multistage problem, or at the endpoint in any of the problems above or during the entire horizon of operation. These path and point constraints:

$$g_k(x_k(t), z_k(t), u_k(t), p, \zeta, t) \leq 0, t \in [t_{k-1}, t_k], k \in K \qquad (9.24)$$

$$g_k^e(x_k(t_k), z_k(t_k), u_k(t_k), p, \zeta, t_k) \leq 0, k \in K \qquad (9.25)$$

are hence added to the model equations in every discrete mode or model stage k. The constraints are usually simple algebraic expressions that can be easily evaluated; however, in some design problem formulations, these constraints are complex and may even comprise a full simulation or optimization problem themselves, such as in scenario-integrated modeling and optimization.[2]

9.2.4.2 Objectives

In addition to constraints, for the representation of design or control specifications one or more objectives are usually formulated, which are supposed to be minimized as part of the design activity. For one objective, usually a suitable criterion to measure economical success is chosen. Examples include such simple criteria as batch time or more complicated ones such as total production cost per unit product including annualized investment and operating costs.

Mathematically, the optimization objectives are formulated in terms of an objective functional Φ. Commonly, two different types of objective functionals are distinguished. The first one, the *Mayer* type, is employed, if the goal is to minimize a functional at final time:

$$\Phi_{Mayer} = M(x(t_f), z(t_f), p, t_f) \qquad (9.26)$$

An example would be to maximize the amount of product D in the methyl acetate example (Figure 9.1), also discussed in the example in Section 9.3.1.5.

The second type, the so-called *Lagrange* type, is formulated as an integral term of a functional L over the operation horizon:

$$\Phi_{Lagrange} = \int_{t_0}^{t_f} L(x(t), z(t), u(t), p, t)\, dt \qquad (9.27)$$

This type of objective functional describes, for example, the minimization of energy consumption, or a measure for the deviation of some process quantity from a certain set point. Note that an objective functional of the Lagrange type always can be converted into a Mayer-type formulation by introducing an additional state variable (see, for example, Betts[19]).

Often, profitability is not the only objective a designer has in mind, as not all of the design decisions impact the profit of the production. For example, not all of the environmental issues influence profit directly. In such cases, one has to deal with a multiobjective problem that reflects the need to trade-off conflicting goals.[43] The objectives can then be stated as a vector of functionals:

$$\Phi = \left[\Phi_1, \Phi_2, \dots, \Phi_n \right]^T \qquad (9.28)$$

The modeling of the objectives is as important as modeling the process dynamics.

9.2.4.3 Optimality

When all the constraints and suitable objective functionals have been formulated, the optimization problem is set up and can be solved to optimality using an appropriate constrained optimization algorithm. The solution of the optimization problem is *optimal* if all constraints are met and if the objective functional is minimal. In the case of multiple objectives, we can construct a single objective by a suitably weighted sum of the individual objectives:

$$\Phi = \sum_{j=1}^{n} \alpha_j \Phi_j \qquad (9.29)$$

or we can determine the Pareto optimal set (e.g., by variation of the weighting factors) and defer the decision on the best compromise to a later point when the Pareto optimal set is known. In fact, the commonly employed economical objective functionals stem from a multicriteria problem as we would like to minimize at the same time production and investment cost. Often, these criteria are conflicting. The cost functionals combining annualized investment and operating costs are exactly a suitably weighted sum of both — very different — cost functionals.

9.2.4.4 Flexibility and Robustness

The previous discussion does not account for uncertainty in the optimization problem. The presence of uncertainties can have an influence on the solution of an optimization problem, because model and process uncertainties can, for example, cause a process to violate the constraints and hence become infeasible, even if the nominal process is feasible. This leads to the definition of flexibility.

We now introduce the notion of the *nominal batch process*, which is characterized by the nominal model with the uncertain parameters ζ (representing uncertain model parameters ϑ and parameterized disturbances $d(t)$) fixed at some nominal value ζ_{nom}. A batch process is *flexible* if the nominal batch process is feasible and if there exist operational degrees of freedom (in particular, controls $u_k(t)$) such that the corresponding operational regimes are feasible for all possible realizations of $\zeta \in C$. In this context, flexibility is a qualitative measure that defines whether model or process uncertainties might cause the violation of constraints.

9.3 OPTIMAL DESIGN

After having established the modeling foundations, we now look at the optimal design problem, where we assume a given production demand and a given model. We distinguish two different cases: (1) design problems with fixed structural decision variables and (2) design problems for which the structural variables are also subject to optimization. The first case is discussed in Section 9.3.1 on the basis of the model and constraint formulations introduced in Sections 9.2.3.1 to 9.2.3.3 and 9.2.4. Design problems with structural degrees of freedom are treated in Section 9.3.2, where we make use of the model types that have been discussed in Sections 9.2.3.4 and 9.2.4.

9.3.1 DESIGN OF BATCH PROCESSES WITH FIXED STRUCTURE

We now consider the optimal design of batch processes with a fixed structure using single- or multistage models, as presented in Sections 9.2.3.1 and 9.2.3.2. The choice of which process variables are considered to be the degrees of freedom of the optimization and which type of objective functional Φ is employed depends on the goals of the design task. In *equipment design* problems, we are interested in optimal time-invariant parameters p, such as an optimal column diameter or a still pot size. These design parameters are regarded here as continuous rather than as discrete quantities. The determination of optimal control profiles, $u_k(t)$, $k \in K$, is frequently performed using an open-loop *trajectory design*. If, in addition, controllers of the process are integrated into the model, optimal closed-loop trajectories are obtained in a so-called *control-integrated design* problem. In the following text, we present how an optimization problem can be formulated and solved for these three different design tasks. We assume that neither uncertain parameters nor disturbances are present within the dynamic models or that nominal values are known; hence, the quantities ζ are suppressed in the following formulations.

9.3.1.1 Mathematical Problem Formulation

The optimization problem formulation employed here is formally based on multistage models (see Section 9.2.3.2). Design problems defined for single-stage models are also covered by simply fixing the number of stages to $n_s = 1$. The objective functional Φ is obtained by summing up individual costs Φ_k formulated for each model stage $k \in K$. Then, we obtain the following optimization problem:

$$\min_{x_k(t), z_k(t), u_k(t), p, t_k} \Phi := \sum_{k=1}^{n_s} \Phi_k(x_k(t_k), z_k(t_k), p, t_k) \tag{P1}$$

$$\text{s.t. Equations 9.9 to 9.11, 9.24, 9.25} \tag{9.30}$$

Note the analogy to a multiobjective problem, in that the objective functional in Problem (P1) is also a summation of individual objectives, as in Equation 9.29; however, here each individual Φ_k objective is associated with a specific stage. In principle, a weighting of these stage objectives different from one is also conceivable. The dynamic optimization problem (P1) can be solved by a number of different approaches, which are discussed in the following text.

9.3.1.2 Solution Strategies

Solution strategies for dynamic optimization problems can be classified into indirect and direct methods. The former class of solution approaches solves the dynamic optimization problem *indirectly*. These methods, also known as *variational methods*, use the first-order necessary conditions from Pontryagin's maximum principle[96] in order to reformulate the problem as a two-point boundary problem; however, the resulting boundary value problems are usually very difficult to solve.

In order to illustrate this solution method we restrict the problem class to single-stage models without disjunctions. We assume a process model given as a system of ordinary differential equations, $x = f(x,u)$, with initial conditions $x(t_0) = x_0$. We further allow path and endpoint constraints (Equations 9.24 and 9.25) and assume an objective functional of the Mayer type (Equation 9.26) for simplicity. Then, the application of Pontryagin's maximum principle gives the Hamiltonian:[35]

$$H = \lambda^T f(x,u) + \mu^T g(x,u) \tag{9.31}$$

and the associated so-called adjoint equations:

$$\dot{\lambda}^T = -\frac{\partial H}{\partial x}, \quad \lambda^T(t_f) = \frac{\partial \Phi}{\partial x}\bigg|_{t_f} + v^T \left(\frac{\partial g^e}{\partial x}\right)\bigg|_{t_f} \qquad (9.32)$$

Here, $\lambda(t) \neq 0$ represents the vector of adjoint states, $\mu(t) \geq 0$ the vector of Lagrange multipliers for the path constraints, and $v \geq 0$ the vector of Lagrange multipliers for the terminal constraints. The solution of this two-point boundary value problem requires forward integration of $x = f(x,u)$ with the initial condition $x(t_0) = x_0$ and backward integration of the differential equation system for λ (Equation 9.32) with the given initial condition at time point t_f.

The necessary conditions for optimality (NCO) are defined by:

$$\frac{\partial H}{\partial u} = \lambda^T \frac{\partial f}{\partial u} + \mu^T \frac{\partial g}{\partial u} = 0 \qquad (9.33)$$

and by:

$$\mu^T g = 0, \quad v^T g^e = 0 \qquad (9.34)$$

If a free end time is allowed, the transversality condition:

$$H(t_f) = (\lambda^T f + \mu^T g)\bigg|_{t_f} = 0 \qquad (9.35)$$

has to be fulfilled, too. The numerical solution of these problems with state path constraints is rather difficult;[66] however, this solution approach gives deep insight into the solution structure that can be exploited in online optimization applications as shown in Section 9.4.2.3.

Alternatively, *direct methods* can be used. These solution methods solve the optimization problem directly. The decision (control) variables $u_k(t)$ as well as the state variables $x_k(t)$, their derivatives $\dot{x}_k(t)$ with respect to time, and the algebraic state variables $z_k(t)$ appear as infinite dimensional quantities in this problem. The problem can be converted into a finite dimensional one, usually a nonlinear programming problem (NLP), which subsequently can be solved by a suitable numerical optimization algorithm. This can be accomplished by discretizing the infinite dimensional quantities using a finite number of parameters. In general, three different direct methods can be distinguished: the control vector parameterization approach, the multiple-shooting approach, and the full discretization approach. These methods differ in how the infinite dimensional quantities are discretized and which of them are degrees of freedom in the optimization algorithm. The control vector parameterization approach and the full discretization method are described in more detail in the following sections. The multiple-shooting approach can be seen as an in-between one that combines features of both of the other approaches. A concise review on multiple shooting can be found elsewhere.[33,85]

9.3.1.3 Solution via Control Vector Parameterization

In the control vector parameterization approach (also termed *sequential* or *single shooting* approach), only the discretized control variables, the time-invariant parameters, and the final time of each stage are degrees of freedom for the optimization.

Conversion into an NLP Problem

The optimization problem (P1) contains an n_{u_k}-dimensional vector of time-dependent control variables u_k on each stage k according to:

$$u_k(t) = [u_1^k(t), \ldots, u_{n_{u_k}}^k(t)]^T \tag{9.36}$$

Usually, the control profiles $u_l^k(t), l = 1, \ldots, n_{u_k}$ are approximated on each model stage $k = 1, \ldots, n_s$ by piecewise polynomial expansions of the form:

$$u_l^k(t) \approx \tilde{u}_l^k(\hat{u}_l^k, t) = \sum_{j \in \Lambda_l^k} \hat{u}_{l,j}^k \phi_{l,j}^k(t), \quad l = 1, \ldots, n_{u_k}, \forall k \in K \tag{9.37}$$

where Λ_l^k denotes the index set of the chosen parameterization functions $\phi_{l,j}^k(t)$, and the vector $\hat{u}_l^k$ contains the corresponding parameters. Each stage time interval $[t_{k-1}, t_k]$ is therefore divided into subintervals defined on each model stage k with grid points $t_{k,j}, j \in \Lambda_l^k$. A typical choice for a parameterization are piecewise constant functions where $\phi_{l,j}^k(t) = 1$, if $t_{k,j} \leq t \leq t_{k,j+1}$, and $\phi_{l,j}^k(t) = 0$, otherwise. However, higher order splines can also be used to parameterize $u_l^k(t)$. The grid points for each $u_l^k, l = 1, \ldots, n_{u_k}$, are contained in the mesh $M_{\Lambda_l^k} := \{t_{k,j} \mid j \in \Lambda_l^k\}$.

In the sequential approach, discretization of the state variables $x_k(t)$ and $z_k(t)$ is done implicitly by means of numerical integration of the initial-value problems defined on the n_s model stages. Hence, the dynamic optimization problem can be transformed into the following NLP for fixed Λ_l^k with the search variable vector $\theta_k := [u^{k,T}, p^T, x_k^T(t_{k-1}), t_k]^T$, where $x_k(t_{k-1})$ denotes the vector of the free initial values of differential state variables in each model stage k:

$$\min_{\theta_k} \sum_{k=1}^{n_s} \Phi(x_k(\theta_k, t_k), z_k(\theta_k, t_k), p, t_k) \tag{P2}$$

$$\text{s.t.} \quad g_k(x_k(\theta_k, t_{k,j}), z_k(\theta_k, t_{k,j}), \theta_k, t_{k,j}) \leq 0, \forall t_{k,j} \in M_{\Lambda^k}, \forall k \in K \tag{9.38}$$

$$g_k^e(x_k(\theta_k,t_k),z_k(\theta_k,t_k),\theta_k,t_k) \leq 0, \forall \, k \in K \tag{9.39}$$

$$l(\dot{x}_1(\theta_1), x_1(\theta_1), z_1(\theta_1), p) = 0 \tag{9.40}$$

Typically, path constraints g_k are enforced only point-wise along the time horizon. For this purpose, we define a unified mesh of all control variables on each model stage k according to $M_{\Lambda^k} := \bigcup_{l=1}^{n_{uk}} M_{\Lambda_l^k}$. Path constraints are then enforced at all time points $t \in M_{\Lambda^k}$, as indicated in Equation 9.3.

Solution of the NLP Problem

Problem (P2) describes a finite dimensional nonlinear programming problem (NLP), where the underlying DAE systems (Equations 9.9 to 9.11) are solved by numerical integration. Due to this fact, the sequential approach corresponds to a *feasible path strategy*, because each iterate of the optimization is feasible with respect to the DAE system. Nevertheless, problem P1 might be infeasible with respect to the constraints found in Equations 9.38 to 9.40 during the iterations. The optimization problem itself can be solved by a suitable NLP solver. Common choices in practice are sequential quadratic programming (SQP) solvers.[63]

In the sequential approach, the dimension of the search variable vector θ_k for the NLP is typically much smaller than the dimension of the DAE system (Equation 9.9); therefore, the computational performance of this approach strongly depends on an efficient evaluation of the objective and constraint function values and also the gradient information of these functions with respect to θ_k. For efficiency, NLP solvers generally require this gradient information, which can be found in three different ways: (1) by integrating the sensitivity equation systems, (2) by backward integration of the adjoint system, or (3) by perturbation. Although advances have been reported,[115] perturbation methods are generally considered to be less accurate and efficient than the other methods. The computational requirements for the adjoint approach[35] are proportional to the number of constraints and independent of the number of decision variables. However, in the presence of path constraints, such as Equation 9.38, this approach becomes almost intractable because a separate adjoint system must be developed for each constraint.[32] Solution of the sensitivity equations provides the state variable sensitivity matrices:

$$S_k^x := \frac{\partial x_k}{\partial \theta_k}, S_k^z := \frac{\partial z_k}{\partial \theta_k} \tag{9.41}$$

which can be used for calculating the derivatives of the objective functional with respect to the degrees of freedom θ_k:

$$\frac{d\Phi_k}{d\theta_k} = \left(\frac{\partial\Phi_k}{\partial x_k}\right)^T S_k^x + \left(\frac{\partial\Phi_k}{\partial z_k}\right)^T S_k^z + \frac{\partial\Phi_k}{\partial\theta_k} \qquad (9.42)$$

In a similar way, the derivatives of the constraint functionals with respect to θ_k can be computed. The sensitivity integration approach can be applied in a straightforward way to problems including path constraints. It is the method of choice in most implementations of sequential approach dynamic optimizers.[106,125] The drawback of this approach is that it can become computationally expensive, because for each degree of freedom an additional sensitivity equation system must be solved. The sensitivity system can be obtained easily from the DAE model. It has special properties that allow an efficient integration together with the DAE system itself.[54] Nevertheless, the major part of computation time used in the sequential approach to dynamic optimization is spent on the sensitivity computation, because the effort grows superlinearly with the number of decision variables for the control profiles. Hence, it is clearly desirable to keep the number of decision variables as small as possible, without losing much accuracy in the approximation of the control profiles. This can be accomplished by using an adaptive strategy for the optimal selection of the discretization meshes $M_{\Lambda_l^k}$ as we will see later.

As shown in Section 9.2.3.2, stage transition conditions (see Equations 9.11 and 9.12) are used to connect process state variables across the stage boundaries. In a similar way, mapping conditions have to be used for the sensitivities S_k^x of the differential states:

$$S_{k+1}^x(t_k) - S_k^x(t_k) = 0 \qquad (9.43)$$

when we have a continuity condition for the differential states such as the one given in Equation 9.11. The algebraic sensitivities (S_k^z) are calculated by a consistent initialization of the sensitivity DAE system.

Discrete–Continuous Hybrid Models

In order to be able to optimize a discrete–continuous hybrid batch process, it is not sufficient to only locate the point in time t^* where the discontinuity takes place and map the differential state variables from one discrete mode to the next. Depending on the switching function φ_k and the mapping condition, sensitivities of differential state variables may be discontinuous even if we assume continuous-state variable profiles enforced by the continuity condition shown in Equation 9.16. If a discrete event takes place at time t^*, the difference of the sensitivities at t_-^* and t_+^* is calculated by:

$$S_k^x(t_+^*) - S_k^x(t_-^*) = -\left(x_k(t_+^*) - x_k(t_-^*)\right)^T \frac{\partial t^*}{\partial\theta_k} \qquad (9.44)$$

Hence, a jump in the sensitivities occurs if the switching time t^* depends on the decision parameters θ_k. The partial derivative $\frac{\partial t^*}{\partial \theta_k}$ is obtained through total differentiation of the switching function φ_k with respect to θ_k as:

$$\frac{\partial t^*}{\partial \theta_k} = \frac{\left(\left(\frac{\partial \varphi_k}{\partial x_k}\right)^T \bigg|_{t_-^*} s_k^x(t_-^*) + \frac{\partial \varphi_k}{\partial \theta_k}\bigg|_{t_-^*}\right)}{\left(\left(\frac{\partial \varphi_k}{\partial x_k}\right)^T \bigg|_{t_-^*} \dot{x}(t_-^*) + \frac{\partial \varphi_k}{\partial t}\bigg|_{t_-^*}\right)} \tag{9.45}$$

For simplicity, Equation 9.45 is derived on the basis of a switching function that is independent of the algebraic state variables $z_k(t)$. This expression reveals an important difference between implicit and explicit discontinuities. With explicit discontinuities, we always have $\frac{\partial t^*}{\partial \theta_k} = 0$ and, thus, continuous sensitivities due to the fact that φ_k is independent of process states and parameters. If, instead, the switching is triggered by variable state or parameter values, discontinuous sensitivities have to be expected. In this context, it is interesting to note that for multistage models the sensitivities are continuous across the stage boundaries (see Equation 9.43) because the corresponding switching function φ_k^{ms} (see Equation 9.14) is independent of the process states and parameters. Consequently, a dynamic optimization problem involving a discrete–continuous hybrid model may become a nonsmooth optimization problem for which smooth NLP methods (such as SQP) are not appropriate.[93] An alternative way to solve these nonsmooth problems is to reformulate the discrete–continuous model as a disjunctive (see Section 9.2.3.4) or mixed-integer model to obtain piecewise smooth models for which special solution techniques are available. An overview of these methods is given later in conjunction with the structural design of batch processes in Section 9.3.2.

Adaptation of Control Vector Grids

Due to the inherently transient nature of batch process operation, optimal control profiles often exhibit large local differences in frequency content over the time horizon. This, in addition to computational efficiency arguments, calls for adaptively chosen, possibly nonequidistant control discretization meshes. In particular, a parameterization that is too coarse on a uniform grid or which locally resolves the control profiles at the wrong place on a uniform grid cannot meet the prespecified accuracy requirements. On the other hand, a parameterization, that is too fine not only leads to inappropriately high computational cost, but can also cause robustness problems, such as oscillatory control profiles; however, it is no trivial matter to generate a problem-adapted mesh *a priori*. A few attempts have been made to incorporate adaptivity into the parameterization scheme with the objective of automatically determining an appropriate mesh. For example, Waldraff et al.[129] applied a grid generation procedure based on curvature information of the optimal solution. Betts and Huffmann[20] proposed grid refinement based on local error analysis of the differential equation where the parameterization of the control variables is directly linked to the approximation error. Vassiliadis et al.[125] introduced the lengths of the discretization intervals of the control variables as additional degrees of freedom into the NLP problem. Binder et al.[24] suggested a framework that automatically generates sequences of nonuniform grids with an increasing degree of resolution. This method is started from a coarse grid with few decision variables so the problem can be solved efficiently and robustly. In every following refinement cycle, grid points are either inserted or deleted based on a multiscale-based refinement criterion. This framework makes use of the fact that the Equation 9.37 offers the choice of separate, nonuniform parameterization grids for each control variable. The refinement loop generates efficient, problem-adapted meshes $M_{\Lambda_l^k}$.

9.3.1.4 Solution via Full Discretization

In contrast to the control vector parameterization and multiple-shooting approaches, full discretization methods *fully* discretize both, state and control variables.[19,22] These methods are also referred to as *simultaneous methods*, because the solutions of the optimization problem and the DAE system are found simultaneously. In particular, the DAE system is solved only once at the optimal point and does not have to be fulfilled on the solution path of the optimizer. Therefore, the full discretization is an *infeasible path strategy* in contrast to the control vector parameterization approach, where the DAE system has to be solved in each optimization iteration.

Conversion into an NLP Problem

For conversion of the dynamic optimization problem into an NLP, the control variables are approximated in a fashion similar to that for the sequential approach:

$$u_l^k(t) \approx \tilde{u}_l^k(\hat{u}_l^k,t) = \sum_{j \in \Lambda_l^k} \hat{u}_{l,j}^k \phi_{l,j}^k(t), \quad l = 1,\dots,n_{u_k}, \forall\, k \in K \qquad (9.46)$$

Additionally, the derivatives of the differential state variables and the algebraic state variables are approximated as:

$$\dot{x}_l^k(t) \approx \dot{\tilde{x}}_l^k(\dot{\hat{x}}_l^k,t) = \sum_{j \in \Upsilon_l^k} \dot{\hat{x}}_{l,j}^k \psi_{l,j}^k(t), \quad l = 1,\dots,n_{x_k}, \forall\, k \in K \qquad (9.47)$$

$$z_l^k(t) \approx \tilde{z}_l^k(\hat{z}_l^k,t) = \sum_{j \in \Xi_l^k} \hat{z}_{l,j}^k \phi_{l,j}^k(t), \quad l = 1,\dots,n_{z_k}, \forall\, k \in K \qquad (9.48)$$

Typically, collocation on finite elements is applied (see, for example, Biegler et al.[22]) by choosing Lagrange polynomials as parameterization functions $\phi_{l,j}^k(t)$. Then, the coefficients $\hat{z}_l^k$ and $\hat{u}_l^k$ correspond to the values of the profiles of $\tilde{z}_l^k$ and $\tilde{u}_l^k$ at the collocation points. This is advantageous, because for example path constraints can be easily implemented. Also, scaling of the original DAE model directly propagates into the NLP problem. The collocation points for all x^k, z_k, and u^k are contained in the mesh M_k. The parameterization functions $\psi_{l,j}^k(t)$ for the differential variables are then chosen as polynomials, such that the coefficients $\hat{x}_l^k$ match the derivative values of the profiles $\tilde{x}_l^k$. To ensure continuity of the differential state variable profiles, continuity equations are enforced at the element boundaries.

Besides using polynomial expansions for approximation of the continuous profiles, other discretization schemes have also been suggested. For example, a multiscale-based discretization approach using wavelets as basis functions has been employed by Binder et al.[27] By substituting Equations 9.46 to 9.48 into problem (P1), the following NLP problem is obtained:

$$\min_{\theta_k} \sum_{k=1}^{n_s} \Phi(\theta_k, t_k) \tag{P3}$$

$$\text{s.t.} \quad f_k(\theta_k, t_{k,j}) = 0, t_{k,j} \in M_k, \forall k \in K \tag{9.49}$$

$$g_k(\theta_k, t_{k,j}) \leq 0, t_{k,j} \in M_k, \forall k \in K \tag{9.50}$$

$$g_k^e(\theta_k, t_k) \leq 0, \forall k \in K \tag{9.51}$$

$$l(\theta_1) = 0 \tag{9.52}$$

with the search variable vector $\theta_k := [\hat{x}^{k,T}, x_k^T(t_{k-1}), z^{k,T}, u^{k,T}, p^T, t_{k+1}]^T$, where $x_k(t_{k-1})$ denotes the vector of free initial values of differential state variables in each model stage k. Note that by the choice of the collocation polynomials the differential–algebraic model equations are essentially evaluated point-wise at the collocation points.

Solution of the NLP problem

The dimension of the search variable vectors θ_k, $k \in K$ is usually much larger than in the sequential approach, because it depends not only on the dimension n_u of the control vector but also on the size of the DAE model $(n_x + n_z)$, which in typical practical applications is much larger than n_u. Whereas the NLP problem (P2) in the control vector parameterization approach is usually of small to moderate size and can be solved with existing general purpose NLP solvers, the NLP problem (P3) of the full discretization method can be quite large and requires specifically tailored solution methods. Commonly, reduced space SQP (rSQP) methods[22] are applied using projected Hessian matrices or their quasi-Newton approximation in order to avoid the need for second-order derivative information.

The key feature of an rSQP strategy is to partition the optimization variables θ_k into dependent variables determined by the constraints, and independent variables. It is important to note that the independent variables are not always comprised of the control variables and parameters; rather, control variables and parameters can be exchanged with algebraic state variables, if unstable modes in the states are detected (see, for example, Biegler et al.[22] for details). Although this decomposition strategy is generally efficient, solution of the reduced quadratic subproblems (QPs) can become a bottleneck if active set methods are used for this purpose. The reason for this is the combinatorial problem of selecting the active set, because bounds on all variables are present. As an alternative to active set strategies, interior point (IP) methods have been developed for the solution of large-scale NLP problems.[58] These methods have been successfully applied in a dynamic optimization context.[38]

Discretization Grid Adaptation

Analogous to the sequential approach, the accuracy, efficiency, and robustness of the solution of a discretized dynamic optimization problem strongly depend on the chosen discretization grids. In the simultaneous approach, this holds for the grids for the control and state variables. In particular, the accuracy of the state profiles depends on these grids, whereas in the sequential approach the accuracy of the state profiles is directly controlled by the numerical integration algorithm. Tanartkit and Biegler[116] and von Stryk[128] suggested a full discretization method in which the optimizer determines the element length or spatial position of the collocation points as additional decision variables. Biegler et al.[22] presented an extension of an interior point optimization strategy that incorporates a strategy for finite-element movement. The multiscale-based full discretization approach of Binder et al.[28–30] allows different, adapted discretization grids for state and control variables by solving a hierarchy of successively refined finite dimensional problems.

9.3.1.5 Applications

In the following, some of the previously introduced concepts are illustrated by means of a case study before we give a brief overview of some practical applications of batch process optimization.

Case Study

Let us again look at the methyl acetate production introduced in Section 9.1. We select a semibatch reactive distillation process like the one shown in Figure 9.1. The structure and size of the process equipment are assumed to be given. The objective of the optimization is to maximize the amount of product D at the end of a fixed batch time ($t_f = 4$ hours). Thus, the final product concentration x_D^{MeAc} should be at least 0.95. The degree of freedom, $u(t)$, that can be manipulated during the operation is the internal reflux ratio, $r = R/V$.

A formulation of this problem according to the definitions introduced in Sections 9.2 and 9.3 yields:

$$\min_{r(t)} \Phi := -D(t_f) \tag{P4}$$

$$\text{s.t.} \quad f(\dot{x}(t), x(t), z(t), r(t)) = 0, t \in [t_0, t_f] \tag{9.53}$$

$$x(t_0) - x0 = 0 \tag{9.54}$$

$$x_D^{\text{MeAc}}(t_f) \geq 0.95 \tag{9.55}$$

$$0 \leq r(t) \leq 1 \tag{9.56}$$

Here, Equations 9.53 and 9.54 represent a DAE model of the semibatch reactive distillation process and the corresponding initial conditions. Note that the stage index k is not stated explicitly as this is a single-stage problem with $n_s = 1$. The model is based on the following assumptions: (1) constant liquid and negligible vapor holdups on the stages and in the condenser/reflux drum, (2) constant molar overflow, (3) constant boil-up rate, (4) ideal gas phase, (5) start-up phase not considered. A Wilson model is used to calculate the activity coefficients for the vapor–liquid equilibrium. Initially, the column is filled with pure methanol. The side feed stream of pure acetic acid is constant over time. The stripping section of the column is comprised of seven reactive equilibrium stages, whereas the upper part of the column has five nonreactive trays. For more details about this separation problem, refer to Lee et al.[81] The model at hand has $n_x = 74$ differential variables and $n_z = 739$ algebraic variables. The problem does not have any state path constraint but does have one endpoint constraint (Equation 9.55) to enforce the desired product purity. The reflux ratio is bounded between its physical limits 0 and 1 (see Equation 9.56).

This problem can be solved by one of the approaches introduced in Section 9.3.1. Here, we chose the control vector parameterization approach; therefore, the control profile $r(t)$ is discretized according to Equation 9.37. Piecewise constant functions are used for this purpose.

The results of this optimization are depicted in Figure 9.8. The left graph shows the optimal profile for the reflux ratio, $r(t)$, discretized into 32 equidistant intervals. The right graph depicts the trajectory for the product concentration x_D^{MeAc}.

The results have a physical interpretation. In the first phase of operation, the reflux ratio is set to the upper bound 1.0, which corresponds to a full reflux in order to increase the composition of the light component in the top of the column. Then it switches to a lower-value profile that increases the amount of product in the product vessel in an optimal way, such that the endpoint purity constraint will be met. This can also be seen from the trajectory of the product concentration, which first rises sharply (due to the full reflux regime) and then increases only slowly to finally reach the desired value. In the last phase, the reflux jumps to a lower value in order to exploit the high concentration of MeAC available in the condenser at the end of the batch.

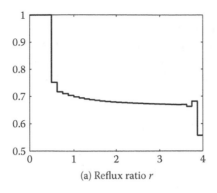

(a) Reflux ratio r

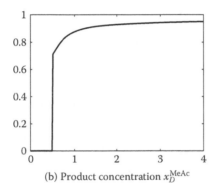

(b) Product concentration x_D^{MeAc}

FIGURE 9.8 Optimal trajectories for reflux ratio and vapor stream.

Practical Applications

Whereas continuous processes have been subject to many rigorous optimization studies, industrial batch and semibatch processes are often still operated using recipes based on heuristics and experience; consequently, only a limited number of studies for applications to laboratory or production scale processes can be found in the literature. Lehtonen et al.[83] reported on the optimization of reaction conditions for an alkali fusion process in which optimal operational profiles have been calculated using unconstrained optimization. The productivity optimization of an industrial semibatch polymerization reactor under safety constraints was treated by Abel et al.[2] Ishikawa et al.[72] determined optimal control profiles for an industrial vapor–liquid batch reactor for the productions of dioctyl phthalate. The process model consists of about 2000 differential–algebraic equations. The optimization problem has been solved using the control vector parameterization approach introduced above. Klingberg[77] presented a study on the dynamic optimization of an industrial batch reactive distillation column in which the model had about 4200 DAEs. Again, this problem has been solved by control vector parameterization. To the authors' knowledge, no reports of the solution of problems involving models of such a size using the full discretization approach can be found in the literature.

A case study of the optimization of a multistage batch process involving a reaction and separation task as well as recycles was presented by Charalambides et al.[41] The multistage model employed was comprised of 420 DAEs and 76 decision variables.

9.3.2 STRUCTURAL DESIGN OF BATCH PROCESSES

Besides flexibility in terms of operational strategies, batch processes allows for a number of different design configurations, such as the middle or multivessel design of a batch distillation process or the equipment configuration of a batch reactor. These structural design alternatives, combined with different possible operating modes, provide great flexibility and economic potential but result in demanding optimization problems due to the combined discrete and continuous nature of the degrees of freedom.[76] Methods that simultaneously address the structural design and determination of operational strategies of transient reaction and separation processes in chemical engineering have been proposed by several authors, including Barrera and Evans,[16] Mujtaba and Macchietto,[91] and Bhatia and Biegler,[21] who integrated discrete decisions into the batch design task. The corresponding approaches address batch design problems with process models of a different level of detail. Allgor and Barton[5] and Sharif et al.[110] formulated the design problem as a *mixed-integer dynamic optimization* (MIDO) problem that incorporates both process dynamics and discrete decision variables in a *superstructure* model.

9.3.2.1 Mathematical Problem Formulation

In Section 9.2.3.4, we introduced multistage models with disjunctive constraints in order to formalize structural design decisions. This superstructure model contains Boolean variables, $Y \in \{\text{True}, \text{False}\}, ^{n_Y}$ that are related to disjunctive sets of differential–algebraic equations and constraints. The corresponding class of optimization problems is termed *disjunctive programming* and is well known in the context of process synthesis of continuously operated processes.[99]

The basic ideas of disjunctive programming have been transferred to the dynamic case.[94] Here, the multistage dynamic optimization problem (P1) is extended by disjunctive constraints, leading to the following mixed-logic dynamic optimization (MLDO) problem:

$$\min_{x_k(t), z_k(t), u_k(t), p, Y, t_k} \Phi := \sum_{k=1}^{n_s} \Phi_k(x_k(t_k), z_k(t_k), p, t_k) + \sum_{i=1}^{n_Y} b_i \quad \text{(P5)}$$

$$\text{s.t. Equations 9.19 to 9.21, 9.24, 9.25} \quad (9.57)$$

$$
\begin{bmatrix}
Y_i \\
q_{k,i}(\dot{x}_k(t), x_k(t), z_k(t), \\
u_k(t), p, t) = 0 \\
t \in [t_{k-1}, t_k], k \in K \\
s_i(\dot{x}_1(t_0), x_1(t_0), z_1(t_0), p) = 0 \\
r_{k,i}(x_k(t), z_k(t), u_k(t), p, t) \le 0 \\
t \in [t_{k-1}, t_k], k \in K \\
r_{k,i}^e(x_k(t_k), z_k(t_k) \\
u_k(t_k), p, t_k) \le 0 \\
k \in K \\
x_{k+1}(t_k) - x_k(t_k) = 0 \\
k \in K_m \\
b_i = \gamma_i
\end{bmatrix}
\vee
\begin{bmatrix}
\neg Y_i \\
B_{k,i}[x_k(t)^T, z_k(t)^T, \\
u_k^T(t), p^T, x_k(t_{k-1})^T]^T = 0 \\
t \in [t_{k-1}, t_k] \\
k \in K \\
b_i = 0
\end{bmatrix}
\tag{9.58}
$$

$$
i = 1, \ldots, n_Y
$$
$$
\Omega(Y) = \text{True}
\tag{9.59}
$$

Note that Equation 9.57 is identical to those of the multistage formulation used in Section 9.3.1.1, whereas the objective function Φ includes an extra term that covers the investment costs b_i induced by a process part or unit i. Whereas the differential equations (Equation 9.19), initial conditions (Equation 9.20), stage transition conditions (Equation 9.21), and path and endpoint constraints (Equations 9.27 and 9.28) contained in Equation 9.57 and the objective function Φ hold globally, there are further equalities $q_{k,i}$, inequalities $r_{k,i}$, $r_{k,i}^e$, and stage transition conditions included in Equation 9.58 that are only enforced if the corresponding Boolean variable Y_i is true. Otherwise, if Y_i is false, a subset of the state and control variables, time-invariant parameters, initial values of differential states, and fixed investment costs (b_i) for nonexisting units are set to zero, as already introduced in Section 9.2.3.4. Commonly, the Boolean variables Y themselves are partly related through propositional logic constraints (Equation 9.59), as explained in Section 9.2.3.4. Usually, a large portion of the combinatorial complexity of the MLDO problem is governed by these logic relationships.

9.3.2.2 Solution Strategies

The optimization problem (P5) can be solved using a number of different approaches. The corresponding solution methods rely either on reformulation of the mixed-logic dynamic optimization (MLDO) into a mixed-integer dynamic optimization (MIDO) problem or on solving the MLDO problem directly.[94] We focus here on approaches based on a MIDO reformulation that have been widely studied in the literature.

Reformulation as MIDO problem

A mixed-integer dynamic optimization problem is obtained by replacing the Boolean variables with binary variables $y \in \{0,1\}^{n_y}$ and by representing the disjunctions (Equation 9.58) using big-M constraints[131] or a convex-hull formulation.[12,99] The use of big-M constraints is illustrated by reformulation of the inequality constraint $r_{k,i}(u_k(t), t) \leq 0$ in (Equation 9.58). When $M_{k,i}$ is a sufficiently large positive constant and with bounds $u_{k,L}$, $u_{k,U}$ for the control variables, we obtain:

$$r_{k,i}(u_k(t),t) \leq M_{k,i}(1-y_i) \tag{9.60}$$

$$u_{k,L} y_i \leq u_k(t) \leq u_{k,U} y_i \tag{9.61}$$

By substituting $y_i = 1$ into Equation 9.60 we see that the constraint $r_{k,i} \leq 0$ is enforced. Otherwise, if $y_i = 0$, $r_{k,i}$ is unconstrained for an appropriate choice of $M_{k,i}$ and a subset* of the variables, $u_k(t)$ may be set to zero as shown in Equation 9.61.

Both reformulation techniques are well known in the context of mixed-integer nonlinear programming (MINLP) for the synthesis of process flow sheets. The basic concepts can also be applied to problems involving dynamic process models, such as the MLDO problem stated above. Thus, by employing big-M constraints for the disjunctions in Equation 9.58, we can state the following set of equations involving binary variables y_i instead of Boolean variables Y_i:

---

* The subset is defined by a part of the matrix $Bk_{,j}$ (see Equation 9.22) that corresponds to the variables $u_k(t)$.

$$-M_{k,i}(1-y_i) \leq q_{k,i}(\dot{x}_k(t), x_k(t), z_k(t), u_k(t), p, t) \leq M_{k,i}(1-y_i)$$

$$t \in [t_{k-1}, t_k], k \in K$$

$$-M_{k,i}(1-y_i) \leq s_i(\dot{x}_1(t_0), x_1(t_0), z_1(t_0), p) \leq M_{k,i}(1-y_i)$$

$$r_{k,i}(x_k(t), z_k(t), u_k(t), p, t) \leq M_{k,i}(1-y_i)$$

$$t \in [t_{k-1}, t_k], k \in K \qquad (9.62)$$

$$r^e_{k,i}(x_k(t_k), z_k(t_k), u_k(t_k), p, t_k) \leq M_{k,i}(1-y_i)$$

$$k \in K$$

$$-M_{k,i}(1-y_i) \leq x_{k+1}(t_k) - x_k(t_k) \leq M_{k,i}(1-y_i)$$

$$k \in K_m$$

$$[x^T_{k,L}, z^T_{k,L}, u^T_{k,L}, p^T_L, x^T_{k,L}]^T \, y_i$$

$$\leq B_{k,i}[x_k(t)^T, z_k(t)^T, u^T_k(t), p^T, x_k(t_{k-1})^T]^T \qquad (9.63)$$

$$\leq [x^T_{k,U}, z^T_{k,U}, u^T_{k,U}, p^T_U, x^T_{k,U}]^T \, y_i$$

$$Ay \leq b \qquad (9.64)$$

As shown by Türkay and Grossmann,[121] propositional logic constraints formulated for the mixed-logic optimization problem can be expressed in terms of linear constraints including binary variables in a MIDO problem formulation. Considering again the example presented in Section 9.4, where a batch reactor is excluded from the set of alternatives when a reactive batch distillation column is chosen (see Equation 9.17), we can state the linear inequality as:

$$1 - y_{reactive-column} + 1 - y_{reactor} \geq 1 \qquad (9.65)$$

or equivalently

$$y_{reactive-column} + y_{reactor} \leq 1 \qquad (9.66)$$

in order to obtain a mathematical representation instead of a logical expression. Note that the inequality (9.65) can be directly deduced from Equation 9.18.

9.3.2.3   Solution Algorithms for MIDO Problems

Direct solution methods for dynamic optimization problems without discrete variables have been discussed in the previous sections. In fact, these methods are capable of solving a broad class of problems, including applications governed by large-scale DAE systems.[2] As mentioned in Section 9.3.1, they convert the time-continuous dynamic optimization problem into a finite-dimensional *nonlinear programming* problem (NLP) by discretization.

By using one of these direct approaches, the mixed-integer dynamic optimization problem (P5) can be converted into an algebraic problem which then forms a mixed-integer nonlinear programming instead of an NLP. Fortunately, a number of solution methods are available for solving MINLP problems.[57] In fact, these algorithms were shown to be suitable for solving problems of a considerable size; however, all of them suffer from the fact that they inherently rely on the convexity of the MINLP problem, which does not hold true for almost all practical applications.

The MINLP algorithms reported in the literature can be divided into two classes: based on enumeration or on decomposition. The former class is comprised of the complete enumeration of the entire discrete decision space (which is only tractable for problems with a small number of discrete variables) and branch-and-bound type of methods.[57,92] The branch-and-bound methods relax the problem (P5) continuously; that is, the binary variables are allowed to take values between 0 and 1. The relaxed problem is then solved in order to provide a lower bound to the original solution. Subsequently, different strategies can be applied to fix subsets of the binary variables in the nodes of a search tree. Parts of the search tree can be cut off if the current solution is greater than a known upper bound to the solution of the original problem or if the continuously relaxed binary variables take discrete values 0 or 1. The (global) optimal solution is found when all binary variables take discrete values $\{0,1\}$ and the relaxed dynamic optimization problems have been solved to (global) optimality in all the nodes. A branch-and-bound solution strategy for MIDO problems was recently proposed by Buss et al.[36]

On the other hand, decomposition methods are based on decomposing the optimization problem into two subproblems that are solved in an iterative manner. *Outer Approximation* (OA)[52,56,78,127] and *Generalized Benders Decomposition* (GBD)[62] are two well-known decomposition algorithms that have been used by Schweiger and Floudas[108] and Bansal et al.[15] to solve MIDO problems of considerable size.

In the OA method, a primal subproblem, which constitutes a dynamic optimization problem with fixed binary variable values, is solved to yield an upper bound, Z_{ub}^L, for the optimization problem (P5). Subsequently, a linear outer approximation of the overall mixed-integer problem is utilized within the master subproblem (a *mixed-integer linear programming* [MILP] problem) to provide a lower bound to problem (P5) and a new set of binary variables for the next primal subproblem. After a finite number of iterations the nondecreasing lower bound and the minimum of all upper bounds $Z_{ub}^L := \min(Z_{ub}^\ell), \ell = 0, ..., L$, approach each other up to a specified error tolerance ε at the optimum value of the original problem (P5), given that the problem under consideration is convex. In practical applications, where the problem generally is assumed to be nonconvex, termination within an ε-tolerance cannot be expected and a heuristic stopping criterion has to be employed instead, such as stopping when there is no decrease in two successive primal solutions Z_{ub}^L. To tackle this problem, an extension of the OA algorithm, termed *augmented penalty*, has been proposed by Viswanathan and Grossmann.[127] Moreover, convexity tests can be applied to validate linearizations accumulated in the master problems (for more information, we refer to Grossmann and Kravanja[64]). These extensions for the nonconvex case have been proven to work reasonably well for a number of problems; however, in either case, there is no guarantee that the global optimum is located in a finite number of iterations. In order to circumvent this restriction, global solution algorithms for nonconvex problems of a relatively small size have been developed.[4] Large-scale problems will not be tractable by these algorithms in the near future due to their high computational burden.

Various decomposition algorithms have been applied to solve MIDO problems of considerable sizes by, for example, Bansal et al.[15] and Schweiger and Floudas.[109] A logic-based decomposition approach that is applied to the MLDO problem stated above has been proposed by Oldenburg et al.[94] In the cases mentioned here, a control vector parametcrization method is used to solve the primal subproblems due to its efficiency and robustness when dealing with large-scale multistage process models. This choice, however, means that the treatment of disjunctive equations $q_{k,i}$ (see Equation 9.58) is more complex, as in the case when a full discretization approach is applied to solve the MLDO problem or its MIDO transformation (see Equations 9.62 and 9.64). For more details, refer to Avraam et al.[9] This restriction would not apply to MIDO approaches based on full discretization as proposed in Avraam et al.,[9] who addressed the design of process operations described by hybrid discrete–continuous dynamic process models with an *a priori* unknown number and sequence of discrete modes.

Allgor and Barton[5] proposed an alternative approach where the MIDO problem is solved by iterating between two design subproblems: recipe design and equipment allocation. The recipe design problem constitutes a dynamic optimization problem which assumes a fixed process structure and thus fixed discrete (binary) decision variables. However, in contrast to the approaches based on MINLP, Allgor and Barton[5] generated equipment allocation subproblems on the basis of *screening models* developed using "domain-specific information gathered from physical laws and engineering insight." These screening models lead to MILP problems with the desired underestimating property. The development of these models is, however, a rather complex task.

9.3.2.4 Applications

As shown in Section 9.3.1.5, a number of batch process design applications have been reported where the process structure has been considered to be fixed. A number of studies in batch process synthesis have been reported. Usually, a superstructure is generated and an optimization problem is formulated employing very simple models based on material balances and recovery or conversion models only. Chakraborty and Linninger,[39] for example, studied a process synthesis problem for waste management. A detailed consideration of the physics or even the dynamics of the process has not yet been considered. These approaches give information on a suitable process structure on a coarse level of detail. To the authors' knowledge, very little work has been done on batch process design problems with structural degrees of freedom and consideration of rigorous dynamics. Sharif et al.[110] formulated a MIDO problem for a two-stage batch distillation system where the number of trays and the equipment sizes were subject to optimization together with the continuous degrees of freedom reflux ratio and vapor flow rate in each batch stage. The MIDO problem was solved using a decomposition approach (OA) based on control vector parameterization.

The optimal structural design of a three-stage batch distillation process for the separation of a quaternary mixture was also considered by Oldenburg et al.,[94] who included a regular and inverse mode of operation and the interconnection between the stages as discrete decisions in a superstructure MLDO model. Here, the goal was to simultaneously determine the optimal sequence and the optimal reflux (or reboil) policy in order to minimize the overall batch time required for separating a quaternary mixture into pure components. The optimal batch process design is shown in Figure 9.9 (right), where the first and second stages are operated regularly to obtain the lightest component at the top and the heaviest component as residue of the still pot at the final time of the respective process stages. The third stage is operated inversely to separate components two and three. When compared to a standard multistage separation (operated with an optimal reflux strategy) shown in the right part of Figure 9.9, the optimal batch process achieves a 10% reduction in batch time. This also translates into energy savings of 10% in this case, because all design options were considered with the same constant reboiler heat duty. The optimization problem has been solved using a logic-based solution approach together with control vector parameterization.

The application of the screening model approach to solve batch design problems has been illustrated by Allgor et al.[7] as well as Allgor and Barton,[6] who presented several batch design case studies involving reaction and separation tasks.

9.3.3 DESIGN UNDER UNCERTAINTY

After the discussion of generating the nominal optimal solution of an optimization problem in the previous sections, the question must be considered how uncertainties can be incorporated into the optimization problems (P1) to (P5). According to the definition in Section 9.2.4.4, optimality in the presence of uncertainties requires flexibility; therefore, any attempt to determine an optimal process (structure, parameters, and recipes) should actively incorporate the uncertainties into the design procedure.

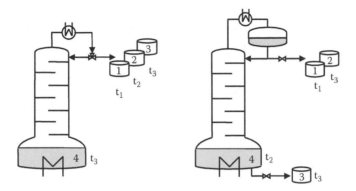

FIGURE 9.9 Three-stage batch distillation process with regular (left), and combined regular and inverse operation (right).

Two approaches can be distinguished. In the first case, we assume that the uncertainty can be completely resolved by measurements available before the batch is started. In this case, the process and its operational strategy can be adapted according to precomputed optimal designs. The resulting problem class is referred to as *parametric programming*,[13] as it provides a solution for each combination of potential uncertain parameters.

In the second approach, called *robust optimization*, we do not have to assume that the uncertainties can be reduced by additional measurements; hence, a single recipe must be calculated incorporating all possible uncertainties leading to rather conservative designs in many cases. A variety of robust optimization problem formulations have been proposed in the literature. An important class of these approaches is based on feasibility and flexibility measures originally introduced by Halemane and Grossmann.[65] These measures were extended by Mohideen et al.,[90] who applied optimization-based design in the presence of uncertainty to dynamic systems. The corresponding optimization problems involve a very large set of DAE equations that are only required to account for the uncertainty of the system. Hence, for practical applications, robust batch process design is an extremely difficult task. Consequently, only a few applications of batch process design under uncertainty have been reported.

In the next section, we focus on two comparatively simple problem formulations for parametric programming and robust optimization, respectively, rather than providing a comprehensive survey of methods for design under uncertainty. In order to keep the notation as simple as possible, only single-stage optimization problems (P1) with $n_s = 1$ are discussed. As above, we assume that all model uncertainties and disturbances are parameterized by the vector ζ.

9.3.3.1 Parametric Programming

In a parametric programming framework, the state and control trajectories $x(t,\zeta)$, $z(t, \zeta)$, and $u(t,\zeta)$ are determined by solving the optimization problem for all the possible values of $\zeta \in C$. The results are stored in a database. If additional information on the parameters ζ becomes available before starting the batch, the corresponding optimal trajectory can be retrieved from the database and applied to the process.

The set of solutions can be determined by collecting the infinite number of optimization problems into the following parametric programming problem:

$$\min_{\substack{x(t,\zeta),z(t,\zeta) \\ u(t,\zeta),p(\zeta),t_f(\zeta)}} \int_{\zeta_{min}}^{\zeta_{max}} \Phi(x(t_f(\zeta),\zeta),z(t_f(\zeta)),p(\zeta),\zeta,t_f(\zeta))\,d\zeta \qquad \text{(P6)}$$

$$\text{s.t.} \quad f(\dot{x}(t,\zeta),x(t,\zeta),z(t,\zeta),u(t,\zeta),p(\zeta),\zeta,t) = 0$$

$$g(x(t,\zeta),z(t,\zeta),u(t,\zeta),p(\zeta),\zeta,t) \leq 0$$

$$g^e(x(t_f(\zeta),\zeta),z(t_f(\zeta),\zeta),u(t_f(\zeta),\zeta),p(\zeta),\zeta,t_f(\zeta)) \leq 0$$

$$l(x(t_0,\zeta),x(t_0,\zeta),z(t_0,\zeta),p(\zeta),\zeta) = 0$$

$$\forall t \in [t_0,t_f(\zeta)], \ \forall \zeta \in [\zeta_{min},\zeta_{max}]$$

Due to the infinite number of constraints a finite approximation of the set C is mandatory. A finite sample ζ_p, $p = 1,...,P$, of all the possible combinations of $\zeta \in C$ is used to solve the problem. Then, the integral in the objective functional of problem (P6) can be replaced by a finite sum. To solve this type of problem, sophisticated techniques exist to generate an appropriate grid in the ζ space.[50] The discretized dynamic optimization problem has a decoupled structure, as each set of the degrees of freedom $u(t,\zeta_p)$ and $p(\zeta_p)$ only influences one set of model equations and constraints. Therefore, the problem can be separated into P independent optimization problems, which are solved by the numerical techniques introduced above.

Note that several alternative formulations can be employed instead of problem (P6). For example, Bansal et al.[14] employed feasibility and flexibility measures in a parametric programming approach as explicit functions of the uncertain parameters.

9.3.3.2 Robust Optimization

In most cases, the uncertainties cannot be removed by additional measurements, in which case all possible uncertainties have to be incorporated into one single design. This is a fundamental difference between robust optimization and parametric programming. In robust optimization, the relative influence of different parameter values and disturbances on the objective function is defined on the basis of probabilities.

Depending on the overall production objective for the process considered, various possibilities exist to account for the uncertainties in the objective function of the optimization. A convex combination of the expected cost $E[\Phi(\cdot,\zeta)]$ and the variance of the cost $Var[\Phi(\cdot,\zeta)]$:

$$\min\ a\,E[\Phi(\cdot,\zeta)]+(a-1)\sqrt{Var[\Phi(\cdot,\zeta)]} \tag{9.67}$$

offers a flexible means of achieving a suitable trade-off between performance and risk.[47] The parameter a $(0 \leq a \leq 1)$ is a constant that weights the level of profit (i.e., the expected value) and variation of the profit (or risk) (i.e., the variance).

If a time-invariant probability distribution $F(\zeta)$ is assumed, the dynamic optimization problem incorporating parametric uncertainties can be formulated as:

$$\min_{\substack{x(t),z(t)\\u(t),p,t_f}} \int_{\zeta_{min}}^{\zeta_{max}} F(\zeta)\,\Phi(x(t_f),z(t_f),p,\zeta,t_f)\,d\zeta \tag{P7}$$

$$\text{s.t.}\quad f(\dot{x}(t),x(t),z(t),u(t),p,\zeta,t)=0$$

$$g(x(t),z(t),u(t),p,\zeta,t)\leq 0$$

$$g^e(x(t_f),z(t_f),u(t_f),p,\zeta,t_f)\leq 0$$

$$l(x(t_0),x(t_0),z(t_0),p,\zeta)=0$$

$$\forall t \in [t_0,t_f],\ \forall \zeta \in [\zeta_{min},\zeta_{max}]$$

when $a = 1$. The solution of problem (P7) is required to satisfy all constraints.

In many cases, such a constraint feasibility cannot be achieved for all possible values of the uncertain parameters ζ. Hence, some constraint softening has to be introduced. As in deterministic optimization, a penalty framework can be employed where the expectation of the constraints is added to the objective function. Two alternative formulations have been established.[46,47] In a first approach, the constraints are satisfied on average by bounding the expectation of the constraint (e.g., to fulfill a required purity constraint on average):

$$E[g(x,z,u,p,\zeta)]\geq 0 \tag{9.68}$$

which, however, does not restrict constraint violation for a single sample of ζ. To remedy this disadvantage, a conditional representation can be used to limit the average constraint violation according to:

$$E[-g(x,z,u,p,\zeta)|g(x,z,u,p,\zeta)<0]\leq b \qquad (9.69)$$

This formulation still ignores the frequency of constraint violation. Often, several expectation-based constraints — such as the two presented — are used simultaneously to compensate for the disadvantages of a single formulation.

In a second approach, the constraint is interpreted as a random variable with an associated probability density. The idea is then to formulate an acceptable level of probability for the constraint violation. The probability (or chance) constraint:

$$P\Big(g(x,z,u,p,\zeta)\leq 0\Big)\geq \rho_0 \qquad (9.70)$$

ensures that all of the process constraints have to be satisfied with the probability ρ_0 simultaneously. Alternatively, individual probabilities can be formulated for each constraint $g_i(\cdot)$:

$$P\Big(g_i(x,z,u,p,\zeta)\leq 0\Big)\geq \rho_i \quad,i=1,\dots,n_g. \qquad (9.71)$$

The resulting optimization problem is easier to solve but has the disadvantage that the worst-case probability level at which a solution satisfies all of the constraints is equal to the product of the levels ρ_i, $i = 1,\dots,n_g$, which is much smaller than any ρ_i.

Regardless of the peculiarities of the formulation, the resulting robust (or stochastic) optimization problems are difficult to solve, particularly if the objective and constraints are nonlinear in the uncertain parameters. All formulations result in semi-infinite programming problems (for a comprehensive overview, refer to Hettich and Kortanek[70]) that require a substantial computational effort to achieve a sufficiently accurate solution. In particular, the decoupled solution approach successfully applied for parametric programs is not possible for robust optimization due to the convolution of probability density and objective functions in problem (P7).

An approximate numerical solution of stochastic optimization problems can be obtained by sampling of the probability density function or by application of discretization techniques.[105,118] The discretization of ζ can be chosen randomly by Monte Carlo sampling or systematically by one of the discretization methods described in Section 9.3. This technique is general and can be applied to almost any type of formulation with nonlinear objectives and constraints but requires a tremendous computational effort.

There exists a variety of different techniques to solve robust optimization problems. A comprehensive review is beyond the scope of this contribution. The interested reader may consult the papers of Zhang et al.,[133] Darlington et al.,[46,47] and Wendt et al.[130] for more information.

9.3.3.3 Applications

Only few practical applications of batch operation under uncertainty can be found in the literature (see Terwiesch et al.[120] for a review). Terwiesch et al.[121] applied robust optimization techniques to a semibatch reaction process analyzing a consecutive reaction system and discussed various objective functions, including risk threshold optimization and quality variance minimization. Their aim was to determine an optimal operation strategy under two different types of uncertainty. Furthermore, the authors briefly analyzed to what extent structural model plant mismatch can be covered as parametric uncertainty. Bernardo and Saraiva[18] proposed a robust optimization framework that is applied to a small-scale case study of batch distillation column. Visser et al.[126] solved an operation problem of a fed-batch bioprocess with a cascaded optimization approach that ensures satisfaction of path constraints in the presence of uncertainties.

Acevedo and Pistikopoulos[3] used a mixed-integer stochastic optimization-based algorithm to solve a process synthesis problem under uncertainty. The problem was formulated as a multiperiod stochastic optimization problem where the objective function represents the cost of the design selected and the expected optimal profit under uncertainty. The process considered was described as steady-state model. A study with similar intent and methodology has recently been reported by Chakraborty and Linninger.[40] The high computational effort to solve optimization problems under uncertainty is typical for all applications; therefore, the application of these techniques is currently restricted to small-size problems.

9.4 OPTIMIZATION-BASED ONLINE CONTROL AND OPERATION

In the previous section, we dealt with a nominal model and either no or known disturbances. The design decisions fixed the process structure and its parameters as well as the sequence of tasks and the time-varying control profiles during each of these tasks in order to achieve maximum profit. The assumptions defining the nominal (design) case usually do not hold anymore during process operation. In reality, various types of uncertainties are unavoidably present. Consequently, the design loses optimality and sometimes even feasibility, if a sufficient degree of flexibility has not been accounted for in the design.

In order to cope with the process and model uncertainties as well as with changing production objectives, we would ideally like to adjust the design momentarily during operation in real time in order to regain feasibility as well as optimality; however, only part of the decision variables available during the design phase can still be manipulated during the operational phase. Typically, the equipment parameters and most of the equipment structure are fixed. In contrast, some stream connections, the sequence of tasks, the control profiles, and the initial conditions can be manipulated in real time, at least in principle. The availability of all or part of these operational degrees of freedom is largely determined by the design of the batch plant and its instrumentation. Subsequently, we do not deal with structural decision variables during operation; hence, the process structure as well as the sequence of tasks are considered to be fixed during operation. However, the techniques presented here extend, at least in principle, to the more general case, where some structural decision variables are also available in real time.

When deviations from the nominal design occur, optimality and feasibility can only be regained during operation if process and model uncertainties can be reduced by means of measurements. In particular, we exploit the feedback of measurements in order to reduce model uncertainty by model adaptation and to estimate disturbances in order to cope with process uncertainty. On the basis of such a model update, we may recompute the optimal values of the operational degrees of freedom $u_k(t)$, p, and t_k in every stage k via *on-line optimization*. Typically, estimates of the states $x_k(t)$ and $z_k(t)$, the disturbances $d_k(t)$, and the model parameters $\vartheta_k(t)$ have to be inferred from the measurements and a process model to implement output feedback with high performance. This combined state, parameter, and disturbance estimation problem has also been called *dynamic data reconciliation*.[26] Therefore, we have to solve not only a modification of the multistage optimal control problem (P1) used during design but also a dynamic data reconciliation problem to adjust the model and to compute state estimates.

The techniques that use measurements in addition to models to reduce uncertainty in optimal batch process operation have recently been referred to as *measurement-based optimization* by Srinivasan et al.[112] We refer the reader to their paper for a review of techniques to achieve optimal batch process using measurements and feedback control.

For further illustration, we reconsider the methyl acetate process (see Figure 9.1). Let us assume that a nominal optimal design has been determined by means of the optimization methods introduced in Section 9.3, resulting in a single-stage batch process with a semibatch reactive distillation column together with optimal values of the initial charge of material in the still as well as optimal feed, reflux, and boil-up policies (see Section 9.3.1.5 for details). Now we are interested in implementing this process but want to account for disturbances in the feed concentration of acetic acid and for uncertainties in the reaction kinetic parameters of the esterification. Measurements are available for the product concentration and the temperature at the top tray of the column. Because the actual control moves are known, the measurements can be used to estimate the kinetic parameters and reconstruct the state vector and actual feed concentration. These estimates can then be used to compute the next control moves to track the optimal trajectory of the process from a prediction model.

In the following section, we first refine the previous considerations and present a general optimization problem, the solution of which would lead to optimal operation of a batch process when confronted by model and process uncertainties through measurement feedback. Section 9.4.2 discusses various solution strategies to tackle the quite complicated optimization problem. Then, some remarks on tailoring algorithms to exploit the peculiarities of online optimization are given. We conclude this section with a discussion of case studies to illustrate the potential of the techniques presented.

9.4.1 MATHEMATICAL PROBLEM FORMULATION

So far, the online optimization approach has been sketched only roughly. Before we turn to the mathematical problem formulation, we have to refine the proposed approach. In particular, we introduce the concept of moving (or receding) horizon optimization. To simplify notation, we consider a sequence of K batches produced in a single-stage batch process modeled by Equations 9.7 and 9.8. The model of the sequence of single-stage batch processes is similar to the multistage model (Equations 9.9 and 9.10) if the stage transition conditions (Equation 9.11) are replaced by initial conditions independent of the states in the previous stage. Obviously, we could also treat the more general case of a sequence of multistage batch processes by introducing a multistage model with a total number of stages given by the product of the number of stages in every individual batch and the number of batches. Note that there would be two qualitatively different types of stage transition conditions, which are either independent or dependent on the states of the previous stage.

We are looking at batch (or stage) k carried out between $t = t_{k-1}$ and $t = t_k$ (Figure 9.10). To simplify the notation, we drop the index k and consider the interval $\Delta_k = [t_{k-1}, t_k] = [t_r, t_f]$. The interval is subdivided by a mesh of equidistant points $t_{c,\ell} = t_r + \ell\Delta t$, $\ell = 1,...,L$. The current time $t_{c,\ell}$ separates Δ_k into two intervals: $\Delta_{r,\ell} = [t_r, t_{c,\ell}]$ and $\Delta_{c,\ell} = [t_{c,\ell}, t_f]$.

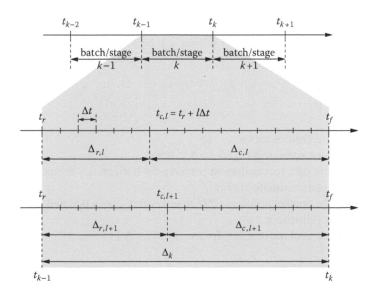

FIGURE 9.10 Moving horizon in one stage or batch.

The (first) reconciliation interval lies in the past of the current time $t = t_{c,\ell}$, whereas the (second) control interval refers to the future. Measurements $\eta_\ell(t)$ are available on the reconciliation interval $\Delta_{r,\ell} = [t_r, t_{c,\ell}]$. For convenience, we do not introduce discrete measurements here. The quantities $\eta_\ell(t)$ can be viewed as interpolations of discrete measurements on $\Delta_{r,\ell}$. These past measurements can therefore be used to solve a reconciliation problem at the current time. In general, reconciliation is not restricted to $\Delta_{r,\ell}$ of stage k. Rather, we may extend it to also include measurements on previous stages $k - \kappa$, $\kappa = 1,2,\dots$. For the sake of a simple presentation, we will not consider this case in detail.

The reconciled model and the resulting state estimates are then used to solve an optimal control problem on the control interval $\Delta_{c,\ell}$. Again, we may not restrict the control interval to the remainder of stage k. Rather, we could extend it to additional future stages $k + \kappa$, $\kappa = 1,2,\dots$. Due to unavoidable uncertainties, the computed controls will only be implemented on the interval $[t_{c,\ell}, t_{c,\ell} + \Delta t]$. After implementation, the current time $t_{c,\ell}$ is shifted by Δt; that is, $t_{c,\ell+1} = t_{c,\ell} + \Delta t$. Hence, the reconciliation interval is extended by Δt, whereas the control interval is decreased by Δt. In some cases, the control interval $\Delta_{c,\ell}$ does not extend to the end of the stage but is chosen significantly shorter. Such a choice is always motivated by the insufficient prediction quality of the model and less significant endpoint constraints at t_f. In such cases, the estimation interval may still grow in length, whereas the control interval is kept at a constant length. The reconciliation and optimal control problems are solved repeatedly and a control move is implemented.

Tacitly, we have assumed that the computations can be carried out in zero time. This is obviously a simplification. More precisely, the reconciliation and control intervals ($\Delta_{r,\ell}$ and $\Delta_{c,\ell}$, respectively) are separated by Δt, which has to be sufficiently long to carry out the computations. For notation simplicity, we will neglect this subtlety. Then, the moving horizon scheme can be stated in detail as follows:

1. Initialize:
   Select Δt.
   Set $\ell = 1$.

2. Compute control move:
   Access measurements $\eta_\ell(t)$ on $\Delta_{r,\ell}$. (P8)
   Solve the data reconciliation problem on horizon $\Delta_{r,\ell}$ to update model
      and state estimate at $t = t_{c,\ell}$.
   Solve the control problem (P9) on interval $\Delta_{c,\ell}$ with updated model and
      state estimate at $t = t_{c,\ell}$.
   Apply control move $u_{c,\ell}(t)$ for $t \in [t_{c,\ell}, t_{c,\ell} + \Delta t]$.

3. Modify horizons:
   $\ell := \ell + 1$
   $t_{c,\ell} := t_{c,\ell} + \Delta t$
   $\Delta_{r,\ell} := \Delta_{r,\ell} + \Delta t$
   $\Delta_{c,\ell} := \Delta_{c,\ell} + \Delta t$

4. Stop or loop back:

If $t_{c,\ell} \neq t_f$, go to 2; otherwise, stop.

The dynamic data reconciliation problem referred to in the algorithm can be formulated as follows:

$$\min_{x_{r,\ell}(t_r),d_{r,\ell}(t)} \Phi_r(y_{r,\ell}(t_{c,\ell}),\eta_\ell(t_{c,\ell}),x_{r,\ell}(t_{c,\ell}),z_{r,\ell}(t_{c,\ell}),p_{r,\ell},d_{r,\ell}(t_{c,\ell}),t_r,t_{c,\ell}) \quad \text{(P8)}$$

$$\text{s.t.} \quad 0 = f_r(\dot{x}_{r,\ell}(t),x_{r,\ell}(t),z_{r,\ell}(t),u_{r,\ell}(t),p_{r,\ell},d_{r,\ell}(t)) \tag{9.72}$$

$$0 = l_r(x_{r,\ell}(t_r),x_{r,\ell}(t_r),z_{r,\ell}(t_r),p_{r,\ell}) \tag{9.73}$$

$$y_{r,\ell}(t) = h_r(x_{r,\ell}(t),z_{r,\ell}(t),u_{r,\ell}(t),p_{r,\ell},d_{r,\ell}(t)) \tag{9.74}$$

$$u_{r,\ell}(t) = U[u_c(\Delta_{c,\ell-1})] \tag{9.75}$$

$$0 \geq g_r\left(x_{r,\ell}(t),\ z_{r,\ell}(t),\ u_{r,\ell}(t),\ p_{r,\ell},d_{r,\ell}(t),t\right)$$
$$t \in [t_r,t_{c,\ell}],\ \ell = 1,...,L \tag{9.76}$$

Analogously, the control problem reads as follows:

$$\min_{u_{c,\ell}(t),t_f} \Phi_c\left(x_{c,\ell}(t_f),z_{c,\ell}(t_f),p_{c,\ell},t_{c,\ell},t_f\right) \tag{P9}$$

$$\text{s.t.} \quad 0 = f_c\left(\dot{x}_{c,\ell}(t),\ x_{c,\ell}(t),z_{c,\ell}(t),u_{c,\ell}(t),p_{c,\ell},d_{c,\ell}(t)\right) \tag{9.77}$$

$$0 = x_{c,\ell}(t_{c,\ell}) - x_{r,\ell}(t_{c,\ell}) \tag{9.78}$$

$$d_{c,\ell}(t) = D[d_{r,\ell}(\Delta_{r,\ell})] \tag{9.79}$$

$$0 \geq g_c\left(x_{c,\ell}(t),z_{c,\ell}(t),u_{c,\ell}(t),p_{c,\ell},d_{c,\ell}(t)\right) \tag{9.80}$$

$$0 \geq g_c^e\left(x_{c,\ell}(t_f),z_{c,\ell}(t_f),u_{c,\ell}(t_f),p_{c,\ell},d_{c,\ell}(t_f)\right)$$
$$t \in [t_{c,\ell},t_f],\ \ell = 1,...,L \tag{9.81}$$

Here, we have assumed that neither the model nor the production and process constraints are the same in the reconciliation and control problems. The notation is the same as in problem P1 for a single stage. Here, we assume that all uncertain quantities vary with time and replace the parameters ζ by the disturbances $d(t)$. The indices r and c refer to quantities in the reconciliation and control problems. In addition, the variables $y_{r,\ell}(t)$ are the predicted outputs for which the measurements $\eta_\ell(t)$ are available. The index ℓ distinguishes the quantities in the problem to be solved at current time $t_{c,\ell}$. Hence, ℓ is the horizon counter on stage k. The Mayer-type objective function stated in problem (P8) is usually replaced by a least-squares objective function including some regularization term $R(\cdot)$ (for more information, refer to Allgöwer et al.[8]). This is achieved by refining the objective [problem (P8)], for example, by:

$$\Phi_r = \int_{t_r}^{t_{c,\ell}} \| y_{r,\ell}(t) - \eta_\ell(t) \| + R(d_{r,\ell}(t), x_{r,\ell}(t_{c,\ell})) \, dt \qquad (9.82)$$

The reconciliation and control problems (P8 and P9) are not independent. The data required to solve the reconciliation problem (P8) on horizon $\Delta_{r,\ell}$ are the control moves $u_c(t)$ applied to the process on $\Delta_{c,\ell-1}$. The solution of the control problem (P9) requires knowledge about the states $x_{r,\ell}(t_{c,\ell})$ at the end of $\Delta_{r,\ell}$ as well as the disturbances $d_{r,\ell}(t)$ on $\Delta_{r,\ell}$, which are computed earlier from the reconciliation problem. The predictor $D[\cdot]$ is forecasting the disturbances $d_{c,\ell}(t)$ on $\Delta_{c,\ell}$ using the disturbance estimates $d_{r,\ell}(t)$ on $\Delta_{r,\ell}$. In the most simple case, constant values for the disturbances are assumed. For example, if the feed concentration in the methyl acetate column has been identified to deviate from the design case, the mean deviation on $\Delta_{r,\ell}$ may be taken as a prediction of the deviation that could occur in the future on $\Delta_{c,\ell}$.

The problem structure of the combined reconciliation and control problem is very similar to the two-stage problem (P1) with $n_s = 2$. The way we have formulated the problem does not account for the two-stage character because two independent objective functions, $\Phi_r(\cdot)$ and $\Phi_c(\cdot)$, have been formulated. A true two-stage formulation would result if both objectives are summed up to a single two-stage objective ($\Phi = \Phi_r + \Phi_c$) which has to be minimized simultaneously subject to the constraints in both problems (P8) and (P9). A coupled solution of both problems does not seem to have been analyzed in the literature for when a higher economical performance (e.g., a smaller value of Φ_c) might be achieved.

Obviously, the problems on the different horizons are not independent due to the coupling introduced by the operator $U[\cdot]$ in Equation 9.75. It links the open-loop optimal control problems on adjacent horizons indexed by $\ell - 1$ and ℓ. In fact, this coupling is responsible for implementing feedback in the overall receding horizon scheme.[53,101] For the methyl acetate semibatch distillation column, $U[\cdot]$ maps the control moves of reflux R and boil-up V realized on horizon $\Delta_{r,\ell}$ into the reconciliation problem (P8). After its solution, new estimates of the states, the feed concentrations, and the reaction kinetic parameters are available and are used to compute the control moves on horizon $\Delta_{c,\ell}$.

From a control perspective, problems (P8) and (P9) constitute an output feedback optimal control problem with a general objective reflecting process economics rather than deviations from reference trajectories. Because there are no operational targets in the sense of set points or reference trajectories as in model predictive control,[101] the problem may also be interpreted from an operational perspective. Hence, the problem can be considered as a generalization of state-of-the-art (steady-state), real-time optimization,[88] which aims at establishing economically optimal transient plant operation.[10,11,68] The solution of this operation support problem would achieve an integration of advanced (predictive constrained) process control and economical optimization in a transient environment. The approach is not only applicable to batch but also to continuous processes in transient regimes, for example during grade or load transitions.

9.4.2 SOLUTION STRATEGIES

The solution of the receding horizon online optimization problem is quite involved; therefore, we are interested in simplifications that are appropriate in certain situations or which cut down computational effort. Subsequently, we first consider in Section 9.4.2.1 the case where measurements are not used during batch k to reduce model or process uncertainties. Instead, measurements from previous batches $k - \kappa$, $\kappa = 1,2,\ldots$ are exploited in the operational strategy. Next, in Section 9.4.2.2, we discuss the direct solution of the online optimization problem (P8 and P9) at high frequency associated with the (largest) sampling time of the measurements. In some cases, the computational burden can be significantly reduced if the necessary optimality conditions of the problem are identified and used to implement the output feedback. Such a strategy, recently suggested by Bonvin and coworkers,[34] is introduced in Section 9.4.2.3. Finally, Section 9.4.2.4 introduces decomposition strategies exploiting differing time scales in the process.

9.4.2.1 Batch-to-Batch Optimization

If measurements are not available in real time, the online optimization problems (P8 and P9) cannot be solved during the actual batch. Nevertheless, some kind of model-based operation is still possible, if measurements (for example of concentration and product quality) become available after completion of the batch. Such a situation often occurs in industrial practice, where a detailed analysis of samples taken during and at the end of the batch are taken to assess the quality of the batch produced. Even though this approach is not online, it fits in the proposed framework of measurement and model-based optimization described in this section.

In the situation described above, a reconciliation problem similar to problem (P8) is solved before the actual batch k of duration Δ_k is started at $t = t_{k-1}$. The objective $\Phi_r(\cdot)$ is comprised of only the deviations between model predictions $y_{r,k-\kappa}(t)$ and measurements $\eta_{k-\kappa}(t)$ in *previous* batches $k - \kappa$, $\kappa = 1,2,\ldots$. Obviously, the control profiles $u_{k-\kappa}(t)$ from the completed batches, $k - \kappa$, $\kappa = 1,2,\ldots$ are also available from historical data to set up the problem. The optimization is subject to the model and process constraints in those previous batches, where measurements are used for reconciliation. The results of the reconciliation problem, particularly the state estimates $x_k(t_k)$ and $z_k(t_k)$, the parameters ϑ_k, and the predicted disturbances $D_k[d_{k-1}(t), d_{k-2}(t),\ldots]$, can be used to subsequently solve an optimal control problem for batch k. The control profile $u_k(t)$ is applied to batch k without the opportunity to compensate any uncertainties during the batch. The base control

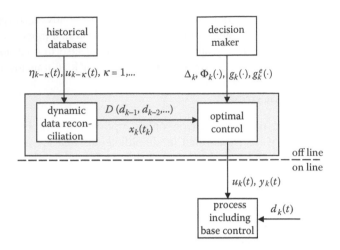

FIGURE 9.11 Batch-to-batch optimization.

system may be used to track the optimal trajectories $u_k(t)$ and $y_k(t)$. This strategy is depicted schematically in Figure 9.11.

Such an iterative offline model reconciliation and batch optimization exploits the repetitive nature of batch processes. Data collected during or after the batch are used to account for parametric uncertainties and disturbances. After a model update, the calculated optimal trajectory can be implemented in a feed-forward sense. This procedure is referred to as *batch-to-batch* or *run-to-run optimization*.[44,90] It is particularly effective if model and process uncertainties are of a repetitive nature and do not change very much from batch to batch.

9.4.2.2 Direct Online Optimization

The most rigorous approach for the implementation of measurement-based optimization is solution of the combined reconciliation and control problem (P8 and P9) on a receding horizon. In this case, a single optimizing feedback control layer is implemented as shown in Figure 9.12. The goal of the optimization is minimization of the economic cost function Φ_c on the control horizon $\Delta_{c,\ell}$, both set by a decision maker on a higher level in the automation hierarchy (e.g., a planner or a scheduler; see Chapter 10). Both problems (P8) and (P9) are solved in real time repeatedly during operation of the batch process at a high sampling rate δ_c.

Successful implementation of such a strategy requires a high-quality process model and disturbance forecast to facilitate the prediction of the states with sufficiently small error until the end of batch k. Note that this strategy differs significantly from output feedback in a model-predictive control scheme,[8] as the economic objective function Φ_c is used directly on the control level and the endpoint constraints are enforced. This approach is also termed *online reoptimization* or *optimization on a shrinking horizon*. The drawback of such an approach is the high computational burden of online optimization that must be mastered

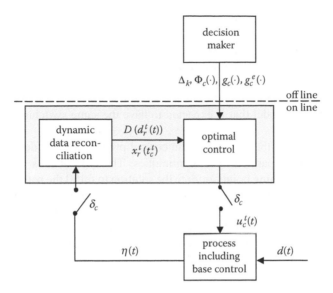

FIGURE 9.12 Direct online optimization.

with high frequency to exploit the information in the sampled measurements quickly, a potential lack of robustness for large-scale and strongly nonlinear models, and lack of transparency for the operators.

In some practical cases, the computational burden can be lowered significantly if the control problem is solved on a short horizon rather than on the horizon $\Delta_{c,\ell}$.[68] Such a simplification requires a transformation of the control problem (P9) into a short horizon optimization problem defined over the horizon $\overline{\Delta}_{c,\ell} = [t_{c,\ell}, t_{c,\ell} + n\Delta t]$ significantly shorter than $\Delta_{c,\ell}$. If properly chosen, the local objective can be fully equivalent to the corresponding full horizon objective. Such a transformation is obviously not possible in all cases, especially, if endpoint constraints have to be met precisely. However, if a representative short horizon problem can be formulated, not only can the computational burden be reduced but the process model can also be rather crude as it is frequently updated by means of online measurements. In particular, the very simple trend models introduced in Section 9.2.1 are very favorable. The resulting scheme can be regarded as an adaptive predictive controller with a suitable parameterization of the model uncertainty. Applications of this direct short-horizon, online optimization are discussed in Section 9.4.4.

9.4.2.3 Tracking of the Necessary Conditions of Optimality

Implementation of the direct approach is computationally demanding and requires the robust solution of online optimization problems. An alternative solution strategy has been suggested recently by Bonvin et al.[34] Instead of tracking and updating an optimal trajectory during the batch by repetitive solution of the reconciliation and control problems (P8 and P9), these authors suggest instead tracking the necessary conditions of optimality (NCO) (see Equations 9.33 and 9.34).

The basic assumption is that the set of active path and terminal constraints of the *real batch plant* is known *a priori* and does not change due to process uncertainty or modified production objectives. This solution structure has to be determined from operating experience or by offline optimization of a (simplified nominal) process model that is expected to have the same switching structure as the real plant.

The structure of the solution of the control problem (P9) is determined by a sequence of arcs separated by discontinuities. The type of arcs as well as the position of the switching points enforce the solution to fulfill the path and endpoint constraints and to establish a compromise between competing phenomena. The optimal solution is characterized by scalar quantities θ^s related to the switching times and by functions $\alpha(\theta^a,t)$ to represent all the arcs. A characterization of the optimal solutions shows that the decisions θ^s and $\alpha(\theta^a,t)$ can be classified as either constraint seeking, $\bar{\theta}^s$, $\bar{\alpha}(\bar{\theta}^a,t)$ (which establish path or terminal constraints) or as sensitivity seeking, $\tilde{\theta}^s$, $\tilde{\alpha}(\tilde{\theta}^a,t)$ which facilitate the compromise between competing objectives). Constraint- and sensitivity-seeking decisions are determined by the NCO, which may depend on nonmeasurable process states and (probably) unknown disturbances.

In summary, NCO tracking has two main features: (1) online adjustment of $\bar{\theta}^s$, $\bar{\alpha}(\bar{\theta}^a,t)$ via feedback control, and (2) run-to-run adjustment of $\tilde{\theta}^s$, $\tilde{\alpha}(\tilde{\theta}^a,t)$ via feedback control. The implementation of such a scheme requires: (1) identification of the structure of the optimal control profile (preferably for the real process) including the number of switching points and the type of arcs between two adjacent switching points; (2) determination of the constraint- and sensitivity-seeking decision variables; and (3) measurement (or estimation) of the constrained quantities (see Figure 9.13).

With this information, it is sufficient to control the NCO to their setpoint zero in order to determine the control variables $u(t)$. This entails an online evaluation of the necessary conditions of optimality without numerically solving an optimization problem online. In some cases, the problem structure is such that no model is required to implement the strategy.[34] An application of this approach is discussed in Section 9.4.4.

9.4.2.4 Decomposition Approaches for Online Optimization

The disadvantages of the direct online solution of the combined reconciliation and control problem (P8 and P9) as introduced in Section 9.4.2.2 can be overcome in many cases by exploiting multiple time-scale contributions in the disturbances $d(t)$ and the significant difference in the dominant time scales of different parts of the batch process. Decomposition seems to be inevitable, particularly if we are considering large-scale processes with many interacting process units with cross-functional integration.[10,87] Spatial or vertical decomposition addresses

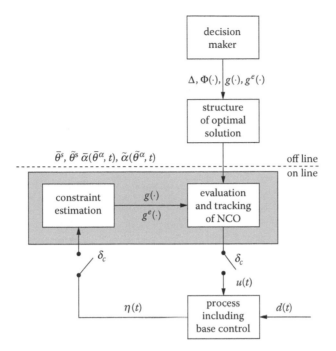

FIGURE 9.13 Tracking of necessary conditions of optimality.

differences in the dynamics of different parts of the process, whereas horizontal decomposition addresses the multiple time scales in the same part of the process. If horizontal decomposition is applied, base control, predictive reference trajectory tracking control, and dynamic economic optimization are typically applied at widely differing sampling rates in the range of seconds, minutes, and hours.[55] This is the case we are focusing on in this section.

According to Helbig et al.,[68] the feasibility of a multiple time-scale decomposition largely depends on the dynamic nature of the disturbances. If, for example, the disturbance can be decomposed into at least two contributions:

$$d(t) = d_0(t) + \Delta d(t) \qquad (9.83)$$

that is, a slow trend, $d_0(t)$, containing slow frequency contributions and an additional zero mean contribution, $\Delta d(t)$, containing high frequencies, then some sort of horizontal decomposition should be feasible. The slow frequency contributions mainly affect the economics of the process and impact the generation of optimal trajectories. In contrast, process economics are generally insensitive to high-frequency contributions of the disturbances; however, the fast disturbances may drive the process off the optimal trajectory.

Figure 9.14 shows a possible structure of an optimization-based operation support system that exploits two time scales in the disturbances. The upper level is responsible for the design of a desired optimal trajectory $x_c(t)$, $u_c(t)$, $y_c(t)$,

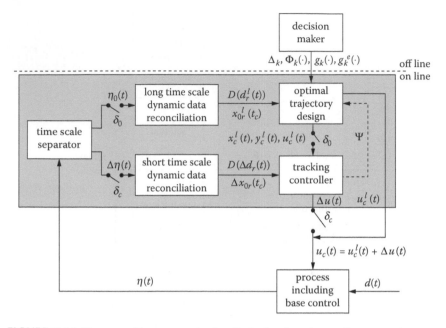

FIGURE 9.14 Decomposition approach of optimization-based operation support.

whereas the lower level is tracking the trajectory set by the upper level. Due to the time-varying nature of the disturbances, $d(t)$, feedback is not only necessary to adjust the action of the tracking controller but also to adjust the optimal trajectory design to compensate for variations in $d_0(t)$ and $\Delta d(t)$, respectively. The control action $u_c(t)$ is the sum of the desired control trajectory $u_c^\ell(t)$ and the tracking controller output $\Delta u(t)$. Reconciliation is also based on the slow and fast contributions $\eta_0(t)$ and $\Delta\eta(t)$, which are generated by a time-scale separation module. Control and trajectory design are typically executed on two distinct sampling intervals δ_c and $\delta_o = n\delta_c$, where an integer $n \gg 1$. The performance of the controller, coded by some indicator ψ,[71] must be monitored and communicated to the trajectory design level to trigger an update of the optimal trajectory in case the controller is not able to achieve acceptable performance. Although this decomposition scheme is largely related to so-called composite control in the singular perturbation literature,[79] the achievable performance will be determined by the way the time-scale separator is implemented.

As compared to the direct solution of the problems (P8) and (P9), the decomposition approach is expected to be computationally less demanding and to provide better transparency to the operators. Further, different methods and models can be used for reconciliation and optimal control on the two time scales.

9.4.3 ALGORITHMS FOR ONLINE OPTIMIZATION

In this section, we sketch algorithmic approaches to (approximately) solve the online optimization problem introduced in Section 9.4.2. We first briefly address the reconciliation problem (P8) in the next section and then return to algorithms for general real-time dynamic optimization problems such as problems (P8) and (P9).

9.4.3.1 Dynamic Data Reconciliation

In principle, two different approaches can be distinguished to solve the reconciliation problem (P8): recursive and receding horizon estimation. The most commonly used representative of the recursive approach is the (extended) Kalman filter (EKF).[75,61] Here, we have to make assumptions on the stochastic nature of the model and measurement uncertainties and neglect the path and terminal constraints. In addition, the reconciliation interval $\Delta_{r,\ell}$ has to be restricted to one sampling time; hence, $\Delta_{r,\ell} = [t_{c,\ell} - \Delta t, t_{c,\ell}]$. The reader is referred to Henson and Seborg[69] for an introductory exposition of these filter techniques.

In practical applications, estimators based on the extended Kalman filter have to cope with a number of limitations.[103] For nonlinear models, the theoretical basis breaks down almost completely because the influence of past data on the current estimate is not properly reflected by the covariance equations anymore. Hence, tuning of the EKF to obtain good estimates may become very difficult. Another drawback of the filter is the difficulty incorporating inequality constraints in a rigorous manner. Especially this last drawback can limit the applicability of these techniques and has been a motivation for development of receding horizon estimation (RHE) methods.[103]

Receding horizon estimation provides a general but computationally demanding alternative for solving the dynamic data reconciliation problem. Any kind of constraints can be incorporated in the problem formulation, and no restrictive assumption on the stochastic properties of model and measurement uncertainty have to be introduced. Here, problem (P8) is solved repetitively on the horizon $\Delta_{r,\ell} = [t_r, t_{c,\ell}]$. In order to reduce the computational effort, the reconciliation horizon is often chosen to be of fixed length $\overline{\Delta}_{r,\ell} = [t_{c,\ell}, t_{c,\ell} - n\Delta t]$ shorter than $\Delta_{r,\ell}$. In this case, a key issue is correct consideration of the past data in order to achieve estimation errors converging to zero in the limit.[100] A recursive version of a receding horizon estimator where the horizon length is restricted to one sampling interval has been reported by Cheng et al.[42]

9.4.3.2 Real-Time Dynamic Optimization Algorithms

Algorithms for the solution of dynamic optimization problems such as (P8) and (P9) were described in detail in Section 9.3. In principle, the same algorithmic framework can be employed to devise online optimization algorithms; however, the algorithms must be tailored specifically in order to meet the constraints on computational time. In particular, the repetitive solution of a sequence of similar optimization problems can be exploited. Some algorithmic ideas will be summarized below.

In order to solve problem (P8) and (P9), the methods described in Section 9.3 are in principle applicable. The advantages of this methods (e.g., handling of constraints, flexibility, robustness) must, however, be weighted against the real-time requirements. In general, a run-time guarantee cannot be given for these methods as the number of NLP iterations can neither be predicted nor limited. Explicit time management and tracking of the status of the numerical algorithm are therefore inevitable. Real-time implementation requires explicit time management to schedule and control the numerical calculations. Obtaining information about the convergence of the current approximate solution and hence its reliability must be supervised.

In the face of stringent real-time requirements, a tailoring of the offline algorithms described in Section 9.3 has to be done. In particular, all the algorithmic extensions take advantage of the fact that a sequence of similar problems has to be solved repetitively. We briefly review some strategies to initialize the solution of the current problem ℓ based on the results from solution $\ell - 1$. For an in-depth treatment of the subject, refer to the survey of Binder et al.[28]

Initialization Strategies

If the data do not change much in the problems on two adjacent horizons $\ell - 1$ and ℓ and if the sampling time Δt is short compared to the dominant time constants of the process, the solution to the optimal control and reconciliation problems will obviously not change much. If we assume that the structure of the parameterization of the decision variables does not change from $\ell - 1$ to ℓ, then the solution on horizon $\ell - 1$ can be shifted by one sampling time to result in a first approximation of the solution on horizon ℓ which can be used to initialize the optimization algorithm. Alternatively, one could reuse the solution on the overlapping part of both horizons to extrapolate the solution on the disjunctive time interval to get a first approximate solution. In addition to reusing the previous solution one can also try to reuse the information on the gradient and in particular on the Hessian, which is expensive to compute.

Multiscale Optimization

Powerful initialization strategies are possible not only between problems on adjacent horizons but also between two approximations within a horizon if multiscale representations of the optimization problems are used. Such a multiscale framework with applications to control and estimation has recently been suggested by Binder and coworkers.[23,25,28–30] In their approach, the optimization problem is projected to a sequence of nested spaces that provide local resolution of the state and control variables. An adaptation strategy allows for a gradual refinement of the optimal solution reusing information from a previous solution step. Tailored numerical algorithms are currently under development to exploit the multiscale properties with high computational efficiency.

Real-Time Iteration

Often these warm start strategies do not sufficiently reduce the computational effort. Diehl[48] and Diehl et al.[49] proposed a drastic simplification. They suggested a sequence of Newton-type iterations to determine the values of the degrees of freedom in the next optimization step. This approach is based on a linearization of all problem functions resulting in a quadratic program of a special structure. For least-squares objectives, the Gauss–Newton matrices can be used as an inexpensive approximation of the exact Hessian. This iterative procedure extends linear time-varying model predictive control and is comparable to a model predictive controller based on system linearization along the currently best predicted trajectory.[60]

9.4.4 Applications

This section concludes with a brief discussion of applications of optimal operation of batch processes. A case study on optimization-based operation of an industrial semibatch reactor is presented. The primary emphasis is on the applicability of online optimization as outlined in Section 9.2 when significant uncertainty is present. A measurement-based approach via tracking of the necessary conditions of optimality proposed in Section 9.4.2.3 is also employed.

Case Study

A schematic of the batch process considered in the case study is given in Figure 9.15.[45] The setup consists of a common stirred tank reactor equipped with a jacket in which a fixed amount of heating or cooling medium is circulated. The reactor temperature can be manipulated by adjusting the temperature of the medium by inserting either hot or cold water into the loop through equal percentage control valves.

The exothermic liquid-phase reaction follows the simple scheme $A + B \rightarrow C + D$, where A and B are the reactants, C is the desired product, and D is a byproduct. In addition, a solvent S and a catalyst Cat are present. The reactor is

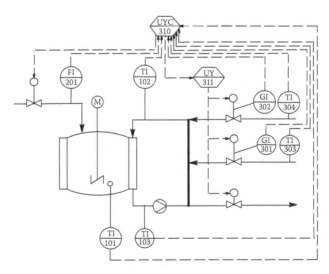

FIGURE 9.15 Reactor schematic.

initially filled with a fixed amount of A, S, and Cat at ambient temperature and pressure. After the initial reactor content has been heated to the required reaction temperature, feeding of a B/S mixture is initiated and the reaction phase begins.

Online available measurements include (see Figure 9.15) the feed rate F_f (FI 201), the reactor temperature T_R (TI 101), the jacket inlet temperature $T_{J,i}$ (TI 102), the jacket outlet temperature $T_{J,o}$ (TI 103), the temperature of the hot stream T_{hot} (TI 303), and the temperature of the cold stream T_{cold} (TI 304).

The control and optimization system utilizes the split-range variable C_{con} and the feed rate F_f as manipulated variables. The first task is to heat the initial reactor content from ambient temperature to the required reaction temperature of 70°C and to keep temperature at this value during the subsequent reaction phase. Feeding may only begin after the reaction temperature has been reached. A total amount of 5000 kg has to be fed during the reaction phase. At the beginning of the reactant feeding phase, the feed rate is constrained by a linear increase of the feed rate, from 100 to 1000 kg/hr within half an hour. The batch ends when B has been converted up to a remaining amount of 50 kg.

The operational objective is to minimize the duration of the reaction phase t_f by determining a suitable feed rate profile, $F_f(t)$. Tight temperature control has to be achieved to avoid additional unwanted side reactions. In order to safely produce a product of desired quality, it is not sufficient to control only the reaction temperature. The feed rate must be chosen in such a way that in case of a cooling system failure adiabatic conversion of the current reactor content will not increase the reaction temperature beyond a specified limit. In order to avoid runaway due to decomposition reactions beginning at that temperature level, a path constraint on the so-called *adiabatic maximum temperature* (T_{ad}) has to be considered (see Equation 9.95, below).[114] The adiabatic end temperature is constrained for safety reasons to be, $T_{ad}(t) \leq 85°C$. The cooling system control action C_{con} has to be between its lower and upper bounds: C_{con}^{min} and C_{con}^{max}, respectively. At the end of the batch, the concentration of B or the amount M_B of B in the reactor has to be below a given value.

The mathematical formulation of the optimization problem is given by:

$$\min_{C_{con}(t),F_f(t)} \Phi := t_f \tag{P10}$$

s.t.   dynamic reactor model

$$C_{con}^{min} \leq C_{con}(t) \leq C_{con}^{max}$$

$$F_f^{min} \leq F_f(t) \leq F_f^{max} \tag{9.84}$$

$$T_{ad}(t) \leq T_{ad}^{max}$$

$$M_B(t_f) \leq M_B^{max}$$

This dynamic optimization problem could readily be solved for the controls $C_{con}(t)$ and $F_f(t)$ if a sufficiently accurate model is available; however, this is not the case in many industrial situations. Therefore, measurement-based optimization has to be employed in some variant to cope with the uncertainty.

In our scenario, we assume that the reaction kinetics are unknown. Further, there are uncertainties in the heat-transfer coefficient and in the valve positions that send the streams of hot or cold medium into the heating or cooling circulation loop. To address this operational problem, we will now take a look at a special case of the direct optimization scheme introduced in Section 9.4.2.2 which is closely related to model predictive control.

Optimizing Adaptive Calorimetric Model Predictive Control

The basic principle of the model predictive control (MPC) scheme for the optimization of semibatch reaction processes is to use available degrees of freedom on the process (such as the feed rate) to reduce the batch time while simultaneously solving a classical temperature control problem. For this purpose, the optimization problem (P10) ("minimize batch time") has to be transformed into a local optimization problem on a finite prediction horizon that can be interpreted as a particular form of the direct optimization scheme described in Section 9.4.2.2.

As shown in a case study of an industrial two-phase polymerization reactor,[67] simple process models based on the principles of reaction calorimetry[107] are sufficient in order to solve the problem. Calorimetric state estimation techniques can be applied in order to permanently adapt crude models to match current process dynamics, thus allowing at least short horizon predictions even in cases of large structural and parametric uncertainties. This strategy gives rise to the term *adaptive calorimetric model predictive control*.

The design of any optimizing MPC scheme consists of three basic elements: First, an appropriate *cost function* has to be chosen which drives the process along different active constraints (see problem P9). For the above type of reaction process, this can be achieved by extending the classical MPC controller objective function for temperature control with an additional term maximizing the feed rate. Second, a suitable calorimetric *estimator* must be derived in order to infer unknown quantities from available measurements (see problem P8). In addition to simple state estimation we have to deal with the uncertain reaction kinetics. For that purpose, the heat of reaction (Q_R) is interpreted as an unknown input in the energy balance:

$$... \frac{dT}{dt} = Q_R + ... \tag{9.85}$$

The unknown heat of reaction (Q_R) may be estimated on the basis of a simple trend model such as:

$$\frac{dQ_R}{dt} = 0 \tag{9.86}$$

Finally, *prediction models* for the estimated variables (e.g., Q_R) have to be formulated to predict the dynamics of the unknown input for the solution of the control problem into the future (see Equation 9.79). Application of the estimation model as the predictor may be restricted by controllability issues in cases where the estimated variable depends on the manipulated variables; therefore, a specific *prediction model differing from the estimation model* is often inevitable.

Estimation of calorimetric state and input is implemented by means of an extended Kalman filter (EKF) with output clipping.[122] The symbols used in the following model and their meaning are compiled in Table 9.1. The numbers in the equations are of appropriate physical dimensions resulting from the conversion of watts (W) to kilowatts (kW) or seconds (s) to hours (hr).

The mass and component balances are:

$$\frac{d M}{d t} = F_f$$

$$\frac{d M_A}{d t} = \frac{Q_R}{\Delta H_R} MW_A \cdot 3.6 \tag{9.87}$$

$$\frac{d M_B}{d t} = w_{B,f} F_f + \frac{Q_R}{\Delta H_R} MW_B \cdot 3.6$$

The energy balance for the reactor content is:

$$Mc_p \frac{d T_R}{d t} = ([Q_R + (\alpha_0 + \Delta\alpha) M / M_0$$

$$(0.5(T_{J,o} + T_{J,i}) - T_R)] \cdot 3600 \tag{9.88}$$

$$+ F_f c_{p,f}(T_f - T_R))$$

The energy balance of the jacket content is:

$$M_J c_{p,c} \frac{d T_{J,o}}{d t} = F_c c_{p,c}\left(T_{J,i} - T_{J,o}\right)$$

$$-\left(\alpha_0 + \Delta\alpha\right) M / M_0 \tag{9.89}$$

$$\left(0.5\left(T_{J,i} + T_{J,o}\right) - T_R\right) \cdot 3600$$

TABLE 9.1
Symbols Used in the Model

Symbol	Unit	Description
C_{con}	%	Split range
C_{cold}	%	Valve position cold stream
C_{hot}	%	Valve position hot stream
c_p	kJ/kg/K	Heat capacity reactor content
$c_{p,f}$	kJ/Kg/K	Heat capacity feed
$c_{p,c}$	kJ/kg/K	Heat capacity cooling medium
$c_{p,cold}$	kJ/kg/K	Heat capacity cold stream
$c_{p,hot}$	kJ/kg/K	Heat capacity hot stream
F_f	kg/hr	Feed rate
ΔH_R	kJ/mol	Reaction enthalpy
MW_A	kg/kmol	Molar weight component A
MW_B	kg/kmol	Molar weight component B
$w_{B,f}$	—	Weight fraction component B
T_R	°C	Temperature of reactor content
$T_{J,i}$	°C	Jacket inlet temperature
$T_{J,o}$	°C	Jacket outlet temperature
T_f	°C	Temperature of feed
T_{hot}	°C	Temperature of hot stream
T_{cold}	°C	Temperature of cold stream
T_{ad}	°C	Adiabatic temperature
M	kg	Mass of reactor content
M_A	kg	Mass of component A
M_B	kg	Mass of component B
M_0	kg	Initial mass of reactor content
M_J	kg	Mass of jacket content
M_M	kg	Mass of medium in cooling utility
F_c	kg/hr	Circulation stream
α_0	kW/K	Heat transfer coefficient
QS_{cold}	kg/hr	Maximum cold stream
$V\,T_{cold}$	—	Coeffcient valve characteristic
K_{VS}^{cold}	—	Coeffcient valve characteristic
K_{V0}^{cold}	—	Coeffcient valve characteristic
QS_{hot}	kg/hr	Maximum hot stream
$V\,T_{hot}$	—	Coeffcient valve characteristic
K_{VS}^{hot}	—	Coeffcient valve characteristic
K_{V0}^{hot}	—	Coeffcient valve characteristic
Q_R	kW	Heat of reaction
Q_{REKF}	kW	Estimated heat of reaction
ΔQ_c	kW	Error term heating/cooling utility
$\Delta \alpha$	kW/K	Error term heat transfer coefficient

The energy balance for the cooling/heating utility is:

$$M_M c_{p,c} \frac{dT_{J,i}}{dt} = \Delta Q_C \cdot 3600. + \dot{M}_c c_{p,c} \left(T_{J,o} - T_{J,i} \right)$$

$$+Q_{S,cold} \left[1 + VT_{cold} \left(\frac{K_{V0}^{cold}}{K_{VS}^{cold}} \exp\left(\log\left(\frac{K_{VS}^{cold}}{K_{V0}^{cold}} \right) C_{cold} \right) \right)^{-2} - 1 \right]^{-0.5}$$

$$\cdot \left(c_{p,cold} T_{cold} - c_{p,c} T_{J,o} \right) \tag{9.90}$$

$$+Q_{S,hot} \left[1 + VT_{hot} \left(\frac{K_{V0}^{hot}}{K_{VS}^{hot}} \exp\left(\log\left(\frac{K_{VS}^{hot}}{K_{V0}^{hot}} \right) C_{hot} \right) \right)^{-2} - 1 \right]^{-0.5}$$

$$\cdot \left(c_{p,hot} T_{hot} - c_{p,c} T_{J,o} \right)$$

The split range for the valve positioning is:

$$C_{cold} = \begin{cases} 100 - 2C_{con}, & \text{if } C_{con} < 50 \\ 0, & \text{if } C_{con} \geq 50 \end{cases} \tag{9.91}$$

$$C_{hot} = \begin{cases} 2(C_{con} - 50), & \text{if } C_{con} > 50 \\ 0, & \text{if } C_{con} \leq 50 \end{cases} \tag{9.92}$$

The trend models for unknown quantities are:

$$\frac{d\,\Delta\alpha}{dt} = 0 \tag{9.93}$$

$$\frac{d\,\Delta Q_C}{dt} = 0 \tag{9.94}$$

$$\frac{d\,Q_R}{dt} = 0 \tag{9.95}$$

In addition to the heat of reaction (see Equation 9.95), two additional sources of uncertainty are included. The uncertainties in the jacket heat transfer are handled by estimating a drifting parameter $\Delta\alpha$ (see Equation 9.93). This corresponds to the assumption that the coefficient α_0 for the initially filled reactor is known rather well. The complex terms in Equation (9.90) arise due to consideration of the valve behavior described by equal percentage valves. Uncertainties in these valve flows lead to a nonzero error term, ΔQ_C, which is assumed to be constant over the prediction horizon (see Equation (9.94)).

The optimization task is to minimize the duration of the reaction phase, which can be accomplished by feeding the reactants as quickly as possible subject to constraints. The following type of cost functional may be applied:

$$\Phi = \alpha_1 \sum_{i=1}^{n} (T_R(t_i) - T_R^{set}(t_i))^2$$

$$+\alpha_2 \sum_{i=1}^{n} (C_{con}(t_i) - C_{con}(t_{i-1}))^2$$

$$+\alpha_3 \sum_{i=1}^{n} (F_f(t_i) - F_f(t_{i-1}))^2 \tag{9.96}$$

$$-\phi_J(\Delta_{c,\ell})$$

$$t_i = t_{c,\ell} + i \cdot \Delta t$$

It contains four terms representing the temperature control task, penalties on control moves, and a free customizable fourth term. In this application, the fourth term depends on the feed rate. It is evaluated continuously as:

$$\phi_J(\Delta_{c,\ell}) = \alpha_4 \int_{t_{c,\ell}}^{t_{c,\ell} + \overline{\Delta}_{c,\ell}} F_f(t) \, dt \tag{9.97}$$

Here, $\overline{\Delta}_{c,\ell} = [t_{c,\ell}, t_{c,\ell} + n\Delta t]$ is the prediction horizon, where Δt is the sampling interval used for control and n is the number of prediction steps (see Section 9.4.2.2).

In the most streamlined version, the prediction model for the MPC includes all equations of the EKF model. If, however, the estimated variable represents an important process variable which itself depends on the manipulated variables (i.e., the heat of reaction), a specific prediction model structure:

$$Q_R = f_Q(\vartheta_Q, \ldots) \tag{9.98}$$

may be proposed for prediction (see Equation 9.2). The initial value of the prediction should match the current estimate, which can be achieved by updating some parameter ϑ_Q or assuming a constant error term over time.

Because the reaction is isothermal, the major prediction aspect with respect to the heat of reaction covers the dependency on feed component B. Assuming the reaction order to be locally of first order with respect to B, the following realization of Equation 9.98 can be formulated:

$$Q_R = \vartheta_Q M_B \tag{9.99}$$

The model parameter ϑ_Q is adapted at the beginning of each prediction step by:

$$\vartheta_Q = Q_{R_{EKF}} / M_B \tag{9.100}$$

In order to handle the safety constraint, the MPC model also contains an equation to predict the adiabatic maximum temperature (T_{ad}):

$$T_{ad} = T_R - \frac{M_B \Delta H_R \cdot 1000}{MW_B M c_p} \tag{9.101}$$

which has to be constrained over the reaction phase.

The results of this online recipe optimization are illustrated in Figure 9.16 and Figure 9.17. The top diagram of Figure 9.16 shows a comparison of the reactor temperature for the simulated reactor (SIM) and the temperature adapted with the extended Kalman filter (EKF). Obviously, the optimizing controller is able to handle the temperature control task. The three diagrams at the bottom of Figure 9.16 show the trajectories of the constrained variables. These are the adiabatic temperature $T_{ad}(t)$ (top), the split range variable $C_{con}(t)$ (middle), and the feed rate $F_j(t)$ (bottom). Active constraints are indicated by intervals between dashed lines. The sequence of active constraints is given by the feed rate limitation imposed by the initial feeding ramp, limitations by the cooling utility and the safety constraint on the adiabatic temperature. The feed rate has been determined in such a way that the local optimal solution under consideration of all constraints is met. Figure 9.17 shows a comparison of the unknown simulated heat of reaction (SIM) and the estimated heat of reaction (EKF) at the top and the unknown simulated adiabatic temperature (SIM) and estimated adiabatic temperature (EKF) at the bottom.

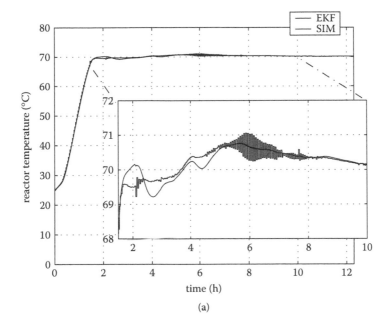

(a)

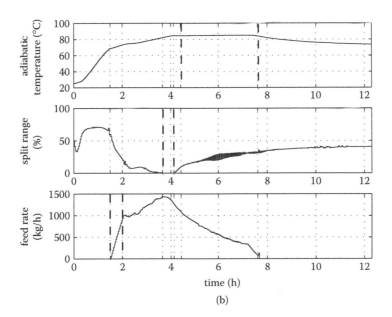

(b)

FIGURE 9.16 Online optimization: (top) controlled reactor temperature, and (bottom) path-constrained feed rate, cooling capacity, and adiabatic temperature.

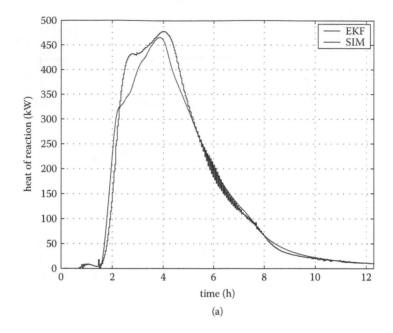

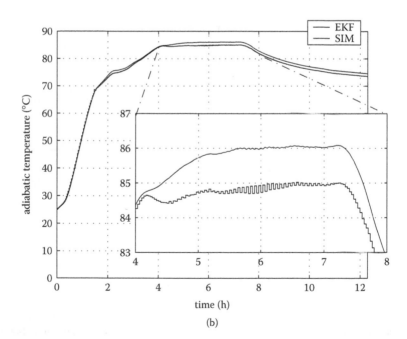

FIGURE 9.17 Online optimization: (top) estimation and prediction error of heat of reaction, and (bottom) adiabatic temperature.

Online Optimization via Feedback Control

From the results of the previous section it is obvious that the optimal solution of problem (P10) is determined by a sequence of active constraints, as shown in the three diagrams at the bottom of Figure 9.16. The sequence of active constraints in the reaction phase, i.e., after reaching the desired reactor temperature at about $t = 1.7$ hr, is given by: (1) F_f^{max}, the feed rate limitation imposed by the initial feeding ramp; (2) C_{con}^{min}, indicating a limitation of the cooling utility; (3) T_{ad}^{max}, the safety constraint; and (4) F_f^{min}, after the desired total amount has been fed. For this particular case study, it can even be proven[113] that the optimal solution for this reaction type is always determined by the constraints. This is a nice result because the optimal solution can be implemented by simple proportional–integral controllers used to adapt the feed rate F_f in order to track the sequence of active constraints: (1) manipulate F_f to the set point given by the ramp in the initial feeding phase, (2) manipulate F_f such that the split range variable is at its lower bound, (3) adjust F_f in order to satisfy the safety constraint, and (4) set the set point for F_f to F_f^{min}.

Recall from Section 9.3 that no model of the process is required to implement this control strategy if all constraints can be measured; however, because the adiabatic temperature cannot be measured online, this constraint has to be estimated. Comparison with Equation 9.97 shows that the only unknown quantity in the expression for the adiabatic temperature is the am ount $M_B(t)$ of component B; therefore, a simple regression model can be used to calculate an estimate $\hat{M}_B$ of M_B and thus T_{ad}. The following regression model is derived from Equations 9.87 and 9.88 under the assumption of perfect temperature control:

$$\frac{d\,\hat{M}_B}{d\,t} = F_f\,w_{B,f} - \vartheta_I \frac{M}{M_0}(0.5(T_{J,o} + T_{J,i}) - T_R)$$

$$-\vartheta_{II}\,F_f\,(T_f - T_R) \qquad\qquad (9.102)$$

$$\hat{M}_B = \vartheta_{III} + \tilde{M}_B$$

where ϑ_I represents a parameter associated with heat transfer, ϑ_{II} represents a parameter related to temperature changes due to the difference in feed and reactor temperature, and ϑ_{III} parameterizes the initial amount of M_B. Recall that the structure of this predictive model was introduced in Section 9.2 by Equations 9.3 and 9.4.

In the first 2 hours of the reaction phase, the constraint on the adiabatic temperature is not active. This interval can be used as reconciliation interval Δ_r. Samples of the reactor content taken during this period in order to determine the current amount M_B of component B can be used to fit the model (Equation 9.102) to measurements $\eta_{r,\ell}$ by solving problem (P8) once for the collected data.

In analogy to Figure 9.16, Figure 9.18 shows the reactor temperature (top) and trajectories of the constrained variables (bottom). Again, the adiabatic temperature, $T_{ad}(t)$ (top); the split range variable, $C_{con}(t)$ (middle); and the feed rate, $F_j(t)$ (bottom) are presented. Active constraints are indicated by intervals between dashed lines. The differences between Figure 9.16 and Figure 9.18 result from regularization with the manipulated variables in the objective function (Equation 9.97) which prohibits drastic control moves.

Practical Applications

A significant number of applications of optimal operation of batch processes have been reported in the literature. Batch-to-batch optimization (see Section 9.4.2.1) has gained a lot of attention in recent years. Clarke-Pringle and MacGregor[44] applied a batch-to-batch optimization methodology for producing a desired molecular-weight distribution (MWD) using an approximate model. A measurement of the MWD at the end of the batch is used to update manipulated variable trajectories for the next batch, thus iterating into a good operating policy. Srinivasan et al.[111] used a parameterization of the inputs, updated on a run-to-run basis using a feedback control scheme that tracks signals that are invariant under uncertainty. The methodology was conceived to improve the cost function from batch to batch without constraint violation. Batch-to-batch control and online single batch control strategies were investigated by Lee et al.[82] for controlling the particle size distribution (PSD) in a precipitation process in a semibatch reactor. A systematic integration of the two strategies has been shown to have a complementary effect on the control performance. This approach is expected to find more industrial applications as the computational complexity of this approach is small and measurements are usually available after completion of a batch at no extra cost.

Direct on-line optimization (see Section 9.4.2.2) has been applied to simulated and experimental systems in various ways. For example, Helbig et al.[67] reported an application with a short prediction horizon. Direct optimization with predictions to the end of the horizon have been studied by Terwiesch,[117] Ruppen et al.,[104] and Abel et al.[1] in the context of semibatch reactors. A very special case of direct online optimization is nonlinear model predictive control with output feedback. In this case, the control objective is to follow a given reference trajectory or a set point. For example, Prasad et al.[97] applied this concept to an industrial styrene polymerization reactor. The main goal was to control product qualities such as average molecular weight and polydispersity as well as the production rate. An EKF was used for estimating the states and disturbances from the measurements, which were available for different sample times. It should be noted, however, that a fixed recipe is assumed in this case as no economical optimization was carried out online.

Tracking the necessary conditions of optimality (see Section 4.2.3) has only recently been introduced. Some examples are presented in the papers of Bonvin et al.[34] and Srinivasan et al.[112] Decomposition approaches (see Section 9.4.2.4) are still under development and have not yet been widely applied. A semibatch reactive distillation process was considered by Kadam et al.,[73] who presented a particular aspect of the concept. Kadam et al.[74] presented an example of a

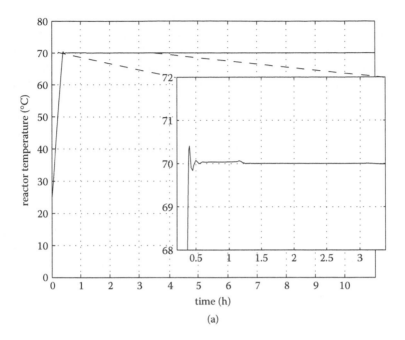

(a)

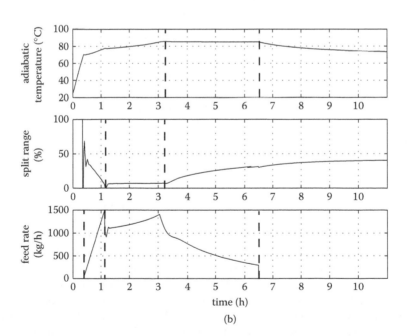

(b)

FIGURE 9.18 Online optimization via feedback control: (top) controlled reactor temperature, and (bottom) path-constrained feed rate, cooling capacity, and adiabatic temperature.

decomposition approach applied to an industrial polymerization process in which they considered an optimal grade change of a continuous process. The online optimization was decomposed into two subproblems which were solved for different sample times. Rigorous dynamic optimization, a Kalman filter, and a linear time-variant model predictive controller were used to implement the various functions.

Despite the significant potential of optimal operation of batch processes employing a dynamic process model and the measurements available after or during a batch, only few real applications have been reported so far. A significant increase in industrial application is predicted in the future.

9.5 SUMMARY

Batch processes are characterized by an inherent dynamic nature that allows them to react to market-driven fluctuations, such as changes in feedstock and product specifications, in a flexible way; however, the unsteady nature and flexibility of batch processes pose challenging design and operation problems. Traditional approaches to the design of batch unit operations include short-cut methods, rules of thumb, and design by analogy. Recent research activities have tried to come up with more fundamental and rigorous design methodologies. Systematic methods for the synthesis and conceptual design of batch processes are a central point of interest in these activities; thus, model-based design and optimization strategies play a key role. They are the topic of this chapter.

Throughout this chapter we have discussed various aspects of batch process modeling with a focus on model application in dynamic optimization. The models governing the process behavior have to be derived from the process knowledge available; however, unavoidable uncertainty and a lack of mechanistic understanding in batch process modeling often discourage potential users from proceeding with model-based techniques. In order to overcome these obstacles, different sources of uncertainty have been classified and suggestions made regarding how these uncertainties can be modeled and parameterized. Models integrated based on fundamental mechanisms and those derived from reasoning or data are referred to as *mixed fundamental–empirical models* and are in many cases adequate to represent various types of uncertain model aspects.

Our analysis of model uncertainty is followed by an introduction to several types of batch process models with increasing levels of complexity. We began our considerations with continuous DAE models describing batch processes that are operated in a single stage. We then generalized this class of models to multistage models describing a batch process involving more than one processing step. In fact, multistage models reveal a strong relationship to the far more complex class of general discrete–continuous hybrid models. These models are capable of describing the dynamic behavior of batch processes with explicit as well as implicit discontinuities. Various examples are given to illustrate the different types of discontinuities that might appear during the operation of batch processes. These examples further illustrate the concepts of these three model classes.

A physicochemical process model defines a simulation problem when all degrees of freedom describing the process structure and operation are fixed. Instead of carrying out simulation experiments with parameter variations, we advocate the use of optimization techniques. Given the immense complexity of batch design, control, and operation problems, these techniques provide a means of searching for optimal parameters in a more organized way than by trial and error; however, such an approach requires some insight into the formulation and solution techniques of optimization problems. We introduced the use of constraints and objective functions for this purpose. The discussion of solution techniques distinguished between cases with a fixed process structure and cases where structural design alternatives are also degrees of freedom. The former case can be formulated as a single- or multistage dynamic optimization problem. The most popular solution techniques for this problem type (i.e., control vector parameterization and full discretization) were revised. In the latter case, the incorporation of structural design alternatives gives rise to a superstructure model formulation involving disjunctions for which special solution techniques such as mixed-integer optimization methods are required. An overview of these methods is also presented in this chapter.

After the discussion of generating nominal optimal solutions of optimization problems we introduced the concept of optimization considering uncertainty explicitly in the problem formulation. Two approaches can be distinguished. In the first case, we assume that the uncertainty can be completely resolved by measurements available before the batch is started, leading to a parametric programming problem that provides a solution for each combination of potential uncertain parameters. If the uncertainties cannot be removed by additional measurements, all possible uncertainties have to be incorporated into one single robust optimization problem. In robust optimization, the relative influence of various parameter values and disturbances on the objective function is defined on the basis of probabilities.

Batch process design and optimization not only involve nominal offline calculated solutions but also must consider the application of those solutions online. Solution of the receding horizon online optimization problem is quite involved; therefore, we discussed several concepts appropriate in certain situations to cut down the computational effort. All approaches exploit measurement feedback, in contrast to the robust optimization formulation, in order to adapt the models used in the optimization to the real plant or to evaluate the necessary conditions for optimality directly. The methods explained differ according to the available online information, the computational effort, and the process knowledge available or necessary. We introduced a case where measurements are not used during the batch to reduce model or process uncertainties in a batch-to-batch optimization scheme. Here, measurements from previous batches are exploited in the operational strategy. Next, we discussed the direct solution of a high-frequency data reconciliation and online optimization problem associated with the (largest) sampling time of the measurements. The computational burden can be significantly reduced if the necessary optimality conditions of the problem can be identified and used to implement an output feedback. Finally, decomposition strategies were introduced to exploit various time scales in the process. We concluded with some remarks on tailoring algorithms to handle the peculiarities of online optimization.

Today, most aspects of model-based design and optimization are by no means mature. Many open questions and points of improvement in this area are subject to research in academia and industry. Among these are aspects of model development and validation, as well as the immense variety of numerical solution techniques, which are continuously being improved. Nevertheless, rigorous models and optimization-based strategies have a high potential for the future and will remain a central focus of research.

REFERENCES

1. Abel, O., Helbig, A., Marquardt, W., Zwick H., and Daszkowski, T., Productivity optimization of an industrial semi-batch polymerization reactor under safety constraints, *J. Proc. Contr.*, 10, 351–362, 2000.
2. Abel, O. and Marquardt, W., Scenario-integrated optimization of dynamic systems, *AIChE J.*, 46(4), 803–823, 2000.
3. Acevedo, J. and Pistikopoulos, E.N., Stochastic optimization based algorithms for process synthesis under uncertainty, *Comp. Chem. Eng.* 22(4–5), 647–671, 1998.
4. Adjiman, C.S., Androulakis, I.P., and Floudas, C.A., Global optimization of mixed-integer nonlinear problems, *AIChE J.*, 46(9), 1769–1797, 2000.
5. Allgor, R.J. and Barton, P.I., Mixed-integer dynamic optimization, *Comput. Chem. Eng.*, 21, S451–S456, 1997.
6. Allgor, R.J. and P.I. Barton (1999). Screening models for batch process development. Part II. Case studies, *Chem. Eng. Sci.*, 54, 4065–4087, 1999.
7. Allgor, R.J., Evans, L.B., and Barton, P.I., Screening models for batch process development. Part I. Design targets for reaction/distillation networks, *Chem. Eng. Sci.*, 54(19), 4145–4169, 1999.
8. Allgöwer, F., Badgwell, T.A. Qin, J.S. Rawlings, J.B., and Wright, S.J., Nonlinear predictive control and moving horizon estimation, in *Advances in Control*, Frank, P.M., Ed., Springer-Verlag, New York, 1999, 391–449.
9. Avraam, M.P., Shah, N., and Pantelides, C.C., Modelling and optimisation of general hybrid systems in the continuous time domain, *Comput. Chem. Eng.*, S221–S228, 22, 1998.
10. Backx, T., Bosgra O., and Marquardt, W., Towards intentional dynamics in supply chain conscious process operations, FOCAPO 98, Snowmass, 1998 (www.lpt.rwth-aachen.de/Publication/Techreport/).
11. Backx, T., Bosgra, O., and Marquardt W., Integration of model predictive control and optimization of processes, in *Proc. IFAC Symposium: Advanced Control of Chemical Processes*, Vol. 1, ADCHEM 2000, Pisa, Italy, 249–260.
12. Balas, E., Disjunctive programming and a hierarchy of relaxations for discrete optimization problems, *SIAM J. Alg. Disc. Meth.*, 6, 466–486, 1985.

13. Bansal, V., Perkins, J.D., and Pistikopoulos, E.N., Flexibility analysis and design of linear systems by parametric programming, *AIChE J.,* 46(2), 335–354, 2000.

14. Bansal, V., Perkins, J.D., and Pistikopoulos, E.N., A unified framework for the flexibility analysis and design of non-linear system via parametric programming, *Eur. Symp. Computer Aided Process Eng.*, 11, 961–966, 2001.

15. Bansal, V., Perkins, J.D., and Pistikopoulos, E.N., A case study in simultaneous design and control using rigorous, mixed-integer dynamic optimization models, *Ind. Eng. Chem. Res.* 41(4), 760–778, 2002.

16. Barrera, M.D. and Evans, L.B., Optimal design and operation of batch processes, *Chem. Eng. Comm.*, 45–66, 1989.

17. Barton, P.I. and Pantelides, C.C. Modelling of combined discrete/continuous processes, *AIChE J.*, 40(6), 966–979, 1994.

18. Bernardo, F.P. and Saraiva, P.M., Robust optimization framework for process parameter and tolerance design. *AIChE J.,* 44(9), 2007–2017, 1998.

19. Betts, J.T., *Practical Methods for Optimal Control Using Nonlinear Programming*, SIAM, Philadelphia, 2001.

20. Betts, J.T. and Huffmann, W.P., Mesh refinement in direct transcription methods for optimal control, *Optim. Control Appl. Meth.*, 19, 1–21, 1998.

21. Bhatia, T.K. and Biegler, L.T., Dynamic optimization in the design and scheduling of multiproduct batch plants, *Ind. Eng. Chem. Res.*, 35, 2234–2246, 1996.

22. Biegler, L.T., Cervantes, A., and Wächter, A., Advances in simultaneous strategies for dynamic process optimization, *Chem. Eng. Sci.*, 24, 39–51, 2000.

23. Binder, T., *Adaptive Multiscale Methods for the Solution of Dynamic Optimization Problems*, Fortschritt-Berichte VDI, Reihe 8, No. 969, VDI-Verlag, Düsseldorf, 2002.

24. Binder, T., Cruse, A., Villas C., and Marquardt, W., Dynamic optimization using a wavelet based adaptive control vector parameterization strategy, *Comp. Chem. Eng.*, 24, 1201–1207, 2000.

25. Binder, T. et al., Introduction to model based optimization of chemical processes on moving horizons, in *Online Optimization of Large Scale Systems*, Grötschel, M., Krumke, S.O., and Rambau, J., Eds., Springer-Verlag, Berlin, 2001, 297–339.

26. Binder, T., Blank, L., Dahmen, W., and Marquardt, W., Towards multiscale dynamic data reconciliation, in *Nonlinear Model Based Process Control*, Berber, R. and Kravaris, C., Eds., Kluwer, Dordrecht, 1998, 623–665.

27. Binder, T., Blank, L., Dahmen, W., and Marquardt, W., *An Adaptive Multiscale Approach to Real-Time Dynamic Optimization*, Technical Report LPT-1999-14, Lehrstuhl für Prozesstechnik, RWTH Aachen, Germany, 1999.

28. Binder, T., Blank, L., Dahmen, W., and Marquardt, W., Iterative algorithms for multiscale state estimation. I. The concept, *J. Opt. Theo. Appl.*, 111(3), 501–527, 2001.

29. Binder, T., Blank, L., Dahmen, W., and Marquardt, W., Iterative algorithms for multiscale state estimation. II. Numerical investigations, *J. Opt. Theo. Appl.* 111(3), 529–551, 2001.

30. Binder, T., Blank, L., Dahmen, W., and Marquardt, W., Multiscale concepts for moving horizon optimization, in *Online Optimization of Large Scale Systems*, Grötschel, M., Krumke, S.O., and Rambau, J., Eds., Springer-Verlag, Berlin, 2001, 341–361.

31. Birge, J.R. and Louveaux, F., *Introduction to Stochastic Programming*, Springer, New York, 1997.

32. Bloss, K.F., Biegler, L.T., and Schiesser, W.E., Dynamic process optimization through adjoint formulation and constraint aggregation, *Ind. Eng. Chem. Res.*, 38, 421–432, 1999.

33. Bock, H.G., Diehl, M., Schlöder, J.P., Allgöwer, F., Findeisen, R., and Nagy, Z., Real-time optimization and nonlinear model predictive control of processes governed by differential-algebraic equations, in *Proc. IFAC Symposium: Advanced Control of Chemical Processes*, ADCHEM 2000, Pisa, Italy, 695–70.

34. Bonvin, D., Srinivasan, B., and Ruppen, D., Dynamic optimization in the batch chemical industry, in *Preprints of Chemical Process Control CPC 6*, Tucson, AZ, 2001.

35. Bryson, A.E. and Ho, Y.-C., *Applied Optimal Control*, Taylor & Francis, Bristol, PA, 1975.

36. Buss, M., von Stryk, O., Bulirsch, R., and Schmidt, G., Towards hybrid optimal control, *at-Automatisierungstechnik*, 48(9), 448–459, 2000.

37. Carver, M.D., Efficient integration over discontinuities ODE simulations, *Math. Comp. Simulat.*, 20(3), 190–196, 1978.

38. Cervantes, A.M., Wächter, A., Tütüncü, R., and Biegler, L.T., A reduced space interior point strategy for optimization of differential algebraic systems, *Comp. Chem. Eng.*, 57, 575–593, 2002.

39. Chakraborty, A. and Linninger, A.A., Plant-wide waste management. 1. Synthesis and multiobjective design, *Ind. Eng. Chem. Res.*, 47, 4591–4604, 2002.

40. Chakraborty, A. and Linninger, A.A., Plant-wide waste management. 2. Decision making under uncertainty, *Ind. Eng. Chem. Res.*, 42, 357–369, 2002.

41. Charalambides, M.S., Shah, N., and Pantelides, C.C., Synthesis of reaction/distillation processes using detailed dynamic models, *Comput. Chem. Eng.*, S19, 167–174, 1995.

42. Cheng, Y.S., Mongkhonsi, T., and Kershenbaum, L.S., Sequential estimation for nonlinear differential and algebraic systems: theoretical development and application, *Comp. Chem. Eng.*, 21(9), 1051–1067, 1997.

43. Clark, P. and Westerberg, A., Optimization for design problems having more than one objective, *Comput. Chem. Eng.*, 7(4), 259–278, 1983.

44. Clarke-Pringle, T.L. and MacGregor, J.F., Optimization of molecular weight distribution using batch-to-batch adjustments, in *Proc. of the American Control Conference*, Philadelphia, 1998, 3371–3375.

45. Cruse, A., Marquardt, W., Helbig, A., and Kussi, J.-S., Optimizing adaptive calorimetric model predictive control of a benchmark semi-batch reaction process, in *IFAC World Congress on Automatic Control*, Barcelona, July 21–26, 2002.

46. Darlington, J., Pantelides, C.C., Rusten, B., and Tanyi, B.A., Decreasing the sensitivity of open-loop optimal solutions in decision making under uncertainty, *Eur. J. Oper. Res.*, 121(2), 343–362, 2000.

47. Darlington, J., Pantelides, C.C., Rustem, B., and Tanyi, B.A., An algorithm for constrained nonlinear optimization under uncertainty, *Automatica*, 35, 217–228, 1999.

48. Diehl, M., Real-Time Optimization for Large Scale Nonlinear Processes, Ph.D. thesis, University of Heidelberg, 2001.

49. Diehl, M., Bock, H.G., Schlöder, J.P., Findeisen, R., Nagy Z., and Allgöwer, F., Real-time optimization and nonlinear model predictive control of processes governed by differential-algebraic equations, *J. Proc. Control*, 12, 577–585, 2002.

50. Diwekar, U. and Kalgnanam, J.R., Robust design using an efficient sampling technique, *Comp. Chem. Eng.*, 20, S389–S394, 1996 .
51. Diwekar, U.M., *Batch Distillation: Simulation, Optimal Design and Control*, Taylor & Francis, New York, 1996.
52. Duran, M.A. and Grossmann, I.E., A mixed-integer nonlinear programming algorithm for process systems synthesis, *AIChE J.*, 32(4), 592–606, 1986.
53. Eaton, J.W. and Rawlings, J.B., Feedback control of nonlinear processes using on-line optimization techniques. *Comp. Chem. Eng.*, 14, 469–479, 1990.
54. Feehery, W., Tolsma, J., and Barton, P.I., Efficient sensitivity analysis of large–scale differential–algebraic systems. *Appl. Numer. Math.*, 25, 41–54, 1997.
55. Findeisen, W., Brdys, M., Malinowski, K., Tatjewski, P., and Wozniak, A., *Control and Coordination in Hierarchical Systems*, Wiley, Chichester, 1980.
56. Fletcher, R. and Leyffer, S., Solving mixed integer nonlinear programs by outer approximation, *Math. Program.*, 66, 327–349, 1994.
57. Floudas, C.A., *Nonlinear and Mixed-Integer Optimization: Fundamentals and Applications*, Oxford University Press, London, 1995.
58. Forsgren, A., Gill, P.E., and Wright, M.H., Interior methods for nonlinear optimization, *SIAM Rev.*, 44(4), 525–597, 2002.
59. Foss, B.A., Lohmann, B., and Marquardt, W., A field study of the industrial modeling process. *J. Proc. Contr.*, 8, 325–337, 1998.
60. Gattu, G. and Zafiriou, E., Nonlinear quadratic dynamic matrix control with state estimation. *Ind. Eng. Chem. Res.*, 31, 1096–1104, 1992.
61. Gelb, A., *Applied Optimal Estimation*, MIT Press, Cambridge, MA, 1974.
62. Geoffrion, A.M., Generalized benders decomposition, *J. Opt. Theory Appl.*, 10(4), 237–260, 1972.
63. Gill, P.E., Murray, W., and Wright, M.H., *Practical Optimization*, Academic Press, London, 1995.
64. Grossmann, I.E. and Kravanja, Z., Mixed-integer nonlinear programming techniques for process systems engineering, *Comput. Chem. Eng.*, 19, S189–S204, 1995.
65. Halemane, K.P. and Grossmann, I.E., Optimal process design under uncertainty, *AIChE J.*, 29(3), 425–433, 1983.
66. Hartl, R.F., Sethi, S.P., and Vickson, R.G., A survey of the maximum principles for optimal control problems with state constraints, *SIAM Rev.*, 37(2), 181–218, 1995.
67. Helbig, A., Abel, O., and Marquardt, W., Model predictive control for on-line optimization of semibatch reactors, in *Proc. of the American Control Conference ACC'98*, Vol. 3, Philadelphia, 1998, 1695–1699.
68. Helbig, A., Abel, O., and Marquardt, W., Structural concepts for optimization based control of transient processes, in *Nonlinear Model Predictive Control*, Vol. 26, Allgöwer, F. and Zeng, A., Eds., Progress in Systems and Control Theory, Birkhäuser Verlag, Basel, 2000.
69. Henson, M.A. and Seborg, D.E., *Nonlinear Process Control*, Prentice-Hall, Englewood Cliffs, NJ, 1997.
70. Hettich, R. and Kortanek, K., Semi-infinite programming: theory, methods and applications, *SIAM Rev.*, 35(3), 380–429, 1993.
71. Huang, B., Minimum variance control and performance assessment of time-variant processes, *J. Proc. Contr.*, 12, 707–710, 2000.

72. Ishikawa, T., Natori, Y., Liberis, L., and Pantelides, C.C., Modelling and optimisation in an industrial batch process for the production of dioctyl phthalate, *Comp. Chem. Eng.*, 21, 1239–1244, 1997.

73. Kadam, J.V. et al., A two-level strategy of integrated dynamic optimization and control of industrial processes: a case study, in *Proc. of European Symposium on Computer Aided Process Engineering 12*, Grievink, J. and van Schijndel, J., Eds., Elsevier Science, Amsterdam, 2002

74. Kadam, J.V. et al., Towards integrated dynamic real-time optimization and control of industrial processes, in *Proc. FOCAPO 2003*, Grossmann, I.E. and McDonald, C.M., Eds., 2003, 593–596.

75. Kalman, R.E., A new approach to linear filtering and prediction problems, *Trans. ASME J. Basic Eng.*, 82, 35–45, 1960.

76. Kim, K.-J. and Diwekar, U.M., New era in batch distillation: computer aided analysis, optimal design and control, *Rev. Chem. Eng.*, 17(2), 111–164, 2001.

77. Klingberg, A., Modelling and Optimization of Batch Distillation. Master's thesis, Department of Automatic Control, Lund Institute of Technology, Sweden, 2000.

78. Kocis, G.R. and Grossmann, I.E., Relaxation strategy for the structural optimization of process flow sheets, *Ind. Eng. Chem. Res.*, 26, 1869–1880, 1987.

79. Kokotovic, P.V., Khalil, H.K., and O'Reilly, J., *Singular Perturbations in Control: Analysis and Design*, Academic Press, San Diego, CA, 1986.

80. Kramer, M.A. and Thompson, M.L., Embedding theoretical models in neural networks in *Proc. of the American Control Conference*, 1992, 475.

81. Lee, J.W., Oldenburg, J., Brüggemann, S., and Marquardt, W., Feasibility of batch reactive distillation, contribution to AIChE Annual Meeting, Reno, NV, 2001.

82. Lee, K., Lee, J.H., Yang, D.R., and Mahoney, A.W., Integrated run-to-run and online model-based control of particle size distribution for a semi-batch precipitation reactor, *Comp. Chem. Eng.*, 26(7–8), 1117–1131, 2002.

83. Lehtonen, J., Salmi, T., Vuori, A., and Haario, H., Optimization of reaction conditions for complex kinetics in a semi-batch reactor, *Ind. Eng. Chem. Res.*, 36(12), 5196–5206, 1997.

84. Leineweber, D.B., Bauer, I., Bock, H.G., and Schlöder, J.P., An efficient multiple shooting based reduced SQP strategy for large-scale dynamic process optimization. Part I, Theoretical aspects, *Comput. Chem. Eng.*, 27, 157–166, 2003.

85. Leineweber, D.B., Efficient Reduced SQP Methods for the Optimization of Chemical Processes Described by Large Sparse DAE Models, Ph.D. thesis, Reihe 3, No. 613, VDI-Verlag, Düsseldorf, 1999.

86. Li, P., Arellano, H., Garcia, H., Wozny, G., and Reuter, E., Optimization of a semibatch distillation process with model validation on the industrial site, *Ind. Eng. Chem. Res.*, 37, 1341–1350, 1998.

87. Lu, J.Z., Multi-zone control under enterprise optimization: needs, challenges and requirements, in *Nonlinear Model Predictive Control*, Allgöwer, F. and Zheng, A., Eds., Birkhäuser, Basel, 2000, 393–402.

88. Marlin, T.E. and Hrymak, A.N., Real-time operations optimization of continuous processes, *AIChE Symp. Ser.*, 93(316), 156–164, 1997.

89. Marquardt, W., Numerical methods for the simulation of differential-algebraic process models, in *Methods of Model Based Control*, Vol. 293, Berber, R., Ed., NATO-ASI Series E: Applied Sciences, Kluwer, Dordrecht, 1995, 42–79.

90. Mohideen, M.J., Perkins, J.D., and Pistikopoulos, E.N., Optimal design of dynamic system under uncertainty, *AIChE J.*, 42(8), 2251–2272, 1996.

91. Mujtaba, I.M. and Macchietto, S., Simultaneous optimization of design and operation of multicomponent batch distillation column: single and multiple separation duties, *J. Proc. Cont.*, 6(1), 27–36, 1996.

92. Nemhauser, G.L. and Wolsey, L.A., *Integer and Combinatorial Optimization*, John Wiley & Sons, New York, 1999.

93. Obertopp, T., Spieker, A., and Gilles, E.D., Optimierung hybrider Prozesse in der Verfahrenstechnik, in *Oberhausener UMSICHT-Tage 1998: Rechneranwendungen in der Verfahrenstechnik*, UMSICHT-Schriftenreihe Band 7, Fraunhofer IRB Verlag, Stuttgart, 1998, 5.1–5.18.

94. Oldenburg, J., Marquardt, W., Heinz, D., and Leineweber, D.B., Mixed logic dynamic optimization applied to batch distillation process design, *AIChE J.*, 48(11), 1900–2917, 2003.

95. Park, T. and Barton, P.I., State event location in differential-algebraic models, *ACM Trans. Model. Comput. Simul.*, 6(2), 137–165, 1996.

96. Pontryagin, L.S., Boltyanskiy, V.G., Gamkrelidze, R.V., and Mishchenko, Y.F., *The Mathematical Theory of Optimal Processes*, Wiley-Interscience, New York, 1962.

97. Prasad, V., Schley, M., Russo, L.P., and Bequette, B.W., Product property and production rate control of styrene polymerization, *J. Proc. Cont.*, 12, 353–372, 2002.

98. Psichogios, D.C. and Ungar, L.H., A hybrid neural-network first principles approach to process modeling, *AIChE J.*, 38, 1499, 1992.

99. Raman, R. and Grossmann, I.E., Modelling and computational techniques for logic based integer programming, *Comput. Chem. Eng.*, 18(7), 563–578, 1994.

100. Rao, C.V. and Rawlings, J.B., Constrained process monitoring: moving-horizon approach, *AIChE J.*, 48(1), 97–109, 2002.

101. Rawlings, J.B., Tutorial overview of model predictive control, *IEEE Control Syst. Mag.*, 20(3), 38–52, 2000.

102. René, D. and Alla, H., *Petri Nets and Grafcet: Tools for Modelling Discrete Event Systems*, Prentice Hall, New York.

103. Robertson, Lee D.K.H. and Rawlings, J.B., A moving horizon-based approach for least-squares estimation, *AIChE J.*, 42(8), 2209–2223, 1996.

104. Ruppen, D., Bonvin, D., and Rippin, D.W.T., Implementation of adaptive optimal operation for a semi-batch reaction system, *Comp. Chem. Eng.*, 22(1–2), 185–199, 1998.

105. Samsalit, N.J., Papageorgiou, L.G., and Shah, N., Robustness metrics for dynamic optimization models under parametric uncertainty, *AIChE J.*, 44(9), 1993–2006, 1998.

106. Schlegel, M., Binder, Cruse, A., Oldenburg, J., and Marquardt, W., Component-based implementation of a dynamic optimization algorithm using adaptive parameterization, in *European Symposium on Computer Aided Process Engineering 11*, Gani, R. and Jørgensen, S.B., Eds., Elsevier, Amsterdam, 2001, 1071–1076.

107. Schuler, H. and Schmidt, C.-U., Calorimetric state estimators for chemical reactor diagnosis and control: review of methods and applications, *Chem. Eng. Sci.*, 47, 899–915, 1992.

108. Schweiger, C.A. and Floudas, C.A., Interaction of design and control: optimization with dynamic models, in *Optimal Control: Theory, Algorithms, and Applications*, Hager, W.W. and Paradolos, P.M., Eds., Kluwer Academic, Dordrecht, 1997, 388–435.

109. Schweiger, C.A. and Floudas, C.A., Process synthesis, design, and control: a mixed-integer optimal control framework, in *Proc. of DYCOPS-5 on Dynamics and Control of Process Systems*, Corfu, Greece, 1998.

110. Sharif, M., Shah, N., and Pantelides, C.C., On the design of multicomponent batch distillation columns, *Comput. Chem. Eng.*, 22(suppl.), S69–S76, 1998.

111. Srinivasan, B., Bonvin, D., Primus, C.J., and Ricker, N.L., Run-to-run optimization via control of generalized constraints, *Control Eng. Pract.*, 9(8), 911–919, 2001.

112. Srinivasan, B., Palanki, S., and Bonvin, D., Dynamic optimization of batch processes. I. Characterization of the nominal solution, *Comp. Chem. Eng.*, 27, 1–26, 2003.

113. Srinivasan, B., Bonvin, D., Visser, E., and Palanki, S., Dynamic optimization of batch processes. II. Role of measurements in handling uncertainty, *Comp. Chem. Eng.*, 27, 27–44, 2003.

114. Stoessel, F., Design thermally safe semibatch reactors, *Chem. Eng. Progr.*, 93, 46–53, 1995.

115. Støren, S. and Hertzberg, T., Obtaining sensitivity information in dynamic optimization problems solved by the sequential approach, *Comp. Chem. Eng.*, 23, 807–819, 1999.

116. Tanartkit, P. and Biegler, L.T., A nested, simultaneous approach for dynamic optimization problems. II. The outer problem, *Comp. Chem. Eng.*, 27(12), 1365–1388, 1997.

117. Terwiesch, P., Cautious online correction of batch process operation, *AIChE J.*, 41(5), 1337–1340, 1995.

118. Terwiesch, P. and Agrawal, M., Robust input policies for batch reactors under parametric uncertainty, *Comput. Chem. Eng.*, 37, 33–52, 1995.

119. Terwiesch, P., Ravemark, D., Schenker, B., and Rippin, D.W.T., Semi-batch process optimization under uncertainty: theory and experiments, *Comp. Chem. Eng.*, 22(1–2), 201–213, 1998.

120. Terwiesch, P., Agrawal, M., and Rippin, D.W.T., Bath unit optimization with imperfect modeling: a survey, *J. Proc. Cont.*, 4, 238–258, 1994.

121. Türkay, M. and Grossmann, I.E., Logic-based MINLP algorithms for the optimal synthesis of process networks, *Comput. Chem. Eng.*, 20(8), 959–978, 1996.

122. de Vallière, P. and Bonvin, D., Application of estimation techniques to batch reactors. II. Experimental studies in state and parameter estimation, *Comput. Chem. Eng.*, 13(1–2), 11–20, 1989.

123. van Can, H.J.L. et al., Understanding and applying the extrapolation properties of serial gray-box models, *AIChE J.*, 44(5), 1071–1089, 1998.

124. Varaiya, P. and Wets, R.J.B., Stochastic dynamic optimization approaches and computation, in *Mathematical Programming: Recent Developments and Applications*, Kluwer, Dordrecht, 1989, 309–332.

125. Vassiliadis, V.S., Sargent, R.W.H., and Pantelides, C.C., Solution of a class of multistage dynamic optimization problems. 1. Problems without path constraints, *Ind. Eng. Chem. Res.*, 33(9), 2111–2122, 1994.

126. Visser, E., Srinivasan, B., Palanki, S., and Bonvin, D., A feedback-based implementation scheme for batch process optimization, *J. Proc. Cont.*, 20(5), 399–410, 2000.

127. Viswanathan, J. and Grossmann, I.E., A combined penalty function and outer-approximation method for MINLP optimization, *Comput. Chem. Eng.*, 14(7), 769–782, 1990.

128. von Stryk, O., Numerische Lösung optimaler Steuerungsprobleme: Diskretisierung, Parameteroptimierung und Berechnung der adjungierten Variablen, VDI-Fortschrittsbericht, Reihe 8, No. 441, VDI-Verlag, Düsseldorf, 1995.

129. Waldraff, W., King, R., and Gilles, E.G., Optimal feeding strategies by adaptive mesh selection for fed-batch bioprocesses, *Bioproc. Eng.*, 17, 221–227, 1997.

130. Wendt, M., Li, P., and Wozny, G., Nonlinear chance-constrained process optimization under uncertainty, *Ind. Eng. Chem. Res.*, 41(15), 3621–3629, 2002.

131. Williams, H.P., *Model Building in Mathematical Programming*, John Wiley & Sons, Chichester, 1999.

132. Zafiriou, E. and Zhu, J.-M., Optimal control of semi-batch processes in the presence of modeling error, in *Proc. of the American Control Conference*, San Diego, CA, 1990.

133. Zhang, Y., Monder, D., and Forbes, J.F., Real-time optimization under parametric uncertainty: a probability constrained approach, *J. Proc. Control*, 12, 373–389, 2002.

10 Batch Process Management: Planning and Scheduling

Karl D. Schnelle and Matthew H. Bassett

CONTENTS

10.1 INTRODUCTION

One key topic in batch process management is planning and scheduling. Typically, production scheduling is tactical and deals with detailed timing of specific manufacturing steps, while campaign planning is more strategic and related to controlling costs over longer periods of time. Both concepts are characterized by extensive data needs, uncertainty, a large decision space (sequencing, timing, product assignment to units, etc.), and the need for good, feasible solutions. Optimal plans and schedules, found through numerical modeling techniques, may not always be required to satisfy the real-world business needs even though they might be worth the effort. Depending on the sophistication of approach used to solve the problem, a feasible plan or schedule may be all that is considered necessary to meet the immediate business needs.

Production scheduling is the short-term look (less than a week to a month) at the requirements for each product to be made. The time scale should fit the needs of manufacturing. Decisions that must be made at this level are:

- How many batches required for each product
- Which equipment to use if multiple units are available
- Start and stop time of each batch on each piece of equipment (the run length)
- Allocation of resources to support the production of those batches (e.g., utilities, operators, raw materials, waste facilities)

Campaign planning, then, is a medium-term look (weeks to months) at a series of batches of one product. Because this planning is done for a longer period of time, performance is measured on metrics averaged over time. The time scale for campaign planning depends on both the business and production structure. Decisions to be made include:

- Production goals, or the total amount of each product in each campaign and the resulting work in-process (WIP, or inventories)
- Which production line (processing train) to use if multiple lines are available in the facility
- Sequence of campaigns on each train
- Day when each campaign starts and stops

An extra dimension to consider in campaign planning is that different intermediates and final products may share production facilities but use distinct processing trains, or they could even share certain equipment units. During the planning process, the link between intermediate campaigns and the final product campaign must be maintained. If equipment is shared among intermediates or products, typically long production campaigns (on the order of weeks or months) of single intermediates are required, and then careful *turnarounds* (with possible cleanouts) are performed. Turnarounds are the time and effort needed to change over, and sometimes clean out, equipment for a new intermediate or product. Because many batch chemicals are low-volume, high-value products, they may be structurally complex and require a large number of processing steps and complex intermediates. Thus, decision making must be sophisticated enough to match the number of required intermediates for each product.

Different companies, and even businesses within the same company, determine the time horizons that campaign planning and production scheduling encompass in their own ways. These differences can be based on the complexity of the process being analyzed or the detail and time frame of the solutions needed. The distinction between campaign planning and production scheduling in batch facilities has become further blurred in recent years as new process systems are being utilized. The computational power of current software solutions allows more detailed decisions to be made (down to the level of each batch on each unit) over a longer time period. Thus, both timing and costs can be driven to very good or even optimal solutions over a longer and longer planning and scheduling horizon.

The long-term view of this decision-making process for batch plant operation may be considered *supply-chain planning* with characteristic time scales of months or years. In this case, supply planning would include selecting what products to make in which years, choosing manufacturing sites, utilizing third party contractors, etc. Long-range capacity and forecasting systems would also be needed. Again, because of current software and hardware capabilities, many of these planning systems overlap with campaign planning and even detailed scheduling. At present, the *strategic planning* activities (multiyear, multilocation, multiproduct, and multiconstraint decisions) are still at too high a level to be automated in the same systems as planning and scheduling. How these different levels of planning are differentiated is influenced by the management structure of the company, so each organization may have its own terminology and planning approach. These topics are discussed further in Chapter 12.

10.1.1 WHY ACCURATE PLANNING AND SCHEDULING ARE NEEDED

Due to the increased stress on profitability in all manufacturing processes and pressure to control costs, reduce inventories, and get more product out the door with the same resources, the job of the planner and production scheduler has become extremely complex. In addition, the use of third-party subcontractors for some intermediates (or final products), reduced lead times, and the increasing global character of suppliers and customers are driving the need for advanced planning and scheduling. Additional complexities are encountered when one or more new products are introduced simultaneously into the marketplace. In addition to the normal *uncertainties* in the day-to-day running of a production facility (e.g., equipment breakdowns, raw material shortages, unavailability of required labor and utilities), new product introductions include greater market uncertainties (demand levels, demand timings, and product pricing) and production uncertainties (cycle times and yields). Software systems or modeling tools are needed that can incorporate many of the uncertainties associated with planning and shorten the time required to make longer-term supply-chain decisions.

For production personnel, planning is necessary to help understand customer demands and react effectively to change. If efficient production schedules and campaign plans are available, management can deal with any changes soon after they occur. In fact, what-if scenarios can be formulated beforehand to help solve issues before they happen if a numerical tool or appropriate model is available. The campaign plan improves communication among production personnel, product managers, and the sales and marketing organization. If a concrete plan is in place and different what-if scenarios are run, production has a better chance to obtain buy-in from the commercial side of the business. The commercial product managers will also have more confidence that they will be able to meet customer demands.

To illustrate that a campaign plan is feasible, a detailed schedule of the tasks within the production facility is beneficial. This schedule can help guide all operations personnel in the daily running of the plant. Plant shutdowns, maintenance, critical raw material deliveries, and waste handling should all be linked to the schedule. Any data acquisition and modeling activity used to construct plans and schedules will result in a better understanding of the process and any potential causes for delays. These data and models contribute to continual process improvement.

10.1.2 WHERE DO PLANNING AND SCHEDULING FIT?

Batch plant planning and scheduling systems sit in between enterprise resource planning (ERP) and manufacturing execution systems (MES). ERP encompasses the older manufacturing resource planning concept known as MRP-II (which encompasses when and where materials are needed, long-range planning, capacity planning, business planning, etc.), as well as forecasting, customer order processing and analysis, finance, local and global logistics, and quality control. ERP can be considered as supply-chain management (SCM) if every operation of the value chain is managed to minimize the cost and time of supplying products to customers. MES acquires, manages, and reports production-related data on all plant activities, such as raw material orders, batch tracking, quality, maintenance, personnel, and inventory levels, as they occur.

Figure 10.1 shows the key information systems that relate to planning and scheduling. The ovals represent systems, and the rectangles represent work processes. Customer orders, or sales data, drive the forecasting process. Historic and forecasted demands are used in supply-chain planning, which in turn feeds the campaign planning process. In some companies, supply-chain planning may be called sales and operations planning (S&OP). Detailed production scheduling requires process data from the MES layer as well as logistics information about transportation and warehousing of raw materials and products. For planning and scheduling to make full use of all available information, these business and process systems must be fully incorporated into a computer-integrated manufacturing (CIM) system. In this case, CIM can be thought of as integrating the

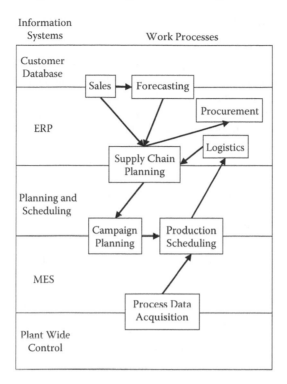

FIGURE 10.1 Key information systems and processes.

flow of all the manufacturing information in Figure 10.1 into one computer system. As noted by Edgar,[1] the chemical processing industry as a whole has not reached this level of integration.

10.1.3 IMPORTANCE OF INTEGRATION

When ERP and MES are integrated, the feasibility of a new forecast can be checked by updating the campaign plans and the production schedules. Rapid feedback of production scheduling issues from production can help the product manager decide on the need for third-party contractors or which customers will be shorted if demands cannot be met. The sequence and timing of batches are also passed to the MES layer so feedback can be obtained quickly on the status of production. Having one planning and scheduling function avoids fragmentation of the planning process that could result in several conflicting plans for production. Challenges do exist before any planning and scheduling systems are successfully integrated and fully utilized. With computer hardware and software capabilities increasing rapidly, information technology is not the barrier so much as human or organizational barriers. Shobrys and White[2] explain how to change these behaviors and give more thoughts on how planning and scheduling systems work together.

10.1.4 How Can You Design a Plant without Planning or Scheduling?

When designing or retrofitting a batch plant, the final design may or may not necessarily be associated with a feasible schedule. In some design optimization cases, constraints are added to handle the planning and scheduling. Lin and Floudas[3] discuss this case with two small example problems. For many applications, these extra constraints may make the problem too complex to solve in a reasonable amount of time. In this case, an iterative approach may be required; if a feasible schedule cannot be found, then certain inputs to the design optimization are adjusted and a new design is found.[4] Heo et al.[5] approached this complexity problem by using a three-step approach for multipurpose plants. Thus, before a new plant is running or even built, planning and scheduling are necessary to verify the feasibility of the design.

10.1.5 Steps for Real-World Planning and Scheduling

The following steps have been modified from Houston and Basu,[6] who were concerned specifically with pharmaceutical pilot-plant management.

1. *Establish the project team.* The personnel involved with inputs to and results from the scheduling and planning process must be brought together as a team. These personnel could include product managers, supply-chain planners, plant schedulers, information services, and mathematical modelers. Extensive modeling expertise, which companies might not have in-house, may be required for the team to implement some of the more complex solution approaches discussed later. However, even a simplified implementation is better than no implementation at all.
2. *Document all assumptions.* Assumptions must be documented so both production and product marketing managers understand the issues and feel part of the same team. Any uncertainties in marketing or production data should be understood, communicated to the team, and managed. The scope should be defined, as well. Misunderstandings will be reduced, and, as personnel leave and join the team, they will understand the issues more quickly and contribute more effectively. These assumptions must be communicated to both planning and scheduling personnel so all decisions are consistent.
3. *Acquire demand data.* Manufacturing and marketing managers must be in frequent contact to keep demand forecasts (amounts and due dates) as up to date as possible. Uncertainties in demand must be acknowledged and the risk managed as well as possible through an integrated team. Also, any downtime at the plant must be communicated with the commercial managers so they can plan accordingly. Demand data will be required if the production strategy is *make-to-order.* The goal of planning

and scheduling is to meet exact due dates for each product, whereas the make-to-stock strategy is based on inventory levels and current capacity.

4. *Plan product campaigns.* The amount of each product in each campaign may be determined from a long-term supply-chain plan or by campaign planning. When campaign amounts over several months are determined, any missed customer demands can be estimated. Difficult to obtain or long lead-time raw materials can be planned for, and any special processing needs or short-term equipment modifications can be identified. With multiple products, either long campaigns of single products or small numbers of batches interspersed between different products may be run. Depending on how planning and scheduling are aggregated, detailed scheduling (step 7) may be integrated into this step to help determine how many batches to run in each campaign. The modeling experts on the team must decide on the appropriate planning algorithms and tools to use based on the business requirements

5. *Check production capacity.* The campaign plan will give manufacturing the campaign start or end date and required amounts. With this information, production must use their current process yields and equipment sizes to verify that the campaigns can be completed on time. Batch sizes, inventory levels, and planned maintenance and shutdowns must be known as well. This capacity is the real-life output rate of the plant based on actual batches. Typically, this is less than the *design capacity*, which is the ideal capacity on paper. The design capacity may never be reached because of scheduling complexities and other obstacles to productivity.

6. *Predict raw material and waste needs.* Once the campaigns are planned, ordering of raw materials can be planned, and any special handling of the waste streams generated can be coordinated with onsite or offsite environmental facilities. The cost to process the campaigns can then be estimated and reported to management.

7. *Schedule production.* If a detailed schedule is not produced in step 4, then the start and stop time of each batch on each equipment unit must be determined. Details on exactly when raw materials, utilities, and operators will be needed and waste streams generated must be known. Consistency must be maintained between the planning and scheduling systems so that the same key assumptions and data are used. The team must choose a suitable scheduling approach to meet the business needs. The appropriate results of supply-chain planning must be passed down to the lower, more detailed systems.

8. *Track actual schedule and costs versus plan.* Areas for continuous process improvement can be identified if actual vs. plan are examined.

9. *Maintain the software and connections to external databases.* The planning and scheduling systems are usually complex software applications that require real-time process data and transactional demand

data. As the manufacturing process is de-bottlenecked or updated and as the customer base or market changes over time, the sophisticated algorithms as well as the data-acquisition systems must be adapted and maintained. Their value is lost and they will fall into disuse if they no longer reflect the current situation.

10.2 DATA AND SYSTEM REQUIREMENTS

Regardless of how a schedule or plan is constructed, certain types of input data are required. Most scheduling applications are data intensive and require some or all of the following types of data:

- Sales and marketing
  - Time horizon, forecast amounts, order due dates, prices
- Operations
  - Manufacturing recipe (precedence of processing tasks, unit operations flowchart, task descriptions, allowable task to unit assignments)
  - Production facility (equipment types, capacities, rates)
  - Inventory (stock-out levels, locations, capacities)
  - Raw materials (availability, amounts, and timing)
  - Intermediates/products (zero wait when stability is an issue, turnaround times)
  - Resources (e.g., amounts or use rates for utilities, operators, waste processing)

10.2.1 IMPORTANCE OF FORECAST DATA

For planning purposes, many batch chemical companies produce a 12-month forecast, focusing more effort on the first 1 to 3 months. Ideally, the forecast should include uncertainties in the demands, but at a minimum an upside and a downside forecast should be estimated. Because of this uncertainty, actual inventory levels may vary widely and some minimum stock-out level or buffer must be set. Production scheduling must handle the risk that customer demands will be unmet or that inventory carrying costs will be too high. The process may be repeated several times a year or even every month. The forecast that is based on customer sales must be disaggregated into the level of product that is manufactured at the facility being scheduled. In other words, if the customer receives a formulated product that is packaged and labeled, then the forecast for that product must be converted into the appropriate chemical ingredients being used in this formulation. This step is required when planning and scheduling the ingredient plant.

10.2.2 IMPORTANCE OF SYSTEM ARCHITECTURE

Forecast data are stored in various places and must be linked to a functional scheduling system that includes a graphical user interface (GUI), model code, and database. Maintaining the software cannot be accomplished without these functions. The GUI is required to help visualize and modify the process flow and the equipment networks and to display results in the form of Gantt charts, graphs of inventory levels and resource usage, etc. Many times, the results of any scheduling system will have to be manually adjusted because industrially sized problems are complex enough that all constraints will never be represented fully. This adjustment should be made user-friendly by interactive tools in the GUI. The GUI should also help manage a number of what-if scenarios that are typically run before the schedule or plan is finalized. Diagnosis is another important function: determining which resources are limiting and where the bottlenecks are, as well as sensitivity analysis of certain input variables on the results.

The model code includes both the means to represent a specific problem in terms of a general framework and the algorithm itself used to find the solution. Unless the system is tuned to solve one problem only, the general framework and algorithm should be robust enough to handle many other problems. Pekny and Reklaitis[7] actually call for three layers of model code: representation, formulation, and computation. Representation describes the problem to be solved, and formulation translates the representation into a format that the computational engine can act on. Computation, then, is the specific algorithm that can generate solutions and solve the problem. This software design lends itself to a modular system architecture, using object-oriented programming (OOP). This type of architecture is required for a stand-alone system as well as one integrated into other information systems.

10.3 CHARACTERISTICS OF BATCH PLANTS

In the manufacturing of chemicals, batch plants have several characteristics that distinguish themselves from continuous plants. The *recipe* (the reaction pathways necessary to produce a chemical) is the first piece of information that is required. The *process flow* (the set of processing tasks required to manufacture a chemical) may not correspond one-to-one to the equipment units in the plant. The tasks may be linked together in a network configuration to portray the precedence constraints, an example of which is shown in Figure 10.2. Combining the recipe

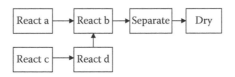

FIGURE 10.2 Example of a recipe.

and process flow together leads to a *state–task network* (STN).[8–9] The STN allows one to see the flow of resources into and out of tasks that may not be obvious from the recipe or process flow individually. The tasks are represented as boxes, as in the previous recipe, but now a circle represents each state (intermediate). For example, the fact that intermediate D is isolated from E is easily seen in the STN shown in Figure 10.3. Pantelides[10] further refined the STN by introducing the *resource–task network* (RTN). The major difference is that, where the STN treats equipment differently than raw materials, the RTN treats them both as resources. Figure 10.4 shows an example of what an RTN might look like for the case when units are dedicated to specific tasks.

More than one unit in the plant may be capable of performing a given task; thus, the first step in production scheduling is the assignment problem: Assign each task to an equipment unit. *Parallel units* may be used for certain tasks to increase capacity or throughput. The network of units needed for the recipe forms a *production line*. The *batch size* will be the amount of intermediate or product coming out of the last unit in the production line. If *batch integrity* is not kept (i.e., batches are mixed somewhere in the line), then the batch size may change inside the production line. The line may be split by *intermediate storage*, which would also cause the batches to lose their identity.

A series of batches of the same product may be run on a production line, representing a *campaign* of that product. Conversely, one batch each of many

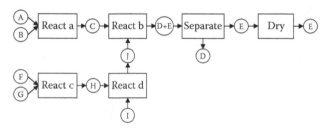

FIGURE 10.3 Example of a state–task network.

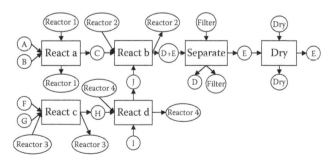

FIGURE 10.4 Example of a resource–task network.

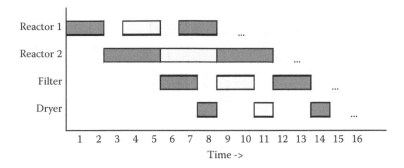

FIGURE 10.5 A simple batch plant.

products could be run on the line. This feature of batch plants leads to the issue of *changeover times* and the expense of transitioning from one product to the next. Equipment cleanouts may be time consuming and expensive if cross-contamination is an issue. A convenient way to illustrate these types of operation is with a *Gantt chart*. Time is usually represented on the *x*-axis, and the *y*-axis contains a list of units or production lines dependent on whether individual batches or campaigns are being depicted. If the top four tasks in Figure 10.2 are to be scheduled for one product, Figure 10.5 would represent the resulting Gantt chart. The first three batches are shown as alternating shaded boxes; the task named React a in Figure 10.2 is performed on Reactor 1, React b on Reactor 2, etc. The batches overlap on each unit so multiple units are running at any given time. The *bottleneck* can be easily identified as Reactor 2. To increase throughput, an additional reactor may be inserted *out-of-phase* for React b. Figure 10.6 shows how this *parallel equipment* can shift the bottleneck from the equipment with a longer processing time, with five batches now illustrated.

10.3.1 MULTIPRODUCT PLANTS

If another product is run on the same equipment, as in Figure 10.6, then the plant becomes *multiproduct*. Figure 10.7 depicts product B being added to the Gantt chart. Four batches of product A are processed, followed by two new batches of B. Thus, a campaign of A is followed by a campaign of B. The four batches of A are processed in the same order as in Figure 10.6, flowing from Reactor 1 to 2a or 2b, to the filter, and then finally to the dryer. Because product B does not require the same processing times, a second Reactor 2 is not needed for B, and the bottleneck shifts. Product B requires all tasks in Figure 10.2. Task React c is done on Reactor 3 and React d on Reactor 4, then the flow follows the same route as for product A. Because scheduling and planning are also considered part of operations research (OR), the term *flowshop* may be applied to multiproduct plants where the products flow through the equipment is mostly in the same order. In fact, the OR literature often uses the term *batch manufacturing* to represent a

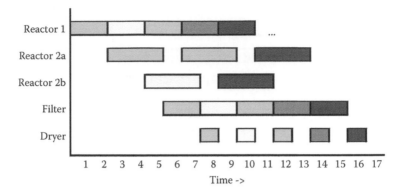

FIGURE 10.6 A batch plant with parallel units.

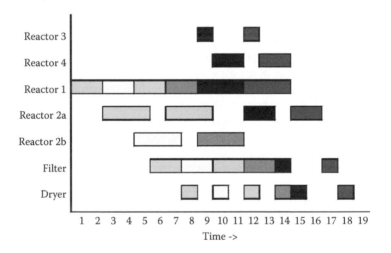

FIGURE 10.7 A multiproduct plant.

flowshop with a medium amount of nonchemical products: 10 to 30 discrete parts, assemblies, etc.[11]

With multiple products in the same equipment, the need for cleanouts and changeovers arises. The time and cost to switch to another product may be sequence dependent. With many products, this constraint may make the sequencing decisions much more difficult. Nevertheless, in the Gantt chart, time blocks of varying lengths may be added between campaigns to block off changeover times.

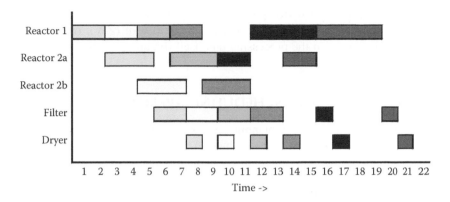

FIGURE 10.8 A multipurpose plant.

10.3.2 MULTIPURPOSE PLANTS

The next step in complexity in batch plants is the *multipurpose* plant, where multiple products are processed through multiple pathways through the equipment. No predefined production line exists for all products. The processing of each product may flow through the plant in various directions. Again, if a second product is added to Figure 10.6, but the order of the equipment is very different from the first product, then we would obtain the Gantt chart in Figure 10.8. Four batches of product A are produced, again beginning in Reactor 1; however, product B starts in Reactor 2a, then moves to Reactor 1, then the filter, and then the dryer. In the OR world, a multipurpose plant may be called a *jobshop*, especially when very short campaigns of many products are produced. A pilot-plant facility would fit this description well.

10.3.3 PERFORMANCE METRICS

When Gantt charts are produced for plant operations, certain performance metrics are easy to demonstrate. The *cycle time* for a product is the time required between successive batches (sometimes called *tact time*); for instance, the introduction of the parallel reactor reduced the cycle time from 3 time units in Figure 10.5 to 2 time units in Figure 10.6. The *residence time* is the time needed to complete a batch on all the equipment (also called *throughput time*).[12] The residence time is 8 time units in both Figure 10.5 and Figure 10.6. Also, the *makespan* is the completion time of the last batch on the last equipment unit (16 time units in Figure 10.6 if the campaign is only five batches). The scheduling *horizon* is the time window in which the entire schedule is built (a week, a month, etc.), depending on order due dates and other business needs.

The Gantt chart helps illustrate visually that intermediate storage is needed to decouple long production lines of many units. If the intermediate is a solid, unlimited intermediate storage (UIS) may be used. For liquids or any intermediates with limited storage, finite storage (FIS) is required. The special case of an unstable intermediate would require zero wait (ZW), or no storage even for a short time in the processing unit itself. Figures 10.5 to 10.8 all show the ZW type.

10.4 PLANNING AND SCHEDULING APPROACHES

The approach necessary for planning and scheduling can vary significantly depending on the complexity of the problem to be solved. Approaches can include manual solution, interactive Gantt charts, simulation, heuristic optimization, and rigorous optimization. Figure 10.9 shows a possible hierarchy for these approaches. Either manual or automatic feedback could exist among different levels of this hierarchy, which could allow solutions to be iteratively improved or modified to better accommodate uncertainty or process upsets. In this section, examples of tools that can be used for each approach as well as possible overlap among the approaches are presented.

In many industrial applications, either the number of planning choices or the time and resources are limited, so a very simple model is designed and used in a matter of a few days or weeks. Typically, a manual spreadsheet or project planning system, such as Microsoft Excel® (Microsoft Corp.; www.microsoft.com/excel) or QuickGantt® (AICOS Technologies AG; www.aicos.com), is used to illustrate the Gantt chart and calculate the material balances, inventory levels, amount of product produced, etc. The use of an interactive Gantt chart can help speed up this manual process by reducing the time required to modify the plan or schedule. When the decisions become too complex or numerous, a more sophisticated technology is sometimes used.

Two distinct classes of sophisticated modeling technology can be used in the scheduling and planning arena: *simulation* and *optimization*. Each has its own

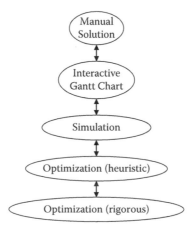

FIGURE 10.9 A possible hierarchy of solution approaches.

advantages. Batch process simulation is usually based on discrete event simulation although Monte Carlo-based spreadsheet simulations can be used for very simple problems. Generally, the equipment network and associated processing tasks are first represented, and then the flow of batches through the plant is simulated. The number of batches per time horizon (capacity driven) or the time to complete a campaign of multiple batches of multiple products (demand driven) is usually the performance metric that is monitored. The downside is that the user must predetermine which tasks are assigned to which units, as well as the order and number of batches if more than one product is being produced. In other words, a trial-and-error approach has to be taken until a good schedule is found. The advantage is that parameter uncertainty may be accounted for, typically unit processing times, as well as equipment and resource availability. Also, dynamic operating conditions can be simulated (e.g., a random 10% of the batches must be reworked due to quality issues). Some examples of spreadsheet-based Monte Carlo simulation tools include @RISK (Palisade Corp.; www.palisade.com) and Crystal Ball® (Decisioneering, Inc.; www.decisioneering.com), and discrete event simulation packages include Arena® (Rockwell Software, Inc.; www.arenasimulation.com) and BATCHES (Batch Process Technologies, Inc.; www.bptech.com). BATCHES is currently the only simulation system built specifically for multi-product batch and semicontinuous processes.

Optimization, on the other hand, is used to determine the best feasible (i.e., optimal) schedule.[13] This is done by first determining a set of *decision variables* that represent the decisions that must be made (e.g., batch size, start time). Together with parameters (constants that define the process, such as processing time and demanded quantity), *constraints* are generated that specify the restrictions on and interactions between the decision variables. A feasible solution is any solution that satisfies all the constraints. In order to determine the optimal solution, an *objective function* that quantifies the consequences of the decision variables is needed.

In general, optimization models are based on deterministic data. The few formulations that attempt to include uncertainty are unable to handle processing time uncertainty. Thus, simulation would be a good tool to evaluate a schedule under uncertainty after an optimization algorithm has established a schedule. Two basic types of optimization models are available: heuristics (directed search methods and knowledge-based systems) and equation-based mathematical programming. Directed search methods include algorithms such as simulated annealing, Tabu search, and genetic algorithms, all of which try to construct a set of feasible schedules and then adjust them iteratively until a better solution is found. Knowledge-based systems originated in the artificial intelligence world and rely on a set of rules to construct the schedule. These rules are developed from knowledge of the planners and schedulers who are familiar with the day-to-day tasks and procedures. In the simplest case, a very simple rule may be used, such as first in, first out (FIFO) or a *product wheel* (a predetermined sequencing order for products).

Mathematical programming has been used in the OR field for many years to plan and schedule discrete parts manufacturing in the jobshop and flowshop setting. Because the decision variables involve sequencing in addition to resource and equipment allocation, binary (0,1) and integer variables are required. Shah[14] points out that mixed-integer nonlinear programming (MINLP) can be used to solve the most general batch or continuous scheduling problems (multipurpose plants). Some of these nonlinearities include learning curves, catalyst decay, and equipment performance; however, the objective function and constraints for many batch scheduling problems can be reformulated so the solution can be found using a mixed-integer linear program (MILP). Solving an MILP is preferred over an MINLP because of the robustness of the available solvers and the generally quicker solution times for the problems. Floudas and Lin[15] give a good overview of the scheduling of multiproduct and multipurpose processes using optimization; they also categorize the approaches into discrete or continuous time formulations.

For detailed control over the formulation and solution of the problem, formulation engines such as GAMS (GAMS Development Corp.; www.gams.com), OPL Studio (ILOG, Inc.; www.ilog.com/products/oplstudio), and Xpress-Mosel (dash optimization; www.dashoptimization.com/modeling_interfaces.html) are available. These tools allow the user to define the equations of the math program themselves and then solve the problem using any number of MILP or MINLP solvers. CPlex (ILOG, Inc.; www.ilog.com/products/cplex) and Xpress-MP (dash optimization; www.dashoptimization.com/products.html) are two of the most commonly used MILP solution engines, while MINOS (Stanford Business Software, Inc.; www.sbsi-sol-optimize.com) and CONOPT (ARKI Consulting & Development A/S; www.conopt.com) are commonly used for NLPs. If standard MILP or MINLP software solvers cannot find a solution in a reasonable amount of time, specialized algorithms must be developed and used. VirtECS Schedule (Advanced Process Combinatorics, Inc.; www.combination.com), MIMI® (Aspen Technology, Inc.; www.aspentech.com), and ModelEnterprise® (Process Systems Enterprise Ltd.; www.psenterprise.com) are examples of third-party tools specifically developed to solve batch planning and scheduling problems using math programming. Siletti and Petrides[16] list a few other similar systems. Use of these tools reduces the dependence on a mathematical modeling expert, but at the loss of flexibility.

A more recent approach to solving planning and scheduling problems is constraint programming (CP), also known as constrained logic programming.[17] CP is a computer programming methodology that solves combinatorial optimization problems using domain reduction techniques. Most CP implementations have been applied to sequencing in the OR arena (e.g., see the work of Baptiste et al.[18]); however, Maravelias and Grossmann[19] applied a hybrid MILP/CP to the scheduling of multipurpose batch plants. ILOG Scheduler (ILOG, Inc.; www.ilog.com/products/scheduler) is an example of a constraint-based scheduling tool.

With all these choices in modeling technology, the appropriate model type will depend on the exact problem being solved. In fact, Reklaitis et al.[20] state that "a single, universal solution approach to all scheduling problems does not exist (contrary to vendor claims) and it is highly unlikely that one will be ever found."

This statement is based on complexity theory and the concept of *NP completeness*, which signifies that a scheduling problem that tries to optimize some performance criterion (e.g., minimize the makespan) based on a set of feasible schedules will have a solution time on a computer processor that may be exponential (nonpolynomial, or NP) to the size of the problem. Thus, large industrial scheduling and planning problems may take an unreasonably long time to solve on the fastest computers available. Also, another issue exists where large differences in solution time and quality between model types could be present for a specific scenario. Pekny and Reklaitis[7] provide more details behind this argument with three strategies to get around this issue:

- Simplify the problem (relax the constraints enough to still produce a reasonable schedule).
- Develop an exact algorithm (a mathematical program) tuned to the type of scheduling problem at hand to get a solution in a reasonable amount of time using the latest computer hardware.
- Develop a heuristic solution using one of the techniques mentioned above and settle on a solution that may be feasible but not near to optimal. In fact, the user would have no guarantee that even a feasible solution could be found.

To evaluate these types of models, Pekny and Reklaitis[7] proposed five features that should be compared: solution quality, usability, extensibility, robustness, and complexity strategy. The *quality* of the solution is an obvious feature to understand before any schedule is implemented in a plant. The result from the system may be infeasible, feasible, suboptimal, or optimal. *Usability* refers to how user friendly the model is, how fast the solution is found, the amount of customization or "knobs" that the user can control to adjust the solution, etc. *Extensibility* and *robustness* are related in that an extensible method allows a new problem to be solved if small changes to the input are made, and a robust method allows major changes that make the problem distinctly different. These changes in the model input could involve parameters (such as the size of an equipment unit or a processing time) or structure (such as an additional unit or a different recipe structure entirely).

10.5 SINGLE-STAGE MULTIPRODUCT PLANTS (SEQUENCING)

The simplest planning and scheduling problems are single-stage plants. A single-stage plant is a plant where all the tasks associated with a product are performed on a single piece of equipment or production line with a number of jobs to be assigned to it. The assignment of jobs to the stage is referred to as *sequencing*. It is also possible to have parallel equipment or production lines that can be sequenced independent of each other for the same products.

The sequence in which the jobs are assigned to the stage can depend on a number of factors such as due date, resource availability, and changeover time and cost. Changeover involves the time and resources required to modify equipment between products. These modifications could include cleaning, calibration, and setup. Changeover costs could also be incurred due to off-spec product and its disposal.

A common example of changeover and the effect of sequencing is a paint production facility. Suppose one needs to formulate black, white, and gray-tinted paints using a single mixer. Obviously, switching directly from black to white requires significant cleaning, while going from black or white to gray or from white to black may not require much, if any, cleaning. So a simple rule of thumb might be to always produce at least one batch of gray paint when switching between black and white paint. This rule of thumb describes a simplistic *product wheel* for these products. Many production facilities use product wheels to help the production schedulers decide on the order of production when issues of due date and resource availability are not overriding. A product wheel in its simplest form describes what order the products should be produced to minimize changeover. These production sequences have either been determined by trial and error or calculated offline and compiled for quick response time on the production floor.[21]

Even though the description of single-stage problems is straightforward, this by no means implies that they are simple to solve. The solution to these problems involves obtaining a solution to a *traveling salesman problem* (TSP), either by enumeration of all possible solutions, via heuristic optimization techniques such as Tabu search,[22,23] or by rigorous math-programming-based optimization.[24] The object of a TSP is to find the shortest and least expensive route that visits a predetermined set of locations. Replacing locations with products and the route with a schedule describes the single-stage problem. The TSP has proven to be a difficult problem to solve, with the addition of due dates and resource availability further complicating the solution of the problem.

10.6 MULTISTAGE MULTIPRODUCT AND MULTIPURPOSE PLANTS (FLOWSHOPS AND JOBSHOPS)

The next increase in complexity of scheduling is when products must progress through multiple stages during production. In the chemical process industries (CPIs), this is by far the most common situation. The tasks in a flowshop are assigned to specific equipment, whereas a jobshop contains flexible equipment that can perform multiple tasks. Multistage plants require sequencing of activities, but now have the added complication that order and timing of activities at one stage are dependent on other stages. So, as we saw in Figure 10.5 earlier, the second batch on Reactor 1 cannot start until the batch in Reactor 2 is finished. For a flowshop, *recurrence relationships* can be developed to handle this dependency. A recurrence relationship simply identifies which situations must be satisfied for another batch to be processed. This includes ensuring the completion of the previous task and the availability of the piece of equipment assigned to the next task. Because the assignment of tasks to equipment is known *a priori* in a flowshop, these relationships are straightforward; however, in a jobshop, it is not necessarily known which equipment will process a task until a schedule has been developed. Introducing *material balances* for the resources created and consumed at each stage (whether stored or not) enforces this dependency between the stages for a jobshop. A material balance simply tracks what comes into a stage vs. what comes out. Over the total time for the schedule, in must equal out.

When dependency between stages has been introduced to a scheduling problem, the issue of how to capture this dependency in the mathematical formulation of the problem is raised. Two approaches are to *discretize* time into "buckets"[25] or to treat it as a continuous variable.[26] Each approach has its own pros and cons. Discretizing time into buckets requires the scheduler to divide the *horizon* (time domain) into uniform or possibly nonuniform pieces (e.g., a day or an hour). This discretization depends on the detail to which one would like to schedule. Enforcing material balances for time-discretized models is done by adding a constraint for every time bucket. The smaller the time buckets, the more constraints there will be and the more difficult the problem will be to solve. Leaving time as continuous means not having to sacrifice accuracy of processing times due to rounding, but continuous time makes enforcing material balances and other time-dependent constraints more complex. The multipurpose nature of the equipment in a jobshop allows for increased utilization of this equipment at the expense of greater planning and scheduling complexity.

10.7 PLANNING PROBLEMS

The issues associated with a planning problem are of a more mid- to long-term nature than those for scheduling, with an emphasis on cost. Some of the issues that are normally addressed by a planning problem include campaign planning, multiplant planning, and inventory planning. Campaign planning determines the sequence in which products will be produced on the same or parallel equipment. This differs from single-stage, multiproduct plant scheduling in that campaign planning does not determine the start and stop times for each task for each batch but only the start and stop time for a block of batches to be run. Campaign planning requires an assumption on the throughput of a facility for each product. Multiplant planning extends campaign planning to cover facilities either at the same geographic location or elsewhere, either in-house or third-party contractors. Inventory planning becomes a significant issue in campaign and multiplant planning. These problems are characterized by a need to store and possibly transport significant quantities of intermediate product or to store excess finished goods. Many companies do strive to minimize inventories through *just-in-time* (JIT) production; however, when production campaigns are planned to satisfy future demands, JIT is not possible. For high-valued products, the inventory that is created can lead to significant costs.

10.8 HOW TO HANDLE UNCERTAINTY

Large uncertainties may exist in four main types of batch planning and scheduling data:

- Sales (demands, forecasts, pricing)
- Process (batch cycle times, yields [unit ratios], changeover times)
- Production (e.g., equipment malfunctions, unavailable raw material)
- New processes and products (ramp-up rates for cycle times and unit ratios)

Sales and marketing data include demand amounts and timing, based on some type of forward-looking forecast or on customer orders (make-to-order). All demands and due dates have a certain amount of uncertainty. In some cases, campaign planning could be affected by pricing if different products can be produced in the facility. Higher priced products may be more beneficial to produce. Usually, supply-chain planning, as described in Chapter 12, is used to resolve the issue. Process data can be uncertain as well. The cycle time of each batch on each unit is not always a deterministic value. Intermediate and final yields or unit ratios may change as processing conditions are optimized in the plant. Equipment breakdowns and unavailable resources cannot be predicted and could affect the schedule and even the campaign plan.

10.8.1 RAPID RAMP-UP OF SPECIALTY CHEMICALS

In recent years, an added complexity to batch process management is the emphasis on decreasing the time required to push new specialty chemicals into the marketplace. Product and process development are now being overlapped to produce initial quantities of a chemical; for products such as agrochemicals or pharmaceuticals, small runs are required to prepare samples for environmental or regulatory studies and experimental trials prior to the product being registered and approved for sale. Pilot-plant studies may also be reduced or even skipped. The driver for this rapid ramp-up is that less financial risk occurs when process development is delayed as long as possible. Also, the probability for regulatory approval and commercial success are better known later in the development process.

When batch chemicals are first manufactured in a regular production facility, ramp-up is the period of time to reach normal production levels, which may be on the order of months. The rate of ramp-up, or learning curve rate, should be estimated to help determine the time and resources required to reach full production.[27] Ramp-up occurs because operators become better acquainted with the process and better trained, because process problems caused by new construction are identified and fixed, and because chemistry problems that were not identified in the accelerated process development are fixed. Many of these improvements are completed in the ramp-up phase and not in process development because of lack of time in development and greatly reduced pilot-plant work. Even if the time to market had not been accelerated, many potential improvements could not have been identified until production is initiated because many interactions are too difficult to define beforehand. Instead of solving all problems on the benchtop or in the pilot plant, the "learn by doing" mode is often employed in the full-scale production facility.[28]

Timing is crucial for agrochemical and pharmaceutical products as well, as the first major campaign should be ready to run as soon as a product becomes approved for sale. If the product becomes registered before expected, then ramp-up must occur even faster. With quicker ramp-up and registration, the product can penetrate the market faster and reach higher sales numbers, and the development costs are recouped quicker. Fast ramp-up signifies both fast gains in productivity (kilograms out the door) and high yields (less raw material consumed per kilogram of product). The result of this higher efficiency is less need for capital investment.

10.8.2 CRITICAL ISSUES

Because of rapid ramp-up, production parameters and uncertainties are difficult to quantify. In addition to uncertainty in process productivity and yields, the other end of the supply chain — the market — is also difficult to predict. Both product demands and prices are uncertain to a great degree; their estimates may be updated as often as several times per year near the end of a 10- to 15-year development cycle. Many individual estimates are done by market segment (usually by location or customer segment) and are then rolled-up into global demands by a product manager. Because of the numerous sources of the various demand and price estimates, some values may be given as "best guess" and any predictions of uncertainty would be difficult to quantify.

10.8.3 POSSIBLE SOLUTIONS

Optimization under uncertainty, in general, was recently reviewed by Sahinidis;[29] however, specific solutions to the planning or scheduling problem that deal with uncertainties have mostly concentrated on demands or processing times. Lee and Malone[30] used Monte Carlo sampling of demands and due dates to try to find a flexible plan using simulated annealing. Uncertainties were also included in the work of Vin and Ierapetritou;[31] however, they adapted a mathematical optimization method to include uncertain demands in order to schedule a multiproduct plant. Honkomp et al.[32] dealt with both uncertain processing times and equipment break-downs. A schedule is fed into a simulator that generates uncertainties. Different types of rescheduling methods are compared through an optimizer, and the expected performance of the new schedules is evaluated. Also, Balasubramanian and Grossmann[33] investigated variable processing times only. They used a branch-and-bound algorithm to schedule a multiproduct plant, then they dealt with demand uncertainty using a multistage stochastic MILP.[34] Also, Bonfill et al.[35] considered financial risk in their scheduling model with uncertain market demands.

10.8.4 Real-Time vs. Offline Applications

Uncertainties often occur because the planning or scheduling is being done dynamically or in real time. Instead of a static or offline application, the model is being used as changes in the manufacturing facility occur (e.g., new orders are received, equipment breaks down); thus, *rescheduling* is required. *Scalability* becomes important in this case because these industrial-sized problems must be solved in a short time.

10.9 PLANNING AND SCHEDULING IN THE FUTURE

Sophisticated planning and scheduling systems have not been 100% successful in an industrial setting at this time. Several reasons exist for the continued use of manual spreadsheets and very simple heuristics. Even if a standard scheduling optimization tool is in place, the user many times will have to adjust the results to fit all the operating constraints that were not built into the system. Most industrial-scale batch plants are complex enough that any automated solution can become quite large. The typical system cannot handle the combination of constrained resources, inventory, uncertain demands, etc., so any resulting plan will not be realistic enough and will have to be manually adjusted. The expertise to build most, or all, of these issues into an automated system is not always available in-house. As Honkomp et al.[36] noted, the use of external contractors drives up the cost and time of implementation, updates, and maintenance.

The issue of complexity and NP completeness points to the development of planning and scheduling systems through the use of *algorithm engineering*. Pekny and Reklaitis[7] stated that, as new software algorithms are written to solve more complex scheduling issues, research must be done to extend the current algorithm approach. As an engineering activity, new planning and scheduling software must be more than just an algorithm. The software must also include a GUI, interfaces to external databases, help aids, and diagnosis and debugging tools to be acceptable to plant engineers and schedulers. As the problem proceeds through the stages of algorithm engineering, different levels of system sophistication are achieved:[37]

- Unsupported (trivial test problems)
- Demonstration (realistic test or prototype with limited scope and some constraints)
- Engineering (realistic problem with useful results, a few key constraints missing that allow automatic use of results)

- Production (realistic problem with results implemented after user interpretation)
- Online (realistic problem with results implemented automatically)

Current state-of-the-art automated systems are somewhere along this research and development timeline. New implementations must be developed so they reach production or online status to be truly useful. Much research is currently being done in improving the efficiency and increasing the scope of current algorithms; the extended bibliography, below, lists some of the most current research. A test set of generic planning and scheduling problems is needed so different solution methodologies may be compared. In fact, Honkomp et al.[36] recently proposed a set of test scenarios for scheduling consumer goods manufacturing to aid in selecting the best solution for a specific plant. These higher-level solutions will require new supporting tools. Grossmann and Westerberg[38] identified several areas that must be developed to make progress in this area: simulation and optimization under uncertainty, dynamic or distributed quantitative–qualitative tools, improved real-time modeling, improved tools to represent heuristics, and more information management and data-mining tools. For large-scale industrial applications, Floudas and Lin[15] noted such future research needs as better mathematical models for short-term and medium-term scheduling; multisite production and distribution scheduling; scheduling of semiconductor operations; uncertainty in processing times, prices, product demands, and equipment breakdown; and, finally, integration of scheduling design, synthesis, control, and planning.[15]

10.10 SUMMARY

One key issue in batch process management that cannot be ignored is planning and scheduling. Production scheduling is the short-term look at the requirements for each product to be made, while campaign planning is a more medium-term look at a series of batches of one product. The long-term view of this decision-making process for batch plant operation is often called supply-chain planning, with characteristic time scales of months or years. Each of these levels of planning requires extensive input data concerning not only sales and marketing but also operations. Because of the complexity of the data needs and of the decisions to be made, the methodology must be linked into other manufacturing information systems, such as an enterprise resource planning (ERP) system or a manufacturing execution system (MES).

The approach used to solve planning and scheduling problems depends on many factors relating to the complexity of the issue, such as amount of detailed input data on hand, amount of uncertainty, type of batch manufacturing facility involved, and modeling expertise and real-time or offline software systems available. Similar modeling approaches may be used for different batch plants — single-stage multiproduct, multiple-stage multiproduct, or multipurpose — but the exact implementation of the approach may differ. For instance, the simplest approach would be a totally manual or spreadsheet solution. An interactive Gantt chart may be used to help automate the planning process and study many more scenarios than possible with a purely manual approach. For more complex problems, off-the-self or custom-built software may either simulate or optimize the batch process using many methodologies, from discrete event simulation to heuristics to mathematical programming.

Even though planning and scheduling tools continue to develop as faster, more efficient algorithms and computers are introduced, many industrial-scale plants are so complex that automated solutions are quite large and cumbersome. As a result, algorithms and software systems are not universally installed, but as research and algorithm engineering continues to progress in the future, more realistic problems will be solved with results being implemented automatically in the batch plant.

EXTENDED BIBLIOGRAPHY

Allahverdi, A., Stochastically minimizing total flowtime in flowshops with no waiting space, *Eur. J. Operational Res.*, 113(1), 101–112, 1999.

Azzaro-Pantel, C., Bernal-Haro, L., Baudet, P., Domenech, S., and Pibouleau, L., A two-stage methodology for short-term batch plant scheduling: discrete-event simulation and genetic algorithm, *Comput. Chem. Eng.*, 22(10), 1461–1481, 1998.

Balasubramanian, J. and Grossmann, I.E., Scheduling optimization under uncertainty: an alternative approach, *Comput. Chem. Eng.*, 27, 469–490, 2003.

Bassett, M.H., Pekny, J.F., and Reklaitis G.V., Obtaining realistic production plans for a batch production facility, *Comput. Chem. Eng.*, 21, S1203–S1208, 1997.

Book, N.L. and Bhatnagar, V., Information models for planning and scheduling of chemical processes, *Comput. Chem. Eng.*, 24(2–7), 1641–1644, 2000.

Burkard, R.E., Fortuna, T., and Hurkens, C.A.J., Makespan minimization for chemical batch processes using non-uniform time grids, *Comput. Chem. Eng.*, 26(9), 1321–1332, 2002.

Cantón, J., Nougués, J.M., Rabiza, M.J., Gonzalez, R., Espuña, A., and Puigjaner, L., Reactive scheduling in real-time operation of batch chemical plants, in *Proc. of Foundations of Computer-Aided Process Operations, FOCAPO 2003*, Coral Springs, FL, 2003, 359–362.

Castro, P., Matos, H., and Barbosa-Póvoa, A.P.F.D., Dynamic modelling and scheduling of an industrial batch system, *Comput. Chem. Eng.*, 26, 671–686, 2002.

Castro, P., Barbosa-Povoa, A.P.F.D., and Matos, H., An improved RTN continuous-time formulation for the short-term scheduling of multipurpose batch plants, *Ind. Eng. Chem. Res.*, 40(9), 2059–2068, 2001.

Castro, P.M., Barbosa-Povoa, A.P., Matos, H., and Novais, A.Q., Simple continuous-time formulation for short-term scheduling of batch and continuous plants, *Ind. Eng. Chem. Res.*, 43, 105–118, 2004.

Chan, K.F. and Hui, C.W., Scheduling batch production using a stepwise approach, *Ind. Eng. Chem. Res.*, 42, 3505–3508, 2003.

Chen, C.-L., Liu, C.-L., Feng, X.-D., and Shao, H.-H., Optimal short-term scheduling of multiproduct single-stage batch plants with parallel lines, *Ind. Eng. Chem. Res.*, 41(5), 1249–1260, 2002.

Crama, Y., Pochet, Y., and Wera, Y., *A Discussion of Production Planning Approaches in the Process Industry*, CORE Discussion Paper 2001/42, www.core.ucl.ac.be/services/psfiles/dp01/dp2001-42.pdf, accessed 2001.

Das, B.P., Rickard, J.G., Shah, N., and Macchietto, S., An investigation on integration of aggregate production planning, master production scheduling and short-term production scheduling of batch process operations through a common data model, *Comput. Chem. Eng.*, 24(2–7), 1625–1631, 2000.

Das, B.P., Shah, N., and Chung, P.W.H., Off-line scheduling a simple batch process production plan using the ILOG scheduler, *Comput. Chem. Eng.*, 22, S947–S950, 1998.

Dennis, D.R. and Meredith J.R., An analysis of process industry production and inventory management systems, *J. Operations Manage.*, 18(6), 683–699, 2000.

Dileepan, P. and Ettikin, L.P., Learning: the missing ingredient in production planning spreadsheet models, *Prod. Invent. Manage.*, 29(3), 32–35, 1988.

Giannelos, N.F. and Georgiadis, M.C., A simple new continuous-time formulation for short-term scheduling of multipurpose batch processes, *Ind. Eng. Chem. Res.*, 41(9), 2178–2184, 2002.

Graells, M., Cantón, J., Peschaud, B., and Puigjaner, L., General approach and tool for the scheduling of complex production systems, *Comput. Chem. Eng.*, 22(Suppl. 1), S395–S402, 1998.

Grossmann, I.E. and Biegler, L.T., Future perspective on optimization, part II, *Comput. Chem. Eng.*, 28(8), 1193–1218, 2004.

Gupta, S. and Karimi, I.A., Scheduling a two-stage multiproduct process with limited product shelf life in intermediate storage, *Ind. Eng. Chem. Res.*, 42(3), 490–508, 2003.

Gupta, S. and Karimi, I.A., An improved MILP formulation for scheduling multiproduct, multistage batch plants, *Ind. Eng. Chem. Res.*, 42(11), 2365–2380, 2003.

Ha, J.-K., Chang, H.-K., Lee, E.S., Lee, I.-B., Lee, B.S., and Yi, G., Intermediate storage tank operation strategies in the production scheduling of multi-product batch processes, *Comput. Chem. Eng.*, 24(2–7), 1633–1640, 2000.

Halim, A.H. and Ohta, H., Batch-scheduling problems through the flowshop with both receiving and delivery in time, *Int. J. Prod. Res.*, 31(8), 1943–1955, 1993.

Harjunkoski, I. and Grossmann, I.E., A decomposition approach for the scheduling of a steel plant production, *Comput. Chem. Eng.*, 25(11–12), 1647–1660, 2001.

Harjunkoski, I. and Grossmann, I.E., Decomposition techniques for multistage scheduling problems using mixed-integer and constraint programming methods, *Comput. Chem. Eng.*, 26(11), 1533–1552, 2001.

Henning, G.P. and Cerdá, J., Knowledge-based predictive and reactive scheduling in industrial environments, *Comput. Chem. Eng.*, 24(9–10), 2315–2338, 2000.

Hong, S. and Maleyeff, J., Production planning and master scheduling with spreadsheets, *Prod. Invent. Manage.*, 28(1), 46–54, 1987.

Huang, W. and Chung, P.W.H., Scheduling of pipeless batch plants using constraint satisfaction techniques, *Comput. Chem. Eng.*, 24(2–7), 377–383, 2000.

Hui, C.-W. and Gupta, A., A novel MILP formulation for short-term scheduling of multistage multi-product batch plants, *Comput. Chem. Eng.*, 24(2–7), 1611–1617, 2000.

Hui, C.-W. and Gupta, A., A bi-index continuous-time mixed-integer linear programming model for single-stage batch scheduling with parallel units, *Ind. Eng. Chem. Res.*, 40(25), 5960–5967, 2001.

Hui, C.-W., Gupta, A., and. van der Meulen, H.A.J., A novel MILP formulation for short-term scheduling of multi-stage multi-product batch plants with sequence-dependent constraints, *Comput. Chem. Eng.*, 24(12), 2705–2717, 2000.

Ierapetritou, M.G. and Floudas, C.A., Short-term scheduling: new mathematical models vs. algorithmic improvements, *Comput. Chem. Eng.*, 22(Suppl. 1), S419–S426, 1998.

Ierapetritou, M.G. and Floudas, C.A., Effective continuous-time formulation for short-term scheduling. 1. Multipurpose batch processes, *Ind. Eng. Chem. Res.*, 37, 4341–4359, 1998.

Ierapetritou, M.G. and Floudas, C.A., Effective continuous-time formulation for short-term scheduling. 2. Continuous and semicontinuous processes, *Ind. Eng. Chem. Res.*, 37, 4360–4374, 1998.

Ierapetritou, M.G. and Floudas, C.A., Effective continuous-time formulation for short-term scheduling. 3. Multiple intermediate dates, *Ind. Eng. Chem. Res.*, 38, 3446–3461, 1999.

Janak, S.L., Lin, X., and. Floudas, C.A., Enhanced continuous-time unit-specific event-based formulation for short-term scheduling of multipurpose batch processes: resource constraints and mixed storage policies, *Ind. Eng. Chem. Res.*, 43(10), 2516–2533, 2004.

Kadipasaoglu, S.N. and Sridharan, V., Alternative approaches for reducing schedule instability in multistage manufacturing under demand uncertainty, *J. Operations Manage.*, 13(3), 193–211, 1995.

Kamburowski, J., Stochastically minimizing the makespan in two-machine flow shops without blocking, *Eur. J. Operational Res.*, 112(2), 304–309, 1999.

Kim, S.B., Lee H.-K., Lee, I.-B., Lee, E.S., and Lee, B., Scheduling of non-sequential multipurpose batch processes under finite intermediate storage policy, *Comput. Chem. Eng.*, 24(2–7), 1603–1610, 2000.

Lamba, N. and Karimi, I.A., Scheduling parallel production lines with resource constraints. 1. Model formulation, *Ind. Eng. Chem. Res.*, 41(4), 779–789, 2002.

Lee, D-Y., Moon, S., Lopes, P-M., and Park, S., Environmentally friendly scheduling of primary steelmaking processes, in *Proc. of Foundations of Computer-Aided Process Operations, FOCAPO 2003*, Coral Springs, FL, 2003, 355–358.

Lee, K.-H., Park, H.I., and Lee, I.-B., A novel nonuniform discrete time formulation for short-term scheduling of batch and continuous processes, *Ind. Eng. Chem. Res.*, 40(22), 4902–4911, 2001.

Lin, X. and. Floudas, C.A., Design, synthesis and scheduling of multipurpose batch plants via an effective continuous-time formulation, *Comput. Chem. Eng.*, 25(4–6), 665–674, 2001.

Lin, X., Floudas, C.A., Modi, S., and Juhasz, N.M., Continuous-time optimization approach for medium-range production scheduling of a multiproduct batch plant, *Ind. Eng. Chem. Res.*, 41(16), 3884–3906, 2002.

Löhl, T., Schulz, C., and Engell, S., Sequencing of batch operations for a highly coupled production process: genetic algorithms versus mathematical programming, *Comput. Chem. Eng.*, 22(Suppl. 1), S579–S585, 1998.

Majozi, T. and Zhu, X.X., A novel continuous-time MILP formulation for multipurpose batch plants. 1. Short-term scheduling, *Ind. Eng. Chem. Res.*, 40(25), 5935–5949. 2001.

McDonald, C.M. and Karimi, I.A., Planning and scheduling of parallel semicontinuous processes 1. Production planning, *Ind. Eng. Chem. Res.*, 36, 2691–2700, 1997.

Méndez, C.A. and Cerdá, J., An efficient MILP continuous-time formulation for short-term scheduling of multiproduct continuous facilities, *Comput. Chem. Eng.*, 26(4–5), 687–695, 2002.

Méndez, C.A. and Cerdá, J., An MILP framework for batch reactive scheduling with limited discrete resources, *Comput. Chem. Eng.*, 28(6), 1059–1068, 2004.

Méndez, C.A., Henning, G.P., and Cerdá, J., An MILP continuous-time approach to short-term scheduling of resource-constrained multistage flowshop batch facilities, *Comput. Chem. Eng.*, 25(4–6), 701–711, 2001.

Méndez, C.A., Henning, G.P., and Cerdá, J., Optimal scheduling of batch plants satisfying multiple product orders with different due dates, *Comput. Chem. Eng.*, 24(9–10), 2223–2245, 2000.

Méndez, C.A. and Cerdá, J., Optimal scheduling of a resource-constrained multiproduct batch plant supplying intermediates to nearby end-product facilities, *Comput. Chem. Eng.*, 24(2–7), 369–376, 2000.

Middleton, D., Stochastic flowshop makespan simplifications, *Int. J. Prod. Res.*, 37(1), 237–240, 1999.

Mockus, L., Vinson, J.M., and Luo, K., The integration of production plan and operating schedule in a pharmaceutical pilot plant, *Comput. Chem. Eng.*, 26, 697–702, 2002.

Murakami, Y., Uchiyama, H., Hasebe, S., and Hashimoto, I., Application of repetitive SA method to scheduling problems of chemical processes, *Comput. Chem. Eng.*, 21, S1087–S1092, 1997.

Nott, H.P. and Lee, P.L., An optimal control approach for scheduling mixed batch/continuous process plants with variable cycle time, *Comput. Chem. Eng.*, 23(7), 907–917, 1999.

Nott, H.P. and Lee, P.L., Sets formulation to schedule mixed batch/continuous process plants with variable cycle time, *Comput. Chem. Eng.*, 23(7), 875–888, 1999.

Oh, H.-C. and Karimi, I.A., Planning production on a single processor with sequence-dependent setups. Part 1. Determination of campaigns, *Comput. Chem. Eng.*, 25(7–8), 1021–1030, 2001.

Oh, H.-C. and Karimi, I.A., Planning production on a single processor with sequence-dependent setups. Part 2. Campaign sequencing and scheduling, *Comput. Chem. Eng.*, 25(7–8), 1031–1043, 2001.

Pinto, J.M., Joly, M., and Moro, L.F.L., Planning and scheduling models for refinery operations, *Comput. Chem. Eng.*, 24(9–10), 2259–2276, 2000.

Richards, K.M., Zentner, M.G., and Pekny, J.F., Optimization-Based Batch Production Scheduling, unpublished paper presented at AIChE Spring Meeting, Houston, TX, 1997.

Rodrigues, L.A., Graells, M., Cantón, J., Gimeno, L., Rodrigues, M.T.M., Espuña, A., and Puigjaner, L., Utilization of processing time windows to enhance planning and scheduling in short term multipurpose batch plants, *Comput. Chem. Eng.*, 24(2–7), 353–359, 2000.

Rodrigues, M.T.M., Latre, L.G., and Rodrigues, L.C.A., Production planning using time windows for short-term multipurpose batch plants scheduling problems, *Ind. Eng. Chem. Res.*, 39(10), 3823–3834, 2000.

Rodrigues, M.T.M., Latre, L.G., and Rodrigues, L.C.A., Short-term planning and scheduling in multipurpose batch chemical plants: a multi-level approach, *Comput. Chem. Eng.*, 24(9–10), 2247–2258, 2000.

Roeterink, H.J.H., Verwater-Lukszo, Z., and Weijnen, M.P.C., A method for improvement potential assessment in batch planning and scheduling situation, in *Proc. of Foundations of Computer-Aided Process Operations, FOCAPO 2003*, Coral Springs, FL, 2003, 301–304.

Roslöf, J., Harjunkoski, I., Björkqvist, J., Karlsson S., and Westerlund, T., An MILP-based reordering algorithm for complex industrial scheduling and rescheduling, *Comput. Chem. Eng.*, 25(4–6), 821–828, 2001.

Ryu, J.-H., Lee, H.-K., and Lee, I.-B., Optimal scheduling for a multiproduct batch process with minimization of penalty on due date period, *Ind. Eng. Chem. Res.*, 40(1), 228–233, 2001.

Safizadeh, M.H. and Ritzman, L.P., Linking performance drivers in production planning and inventory control to process choice, *J. Operations Manage.*, 15(4), 389–403, 1997.

Sand, G., Engell, S., Märkert, A., Schultz R., and Schulz, C., Approximation of an ideal online scheduler for a multiproduct batch plant, *Comput. Chem. Eng.*, 24(2–7), 361–367, 2000.

Sanmarti, E., Espuña, A., and Puigjaner, L., Batch production and preventive maintenance scheduling under equipment failure uncertainty, *Comput. Chem. Eng.*, 21, 1157–1168, 1997.

Sanmarti, E., Puigjaner, L., Holczinger, T., and Friedler, F., Combinatorial framework for effective scheduling of multipurpose batch plants, *AIChE J.*, 48(11), 2557–2570, 2002.

Schilling, G. and Pantelides, C.C., Optimal periodic scheduling of multipurpose plants, *Comput. Chem. Eng.*, 23(4–5), 635–655, 1999.

Sen, T. and Dileepan, P., A bicriterion scheduling problem involving total flowtime and total tardiness, *Inform. Optimization Sci.*, 20(2), 155–170, 1999.

Subrahmanyam, S., Pekny, J.F., and Reklaitis, G.V., Design of batch chemical plants under market uncertainty, *Ind. Eng. Chem. Res.*, 33, 2688–2701, 1994.

Taillard, E., Benchmarks for basic scheduling problems, *Eur. J. Operational Res.*, 64(2), 278–285, 1993.

Wang, K., Löhl, T., Stobbe, M., and Engell, S., A genetic algorithm for online scheduling of a multiproduct polymer batch plant, *Comput. Chem. Eng.*, 24(2–7), 393–400, 2000.

Yee, K.L. and Shah, N., Improving the efficiency of discrete time scheduling formulation, *Comput. Chem. Eng.*, 22(Suppl. 1), S403–S410, 1998.

Zhu, X.X. and Majozi, T., Novel continuous time MILP formulation for multipurpose batch plants. 2. Integrated planning and scheduling, *Ind. Eng. Chem. Res.*, 40(23), 5621–5634, 2001.

REFERENCES

1. Edgar, T.F., Process information: achieving a unified view, *Chem. Eng. Prog.*, 96, 51–57. 2000.
2. Shobrys, D.E. and White, D.C., Planning, scheduling, and control systems: why they cannot work together, *Comput. Chem. Eng.*, 26, 149–160, 2002.
3. Lin, X. and Floudas, C.A., Design, synthesis and scheduling of multipurpose batch plants via an effective continuous-time formulation, *Comput. Chem. Eng.*, 25(4–6), 665–674, 2001.

4. Schnelle, K.D., Preliminary design and scheduling of a batch agrochemical plant, *Comput. Chem. Eng.*, 24(2–7), 1535–1541, 2000.

5. Heo, S.K., Lee, K.H., Lee, H.K., Lee, I.B., and Park, J.H., A new algorithm for cyclic scheduling and design of multipurpose batch plants, *Ind. Eng. Chem. Res.*, 42(4), 836–846, 2003.

6. Houston, R.B. and Basu, P.K., Properly plan production in a pharmaceutical pilot plant, *Chem. Eng. Prog.*, 96, 37–45, 2000.

7. Pekny, J.F. and Reklaitis, G.V., Towards the convergence of theory and practice: a technology guide for scheduling/planning methodology, in *Proc. of the Third Int. Conf. on Foundations of Computer-Aided Process Operations*, AIChE Symposium Series No. 320, American Institute of Chemical Engineers, New York, 1998, 91–111.

8. Kondili, E., Pantelides, C.C., and Sargent, R.W.H., A general algorithm for short-term scheduling of batch operations. Part I. Mathematical formulation, *Comput. Chem. Eng.*, 17, 211–227, 1993.

9. Shah N., Pantelides, C.C., and Sargent, R.W.H., A general algorithm for short-term scheduling of batch operations. Part II. Computational issues. *Comput. Chem. Eng.*, 17, 229–244, 1993.

10. Pantelides, C.C., Unified frameworks for optimal process planning and scheduling, in *Proc. of the Second Int. Conf. on the Foundations of Computer-Aided Process Operations*, CACHE Publications, Crested Butte, CO, 1994, 253–274.

11. Portougal, V. and Robb D.J., Production scheduling theory: just where is it applicable? *Interfaces* 30(6), 64–76, 2000.

12. Sandras, W.A., *Just in Time: Making It Happen*, John Wiley & Sons, New York, 1989, 91.

13. Biegler, L.T. and Grossmann, I.E., Retrospective on optimization, *Comput. Chem. Eng.*, 28(8), 1169–1192, 2004.

14. Shah, N., Single and multisite planning and scheduling: current status and future challenges, in *Proc. of the Third Int. Conf. on Foundations of Computer-Aided Process Operations*, AIChE Symposium Series No. 320, American Institute of Chemical Engineers, New York, 1998, 75–90.

15. Floudas, C.A. and Lin, X., Continuous-time versus discrete-time approaches for scheduling of chemical processes: a review, *Comput. Chem. Eng.*, 28(11), 2109–2129, 2004.

16. Siletti, C.A. and Petrides, D., Overcoming the barriers to batch process scheduling, in *Proc. of Foundations of Computer-Aided Process Operations*, FOCAPO 2003, Coral Springs, FL, 2003, 375–378.

17. Van Hentenryck, P., *Constraint Satisfaction in Logic Programming*, Logic Programming Series, The MIT Press, Cambridge, MA, 1989.

18. Baptiste, P., Le Pape, C., and Nuijten, W., *Constraint-Based Scheduling: Applying Constraint Programming to Scheduling Problems*, Kluwer Academic, Boston, MA, 2001.

19. Maravelias, C.T. and Grossmann, I.E., A hybrid MILP/CP decomposition approach for the continuous time scheduling of multipurpose batch plants, *Comput. Chem. Eng.*, 28(10), 1921–1949, 2004.

20. Reklaitis, G.V., Pekny, J.F., and Joglekar, G.S., Scheduling and simulation of batch processes, in *Handbook of Batch Process Design*, Blackie Academic, London, 1997, 25–60.

21. Bassett, M., Minimizing production cycle time at a formulations and packaging plant, *AIChE Symp. Ser.*, 94(320), 267–272, 1998.

22. Lin, B. and Miller, D.C., Tabu search algorithm for chemical process optimization, *Comput. Chem. Eng.*, 28(11), 2287–2306, 2004.

23. Glover, F., Tabu search: a tutorial, *Interfaces*, 20(4), 74–94, 1990.

24. Carpaneto, G., Dell'Amico, M., and Toth, P., A branch-and-bound algorithm for large-scale asymmetric traveling salesman problems, in *Technical Report Dipartimento di Economica Politica Facolta' di Economica E Commercio*, Universita' di Modena, Italy, 1990.

25. Kondili E., Pantelides, C.C., and Sargent, R.W.H., A general algorithm for scheduling batch operations, in *Third Int. Symp. on Process System Engineering*, Sydney, Australia, 1988, 62–75.

26. Orçun, S., Altinel I.K., and Hortaçsu, Ö., General continuous time models for production planning and scheduling of batch processing plants: mixed integer linear program formulations and computational issues, *Comput. Chem. Eng.*, 25(2–3), 371–389, 2001.

27. Smunt, T.L. and Watts, C.A., Improving operations planning with learning curves: overcoming the pitfalls of 'messy' shopfloor data, *J. Operations Manage.*, 21(1), 93–107, 2003.

28. Pisano, G. P., *The Development Factory*, Harvard Business School Press, Boston, MA, 1997.

29. Sahinidis, N.V., Optimization under uncertainty: state-of-the-art and opportunities, *Comput. Chem. Eng.*, 28(6–7), 971–983, 2004.

30. Lee, Y.G. and Malone, M.F., A general treatment of uncertainties in batch process planning, *Ind. Eng. Chem. Res.*, 40(6), 1507–1515, 2001.

31. Vin, J.P. and Ierapetritou, M.G., Robust short-term scheduling of multiproduct batch plants under demand uncertainty, *Ind. Eng. Chem. Res.*, 40(21), 4543–4554, 2001.

32. Honkomp, S.J., Mockus, L., and Reklaitis, G.V., A framework for schedule evaluation with processing uncertainty, *Comput. Chem. Eng.*, 23, 595–609, 1999.

33. Balasubramanian, J. and Grossmann, I.E., A novel branch and bound algorithm for scheduling flowshop plants with uncertain processing times, *Comput. Chem. Eng.*, 26(1), 41–57, 2002.

34. Balasubramanian, J. and Grossmann, I.E., Approximation to multistage stochastic optimization in multiperiod batch plant scheduling under demand uncertainty, *Ind. Eng. Chem. Res.*, 43(14), 3695–3713, 2004.

35. Bonfill, A., Bagajewicz, M., Espuña, A., and Puigjaner, L., Risk management in the scheduling of batch plants under uncertain market demands, *Ind. Eng. Chem. Res.*, 43(3), 741–750, 2004.

36. Honkomp, S.J., Lombardo, S., Rosen, O., and Pekny, J.F., The curse of reality: why process scheduling optimization problems are difficult in practice, *Comput. Chem. Eng.*, 24(2–7), 323–328, 2000.

37. Pekny, J.F., Algorithm architectures to support large-scale process systems engineering applications involving combinatorics, uncertainty, and risk management, *Comput. Chem. Eng.*, 26, 239–267, 2002.

38. Grossman, I.E. and Westerberg, A.W., Research challenges in process systems engineering, *AIChE J.*, 46(9), 1700–1703, 2000.

11 Monitoring and Control of Batch Processes

Sten Bay Jørgensen, Dennis Bonné, and Lars Gregersen

CONTENTS

11.1 INTRODUCTION

Providing a reasonably accessible exposition of batch process monitoring and control is the purpose of this chapter. It is aimed at readers with a background in batch processing who wish to broaden their perspective in the monitoring and control technology area. The review literature on batch process control is limited and tends to view developments from a control technology perspective; however, several interesting methods and approaches have recently appeared and, because these tend to converge toward a common technology, there seems to be room for a presentation with the above-mentioned aim but which is based on a process-property perspective rather than a control perspective. The purpose of the process-property perspective is to provide a common framework for presenting and discussing the different methods in a clear manner.

Batch processing management involves decision making at different layers of the plant operation hierarchy. Such decisions include planning, scheduling, monitoring, and control. Therefore, it is important to address the relevant layers for solving a batch control problem. Having data available from the batch process enables monitoring, minimizes disturbances, and reduces the influence of uncertainty. In this chapter, we assume that such data are available at suitable sampling rates, which can be high for simple measurements or infrequent for more complex quality-related measurements.

Batch processes are used for three main reasons:

- Dividing a process into batches allows confinement of any undesired properties or byproducts. For pharmaceutical and food industries, this property is essential and regulated by the authorities.
- Separating a processing sequence into batches makes it possible to produce a large number of different products and grades on the same equipment, perhaps using different raw materials or recipes.
- Some operations are better suited for batch or fed-batch processing than for continuous processing (e.g., to circumvent substrate inhibition for microbial cultivations).

In all instances it is desirable to achieve reproducible operation and tight product specifications; however, batch processing is subject to variations in raw material properties, in start-up initialization, and in other disturbances during batch execution. These different disturbances introduce variations in the final product quality. Compensating for these disturbances has been difficult due to the nonlinear and time-varying behavior of batch processing and due to the fact that reliable on- or inline sensors for monitoring final product quality are rarely available. However, recent developments in modeling and control technology have triggered developments that have enabled mature first-of-batch monitoring and also monitoring of batch control, thus it is appropriate to review these developments here.

To set the stage for process-property-based monitoring and control, we must define the context of batch processing. The following text provides definitions of elements of batch processing. These definitions then lead into how these elements are subsequently treated as their state of art is presented in this chapter. Each batch process may be defined as a series of operational tasks (i.e., mixing, reaction and separation). Within each task, a set of subtasks (e.g., heating and cooling, charging and discharging) is handled. In some cases, more than one feasible set of operational tasks can produce the specified products; consequently, an optimal sequence of tasks and subtasks with a defined objective needs to be determined. This set of operational tasks is labeled the *optimal batch operations model*. Thus, the batch operations model combines the batch processing tasks normally specified in a recipe with the batch equipment under availability and other resource constraints. In this chapter, a single production line is considered, as multiple lines typically are handled at the plant scheduling level. When dealing with multiple production lines and sharing of common resources, scheduling becomes a major logistical issue. In industrial practice, several attempts have been made during the past 15 years to standardize batch control systems to deal with these issues at various levels of the plant and its equipment. Clearly, the most successful has been the S88 (or SP88) initiative developed by the Instrument Society of America (ISA),[27] which now is in the IEC batch control standards[25,26] (part 1, model terminology and functionality; part 2, data structures and language guidelines). These attempts seem to be rather successful as most commercial batch control systems comply with the S88 standard. In terms of modeling batch operations, a programming language for sequential control applications has been developed by the IEC which is known as *Grafcet*.[23,24] This language has also been adopted by several batch control system vendors, mainly at the programmable logic controller (PLC) level, while *Grafchart* has been defined by Årzén[2] for modeling at higher levels of the control hierarchy.

This chapter is mainly concerned with batch operation of single production lines, which represents a relevant and essential research problem as chemical batch operation in practice is exposed to disturbances or uncertainties that make it difficult to ensure reproducible and, to a lesser extent, optimal operation. If disturbances did not occur during execution, batch operation would be most reliable; however, direct implementation of the batch operations model on the process constitutes what is often called *open-loop or feedforward process control*, as it provides a feedforward change in the manipulated variable to obtain a desired process trajectory. Calculation of the desired change in the manipulated variable is based on model knowledge or process experimentation; consequently, feedforward action is very prone to uncertainty. Such uncertainty may arise from model inaccuracy or from uncertainty about process disturbances. Thus, in practice it is not possible to implement a batch operations model completely as specified due to variability in, for example, raw material composition, other operating conditions, or equipment availability constraints.

In order to protect batch manufacturing from the influence of disturbances or uncertainty (e.g., in timing, raw material composition, or availability of other resources), it is highly desirable to use simple process monitoring and fault detection, which may become the basis for subsequent implementation of a slow feedback in the form of manual process control. *Process monitoring* provides information derived from measurements on the progression of the processing compared to the nominally designed behavior. *Fault detection* takes place if the processing deviates significantly from a desired standard trajectory. Tools for process monitoring and fault detection enable the operator or plant engineer to request information regarding which measurements will reveal the most useful information regarding these deviations in order to isolate the origin of the fault. After obtaining such information, it is normally left to the plant personnel to analyze the underlying cause of the observed deviations. This step is known as *fault isolation*. Thus, process monitoring is in fact a preamble to implementing manual feedback, obviously in a very slow loop as the operating personnel are involved in the decision making and in implementing measures to counteract the fault. However, monitoring and fault detection constitute a very effective technology to provide more consistent product quality due to taking corrective actions between batches.

A much faster feedback effect may be obtained by using process control by direct feedback loops in one of its many forms to reduce the influence of uncertainty on batch operation. Batch process control can take many forms, as described later in this chapter. The control action can be based upon various types of information, most often related through a process model. For multivariable processes, a model basis is essential for providing reliable feedback; however, to provide insight on the ways in which feedback may be introduced, it is useful to consider the key characteristics of batch vs. continuous processing. These characteristics lead to the definition of the batch control problem. The following definitions also comply with the above-mentioned standards.

11.1.1 THE BATCH CONTROL PROBLEM

Batch processing as outlined above consists of combining a recipe with available equipment to realize a desired batch trajectory as closely as possible. When a batch is completed, the equipment is made ready to produce a new recipe. If the same operations model is used, the batch process may be repeated several times. Thus, batch processing is very similar to periodic processing. For consecutive batches, which aim at producing the same product the key characteristic is thus a two time dimensional behavior of batch (and periodic) operation. These time dimensions (i.e., time within a batch and batch number or index) are depicted in Figure 11.1. These characteristic time dimensions lead to the definition of two types of time variation: *intrabatch* variation, where the states of the process follow a set of trajectories supposedly coinciding with the desired trajectories, and *interbatch* time variation, where attempts are made to reproduce the initial conditions in subsequent batches. Consequently, the disturbances may be similarly classified. *Intrabatch disturbances* occur during batch execution (e.g., as a deviation of the initiator or substrate concentration from the nominal value); such

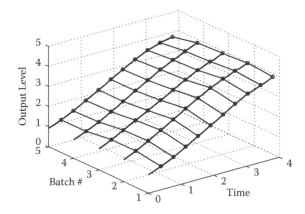

FIGURE 11.1 The three dimensions of batch operation data: output level, time within a batch, and batch number or index. The filled circles represent sample times.

intrabatch disturbances are batchwise uncorrelated. *Interbatch disturbances* occur between batches (e.g., as variations in initial concentration values); hence, interbatch disturbances are batchwise correlated.

For a periodic process, the endpoint of one cycle simply is the starting point of the next cycle; thus, the interbatch variability becomes *intercycle variability* for periodic processes. In the following text, the operations form is referred to as batch as a generalization. The above relatively simple concepts turn out to be very useful both for modeling and for control of batch processes. On the basis of the intra- and intercycle variations, the batch control problem can be defined as follows: The *batch control problem* is implementation of the optimal batch operations model to enable rejection of inter- and intrabatch disturbances. Thus, the key characteristics of the batch (or periodic) control problem are the time-varying characteristics. These time-varying characteristics apparently precluded theoretical progress within batch and periodic process control for many years, until computational power allowed the development of feasible solutions. Such solutions offer significant benefits and can be effectively used to reduce the effects of uncertainties embedded in batch operation. The above definition of the batch control problem is based on an optimal batch operations model that depends on how the optimization objective function is defined. Most often, the objective will be to ensure optimal productivity given a set of equipment and operational constraints, but many other considerations can come into play such as in the pharmaceutical industry, where the overriding concern is reproducible operation. Furthermore, potential deviations from normal operation traditionally render operation scheduling and plant and operation model design conservative; consequently, to reduce this conservatism of traditional plant design and operations model design, it is necessary to apply feedback control. The main benefit of feedback control is that it improves the ability to reject disturbances from nominal operation; hence, implementation of feedback control will increase the reliability or reproducibility and furthermore also enable optimization of the batch process operation.

The purpose of this chapter is to present the state of art of batch process monitoring and control, where the key trend is to implement the batch operations model by extracting knowledge using some form of model and subsequently using this knowledge for control design. Issues related to implementing the (optimal) batch operations model are the thread running through this chapter. The presentation begins with a brief introduction to the development of operations and process models. In the latter case, the focus is on empirical process model development. The key motivation for using models is that a representation of available knowledge about a batch process is essential to implement control of the production, and control is essential for rejection of disturbances. This control may also develop into optimization control; however, optimization may also be achieved based partially on experiments (i.e., where the explicit model capturing step is bypassed). The following section provides a discussion of batch monitoring before reviewing methods for the control of batch processes. In particular, we present methods based on the application of time-series models, although aspects of control using monitoring and solution models also are presented, and we touch briefly on aspects of control based directly on experiments. Finally, time-series models for an industrial bioreactor cultivation are presented, and a simulated control example based on time-series modeling is provided. This latter example demonstrates some of the promising capabilities of the model-based batch control methodologies.

11.2 BATCH OPERATIONS AND PROCESS MODELING

The modeling task may be decomposed into two distinct types: modeling process behavior vs. modeling external actions imposed on the process and its environment by disturbances, operating procedures, or other control actions.[50] The process behavior model usually is represented as a process model, while the model for the external actions imposed on the physical system is a batch operations model. In this section, both model types are briefly treated.

11.2.1 DEVELOPMENT OF AN OPERATIONS MODEL

An operations model may be developed from process knowledge (e.g., using a process model to generate a feasible operations model). This model development may be achieved by conceptual design and synthesis via heuristic usage of computer-aided process engineering (CAPE) tools to exploit basic process limitations. This procedure may provide both the task sequence and a feasible operations model, with definitions of the important process variables for each sub-task.[18] The near-optimal feasible operations model may further be optimized offline using optimization to produce an optimal batch operations model. When a model is not available, at least two different approaches are possible. An operations model may be generated directly from multifactorial experiments; in fact, this is the procedure often employed in industrial practice during scale-up from laboratory to pilot and production scale. However such a procedure requires extensive and costly experiments and suffers from potential combinatorial explosion as the number of degrees of operational freedom increases. An alternative procedure is to consider the operations model as the solution model for the optimal control problem, where the solution model may be inferred from structural knowledge of the optimization problem at hand. This solution model is uncertain, but the uncertainty of the solution may (at least partially) be compensated for by using decentralized control (i.e., several single-loop controllers). (Srinivasan and Bonvin, 2004)

11.2.2 PROCESS MODELING

Ideally, mathematical models should be based on all available information regarding the underlying phenomena. However, first-principles engineering models are relatively time consuming to develop for the wide operation range covered by many batch processes, and often they have only limited validity over this wide operation range. When some underlying phenomenon is not well understood, alternative approaches may be necessary; for example, if reaction kinetics are not completely understood, a model can be developed using an approach that permits combining available first-principles engineering knowledge with process data to elucidate the functional form of specific phenomena models. Recently, such an inverse modeling method based on gray-box stochastic modeling has been reported for batch processes.[33] If the level of *a priori* knowledge is even lower, one may have to resort to purely data-driven approaches, either for monitoring (i.e., correlation models) or for time-series modeling; both approaches seem to be powerful for batch processes.[4,17] The development of models from data is described in the following section, as such models may have wide application and appeal to many industries where the process dynamics are important.

11.2.3 DATA-DRIVEN PROCESS MODELING

For processes where limited fundamental knowledge has precluded proper first-principles modeling, as is the case for many bioreactors, an effort has been made to develop models directly from operating data. Such models were initially developed for monitoring purposes but have recently been demonstrated to be most useful for control also. We first discuss how monitoring models are developed and subsequently how time-series models can be developed from data. The time-series models are particularly interesting because these models can use control theory developed for time-invariant and time-varying processes. Furthermore, the methodology presented here uses the additional batch dimension to provide a fast data-driven alternative to first-principles modeling of batch processes.

11.2.3.1 Modeling for Monitoring

Data from batch and fed-batch processes can conveniently be stored in a three-way matrix $\mathbf{X}$ $(I \times J \times K)$, where I is the number of batches, J is the number of variables, and K is the number of samples from each batch. The size can vary by orders of magnitude depending on the process duration, available on- and offline measurements, and input variables. Note that these variables all form part of the J variables. The matrix $\mathbf{X}$ can be unfolded to a two-way matrix (see Figure 11.2). This two-way matrix is called $(\mathbf{X}$ $(I \times KJ))$. For each fermentation (a row in $\mathbf{X}$), a set of quality measures is recorded and stored in matrix $\mathbf{Y}$. One such measured variable could be the final product concentration, but other measures could also be used (e.g., productivity). Each column of $\mathbf{X}$ corresponds to a certain variable at a certain point in time. Note that the time dependency can be removed by substituting time with any indicator variable that expresses or quantifies advancement of the process (e.g., accumulated base addition to a yeast cultivation).[29,37,64] If the process is carried out following a predetermined batch operations model, it is expected that the trajectories of the measurements are similar and that the mean value of a variable at a certain point in time can be used as a reference value for future executions. The goal of monitoring is to observe and eliminate the cause of deviations from this reference value in future batches. Thus, to facilitate the analysis the columns are centered and scaled to unit variance.

The matrix $\mathbf{X}$ is rather large, but the columns of $\mathbf{X}$ are not independent. They describe similar events in the process and the dimension of the space spanned by $\mathbf{X}$ is usually very low. Thus, by using a multivariate statistical technique to reduce the dimensionality of the variable space, the problem of describing the process becomes more manageable. *Principal component analysis* (PCA) is frequently used for this purpose and is recommended if no quality variables are available. When quality variables are available, one can use *principal component regression* (PCR) or preferably *projection to latent structures* (PLS), which is a linear regression method that optimally utilizes the information in $\mathbf{X}$ and $\mathbf{Y}$ at the same time.[28] In general, when these methods are applied on several measurements, a multivariate approach is used to map the behaviors into the relevant subspace; if the multivariate aspects are stressed, this may be indicated by the use of a capital M (e.g., MPCA).

Assuming that one or more quality variables are available for the described process, a PLS model can be developed. PLS is defined by a bilinear model that is used to project the relationship onto lower dimensional subspaces:

$$\begin{aligned} \mathbf{X} &= \mathbf{TW}^{\top} + \mathbf{E} \\ \mathbf{Y} &= \mathbf{UQ}^{\top} + \mathbf{F} \end{aligned} \qquad \mathbf{u}_a = b_a \mathbf{t}_a \qquad (11.1)$$

by maximizing the covariance between $\mathbf{u}_a$ and $\mathbf{t}_a$ where the number of components A (number of columns in $\mathbf{X}$) is chosen such that $\mathbf{E}$ and $\mathbf{F}$ are small in some sense. The data, in other words, are reduced to a number of scores (either $\mathbf{T}$ or $\mathbf{U}$) that

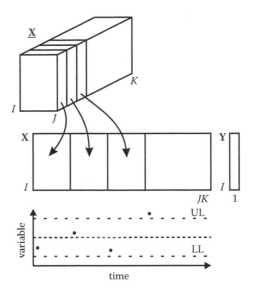

FIGURE 11.2 Unfolding of a three-way matrix to form a two-way matrix. X contains the online variables and Y some quality measure of the process (e.g., final product concentration). The principle behind process chemometrics is shown in the lower part of the figure for a single variable. Every time a new measurement is obtained it will be compared to the expected level. If the deviation is outside (i.e., above the upper limit [UL] or below the lower limit [LL]), the process is behaving abnormally and the process operator should take action.

lie in a low-dimensional space of the data but describe a large fraction of the variation of the data. Equation 11.1 can be rewritten as:

$$\mathbf{Y} = \mathbf{XB} + \mathbf{F}^{*}. \tag{11.2}$$

This expression, in many cases, is easier to work with. The regression parameter matrix is given by:

$$\mathbf{B} = \mathbf{W}(\mathbf{P}^{\top}\mathbf{W})^{-1}\mathbf{Q}^{\top}, \tag{11.3}$$

where the loading matrices $\mathbf{W}$, $\mathbf{P}$, and $\mathbf{Q}$ are determined by the PLS algorithm.[22,40] $\mathbf{Y}$ can be predicted using:

$$\hat{\mathbf{Y}} = \mathbf{XB} = \mathbf{XW}(\mathbf{P}^{\top}\mathbf{W})^{-1}\mathbf{Q}^{\top} \tag{11.4}$$

$$= \mathbf{TQ}^{\top}. \tag{11.5}$$

The model can be used for calculating a vector **t** (a t-score) for a new data set, $\mathbf{X}_{new}$:

$$\mathbf{t}_{new} = (\mathbf{X}_{new}\mathbf{W}(\mathbf{P}^\top\mathbf{W})^{-1})^\top. \qquad (11.6)$$

This expression can be used only when all data from a batch process are available. For online purposes, a full **X** matrix has to be constructed. In this chapter, **X** will be constructed by using all of the available information collected up to the current time, and the remaining part of **X** will be filled in with the most recently obtained measurement. This method results in good fault detection properties and reliable estimation of **Y** as well. This way of filling **X** corresponds to predicting what would happen if a fault is allowed to remain unchanged for the remaining duration of the batch and constitutes a method to evaluate the seriousness of faults. This procedure is often justifiable because the process dynamics become increasingly slow as the batch nears completion.

11.2.3.2 Time-Series Modeling

The periodic nature and finite horizon of batch processes from which observations are collected from a grid of sample points in time make it possible to model the evolution between two consecutive sample points in a batch with local *linear time-invariant* (LTI) models. Such local models are valid between two consecutive points in the sample grid and are known as *grid-point* models. In this fashion, both the time variation within the characteristic regions and the transitions between these may be approximated with a grid of grid-point models. Thus, such a grid-point model set gives a complete description of a batch. The finite horizon of batch processes means that the model set will be finite. The periodic way in which the same recipe is repeated batch after batch means that several measurements from individual sample points are available for identification. That is, the time evolution of a process variable is measured or sampled at specific sample points during the batch operation, and as the batch operation is repeated several measurements are collected from every sample point. With multiple data points or measurements from one specific sample point interval, a multivariable grid-point model can be identified for that sample point interval.

Batch processes, then, are modeled with sets of dynamic grid-point LTI models. Such a set of grid-point LTI models could also be referred to as a *linear time-varying* (LTV) batch model. These grid-point LTI models can be parameterized in a number of ways for example, as autoregressive models with exogenous inputs (ARX), finite impulse response (FIR) models, or state space (SS) models. Here, we have chosen to use an autoregressive moving average model with exogenous inputs (ARMAX) parameterization. This choice of parameterization offers a simple multivariable system description with a moderate number of model parameters.

Let N be the batch length in terms of number of samples, and define the input **u**, output **y**, shifted output $\mathbf{y}^0$, and disturbance **w** profiles as:

$$\mathbf{u} = \begin{bmatrix} u'_0 & u'_1 & \cdots & u'_{N-1} \end{bmatrix}', \quad \mathbf{y} = \begin{bmatrix} y'_1 & y'_2 & \cdots & y'_N \end{bmatrix}'$$

$$\mathbf{y}^0 = \begin{bmatrix} y'_0 & y'_1 & \cdots & y'_{N-1} \end{bmatrix}, \quad \mathbf{w} = \begin{bmatrix} w'_1 & w'_2 & \cdots & w'_N \end{bmatrix} \tag{11.7}$$

Note that the model variables y and u may well be basis functions of the process variables, and not all initial conditions $\mathbf{y}^0$ are measurable or physically meaningful (e.g., off-gas measurements). Thus, the ARX model set may be expressed in matrix form:

$$\bar{\mathbf{y}} - \mathbf{y} = -\mathbf{A}(\bar{\mathbf{y}}^0 - \mathbf{y}^0) + \mathbf{B}(\bar{\mathbf{u}} - \mathbf{u}) + \mathbf{w} \tag{11.8}$$

where the bar denotes the reference profiles and $\mathbf{A}$ and $\mathbf{B}$ are banded lower block triangular matrices. If it is assumed that $n_A(t) = n_A = i$, then $\mathbf{A}$ has the following structure:

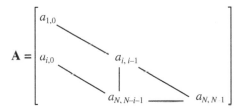

The disturbance profile $\mathbf{w}$ is a sequence of disturbance terms caused by bias in the reference input profile $\bar{\mathbf{u}}$, the effect of process upsets, modeling errors from linear approximations, and errors due to erroneous approximations of transition times between sets of active constraints. This means that disturbance $\mathbf{w}$ contains contributions from batchwise persistent disturbances, such as recipe or input bias, model bias, and erroneous sensor readings, as well as from random disturbances that occur with no batchwise correlation. It thus seems reasonable to model the disturbance profile $\mathbf{w}$ with a random-walk model with respect to batch index k:

$$\mathbf{w}_k = \mathbf{w}_{k-1} + \Delta \mathbf{w}_k \tag{11.9}$$

where the increment disturbance profile $\Delta \mathbf{w}_k$ is modeled with a moving average (MA) model with respect to time:

$$\Delta w_{k,t} = v_{k,t} + c_{t,t-1} v_{k,t-1} + \ldots + c_{t,t-n_C(t)} v_{k,t-n_C(t)} \tag{11.10}$$

with model order $n_C(t) \in [0,\ldots,t-1]$:

$$n_C(t) = \max\{\mathbf{n}_C(i,j,t) \mid i,j = 1,\ldots,n_y(t)\} \tag{11.11}$$

In matrix form, the disturbance model is expressed as:

$$\Delta \mathbf{w}_k = \mathbf{C} \mathbf{v}_k \tag{11.12}$$

where $\mathbf{C}$ is a banded lower block triangular matrix, and the sequence $\mathbf{v}_k = [v_{k,1},$ $v_{k,2}, \ldots, v_{k,N}]'$ represents batchwise nonpersistent disturbances that are assumed to be zero mean, independent, and identically distributed. Considering the difference between two successive batches, we obtain:

$$\Delta \mathbf{y}_k = \mathbf{y}_k - \mathbf{y}_{k-1}$$
$$= -\mathbf{A}(\mathbf{y}_k^0 - \mathbf{y}_{k-1}^0) + \mathbf{B}(\mathbf{u}_k - \mathbf{u}_{k-1}) + \mathbf{w}_k \tag{11.13}$$
$$= -\mathbf{A}\Delta \mathbf{y}_k^0 + \mathbf{B}\Delta \mathbf{u}_k - \mathbf{C}\mathbf{v}_k$$

We have obtained a batch ARMAX (Equation 11.13) that is independent of the reference profiles ($\bar{\mathbf{y}}, \bar{\mathbf{u}}$) and batchwise persistent disturbances. With such a batch ARMAX model, the path is paved for multivariable, model-based monitoring, control, optimization, and, of course, pure simulation.

11.2.4 APPLICATION-SPECIFIC MODELS

Depending on the task to which the batch ARMAX model (Equation 11.13) is to be applied, it is convenient to convert the batch ARMAX model into different representations. If the task at hand is to predict (or simulate) the behavior of a batch before it is started, the following form is convenient:

$$\Delta \mathbf{y}_k = -\mathbf{A}\Delta \mathbf{y}_k^0 + \mathbf{B}\Delta \mathbf{u}_k - \mathbf{C}\mathbf{v}_k$$
$$= \mathbf{H}\Delta y_{k,0} - \mathbf{G}\Delta \mathbf{u}_k + \mathbf{F}\mathbf{v}_k \tag{11.14}$$

Note that the disturbance matrix $\mathbf{F}$ models the propagation of batchwise nonpersistent disturbances — including batchwise nonpersistent model–plant mismatch. The initial condition, $\Delta y_{k,0}$, can be considered as either an input/control variable or a disturbance. The distinction between the two possibilities will of course depend on the information regarding and control of the outputs prior to a batch. The initial output deviation from the reference is also modeled as a random walk with respect to batch index.

Equation 11.14 is also convenient for the task of classification or monitoring (e.g., normal or not) a batch after it has been completed. Furthermore, Equation 11.14 can be used to determine open-loop optimal recipes in the sense of optimizing an objective for the batch. If such an objective is to minimize the deviations $\mathbf{e} = [e_1, e_2, \ldots, e_N]$, from a desired trajectory $\bar{\mathbf{y}}$, then Equation 11.14 can be modified to obtain:

$$\mathbf{e}_k = \bar{\mathbf{y}} - \mathbf{y}_k$$

$$= \mathbf{e}_{k-1} - \mathbf{H}\Delta y_{k,0} + \mathbf{G}\Delta\mathbf{u}_k - \mathbf{F}\mathbf{v}_k \tag{11.15}$$

We must make two important points about the trajectory tracking model (Equation 11.15). First, because the error profile, $\mathbf{e}_k$, in batch k depends on the error profile, $\mathbf{e}_{k-1}$, from batch $k - 1$, the effects of the batchwise persistent disturbances are integrated with respect to the batch index. This means that a properly designed controller can reject the effects of the batchwise persistent disturbances asymptotically with respect to batch index (e.g., removing the effects of recipe and model bias). Second, given the above-mentioned asymptotic behavior and because the control actions generated by such a controller are deviations from the control/input profile realized in the previous batch, the control actions due to batchwise persistent disturbances will converge asymptotically to zero with respect to the batch index. It could be said that the controller learns to reject the batchwise persistent disturbances; that is, the resulting controller is an *iterative learning control* (ILC) scheme. A more precise way of stating this is that both output and input errors are modeled using integrators with respect to the batch index. The trajectory tracking model (Equation 11.15) is similar to that of Lee et al.,[35] but the representations differ significantly because Equation 11.15 includes the effect of the initial conditions ($\mathbf{H}\Delta y_{k,0}$) and disturbance propagation ($\mathbf{F}\mathbf{v}_k$). Another important difference is that Equation 11.15 does not have a double dependence on the batchwise persistent disturbances; that is, the trajectory tracking model representation (Equation 11.15) includes only the batchwise persistent disturbances as represented by $\mathbf{e}_{k-1}$, not as both the part of $\mathbf{e}_{k-1}$ caused by the batchwise persistent disturbances and the batchwise persistent disturbances themselves.

Equation 11.14 and Equation 11.15) of the batch ARMAX model above are applicable to offline or interbatch applications. For online estimation, monitoring, feedback control, and optimization, however, it is convenient to use a state–space realization of the batch ARMAX model. To achieve such a realization, it is necessary to simplify the batch ARMAX model structure with the assumption that the number of outputs is constant: $n_y(t) = n_y$ for $t = 1,\dots,N$. In an observer canonical form, which is structurally a minimal realization, the state–space realization is given as:

$$x_{k,t} = A_t x_{k,t-1} + B_t \Delta u_{k,t-1} + E_t v_{k,t}$$

$$\Delta y_{k,t} = C x_{k,t} \tag{11.16}$$

The SS model matrices A_t, B_t, E_t, and C contain the corresponding block columns in the batch ARMAX model ($\mathbf{A}$, $\mathbf{B}$, $\mathbf{C}$).

11.2.5 IDENTIFICATION

With the batch ARMAX model (Equation 11.13) derived above, parameterization of the batch model is in place, but the model orders and model parameters still have to be determined from process data. One major drawback of the proposed parameterization is the immense dimensionality of the resulting set of models. In practice, this immense dimensionality will render any standard least-squares (LS) identification problem deficient, and the resulting model will generalize poorly. It turns out, however, that the identification problem improves as the grid-point models become progressively analogous with the grid-point models in their neighborhood. In fact, it is this correlation between the forced similarity of neighboring grid-point models and the predictive capability of the model set that forms the basis for the identification framework presented below.

The first step in any modeling work should, of course, be defining the modeling objective — what problems are to be solved, are they to be solved through monitoring or control, how is the most value added with the least effort, etc. Once the modeling objective is well defined, the appropriate inputs and outputs can be chosen as basis functions of either existing process variables or new process variables introduced as new actuators or sensors. Should the modeling objective be monitoring, control, or optimization of product quality variables measured only sparsely offline, these measurements, when resampled as necessary, can be included as online measurements. If all the required inputs and outputs are chosen among the existing process variables, then historical process data can be used for the model identification; otherwise, the necessary new actuators and sensors are installed on the process and data from a few subsequent batch runs are collected.

11.2.6 PARAMETER ESTIMATION

Several suggestions to how sets of LTI or periodic LTV models should be identified from data can be found in the literature.[63] Some coefficient shrinkage or subspace methods are employed to improve the conditioning of the identification problem and hence reduce the variance of the model parameter estimates. Comparisons of some of these methods may be found in Hastie et al.[21] In a modeling framework similar to the one employed here, Dorsey and Lee[13] proposed estimating both a set of FIR models and a LTI SS model using principal component analysis (PCA) and N4SID (van Overshee and de Moor)[66] but such a set of nonparametric FIR models is obviously not parsimonious and would exhibit poor predictive capabilities.[34] Simoglou et al.[57] suggested estimating a set of independent, overlapping local LTI SS models using *canonical variant analysis* (CVA). Here, however, we propose estimating a set of interdependent grid-point LTI ARMAX models using a novel interpretation of Tikhonov regularization.

The batch ARMAX model (Equation 11.13) can be formulated as a pseudo-linear regression model:

$$\Delta \mathbf{y}_k = \Delta \mathbf{x}_k \theta + \mathbf{v}_k \qquad (11.17)$$

where $\Delta\mathbf{x}_k = \Delta\mathbf{x}_k(\Delta\mathbf{y}_k^0, \Delta\mathbf{u}_k, \mathbf{v}_k)$ is a structured regressor matrix with past outputs, inputs, and disturbances, and $\theta = \theta\,(\mathbf{A}, \mathbf{B}, \mathbf{C})$ is a batch ARMAX model. Based on the linear regression model, we find that:

$$\Delta\hat{\mathbf{y}}_k = E\{\Delta\mathbf{y}_k\}$$

$$= \Delta\hat{\mathbf{x}}_k\theta \tag{11.18}$$

with $\Delta\hat{\mathbf{x}}_k = \Delta\mathbf{x}_k(\Delta\hat{\mathbf{y}}_k^0, \Delta\hat{\mathbf{u}}_k, \hat{\mathbf{v}}_k)$ and where $\hat{\mathbf{v}}_k$ is an estimate of the disturbance or a one-step-ahead (OSA) prediction error profile. A multivariate weighted least-squares (wLS) estimate of the model parameters in the linear regression model (Equation 11.17) is found by solving the following minimization problem:

$$\min_{\theta}\left[J_{\text{wLS}}\right]$$

$$s.t. \quad J_{\text{wLS}} = \left\|\Delta\hat{\mathbf{y}}_k - \Delta\hat{\mathbf{x}}_k\theta\right\|_W^2$$

$$= \left(\Delta\hat{\mathbf{y}}_k - \Delta\hat{\mathbf{x}}_k\theta\right)' W\left(\Delta\hat{\mathbf{y}}_k - \Delta\hat{\mathbf{x}}_k\theta\right) \tag{11.19}$$

$$= \sum_{t=1}^{N}\left(\Delta\hat{y}_{k,t} - \Delta\hat{x}_{k,t}\theta_t\right)$$

$$W_t\left(\Delta\hat{y}_{k,t} - \Delta\hat{x}_{k,t}\theta_t\right)$$

where $\|\cdot\|$ denotes the 2-norm, and W is a block diagonal weighting matrix with symmetrical, positive definite, block elements W_t. Note that θ_t are the model parameters of the local grid-point models and that they are mutually independent in the estimation problem (Equation 11.19).

To reduce the variance of model parameter estimates without introducing bias, as much data (of sufficiently high quality) as possible should be used for the model parameter estimation. Typically, the available data set of $N_B = N_B^{\text{est}} + N_B^{\text{val}} + N_B^{\text{test}}$ batches is spilt up into sets of N_B^{est} batches for model parameter estimation, N_B^{val} batches for model validation, and N_B^{test} batches for model testing. The linear system (Equation 11.18) is thus augmented as:

$$\mathbf{Y} = \begin{bmatrix} \Delta\hat{\mathbf{y}}_1 & \Delta\hat{\mathbf{y}}_2 & \cdots & \Delta\hat{\mathbf{y}}_{N_B^{\text{est}}} \end{bmatrix}$$

$$= \begin{bmatrix} \Delta\hat{\mathbf{x}}_1 & \Delta\hat{\mathbf{x}}_2 & \cdots & \Delta\hat{\mathbf{x}}_{N_B^{\text{est}}} \end{bmatrix}\theta \tag{11.20}$$

$$= \mathbf{X}\theta$$

Equation 11.20, however, will most likely still be rank deficient, and solving it in a wLS sense would still produce model parameter estimates with excessive variance. Such excessive model parameter variance would yield models with poor predictive capabilities.[34] Thus, to improve the predictive capabilities of an estimated model, additional measures must be taken to further reduce the variance of the estimated model parameters.

A possible approach to reducing the variance of model parameter estimates is to require that the estimated model possesses some desired model properties. One such model property could be that neighboring grid-point models are analogous in the sense that they exhibit similar behavior. In fact, without this property, the model would be a *set* of independent models and would not constitute a *grid* of interdependent models. Enforcing model properties, however, inevitably introduces bias into the model parameter estimates; thus, there will be a trade-off between the bias and variance of the model parameter estimates, and this trade-off will determine the predictive capabilities of estimated models. One coefficient shrinkage-based parameter estimation method that can incorporate model properties into wLS estimates is Tikhonov regularization (TR). The derived regression method[20] based upon input direction selection called could in principle be an alternative to TR, but in practice the truncated SVD would be computationally infeasible. We thus proposed estimating the model parameters by solving the TR problem:

$$\hat{\theta}_{TR}(\hat{V}, W, \Lambda) = \arg\min_{\theta} \left[J_{wLS}^{TR} \right]$$

$$(11.21)$$

$$s.t. \quad J_{wLS}^{TR} = \left\| Y - X\theta \right\|_W^2 + \left\| L\theta \right\|_{\Lambda^2}^2$$

where $\hat{V} = [\hat{v}_1', \hat{v}_2', \ldots, \hat{v}_{N_B^{est}}']'$, W is a block diagonal matrix with block elements W, the structured penalty matrix (L) maps the parameter vector (θ) into the desired parameter differences, and the diagonal weighting matrix Λ weights the parameter differences, such that the penalty $\Lambda L\theta$ is a column vector of weighted differences between parameters in neighboring grid-point models. The penalty matrix L consists of five submatrices, L_m:

$$L = [L_1' \quad L_2' \quad L_3' \quad L_4' \quad L_5']'$$

$$(11.22)$$

each of which is individual weighted by block diagonal weighting matrices Λ_m:

$$\Lambda = \text{diag}([\Lambda_1 \quad \Lambda_2 \quad \Lambda_3 \quad \Lambda_4 \quad \Lambda_5])$$

$$(11.23)$$

This set is weighted with an individual scalar weight, so we have a total of $5N_c$ weights $\lambda_{m,i,j}^n$.

Each of the submatrices L_m for $m = 1, \ldots, 5$, penalizes violations of specific model (parameter) properties:

- $L_1\theta$ approximates the first-order time derivative of the parameters θ. It thus incorporates the local model interdependency by penalizing the model parameters time evolution.
- $L_2\theta$ approximates the second-order time derivative of the parameters θ. It thus incorporates the local model interdependency by penalizing nonsmoothness of the time evolution of the model parameters.
- $L_3\theta$ approximates the first-order time derivative of the impulse response of the local models θ_t. It thus enforces dampened impulse responses on the local model parameter estimates by penalizing the time evolution of the impulse responses.
- $L_4\theta$ approximates the second-order time derivative of the impulse response of the local models θ_t. It thus enforces smooth impulse responses on the local model parameter estimates by penalizing nonsmoothness of the impulse responses.
- $L_5\theta$ penalizes the variance of the parameter estimates $\hat{\theta}$ and thus enforces minimum variance estimates — that is, $L_5 = I$.

The estimated parameter vector $\hat{\ddot{\theta}}_{TR}(\dot{V}, W, \Lambda)$ is a function of the weighting matrix Λ that determines the coefficient shrinkage and hence the trade-off between bias and variance. This means that the regularization matrix Λ can be used to tune the predictive capabilities of the model estimate. Through the particular choice of penalty matrix L, the regularization matrix Λ also determines the interdependency between the grid-point models in the model grid. Note, if $L'\Lambda^2L$ is designed to be positive definite, then a unique solution to the estimation problem (Equation 11.21) is guaranteed. In general, if the Hessian matrix H:

$$H = X'WX + L'\Lambda^2L \qquad (11.24)$$

is positive definite, then Equation 11.21 has a unique solution.

As their names indicate, the models developed in this section can be used for monitoring and for online control to reject batch-to-batch and intrabatch disturbances. These applications are described in the following sections, first for monitoring to illustrate that process data actually do contain a significant amount of information about the actual operation and subsequently for control to reject both inter and intrabatch disturbances.

11.3 MONITORING

11.3.1 Development of a Monitoring Model

A model for monitoring normal process operation is developed below using an industrial case as an example. Models for monitoring other types of behavior can be developed in an analogous manner. First, the available data are classified into different types of process behavior using MPCA. After selecting the desired class of behavior to be modeled, the batches representing this type of behavior are subdivided into two groups; one group, usually the larger, is used for model development, and the other is used for cross-validation of the model (i.e., investigating the reliability of the model). Industrial cultivation results are presented to illustrate modeling for monitoring. A model has been developed by carefully selecting data sets from the historical database that reflect the normal desired operation of the fermentations. This has been done by first discarding any batch that has substantial undesired or unusual behavior compared to the desired batch behavior (e.g., because of experiments in unfavorable operating regions or infections). Batches that are very short or very long are also discarded. The data suitable for the model development are truncated such that 114 time samples are included in the model. As mentioned, the dynamics of the process becomes slower toward the end of the batch, and few corrective measures can be taken if a fault occurs near the end of a batch. An initial model is estimated, and batches are removed from the modeling data set if they are outside a 99% confidence bound in a score plot using ellipses as confidence bounds. The procedure is repeated iteratively until almost all the remaining batches fall within the 99% confidence bound. It is important here to identify *why* a batch does not lie within the confidence bound in order to make sure that only batches that are really not conforming are eliminated from the normal data set.

After this reduction, 25 data sets are used for model development and 13 for validation of the model. Using the prediction error sum of squares (PRESS) as our validation criterion, two components are found to be sufficient for describing the relationship between X and Y. The obtained model uses only 28% of the information in X but explains 80% of the variation of Y. The low percentage of used variation in X is due to the inclusion of controlled variables that have low variation (e.g., pH and temperature). These variables are *known* to have a large influence on the product formation which is why they are controlled. If the influence of the controlled variables on the product formation is to be modeled by the PLS model, these variables must be perturbed in designed experiments. These experiments have not been performed because doing so would be expected to lead to a decrease in product formation, and gross variation in these variables is not encountered due to the control. If the PLS model was to be used entirely for prediction purposes and if it can be assumed that control is perfect, then the corresponding model performance could be improved by excluding the controlled variables from the model. In our case, fault diagnosis of the controlled variables is of interest, hence these controlled variables remain in the model.

11.3.1.1   Online Estimation of Final Product Concentration

Using Equation 11.4, Equation 11.5, and Equation 11.6, the final product concentration can be predicted in real time. Figure 11.3 shows the performance of this method for a low-producing fermentation. The data from this batch were used for model validation but not for the modeling itself and are used in the following sections. The final product concentration is 0.40, whereas the average

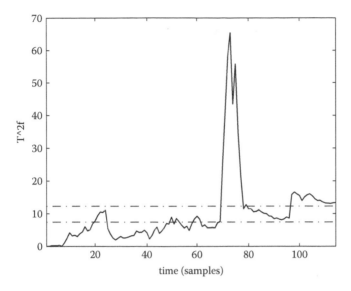

FIGURE 11.3 Prediction of the final product concentration. The dotted line indicates the actual product concentration as it was measured at the end of the batch. If the large fault at $t = 70$ was allowed to persist throughout the fermentation, the product concentration was estimated to be much lower than the one actually obtained.

value for the batches used for model development was 0.46. Figure 11.3 shows that the model is able to predict the final product concentration to within 10% during most of the fermentation except during the interval from 70 to 80, where there is a large deviation due to a process fault. The accuracy of the estimation is similar to the accuracy provided by the lab when chemical analyses are performed. This deviation can be interpreted further by fault diagnosis, as described in the following section.

11.3.1.2 Fault Diagnosis

Fault diagnosis is most useful not only for the detection of faults as a batch progresses but also for revealing whether or not a specific batch belongs to the desired or normal behavior (e.g., whether the particular batch should be included in the set of data used for modeling a particular desired behavior). Fault diagnosis consists of three steps: fault detection, isolation, and identification (FDII). The methods presented in this section will readily detect faults and isolate the measurements that are behaving abnormally, and the methods may also facilitate identification of the fault (i.e., determine the physical origin of the fault in the process).

For fault detection, two statistics, the T_f^2 and the standard prediction error, can be calculated. The T_f^2 statistic (based on the Hotelling T^2 statistic[39]) is calculated using the scores:

$$T_f^2 = \mathbf{t}_{new}^\top \mathbf{S}^{-1} \mathbf{t}_{new} \sim \frac{A(I^2 - 1)}{I(I - A)} F_{A,I-A}, \qquad (11.25)$$

where $\mathbf{S}$ is the covariance matrix of the t-scores contained in the matrix $\mathbf{T}$ calculated during the model development,[62] I is the number of batches used for modeling, and A is the number of components. F represents the F distribution. The T_f^2 statistic reveals faults that can be described by the model.

The squared prediction error (SPE) is calculated by:

$$SPE_k = \sum_{r=(k-1)J+1}^{Jk} \mathbf{e}_r^\top \mathbf{e}_r, \qquad (11.26)$$

where $\mathbf{e}_r$ is the rth column of the matrix $\mathbf{E} = \mathbf{X}_{new} - \mathbf{t}_{new}\mathbf{P}^\top$. In the simple case where we are considering only one new batch, $\mathbf{E}$ is a row vector, $\mathbf{e}_r$ is a scalar, and $\mathbf{e}_r^\top \mathbf{e}_r$ condenses into $\mathbf{e}_r^2$. The distribution of the SPE can be approximated by a weighted χ^2 distribution, $SPE_k \sim (v_k / 2m_k)\chi^2_{2m_k^2/v_k}$, where m_k and v_k are the mean and variance of the SPE obtained for the data set used for the model development at time instant k.[46] The SPE will show if a totally new event is occurring in the process. This measure includes, for example, unusual variation of the controlled variables stabilized by simple control.

A fault is detected whenever the T_f^2 statistic or the SPE exceeds a confidence limit. The 95% limit is usually taken to be a warning level only, and action is taken when the statistic exceeds a 99% limit.

The T_f^2 statistic in Figure 11.4, plotted as a function of time for the same batch as in Figure 11.3, shows that this particular batch has a large deviation from $t = 70$ to $t = 80$. The slow drift of the T_f^2 that can be noted is difficult to detect looking at the raw measurements. It has already been indicated (in Figure 11.3) that this process drift results in a much lower than average product concentration at the end of the fermentation. Figure 11.3 illustrates the importance of this type of monitoring for predicting the consequences of deviations. The SPE in Figure 11.5 indicates that this process is deviating from the average process almost throughout the entire fermentation.

Contribution plots, which indicate the variables that are contributing the most to the T_f^2 statistic or the SPE, can easily be constructed and can thus be used in the fault identification.[42,47] Contribution plots can be used to find the change in the contribution from one point in time to another or the contribution plot can be used to find the deviation of the current batch when compared to the normal batch behavior described by the model. Here, we choose to look at the fault that has been detected around $t = 70$. Figure 11.6 shows the change in contribution of the variables from a point in time just before the fault could be detected in the SPE and T_f^2 plots ($t = 67$) to the point where the fault is at its highest ($t = 73$). The figure shows that there is a large change in the contribution of the CO_2

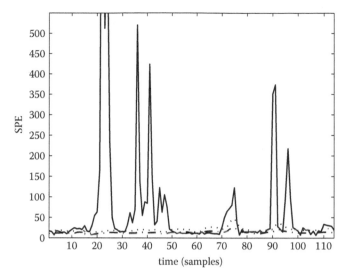

FIGURE 11.4 T_f^2 statistic; dash and dotted lines indicate 95% and 99% confidence limits, respectively.

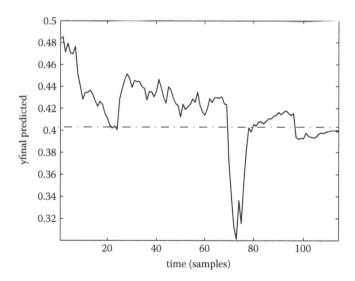

FIGURE 11.5 Squared prediction error (SPE). Maximum peak value when $t = 20$ is about 1400. The dash-dotted line is the 95% confidence limit, and the dotted line is the 99% confidence limits.

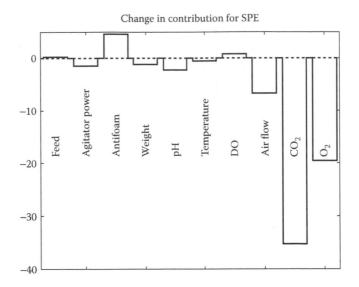

FIGURE 11.6 Contribution plot showing the change in the process from $t = 67$ to $t = 73$. It is seen that the variables CO_2 and O_2 contribute most to the fault and that they are lower than normal. From the plot of the predicted final product concentration (Figure 11.5) it can be seen that the fault has a negative effect on the quality.

and O_2 measurements. Thus, the task of isolating faulty measurement has been reduced to following only a few plots that display if a fault has occurred and subsequently investigating the contribution plots to identify which variables contributed most to the fault.

11.3.1.3 Score Plots

The process can also be monitored using the scores in a so-called *score plot*. Usually, the number of components is low (two to three) so a single plot is usually sufficient to display the state of the process. If the model contains more than two components, one can either construct three-dimensional plots or make several two-dimensional plots to show the variation. The relation between the score plot and the squared prediction error is illustrated in Figure 11.7 for an abnormal sample point in time.

A score plot can be used to monitor the process. When the variable space is compressed using either PCA or PLS, the process behavior can be monitored in a low-dimensional space using simple plots.[46,47] Because the example model contains only two components, the score plot in Figure 11.8 is sufficient to monitor the major variations of a normally operating batch. The figure shows the variation of the process in the reduced space of the two components. The score plot describes the current state of the process and allows the operator to follow and interpret the development of the process. It must be emphasized that the scores usually lack any direct physical meaning, as they combine many physical

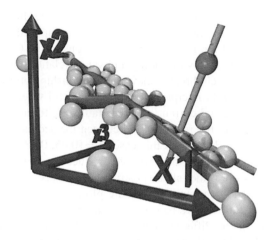

FIGURE 11.7 Three-dimensional score plot. The score plot illustrates the quantification of an abnormal sample instance, both in terms of distance in the two-dimensional plane to its origin as measured by the T^2 statistics and in terms of the level of unmodeled behavior, as indicated by the squared prediction error (i.e., the distance above the two-dimensional plane).

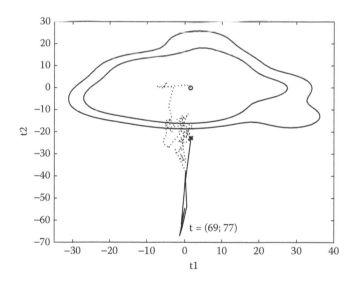

FIGURE 11.8 Score plot for a faulty batch illustrating the development of the process in a reduced space. The beginning of the fermentation is marked with a "°". The time interval [69,77] (starting with a "*") is shown as solid lines. The score t_1 varies, primarily when large oscillations in the temperature occur; t_2 varies primarily when the CO_2 and O_2 measurements change.

effects. One way of interpreting the score plot is to ascribe different phenomena to specific timewise movement of the scores. For example, the example model shows large variation of the t_2 score when deviations in the O_2 uptake and CO_2 production occur. Another way of finding a physical relationship is to investigate the loading matrices, which directly show the relationship between the measurements and the scores. Both interpretations can be useful when the behavior of a batch is to be described and current or future faults are to be eliminated.

Equation 11.5 shows the relationship between the scores and the dependent variable y as $\hat{\mathbf{Y}} = \mathbf{TQ}^\top$. Because in this case $\mathbf{Q} = [0.05, 0.08]$, it can be inferred that batches will have a higher than average product concentration at the end of the batch when the score values all are positive (i.e., the scores are moving around in the upper right-hand quadrant of the score plot).

Score plots can furthermore be used as a fingerprint of the batch. Instead of investigating plots of the different measured variables, we can use a score plot for an entire batch to investigate if something unusual has happened during the fermentation. Figure 11.9 shows such a score plot of a well-behaved batch. The scores stay in this figure close to the point (0,0) which indicates that this batch did not have any faults that affected the product concentration. It would have been much more difficult to interpret the original measurements due to their time-varying nature.

When a new batch is monitored we can then investigate if the batch is operating within the window of the desired behavior given by the model. If the batch deviates significantly, a fault has occurred, and corrective action must be

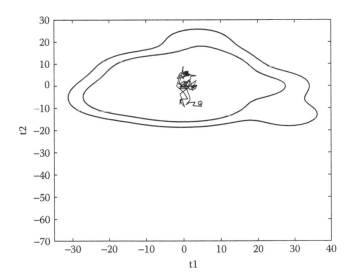

FIGURE 11.9 Score plot for a well-behaved batch. The scores remain near the point (0,0), suggesting that the product concentration will end on the average value of the batches that were used to form the model. Confidence limits on both figures are kernel density estimates.

taken. As the model has been built on historical data where process faults have not been treated, faults unfortunately may be allowed to persist. The goal is that, when fault diagnosis has been implemented and the process variation has been reduced, a new and better model for monitoring should be developed.

11.4 BATCH CONTROL PROBLEM SOLUTIONS

In the following sections, various implementations of batch operations models are reviewed and methods based on time-series models are discussed in more detail.

11.4.1 BATCH CONTROL REVIEW

Despite the fairly difficult nature of the nonlinear and time-varying control problem of batch processes, a number of approaches to online control of batch processes have been presented in literature. This is consistent with the significant attention drawn to batch manufacturing in other areas (e.g., robotics and semiconductor manufacturing).[43] With regard to the control of batch chemical processing, the suggested methods may be distinguished between those using first-principles engineering models and those using empirically based models. For first-principles models, two methods have been advocated, one based on nonlinear differential geometric control and the other on online optimization.

In the differential geometry approaches,[30–32] a nonlinear model is used to perform a feedback transformation that renders the transformed system linear, such that linear control theory can be applied. In the online optimization approaches, the optimal batch operations models are periodically updated during the batch execution at selected instances to optimize some quality or performance measure.[10,11,53] In these methods, sequential quadratic programming may be used to determine the control actions; however, the required first-principles engineering models are only seldom available for many chemical engineering processes. An interesting approach to circumvent this problem was suggested by Lübbert and Jørgensen,[38] who took a hybrid modeling approach to overcome the shortcoming of simple unstructured biomass models for biotechnical processes. Using such a hybrid approach it has been possible to achieve optimizing control in a number of cases.[56]

More general empirical model knowledge has been applied by a number of groups to achieve methods for control of batch processes since it was recognized that usage of detailed first-principles models for control and optimization often is unrealistic in industrial applications. Bonvin et al.[7] and Srinivasan et al.[60,61] determined a sequence of approximate input trajectories using approximate models and subsequently used batch plant measurements to refine the input trajectories through tracking the necessary conditions for optimality (NCO). Their procedure can track both path and terminal conditions, thus leading to a combined intra- and interbatch control. Given the limited availability of models, they aim at experimental adaptation (optimization) of the batch operations model through tracking the NCO. The resulting procedure uses the interesting concept of a model of the optimal solution, hence their resulting optimal batch operations model is referred to as the *solutions model*.[59] Another empirical approach was taken by Åkesson et al.,[1] who utilized the knowledge that optimal batch operation consists of a sequence of operations wherein each sequence is constrained by some physical aspect of the process system. This information is revealed through perturbing selected process actuators to determine the currently most constrained variable. This method was implemented on a substrate- and oxygen-supply-limited bioreactor to provide near-optimal operation. The principle has also been extended to a heat-transfer-limited operation by de Maré et al.[12]

The above works may be viewed as primarily addressing problems related to implementing the batch operations model. Several of the studies mentioned do, however, depend on the availability of a process model, which indeed may be available for some processes but in other cases may be difficult or expensive to develop. Therefore, significant attention should be devoted to developing methods for modeling from data and investigating how such methods can exploit control developments for batch process control.

One such data-driven approach is based on experiences gained from monitoring, wherein predictions of end-of-run properties are made.[15,16,65] These predictions are used for control of end-of-run quality properties, where the control action is restricted to only a few changes in the manipulated variable during the batch operation. However, if the operation requires a more frequent adjustment, then the manipulated variable trajectories may be discretized into a limited number of segments, within which the manipulated variables follow a zero or first-order hold.[8,17,54,55] Such a staircase approach, however, may be undesirable in practice when a smoother progression may be desirable, simply because a too coarse staircase may give rise to undesirable high-frequency disturbances. An advantage of the monitoring methodology is the inherent mapping of the high-dimensional variables into a low-dimensional variation space of a few latent variables, which are determined using principal component analysis (PCA) or partial least squares (PLS) methods. A further advantage of these methods is that all measured variable trajectories can be mapped into the latent variable space, so all control calculations can be performed and then mapped back into the manipulated variables.

Another data-driven approach is also inspired by the ability of monitoring methods to predict endpoint qualities by utilizing monitoring models; however, this inspiration led to the realization that it should be possible to reconstruct approximate time-series models from available process measurement data. Such a reconstruction was realized by Gregersen and Jørgensen[19] and Bonné and Jørgensen,[4,6] who utilized a large number of local models on a grid of time points to approximate the measured process behavior as described in the modeling section of this chapter. These models are referred to as grid of linear models (GoLMs). The resulting model may be represented as being linear time invariant between batches and linear time varying within a batch, thus relatively well-known methods can be used to control these systems, even though the models contain a high number of parameters. A key point in the identification of the models is the usage of regularization to ensure a well-conditioned estimation problem, as described above. Thus, the methodology is essentially equivalent to the projection methods to lower dimensional spaces as mentioned during our discussion of monitoring models.

The development of numerous solutions to the batch control problem has been reported in recent years; therefore, it seems worthwhile to view the developments from the perspective of their implementation in the control hierarchy.

11.4.2 BATCH CONTROL PROBLEM

Conceptual design and synthesis of batch processes[18] provides both the task sequence and a feasible operations model, as well as definitions of the important process variables for each subtask. The feasible operations model may be optimized offline to produce an optimal batch operations model. Implementation of this operations model may be achieved by defining control loops to ensure that the desired trajectory is followed; however, this implementation may be utilized in many different ways, depending on the models used and approaches taken to use experimental information. This section describes an operations model and discusses ways to implement such operations models. Two extreme ways to implement an operations model can be defined, with several hybrid forms in between.

One implementation, in which the prescribed actions of the operations model are implemented directly on the plant for the precalculated duration, is termed *open-loop* (or offline) *optimization*. For this type of operation, the uncertainty caused by disturbances will result directly in variable product quality and productivity; however, in practice, this open-loop implementation has been one of the most utilized. Consequently, there is ample room for improvement by using one of the solutions described below.

In another type of implementation, the actions are downloaded as setpoint trajectories to a set of controllers which then change the manipulated variables such that the controlled variables can follow the prescribed trajectories with the designed performance. This implementation is termed *closed-loop* (or online) *optimization*. Various examples of closed-loop optimizing control are investigated below.

11.4.3 CLOSED-LOOP OPTIMIZING CONTROL

When feedback control is applied to reduce the influence of the process and market disturbances on the batch operations model (i.e., to reject both inter- and intrabatch disturbances), several different implementations are possible. These may be classified according to a combination of the number and type of control layers between the optimization layer and the process and of the control design methodology.

11.4.3.1 Single-Variable Control Layer

When the setpoints or their trajectories are downloaded directly to a single-variable control layer, an improved ability to reject disturbances is obtained. The usage of feedback for implementation of the batch operations model improves the achievable performance to the extent that disturbances may be rejected by the selected control loops in the single-variable layer. The batch operations model will remain optimal if such disturbance rejection is perfectly achievable. In the case of continuous plants, it is in some cases possible to develop a control structure that is insensitive to uncertainty such that the setpoints may be kept constant even during disturbances. The case where this is possible is known as a *self-optimizing control structure*.[58] For batch plants, however, the setpoints usually will be time varying, as specified by the batch operations model.

The control loops are related to the synthesis of a batch operation through the degrees of freedom for the process operation. Some of these degrees of freedom are used by the process actuators in the control loops. Thus, the decision as to which degrees of freedom to provide for the control loops is very important, as is which measurements to combine or pair with each degree of freedom into a control loop. The combined set of control loops for a processing task is a *control structure*. It is clear that the available degrees of freedom change as the status of the constraints change during the batch operation. Thus, when operating on one set of constraints, the corresponding degrees of freedom are not available as control handles, whereas when operation switches to a new set of constraints a new set of degrees of freedom becomes available. Batch operation can be viewed as execution of a sequence of control structures on a plant with the purpose of moving the plant in such a way that available resources are utilized maximally under the given conditions. Clearly, the timing or execution of this sequence becomes important for achieving optimum performance (i.e., realizing the batch operations model). From the perspective of the control hierarchy, a coordinating layer above the single-loop control layer can be introduced to handle the coordination or switching between the different multiloop control structures, each of which usually is implemented as several single loops.

The principal control synthesis problem, then, is to develop basic control structures for the different phases during a batch and to develop procedures for switching between the different control structures. One interesting systematic methodology for implementing the optimal batch operations model solution is under development. This methodology determines a set of controllers that are designed to track the necessary conditions of optimality (NCO), thereby enabling direct adaptation to uncertainty.[59–61] The resulting sequence of control structures and their switching times implements the batch operations model. Each controller or switching time is related to satisfying one specific nominal condition of optimality. This methodology may provide useful insight into the optimal batch operations model and its relation to the process design. Several aspects are still under development so a systematic procedure using this methodology is not yet available. Currently, it is assumed that the structural aspects of the solution model are invariant to uncertainty. Furthermore, the operations model is not unique, and particular choices strongly influence the control structure. For these reasons, it may be relevant to investigate what the introduction of a full multivariable layer in addition to the coordination layer offers. This question is obviously relevant because the batch control problem implies coordination of several different control loops under the influence of process constraints. In reality, this problem clearly is multivariable at the outset.

At one extreme, one could consider the case where it is not possible to achieve a time separation between the optimization and single-variable control layers, thus a tight integration between the optimization and control layers would be desirable. Achieving this requires solving the Hamilton–Jacobi–Bellmann equation, which is computationally infeasible; instead, a repeated optimization may be executed. Such a repeated optimization may be carried out by solving a finite-horizon optimization problem at each sample interval in the multivariable control layer,[14,48,52] which would directly imply a multivariable layer. A different approach has been taken by Cruse et al.,[11] who solved an output feedback optimal control problem by minimizing a general objective reflecting process economics rather than deviations from a batch operations model trajectory. Thus, this procedure can be considered an online optimization of the batch operations model. This approach is a generalization of state-of-the-art, steady-state, real-time optimization aimed at establishing economically optimal transient plant operation.[3] Here, however, our attention is focused on the introduction of a multivariable control layer.

11.4.3.2 Multi- and Single-Variable Control Layers

In fact, a multivariable layer may often be most suitable for implementation of the batch operations model simply due to the required switching between control structures in order to exploit different resource and equipment constraints. The multivariable aspects may be relevant for tracking the setpoint profiles from the optimization layer, where it may be appropriate to keep a careful balance between different actuators to maintain the plant on a desired trajectory. During tracking of setpoint trajectories, the multivariable aspects have two manifestations, where one is related to rejection of intrabatch disturbances and the other to rejection of interbatch disturbances. Preventing intrabatch disturbances from having a deteriorating effect on the ability to follow the desired time-varying optimal batch operations model will in general have to be handled as a time-varying control problem. Thus, one method that could be most suitable is model predictive control, provided a model is available. Model predictive control has the clear additional advantage that it smoothly switches between control structures as constraints become active or inactive. Preventing interbatch disturbances from having a deteriorating effect on the ability to reproduce a batch of the same type, whenever this might occur, may be viewed as a task of learning as much as possible from the previous batch round, including learning from the disturbances actually encountered. For this purpose, iterative learning control is convenient.

11.4.3.3   Iterative Learning Control

Iterative learning control, or ILC, is a well-established technique that can be used to overcome some of the traditional difficulties associated with ensuring control performance for time-varying systems.[43] Specifically, ILC is a technique for improving the transient response and tracking performance of processes, equipment, or systems that execute the same trajectory, motion, or operation over and over, starting from essentially the same initial conditions each time. In these situations, iterative learning control can be used to improve the system response. The approach is motivated by the observation that, if the system controller is fixed and if the process operating conditions are the same each time it executes, then any errors in the output response will be repeated during each operation. These errors can be recorded during operation and can then be used to compute modifications to the operations model (i.e., the input signal that will be applied to the system during the next operation, or trial, of the system). In iterative learning control, refinements are made to the input signal after each trial until the desired performance level is reached. Note that the word *iterative* is used here because of the recursive nature of the system operation, and the word *learning* is used because of the refinement of the control input based on past performance executing a task or trajectory. Iterative learning control depends on the design of algorithms to update the control input. Because the control input in general is multivariable, then iterative learning control should take place at the multivariable layer or higher up in the control hierarchy. In the semiconductor industry and in robotics, run-to-run control has been applied extensively.[43] Most often, these applications focus upon the interbatch disturbances and do not handle the intrabatch aspects. In chemical plants, the batch duration often is long, thus significant benefit may be gained by handling intrabatch disturbances and it may be worthwhile to combine the multivariable and iterative learning controls. Such an integration has been investigated by Lee et al.,[35] and Bonné and Jørgensen.[5]

In this chapter, a multivariable layer with model predictive control and iterative learning control is used to illustrate the possible benefits for implementation of the batch operations model. The purpose of the demonstration is to reveal the straightforward application of MPC plus ILC when an identified model representation is available. First, however, the benefits of this combination are illustrated and discussed.

11.4.4   MODEL PREDICTIVE AND ITERATIVE LEARNING CONTROL

Approximating nonlinear batch processes with sets of linear time-series models provides a model that facilitates application of linear-model-based control. One such model-based control algorithm is *model predictive control* (MPC), which has been comprehensively covered in the literature.[9,41,44,45,51] Model predictive control based on model-based predictions of future process behavior will provide an optimal sequence of actuator signals which will bring the process to a desired operation point within a given time horizon. The actuator sequence will be optimal in the sense that it will optimize a specified control objective. This control objective could be minimum deviation from a desired (optimal) batch operations model, minimum operation costs, or maximum production rate. Further motivation for application of MPC is that it offers optimal specification tracking while considering multivariable dynamic correlations and actuator limitations together with safety and quality constraints on the process variables and potentially also on equipment availability. As depicted in Figure 11.10, the multivariable and constraint handling abilities of MPC may be built on an existing basic (proportional–integral–derivative, or PID) control system. When the final product quality depends on several batch units operated in parallel, the multivariable facets of MPC may also be applied for coordinative control of the separate batch units, keeping the final product quality at its desired level as illustrated in Figure 11.11.

For batch processes the MPC control objective is typically specified as a trade-off between minimum expected deviation ($\hat{\mathbf{e}}_{k,N|t}$) from the desired operational path given measurements up to time t and minimum controller intervention ($\Delta\mathbf{u}_{k,t}$) in the remaining part of the batch:

$$\min_{\Delta u_{k,t}}[\hat{\mathbf{e}}_{k,N|t}^T\mathbf{Q}\hat{\mathbf{e}}_{k,N|t} + \Delta\mathbf{u}_{k,t}^T\mathbf{R}\Delta\mathbf{u}_{k,t}]$$

$$s.t. \quad \hat{\mathbf{e}}_{k,N|t} = F(\hat{\mathbf{e}}_{k,t|t}, \Delta\mathbf{u}_{k,t}),$$

(11.27)

where $\mathbf{Q}$ and $\mathbf{R}$ are weighting matrices. The control objective is optimized at every time step t when only $\Delta u(t)$ is implemented. When utilizing the batch-to-batch transition model (Equation 11.15), the resulting control algorithm may be represented by the block diagram shown in Figure 11.12. The quadratic objective in Equation 11.27 can be formulated as a quadratic programming (QP) problem:

$$\min_{\Delta u_{k,t}}\left[\frac{1}{2}\left\{\Delta\mathbf{u}_{k,t}^T\mathbf{H}\Delta\mathbf{u}_{k,t} + 2\mathbf{f}^T\Delta\mathbf{u}_{k,t}\right\}\right]$$

$$s.t. \quad \mathbf{A}\Delta\mathbf{u}_{k,t}\leq\mathbf{B},$$

(11.28)

which can be solved with commercial QP-solvers. By proper configuration of $\mathbf{A}$ and $\mathbf{B}$, both the process variables and the actuators may be constrained.

Considering the repetitive fashion in which batch processes are usually operated, control of these processes will improve when the experience gathered during operation of the previous batches (for a specific grade or product) is utilized before and during operation of a new batch. Having completed a batch run with control, we could then pose the question: What were the lessons learned?

- Lesson 1. We learned how disturbances encountered during the batch run were rejected. This knowledge about the control profile can be used to improve the nominal control profile, giving better rejection of disturbances in the next batch run.

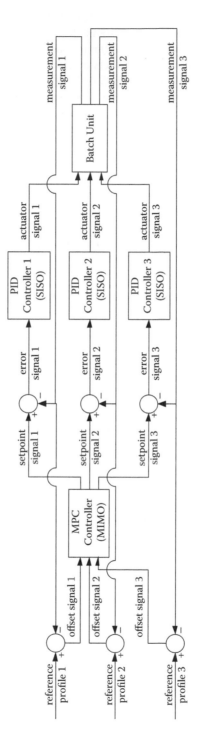

FIGURE 11.10 Conceptual sketch of how the multivariable and constraint handling features of model predictive control (MPC) may be utilized as an add-on to an existing decentralized basic control system.

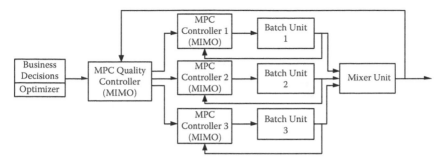

FIGURE 11.11 Conceptual sketch of mixed product quality tracking with coordinative control.

- Lesson 2. We learned which disturbances to expect in the next batch and when. This knowledge about the disturbance profile can be used to improve our prediction of the behavior in the next batch.
- Lesson 3. We learned how well the behavior was predicted. This knowledge about the predictive capabilities can be used to improve both the performance and the robustness of the controller for a batch run as a whole (i.e., to obtain better but also safer disturbance rejection).

These three lessons are depicted in Figure 11.12 and Figure 11.13. This concept of benefiting from experience gathered during operation of past batches is what is captured in iterative learning control, as illustrated in Figure 11.13. Thus far, ILC has been applied between batches as the only control scheme, offering no corrective actions during operation of a single batch; however, merging ILC and MPC offers the learning capabilities of ILC in a MPC framework.[5,35] This merger

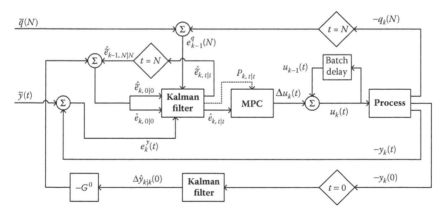

FIGURE 11.12 Block diagram of the model predictive control (MPC) imbedded iterative learning control (ILC) algorithm. The block diagram illustrates how the initial estimate of the error profile and the nominal input profile are refined from batch to batch as well as how the prediction error covariance (P) is used for controller tuning.

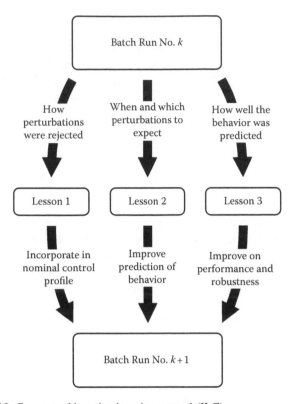

FIGURE 11.13 Concept of iterative learning control (ILC).

provides a MPC framework that asymptotically rejects the effects of batchwise persistent disturbances, including the effects of model bias introduced by the linear process approximation. The time-series modeling framework presented in this chapter fully supports an iterative learning MPC framework.

11.5   RESULTS

To illustrate the applicability of the proposed data-driven time-series modeling approach, two fed-batch fermentation processes are considered. First, the modeling framework is applied to data from an industrial case. Subsequently, the performance improvement that may be achieved using the data-driven predictive time-series models for intra- and interbatch control is demonstrated on a simulated cultivation example.

11.5.1 Modeling for Prediction and Control

11.5.1.1 Industrial Pilot Plant

In collaboration with an industrial partner, historical data from pilot-plant, fed-batch fermentors were investigated for the purpose of modeling.[4] For the sake of simplicity, only online carbon dioxide evolution rate (CER), oxygen uptake rate (OUR), substrate feed rate, and offline product activity measurements were considered. The substrate feed rate was selected as the input or manipulated variable, and the outputs (i.e., the carbon dioxide evolution rate, oxygen consumption rate, and product activity measurements) were correlated to the input in two scenarios:

- Model 1: Using data from ten fed-batch runs with high diversity and cross-validation on data from three different fed-batch runs
- Model 2: Using data from two fed-batch runs with very similar behavior and cross-validation on data from one fed-batch run

The first model showed very good fit of the validation data; therefore, these are not shown here. For the second case, it can be seen in Figure 11.14 that a

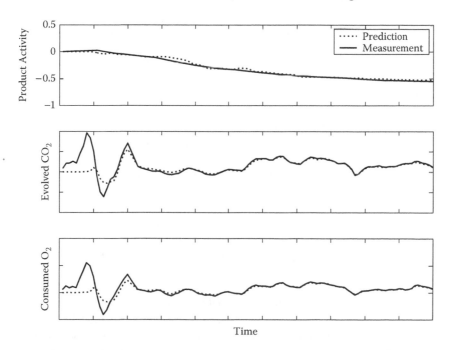

FIGURE 11.14 Model validation for an industrial pilot-plant fermentation. The model convincingly predicts a decrease in product activity from one batch to the next corresponding to 50%.

50% decrease in product activity for the validation batch could be predicted using model 2 and that the experimental costs could have been avoided. Note that the poor predictive performance in the initial phase of the batch depicted in Figure 11.14 is caused by a batch phase preceding the fed-batch phase during which very limited information was available because no inputs were used except for the starting values (i.e., initial conditions). Note also that the available data for this case were very limited and that more data or even designed experiments should be carried out in which the inputs are perturbed at points in time where control action might be desired.

11.5.1.2 Simulated Production of Yeast

To simulate fed-batch cultivation of yeast, a biochemically structured model developed by Lei et al.[37] was used to represent the process. This model describes aerobic growth of *Saccharomyces cerevisiae* on glucose and ethanol and focuses on the pyruvate and acetaldehyde branch points in the metabolic pathways, where overflow metabolism occurs when the growth changes from oxidative to oxido–reductive. In the designed process operation, it was assumed that the cultivation was fed with glucose as substrate while the feed flow rate was manipulated as the input variable. In the simulation, it was assumed that the ethanol (EtOH) concentration, oxygen uptake rate (OUR), and carbon dioxide evolution rate (CER) were measured online without time delay. From the online measurements of OUR and CER, the respiratory quotient (RQ) was calculated (RQ = CER/OUR). To characterize the performance of this fed-batch cultivation, the following dimensionless measures were defined:

- Quality final yeast biomass concentration – final ethanol concentration (∞1000) and substrate concentration ($\times$200), normalized by the concentration unit (1 g/L).
- Yield of produced biomass (yeast) divided by the fed substrate on a mass basis.

The weights on the ethanol and substrate concentration in the quality measure could be based on costs of downstream processing relative to the sales price of yeast, but here they are set such that variations in yeast, ethanol, and substrate concentrations are represented equally in the quality variation.

For simplicity the cultivation was modeled with EtOH, OUR, CER, and RQ as output variables without any other quality-related variables. RQ was treated as a measurement to test if the dynamic behavior could be modeled with the proposed modeling approach. To study the applicability of the proposed modeling approach to processes where initial conditions are not obtainable, the dynamics in $\mathbf{G}_{ini}$ were not included in the modeling. A set of data-driven models was identified in the following scenario:

- The measurement noise was approximated by white noise with a standard deviation equal to one third of the maximum noise level. Maximum measurement noise: EtOH, ±3%; OUR, ±2%; CER, ±2%; RQ, ±2%; u, ±1%, where the individual noise levels are based on laboratory experiments of industrial data.
- The initial conditions were uniformly perturbed by a maximum of ±10% in every batch.
- The identification was based on data from ten realizations of normal runs.
- The identification was in this case optimized by testing possible combinations of the regularization weights $\lambda_1 * I = L_1$ and $\lambda_2 * I = L_2$ only. The combination that minimized the mean (in terms of the set of validation batch data sets) prediction error when the estimated model set is cross-validated on independent data is selected. The optimal regularization weights are shown in Table 11.1 and Table 11.2, along with their respective mean prediction fits.
- The model validation was based on three other normal batch data sets.

TABLE 11.1
Optimal Regularization Weights and Relative Prediction Fit for Noise-Free Identification

Output Variable	λ_1	λ_2	Fit
ΔEtOH	1.18	0.04	0.0157
ΔOUR	0.05	0.01	0.0192
ΔCER	0.05	0.01	0.0178
ΔRQ	0.02	0.01	0.0163

TABLE 11.2
Optimal Regularization Weights and Relative Prediction Fit for Noisy Identification

Output Variable	λ_1	λ_2	Fit
ΔEtOH	1.18	0.05	0.0157
ΔOUR	0.05	0.01	0.0242
ΔCER	0.05	0.01	0.0205
ΔRQ	0.26	0.02	0.0239

To facilitate the identification task a fairly exciting input sequence was chosen. The input sequence was constructed by adding a pseudo-random binary sequence (PRBS) onto the nominal input trajectory. At each sample, the PRBS was uniformly scaled by a maximum of ±10% of the nominal trajectory. Note that input sequences are not system friendly and thus seldom applicable in practice.[49] Because the basic assumption of this interbatch modeling approach is that trajectories vary only slightly from batch to batch, the validation batch data sets were generated with input trajectories similar to those used for identification (i.e., data sets similar to the identification data but not used for the identification). From Table 11.1, it can be seen that quite reasonable models can be obtained from noise-free data for all the outputs with relative prediction fits less than $2\infty10^{-2}$. Furthermore, as indicated in Table 11.2, models of similar quality can be obtained from noisy data without prefiltering. One of the three model cross-validations from the noisy identification is shown in Figure 11.15.

The control simulations were designed to accentuate the ILC feature and study the relation between trajectory control and reproducibility. The performance of a single input/multiple output (SIMO) ILC algorithm without quality control was investigated by randomly perturbing the initial conditions a maximum of ±10% in every batch run and introducing a persistent bias in the input trajectory

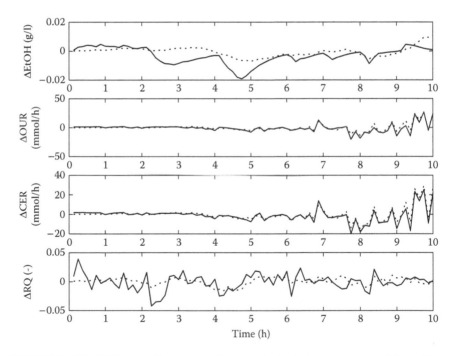

FIGURE 11.15 Validation of noisy identification in batch-incremented variables. True batch difference: solid line; predicted batch difference: dotted line.

by lowering the feed substrate concentration 10%. The benchmark against which the performance was tested was open-loop operation with no control. From the noisy SIMO tracking in Figure 11.16, it is apparent that the control algorithm reduces the summed squared error sequence (SSES) $\mathbf{e}_k^T \mathbf{e}_k$ approximately 80% in the first batch run. Furthermore, after having been trained on the first three batch runs, the control algorithm reduced SSES more than 94% in subsequent batch runs except for batch run 6. By coincidence, the direction of the initial perturbation in batch run 6, which corresponds to starved yeast, was unique to batch run 6 and particularly difficult to handle by control. However, the performance of the control algorithm in batch run 7 does not suffer from the performance drop. Note, however, that, still, almost 60% of the total SSES was rejected by control in batch run 6. Although the SIMO tracking did not ensure reproducible operation in terms of quality, the quality was improved by output tracking alone (i.e., without including a quality measure in the control objective). It is also evident that the yield decreased slightly when control was imposed. This decrease was caused by aggressive controller movements that resulted in feed pulsing, which induced production of ethanol and carbon dioxide. Thus, the cost of a much improved output tracking performance was an insignificantly lower yield.

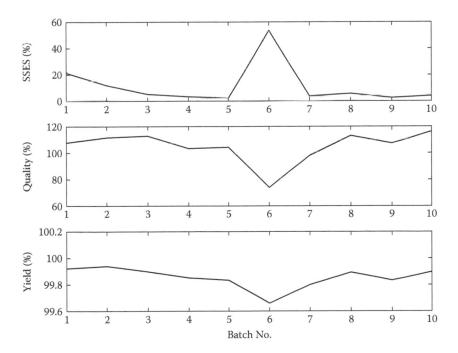

FIGURE 11.16 The batchwise evolution of summed squared error sequence, quality, and yield in noisy tracking of EtOH, OUR, CER, and RQ. No control corresponds to 100%.

11.6 DISCUSSION AND CONCLUSIONS

The chapter formulates the batch control problem as implementation of the optimal batch operations model. A key ingredient in this implementation is application of knowledge of the behavior of the plant which is represented in the form of models. Such knowledge may originate from first-principles knowledge or from data. Data-driven modeling is emphasized in this chapter. One type of data-driven modeling is modeling for monitoring batch processes. It has been demonstrated that data-driven modeling is feasible for the development of time-series models for batch processes leading to LTV models within the batch and LTI models from batch to batch. The moderate number of parameters in the locally linear time-series models are estimated through extensive use of regularization. The models have shown remarkable ability to predict batch cultivations before they have actually been carried out.

Both of the two fundamentally different data-driven modeling approaches presented in this chapter have been shown to possess the property of being able to predict the entire batch process behavior in pure simulation (i.e., before the batch is started). Time-series-based predictive models may be developed from a relatively low number of batches due to the extensive use of regularization; hence, such time-series models hold promise for the development of monitoring tools and learning model predictive control of batch processes. For a new production or new strain, such development could conceivably be based on data from pilot-plant optimization runs. Such data could also enable development of models for other types of products produced using related methods (e.g., cultivation of the same strains of microorganisms).

Monitoring of batch processes is most useful for detection and diagnosis of faults in batch process operation; however, another important use of monitoring is the verification of process performance relative to the behavior represented in the plant model used for control design. When such verification has been obtained, it is reasonable to proceed to further implementation of the optimal batch operations model. In practice, such verification should be carried out at every sample instant before more advanced control action is executed.

We reviewed recent progress in the control of batch processes by presenting the formulation of the batch control problem. Different implementations of the optimal batch operations model were discussed in relation to the control hierarchy and to disturbances affecting the operation. Implementations based on a single-variable control layer designed to reject disturbances were shown to require a coordination layer to ensure proper timing when switching between the different control structures in the single-variable control layer. A methodology to systematically develop these controllers and the coordinated switching is under development based on satisfying the necessary conditions for optimality. This method depends on knowledge of the solutions model (i.e., knowledge of how the optimal solution influences the necessary conditions for optimality). Another type of implementation employs a multivariable control layer with model predictive control that can ensure a smooth transition between control structures to track the process constraints optimally and can also ensure coordinated utilization of actuators to handle multivariable issues, including constraints. A straightforward implementation will enable rejection of intrabatch disturbances. In addition, learning control may be incorporated to enable rejection of interbatch disturbances and ensure optimization of the batch operations model.

The intent of this chapter was to demonstrate that the area of batch control is finally maturing and that several methods are under development for practical implementation of the batch control problem. Several applications of these methods have also been reported. Clearly, many aspects of the various methods require further development; however, if proper validation is carried out then several of the methods may be used in their current state of development. It is noteworthy that most methodologies are based on qualitative or data-driven modeling methods.

Given the methods presented here, it can be expected that batch monitoring and control will lead to significant improvements in batch operations and their optimization, improvements that currently are being exploited in the multivariable control of many continuous plants. Due to the extensive use of open-loop optimization in the implementation of the time-varying optimal batch operations model, however, it is expected that the achievable benefits of closed-loop optimization are substantially greater than those seen for continuous operation.

REFERENCES

1. Åkesson, M., Hagander, P., and Axelsson, J., Probing control of fed-batch cultures: analysis and tuning, *Control Eng. Practice*, 709–723, 2001.
2. Årzén, K.-E.R. and Olsson, J.Å., Grafchart for procedural operator support tasks, in *Proceedings of the 15th IFAC World Congress, Barcelona, Spain*, Pergamon Press, New York, 2002.
3. Backx, T., Bosgra, O., and Marquardt, W., Integration of Model Predictive Control and Optimization of Processes, in *Proceedings of the IFAC Symposium: Advanced Control of Chemical Processes, ADCHEM 2000*, Vol. 1, Pergamon Press, New York, 2000, 249–260.
4. Bonné, D. and Jørgensen, S., Data-driven modeling of nonlinear and time-varying processes, in IFAC symposium on system identification and parameter estimation, *SYSID*, Hoff, P.V., Ed., Pergamon Press, New York, 2003, 1655–1660.
5. Bonné, D. and Jørgensen, S., Iterative learning model predictive control of batch processes, in *BatchPro: Symposium on Knowledge-Driven Batch Processes*, Kipparisidis, C., Ed., CERTH, Thessaloniki, 2004, 67–72.
6. Bonné, D. and Jørgensen, S.B., Development of model-based iterative learning control of batch processes, in *European Control Conference 2001*, de Carvalho, J.L.M., Fontes, F.A.C.C., and de Pinho, M.D.R., Eds., EUCA, 2001.
7. Bonvin, D., Srinivasan, B., and Ruppen, D., Dynamic optimization in the batch chemical industry, *AIChE Symp. Ser.*, 326, 255–273, 2002.
8. Chin, I.-S., Lee, K.S., and Lee, J.H., A Unified Framework for Control of Batch Processes, paper presented at AIChE Annual Meeting '98, Miami, FL, 1998.
9. Clarke, D.W., Mohtadi, C., and Tuffs, P.S., Generalized predictive control. Part I. The basic algorithm, *Automatica, (2)*, 23, 137–148, 1987.

10. Crowley, T. and Choi, K., Experimental studies on optimal molecular weight control in a batch-free radical polymerization process, *Chem. Eng. Sci.*, 53, 2769–2790, 1998.

11. Cruse, A., Marquardt, W., Oldenburg, J., and Schlegel, M., *Batch Process Modeling and Optimization*, 2004.

12. de Maré, L., Velut, S., Briechle, S., Wennerberg, C., Cimander, C., Ramchuran, S., Turnert, P., Silfversparre, G., Holst, O., and Hagander, P., Temperature limited fed-batch cultivations with a probing feeding strategy for *Escherichia coli*, in *Computer Applications in Biotechnology*, Van Impe, M.P.J.F., Ed., Pergamon Press, New York, 2004.

13. Dorsey, A.W. and Lee, J.H., Subspace identification for batch processes, in *Proceedings of the 1999 American Control Conference*, Vol. 4, 1999, 2538–2542.

14. Eaton, J. and Rawlings, J., Feedback control of nonlinear processes using on-line optimization techniques, *Comput. Chem. Eng.*, 14, 469–479, 1990.

15. Flores-Cerrillo, J. and MacGregor, J., Control of particle size distribution in emulsion semi-batch polymerization using midcourse correction policies, *Ind. Eng. Chem. Res.*, 41, 1805–1814, 2002.

16. Flores-Cerrillo, J. and MacGregor, J., Within-batch and batch-to-batch inferential-adaptive control of batch reactors: a PLS approach, *Ind. Eng. Chem. Res.*, 42, 3334–3345, 2003.

17. Flores-Cerrillo, J. and MacGregor, J.F., Control of batch product quality by trajectory manipulation using latent variable models, *J. Process Control*, 14, 539–553, 2004.

18. Gani, R. and Papaeconomou, I., *Conceptual Design/Synthesis of Batch Processes*, Elsevier Science, Amsterdam, 2004.

19. Gregersen, L. and Jørgensen, S.B., Identification of linear models for batch control and optimisation, in *6th IFAC Symposium on Dynamics and Control of Process Systems*, Stephanopoulos, G., Lee, J.H., and Yoon, E.S., Eds., 2001, 269–274.

20. Hansen, P.C., *Rank-Deficient and Discrete Ill-Posed Problems*, Polyteknisk Forlag, Lyngby, 1996.

21. Hastie, T., Tibshirani, R., and Friedman, J., *The Elements of Statistical Learning*, Springer-Verlag, New York, 2001.

22. Höskuldsson, A., *Regression Methods in Science and Technology*, Vol. 1. Thor Publishing, Denmark, 1996.

23. IEC, *Preparation of Function Charts for Control Systems*, Tech. Rep. IEC 60848, International Electrotechnical Commission, Geneva, Switzerland, 1988.

24. IEC, *Programmable Controllers*. Part 3. *Programming Languages*, Tech. Rep. IEC 61131, International Electrotechnical Commission, Geneva, Switzerland, 1993.

25. IEC, *Batch Control*. Part 1. *Models and Terminology*, Tech. Rep. IEC 61512-1, International Electrotechnical Commission, Geneva, Switzerland, 1997.

26. IEC, *Batch Control*. Part 2. *Data Structures and Guidelines for Languages*, Tech. Rep. IEC 61512-2, International Electrotechnical Commission, Geneva, Switzerland, 2001.

27. ISA, *Batch Control*, Tech. Rep. ISA S88.01, Instrument Society of America, Research Triangle Park, NC, 1995.

28. Jackson, J.E., *A User's Guide to Principal Components*, John Wiley & Sons, New York, 1991.

29. Kourti, T., Nomikos, P., and MacGregor, J., Analysis, monitoring and fault diagnosis of batch processes using multi-block and multi-way PLS, *J. Process Control*, 5, 277–284, 1995.

30. Kozub, D. and MacGregor, J., Feedback control of polymer quality on semi-batch copolymerization reactors, *Chem. Eng. Sci.*, 47, 929–942, 1992.

31. Kravaris, C. and Soroush, M., Synthesis of multivariate nonlinear controllers by input/output linearization, *AIChE J.*, 36, 249–264, 1990.

32. Kravaris, C., Wright, R., and Carrier, J., Nonlinear controllers for trajectory tracking in batch processes. *Comput. Chem. Eng.*, 13, 73–82, 1989.

33. Kristensen, N., Madsen, H., and Jørgensen, S., A method for systematic improvement of stochastic grey-box models, *Comput. Chem. Eng.*, 28, 1431–1449, 2004.

34. Larimore, W.E., Statistical optimality and canonical variate analysis system identification, *Signal Process.*, 52(2), 131–144, 1996.

35. Lee, J.H., Lee, K.S., and Kim, W.C., Model-based iterative learning control with a quadratic criterion for time-varying linear systems, *Automatica*, 36(5), 641–657, 2000.

36. Lei, F., Dynamics and Nonlinear Phenomena in Continuous Cultivations of *Saccharomyces cerevisiae*, Ph.D. thesis, Technical University of Denmark, 2001.

37. Lei, F., Rotbøll, M., and Jørgensen, S.B., A biochemically structured model for *Saccharomyces cerevisiae*, *J. Biotechnol.*, 88(3), 205–221, 2001.

38. Lübbert, A. and Jørgensen, S., Bioreactor performance: a more scientific approach for practice, *J. Biotechnol.*, 85, 187–212, 2001.

39. Mardia, K.V., Kent, J.T., and Bibby, J.M., *Multivariate Analysis*, 2nd ed., Academic Press, San Diego, CA, 1980.

40. Martens, H. and Næs, T., *Multivariate Calibration*, John Wiley & Sons, New York, 1989.

41. Mayne, D.Q., Rawlings, J.B., Rao, C.V., and Scokaert, P.O.M., Constrained model predictive control: stability and optimality, *Automatica*, 36, 789–814, 2000.

42. Miller, P., Swanson, R.E., and Heckler, C.F., Contribution plots: a missing link in multivariate quality control, in *37th Annual Conference*, ASCQ, New York, 1993.

43. Moore, K., Dahleh, M., and Bhattacharyya, S., Iterative learning control, *J. Robotic Syst.*, 9, 563–594, 1992.

44. Morari, M. and Lee, J.H., Model predictive control: past, present and future. *Comput. Chem. Eng.*, 23, 667–682, 1999.

45. Muske, K.R. and Rawlings, J.B., Model predictive control with linear models, *AIChE J.*, 39(2), 262–287, 1993.

46. Nomikos, P., Statistical Process Control of Batch Processes, Ph.D. thesis, McMaster University, Hamilton, Canada, 1995.

47. Nomikos, P. and MacGregor, J.F., Multivariate SPC charts for monitoring batch processes, *Technometrics*, 37(1), 41–59, 1995.

48. Palanki, S., Kravaris, C., and Wang, H., Synthesis of state feedback laws for endpoint optimization in batch processes, *Chem. Eng. Sci.*, 48, 135–152, 1993.

49. Pearson, R.K., Input sequences for nonlinear modeling: nonlinear model based process control, in *Proceedings of the NATO Advanced Study Institute*, 1998, 599–621.

50. Puigjaner, L., Espuna, A., and Reklaitis, G., *Framework for Discrete/Hybrid Production Systems*, Elsevier Science, Amsterdam, 2002, 663–700.

51. Rao, C.V. and Rawlings, J.B., Linear programming and model predictive control, *J. Process Control*, 10, 283–289, 2000.

52. Rawlings, J., Tutorial overview of model predictive control, *IEEE Contr. Syst. Mag.*, 20(3), 38–52, 2000.

53. Ruppen, D., Bonvin, D., and Rippin, D., Implementation of adaptive optimal operation for a semi-batch reaction system, *Comput. Chem. Eng.*, 22, 185–199, 1997.

54. Russel, S., Kesavan, P., Lee, J., and Ogunnaike, B., Recursive data-based prediction and control of batch product quality, *AIChE J.*, 44, 2442–2458, 1998.

55. Russell, S., Robertson, D., Lee, J.H., and Ogunnaike, B., Control of product quality for batch nylon 6,6 autoclaves, *Chem. Eng. Sci.*, 53, 3685–3702, 1998.

56. Schubert, J., Simutis, R., Dors, M., Havlik, I., and Lübbert, A., Bioprocess optimization and control: application of hybrid modelling, *J. Biotechnol.*, 35, 51–68, 1994.

57. Simoglou, A., Martin, E.B., and Morris, A.J., Statistical performance monitoring of dynamic multivariate processes using state space modelling, *Comput. Chem. Eng.*, 26(6), 909–920, 2002.

58. Skogestad, S., Plantwide control: the search for the self-optimizing control structure, *J. Process Control*, 10, 487–507, 2000.

59. Srinivasan, B. and Bonvin, D., Dynamic optimization under uncertainty via NCO tracking: a solution model approach, in *BatchPro: Symposium on Knowledge-Driven Batch Processes*, Kipparisidis, C., Ed., CEPERI, Thessaloniki, 2004, 17–35.

60. Srinivasan, B., Palanki, S., and Bonvin, D., Dynamic optimization of batch processes. I. Characterization of the nominal solution, *Comput. Chem. Eng.*, 27, 1–26, 2003.

61. Srinivasan, B., Bonvin, D., Visser, E., and Palanki, S., Dynamic optimization of batch processes. II. Role of measurements in handling uncertainty, *Comput. Chem. Eng.*, 27, 27–44, 2003.

62. Tracy, N.D. and Young, J.C., Multivariate control charts for individual observations, *J. Qual. Technol.*, 24(2), 88–95, 1992.

63. Verhaegen, M. and Yu, X., A class of subspace model identification algorithms to identify periodically and arbitrarily time-varying systems, *Automatica*, 31(2), 201–216, 1995.

64. Vicente, A., Castrillo, J.I., Teixeira, J.A., and Ugalde, U., On-line estimation of biomass through pH control analysis in aerobic yeast fermentation systems, *Biotechnol. Bioeng.*, 58(4), 445–450, 1998.

65. Yabuki, Y., Nagasawa, T., and MacGregor, J., An industrial experience with product quality control in semi-batch processes, *Comput. Chem. Eng.*, 24, 585–590, 2000.

66. van Overshee, P. and de Moor, B., N4SID: subspace algorithms for the identification of combined deterministic–stochastic systems, *Automatica*, 30, 1, 75–93, 1994.

12 Supply-Chain Management

Nilay Shah and Conor M. McDonald

CONTENTS

12.1 INTRODUCTION

12.1.1 DEFINITION OF SUPPLY CHAIN

A supply chain encompasses a firm's or several firms' facilities and resources and the relationships between the facilities and resources required to plan, manage, and execute:

- The sourcing of materials
- The making of intermediate and final products
- The delivery of end products to customers

The Supply-Chain Council (www.supply-chain.org) defines the supply chain as "All interlinked resources and activities required to create and deliver products and services to customers." (See Figure 12.1.)

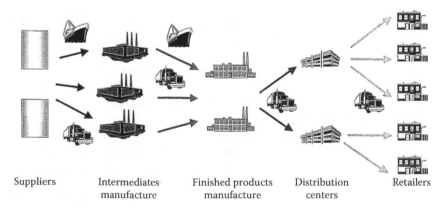

| Suppliers | Intermediates manufacture | Finished products manufacture | Distribution centers | Retailers |

FIGURE 12.1 Example of a supply chain.

12.1.2 SUPPLY-CHAIN MANAGEMENT

The term *supply-chain management* (SCM) has been broadly used in the literature with various meanings due to the development of the philosophy from various perspectives: purchasing and supply, transportation and logistics, marketing, and level of coordination. Though different in origin, these theories have now merged into a holistic and strategic approach to *operations, materials, and logistics management*. A definition given by theSupplyChain.com (http://www.thesupply-chain.com) is:

> SCM is a strategy where business partners jointly commit to work closely together, to bring greater value to the consumer and/or their customers for the least possible overall supply cost. This coordination includes that of order generation, order taking and order fulfillment/distribution of products, services or information. Effective supply chain management enables business to make informed decisions along the entire supply chain, from acquiring raw materials to manufacturing products to distributing finished goods to the consumers. At each link, businesses need to make the best choices about what their customers need and how they can meet those requirements at the lowest possible cost.

In other words, the purpose of SCM is to deal effectively with external strategic changes, such as globalization, and operational uncertainties, such as demand fluctuations, in order to take advantage of any new opportunities and to drive down the overall supply costs. This is achieved through effective management of production, inventories, and distribution.

Supply-chain management is sometimes confused with logistics. The Council of Logistics Management supplies this definition to make the distinction clear:[1]

> Logistics is that part of the supply chain process that plans, implements and controls the efficient, effective flow and storage of goods, services, and related information from point-of-origin to the point-of-consumption in order to meet customer requirements.

Anderson et al.[2] identified seven key principles of supply-chain management:

1. Segment customers based on the service needs of distinct groups and adapt the supply chain to meet the needs of these segments profitably. For example, "type A" customers might be blue-chip companies that buy large quantities of product often. Stocks should always be available for their products. Type B might be medium-size companies that buy smaller amounts often. Stocks should be available, say, 95% of the time, and material might be made available within a week if they are not immediately available. Type C might be independent distributors (who might hold their own stocks). Stock should be available, say, 90%

of the time for them and available within 2 weeks if not immediately available.

2. Customize the supply-chain and logistics network to the service requirements and profitability of customer segments; for example, it is not necessary to use air freight for type C customers.

3. Monitor market signals and align demand planning accordingly across the entire supply chain (i.e., make all information available upstream) to ensure consistent forecasts and optimal resource allocation.

4. Differentiate the product as close to the customer as possible (a good example is paint mixing kits at do-it-yourself stores) and speed up conversion across the supply chain.

5. Manage sources of supply strategically to reduce the total cost of owning materials and services; for example, it is not necessary to hold vast stocks of raw materials if a supplier can provide daily deliveries reliably.

6. Develop a supply-chain-wide technology strategy that supports multiple levels of decision making and gives a clear view of the flow of products, service, and information. Visibility is important if good-quality decisions are to be made.

7. Adopt channel-spanning performance measures to gauge collective success in reaching the end user effectively and efficiently. Often, performance measures traditionally used at individual nodes of the chain are in conflict with each other.

The main elements of supply-chain management, using the Supply Chain Operations Reference (SCOR) model (see www.supply-chain.org) are:

- *Plan.* Strategic or tactical — work out how the enterprise is going to meet demand for the product.
- *Source.* Select suppliers and ensure raw material availabilities; define relationships and develop inbound logistics procedures.
- *Make.* Coordinate the interlinked manufacturing activities often involving complex balancing of resources to manufacture the products and scheduling flexible facilities (see Chapter 10 of this book).
- *Deliver.* Coordinate activities from manufacturing site to customers through the logistics (storage and transportation) infrastructure.
- *Return.* Receive faulty products from customers (best avoided but occasionally unavoidable).

12.1.3 PROCESS INDUSTRY SUPPLY CHAINS

Process companies often sit in the middle of wider supply chains and as a result traditionally perform differently from companies operating at the final consumer end of the chain. Figure 12.2 indicates where products of the European chemical industry end up. In our experience, supply-chain benchmarks for the process

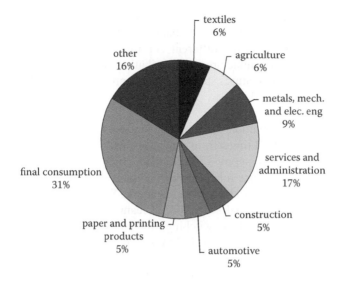

FIGURE 12.2 Process industry supply chains. (Data obtained from www.cefic.be.)

industries do not measure up well when compared with other sectors (e.g., automotive, computers). Examples of such benchmarks in the batch processing sector include:

- Stock levels in the entire chain ("pipeline stocks") typically amount to 30 to 90% of annual demand, and the chain usually contains 4 to 24 weeks' worth of finished good stocks.
- Supply-chain cycle times (defined as elapsed time between material entering as raw material and leaving as product) tend to lie between 1000 and 8000 hours, of which only 0.3 to 5% involve value-adding operations.
- Material efficiencies are low, with only a small proportion of material entering the supply chain ending up as product (particularly fine chemicals and pharmaceuticals, where this figure is 1 to 10%).

Further benchmarks are provided in Section 12.6.

Process industry supply chains, involving manufacturers, suppliers, retailers, and distributors, are therefore striving to improve efficiency and responsiveness. For world-class performance, both the network and the individual components must be designed appropriately, and the allocation of resources over the resulting infrastructure must be performed effectively. The process industries have been hampered in this quest by both intrinsic factors (e.g., the need to influence processes at the molecular level, wide distributions of asset ages) and technological factors (e.g., availability of tools for supply-chain analysis). Many of the reasons for poor performance relate to details of process and plant design and the prevailing economic orthodoxies when key decisions have been made. It is often difficult to effect large improvements simply by changing logistics and transactional processes; fundamental changes at the process and plant level and at the interfaces between the different constituents of the value chain, from product discovery to manufacture and distribution, are often required.

Additionally, some features specific to batch processes that influence supply-chain performance include:

- Multistage production of complex chemicals is often the norm. This implies long lead times from raw materials to products, which makes it difficult for the supply chain to track volatile demands efficiently.
- Plants often produce relatively low volumes of each product and must produce multiple products to be economic. Such flexible plants require effective planning and scheduling. Cleaning or changeover activities may be required when switching products which leads to campaign operation, which in turn again limits responsiveness.

These two features mean that a quick response is difficult, as is make-to-order production, which places pressure on forecasting and planning and limits supply-chain performance. Some companies get around this limitation by making final products (e.g., formulations) to order but making intermediates to forecast.

The process industries will also face new challenges in the future, including:

- Moving from a product-oriented business to a service-oriented business that provides life-cycle solutions for customers
- More dynamic markets and greater competition, with shorter product life-cycles
- Mass customization (trying to deliver specialty products at commodity costs)
- The need to evaluate, report, and improve sustainability and environmental and social impacts throughout the supply chain and aiming to anticipate and respond to future regulation and compliance requirements (e.g., recovery and recycling of consumer products at end of use)

12.1.4 DIFFERENT VIEWS OF THE PROCESS INDUSTRY SUPPLY CHAIN

First of all, it makes sense to define what is meant by the process industry supply chain. Most companies, and indeed researchers, tend to employ a company-centric view of the supply chain, where the supply chain is seen as consisting of the enterprise in question as a central entity, possibly together with some peripheral partners, typically first-tier suppliers and customers.[3] These views involve the integration of production and logistics planning across the enterprise, value-chain management, global network planning, and investment appraisal. Much less work has been reported on the "extended" supply chain, where the view is much broader in that it encompasses the suppliers' suppliers and the customers' customers. This is almost certainly due to:

- The relative youth of the discipline and the fact that considerable benefits can be achieved simply by the use of company-centric views of the supply chain
- A wariness of supply-chain "partners" and a lack of data sharing

12.1.5 TYPICAL SUPPLY-CHAIN PROBLEMS

Supply-chain problems may be divided into three categories:

- Supply-chain infrastructure (network) design and strategy
- Supply-chain planning
- Supply-chain operations and execution

The first two categories are essentially offline activities associated with establishing the best way to configure and manage the supply-chain network. The last category involves deciding how to operate the network to respond best to the external conditions faced by the supply chain.

To different people, "supply-chain problems" can mean different things. The family of problems in these three categories can be mapped onto a two-dimensional domain as shown in Figure 12.3. The vertical dimension is self-explanatory. The horizontal dimension intends to depict the customer-facing part of the supply chain at the rightmost end and the provision of primary resources at the leftmost end; the instances displayed are indicative and will differ from company to company (this example reflects a pharmaceutical supply chain). From the perspective of the firm, classes of problem may be defined by regions 1 to 15 or combinations of regions. Examples include:

- Redesign of the logistics network (regions 4 and 5) (i.e., a strategic activity looking primarily at warehouses and customers)
- Campaign planning at a primary manufacturing site (region 7)
- Real-time supply-chain management and control (regions 11 to 15)
- Negotiation of long-term supply contracts (region 1)
- Long-term manufacturing capacity planning and value-chain management (regions 1, 2, 3)

This chapter reviews important classes of problems in this domain, describes techniques reported for dealing with such problems, and highlights interesting applications using industrial case studies. Both "hard" technical and "soft" managerial and business process issues are described. The chapter then concludes with a view on future developments and challenges.

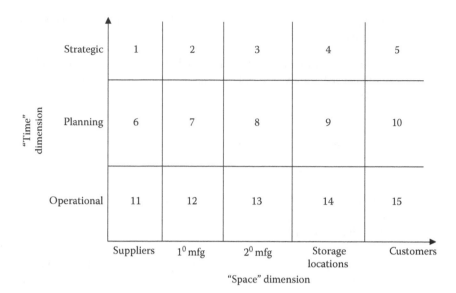

FIGURE 12.3 Supply-chain problem space.

12.2 SUPPLY-CHAIN NETWORK DESIGN AND STRATEGY

The concept of supply-chain network design is very broad and means different things to different enterprises; however, it generally refers to a strategic activity that will lead to making one or more of the following decisions:

- Where to locate new facilities (e.g., production, storage, logistics)
- Significant changes to existing facilities (e.g., expansion, contraction, closure)
- Sourcing decisions (which suppliers and what supply base to use for each facility)
- Allocation decisions (e.g., what products should be produced at each production facility, which markets should be served by which warehouses)

These decisions aim in some way to increase shareholder value. This means that models are employed to try to exploit potential trade-offs. These models may consider:

- Differences in regional production costs
- Distribution costs of raw materials, intermediates, and products
- Differences in regional taxation and duty structures
- Exchange rate variations

- Manufacturing complexity and efficiency (related to the number of different products being produced at any one site)
- Network complexity (related to the number of different possible pathways from raw materials to ultimate consumers)

Most companies do not aim to quantify the latter two explicitly but rather employ policies (e.g., single sourcing of customer zones, exclusive product–plant allocation) to simplify operations to the desired degree.

A relatively rare instance of this class of problems is the "greenfield" design of a new supply chain where no significant assets exist at the time of the analysis (e.g., design of a future hydrogen infrastructure). A more common instance occurs when part of the infrastructure already exists and a retrofit activity is being undertaken, during which products may be reallocated between sites, manufacturing resources may be restructured, the logistics network may be restructured, etc.

Models may be steady state or dynamic and may be deterministic or deal with uncertainties (particularly in product demands). Research in this field started very early on, with location-allocation problems forming part of the early set of "classical" operations research problems; see, for example, Geoffrion and Graves,[4] who considered the problem of distribution system layout and sizing and distribution center-customer allocation. It was recognized early on that systematic, optimization-based approaches should be used, and that common-sense heuristics might lead to poor solutions.[5] These early models tended to focus on the logistics aspects. Clearly, much more benefit could be achieved by simultaneously considering the production aspects.

An early example of a production–distribution network optimization study in the process industries was given by Brown et al.,[6] who considered the biscuit division of Nabisco. Their model involves the opening or closing of plants, the assignment of facilities to plants, and the assignment of production to facilities. The production model is based on the relative product–facility "yields." A thorough review of the work in this area was presented by Vidal and Goetschalckx,[7] who categorized previous work according to a number of characteristics, including:

- Treatment of uncertainties and dynamics and production and supplier capacity
- Ability to include single-sourcing restrictions
- Customer service and inventory features
- "International" features (e.g., taxes, duties)
- Number of echelons considered (see below)
- Cost nonlinearities, model size, and solution techniques

They concluded that features that are not well treated include stochastic elements, accurate descriptions of manufacturing processes (and hence capacity), international aspects, extended and multiple-enterprise networks, and solution techniques.

In general, the works reviewed above use fairly simple representations of capacity and treat all data as deterministic. Given that many of the plants under consideration are flexible and multipurpose with a wide product slate, a better representation of capacity and demand uncertainty is required for more accurate solutions. In general, the infrastructure design problem is posed as a mathematical optimization problem. The key variables include:

- Binary variables to represent the establishment or not of nodes at candidate locations
- Binary variables representing links between nodes (e.g., from factories to warehouses)
- Binary variables representing allocations of products to factories
- Binary variables reflecting the choice of manufacturing resources at each site
- Continuous variables reflecting the capacities of storage locations
- Continuous variables reflecting the production in each product over each period at each manufacturing site
- Continuous variables reflecting the inventories of each material at each node over each period
- Continuous variables reflecting the total flow of each material between every pair of nodes over each period

The key constraints include:

- Logical constraints ensuring that links only exist between nodes that are selected to exist
- Constraints on the total flow between nodes
- Resource constraints at each manufacturing site, ensuring that the production of materials over each period is constrained by the resources available
- Material balances for each material at each node, reflecting contributions due to production, receipts, and deliveries
- Warehouse capacity constraints

The optimization procedure requires an objective function that usually takes the form of cost minimization or net present value (NPV) maximization. It includes terms such as:

- Fixed infrastructure costs (fixed costs of establishment of facilities at candidate locations)
- Variable infrastructure costs (costs that depend on the scale of the facilities and their resource configuration)
- Manufacturing costs
- Material handling costs

- Transportation costs
- Sales revenues

Much work has been reported in the academic literature. Kallrath[8] addressed the issue of process and plant representation and described a tool for simultaneous strategic and operational planning in a multisite production network. Counterintuitive but credible plans were developed that resulted in cost savings of several millions of dollars. Sensitivity analyses showed that the key decisions were not too sensitive to demand uncertainty.

Sabri and Beamon[9] have also developed a combined strategic–operational design and planning model, with two interesting features: a multi-objective optimization procedure is used because of the difficulty of trading off very different types of objectives, and uncertainties in lead times as well as demands are treated. Tsiakis et al.[10] showed how demand uncertainty can be introduced in a multiperiod model. They argue that future uncertainties can be captured well through a scenario tree, where each scenario represents a different discrete future outcome (see next section). These scenarios should correspond to significant future events rather than just minor variations in demand. In their multipurpose production model, flexible production capacity is allocated among different products to determine the optimal layout and flow allocations of the distribution network.

All of the above works rely on the concept of fixed *echelons*; that is, they assume a given fundamental structure for the network in terms of the echelons involved (e.g., suppliers, manufacturing plants, warehouses, distribution centers, customers). Thus, a rather rigid structure is imposed on the supply chain, and the design procedure focuses on determination of the number of components in each echelon and the connectivity between components in adjacent echelons. However, changes in the fundamental structure of the network (e.g., the introduction of additional echelons or removal or partial bypassing of existing ones) may sometimes lead to economic benefits that far exceed what can be achieved merely by changing the number of components or the connectivity within an existing structure. Tsiakis et al.[11] extended this body of work by developing a general framework that integrates the different components of a supply chain without any *a priori* assumption as to the fundamental structure of the network. The framework uses the concept of a flexible, generalized production/warehousing (PW) node. These PW nodes can be located at any one of a set of candidate locations; they produce one or more products using one or more shared resources, hold inventories of the above products as well as any other material in the network, and exchange material with other PW or external nodes. The functions of these nodes, therefore, are not specified *a priori* nor is any flow network superimposed; rather, the node functionalities (production, storage, or both) and the flows between nodes are determined as part of the optimization. This approach tends to result in leaner networks, where storage capacity is established only where necessary. The flexible network structure also provides more scope for exploiting economies of scale in transportation.

12.2.1 MODELING UNCERTAINTY THROUGH SCENARIOS

Process industry companies operate a wide variety of assets, with widely varying ages and expected lifetimes. At any given time, the decisions relating to investment in infrastructure include how best to configure assets at existing sites and whether or not to establish new sites. These are tied in with production and inventory planning. The main issue associated with investment planning is that capacity-related decisions have impacts far beyond the time period over which confidence in data exists. Hence, decisions must be made in the face of significant uncertainty relating in particular to the economic circumstances that will prevail in the future.

12.2.1.1 Traditional Approaches

Most companies approach capacity planning through the use of NPV-based analysis. The data used in this kind of analysis are essentially the *expected* values of key parameters (e.g., product demands and life-cycles, product and raw material prices). The NPV approach is used to define the benefits of a *nominal* plan. This nominal plan is then subjected to sensitivity analysis where each of the key parameters is perturbed and its effect on the project assessed. Any problems identified through the sensitivity analysis are removed as far as possible. A slightly more sophisticated approach develops separate optimal plans for a number of possible scenarios. It then selects the plan that relates to the most likely scenario or performs reasonably across all scenarios. These approaches have a number of serious problems:

- Risks are not clearly identified and evaluated.
- It is assumed that all the decisions must be made up front; no provision is made for contingent decisions that may be made at a later stage in light of additional information that may become available by then. Even less so does it consider a proper accounting for the interactions between present and future decisions; for example, whether or not extra plant capacity will have to be added in 5 years' time is a decision that does not actually have to be made right now; rather, it can wait until future information on product demand becomes available. On the other hand, decisions on whether or not to actually establish the plant and on its initial capacity may have to be taken immediately. However, these immediate decisions may depend on the flexibility to add extra capacity at later stages.
- The future scenarios are very often not thought through clearly and in a systematic manner that covers all major eventualities.
- In practice, the "best" plan is often not optimal with respect to any one scenario but reflects a compromise between risk and expected return; hence, this approach will not identify the best plan.

12.2.1.2 The Scenario-Based Approach to Infrastructure Planning

The scenario-based approach takes systematic account of the uncertainty inherent in medium- and long-term decision making in capacity planning and the associated capital investments. Coupled with a sufficiently detailed model of the manufacturing processes involved, it allows a rational evaluation of the complex trade-offs between current and future decisions, taking account of all relevant capital and operating costs. The scenario-based approach divides the planning horizon into a number of stages. The boundaries between successive stages reflect points in time where important uncertainties will be resolved. These often relate to important external events (e.g., economic, geopolitical) or important internal events (e.g., launches of new products, patent losses on existing products). Normally, the number of scenarios is kept small, the aim being to capture only the most important events. Schoemaker[12] explains how to construct scenarios.

12.2.1.3 An Illustrative Example

Suppose, for example, that a company produces a particular polymer. A major consideration in planning investment decisions is the market demand for this product:

- In the near future (i.e., for the first 2 years), the level of demand (D) is fairly steady.
- After about 2 years, there is a 50% chance that legislation will be introduced that will make the polymer an important component of the automobile industry. This will increase demand by 50%.
- After a further 2 years, however, there is a 50% probability that a similar polymer will be produced by a competitor. If this happens, this will take up about half the increased demand from the automobile sector.

The above situation gives rise to a multistage scenario tree as illustrated in Figure 12.4, where each branch is characterized by:

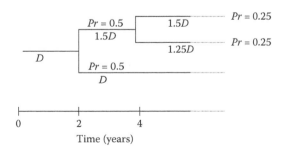

FIGURE 12.4 Scenario tree for polymer demand.

- A probability of occurrence;
- An expected annual level of demand.

We now have three important sets of decisions associated with each decision-making time period and each branch on the tree:

- Whether or not to invest in new capacity and, if so, how much
- How much material to produce
- How much inventory to carry over to the next period

In some decision-making models, the amounts produced over each branch are assumed to be equal to demand; this effectively removes the inventory decisions and the additional flexibility of action available to the decision maker. If a maintenance strategy is important, then maintenance decisions and their effect on equipment availability would be included.

The main advantages of these types of models include the following:

- Although all possible decisions are considered, only the decisions that must be made immediately (e.g., the initial capacity investment and production plan for the near future) have to be made; other decisions are automatically deferred until more information becomes available.
- Risk measures can be introduced and enforced.
- More appropriate optimization metrics can be used (e.g., expected NPV).

12.2.1.4   Data Required by the Scenario-Based Approach

The scenario-based approach requires the following data:

- A manufacturing model of the processes that is of sufficient detail to predict:
  - The relationship between the availability of manufacturing resources (e.g., production equipment, utilities) and the production capacity of the plant
  - The influence of the main operating decisions (e.g., throughput) on operating cost and on the important (e.g., operability and environmental) constraints that the plant operation has to satisfy
- Major current and future uncertainties confronting the plant and corresponding scenario tree data (e.g., market sizes, expected sales value)
- Initial plant configuration
- Possible alternatives for investment in additional plant equipment and associated data (e.g., capital costs, internal rates of return)
- Where relevant, the expected degradation of equipment over its lifetime

12.2.1.5 Objectives of the Scenario-Based Approach

The objective of the scenario-based approach is to optimize some performance measure that explicitly takes account of the inherent uncertainty as expressed in terms of the postulated scenarios. A popular objective is maximization of the *expected* NPV; however, pursuing this objective may result in decisions that are excessively risky — for example, involving a very considerable downside under certain reasonably probable scenarios. It is therefore advisable to impose additional constraints reflecting risk. For example, it may be demanded that the probability of an NPV less than a certain (acceptable) value does not exceed 0.1. Thus, the scenario-based approach provides a powerful tool for the decision maker to make informed decisions on the trade-offs between potential reward vs. risk. From the mathematical point of view, the scenario-based approach results in a mathematical programming optimization problem. The complexity of the latter depends primarily on:

- The complexity of the underlying process model; for example, nonlinear models are more complex than linear ones, and models involving discrete decisions (as is often the case with flexible multipurpose plants) are even more complex
- The number of postulated scenarios

12.2.1.6 Results Produced by the Scenario-Based Approach

The scenario-based approach determines simultaneously:

- The optimal capital investment in new capacity that must be put in place by the start of each planning stage under each and every scenario
- The optimal operating policy (e.g., plant throughput) and sales volume throughout each planning stage of every scenario
- The inventory amounts that should be carried over from one stage to the next in each branch of the scenario tree
- Where relevant, the maintenance strategies for different items of equipment.

Note that, while the (postulated) product demand during any particular stage provides an upper bound on the (optimal) sales volume, the two are not necessarily equal. Moreover, the sales volume during a certain stage is not necessarily equal to the production rate given the possibility of inventory carryover from earlier stages.

12.2.1.7 A Case Study from the Pharmaceutical Industry

The pharmaceutical industry has a pipeline of potential new products. Some may be close to launch and some may be in the early stages of clinical trials. Considerable uncertainty exists as to whether products in clinical trials will ever be

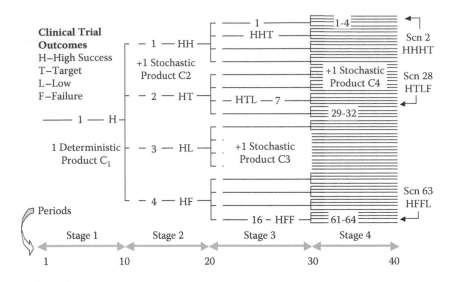

FIGURE 12.5 Scenario tree for pharmaceuticals case study.

launched. Due to the low production volumes, it is unusual to dedicate plants to products; multipurpose plants are used instead. The key problem, then, is to develop a capacity expansion plan for a multipurpose plant given the nature of products in current production and potential products in the pipeline.

The scenario tree for the case of four products is shown in Figure 12.5. Product C1 is in manufacture already; product C2 will be the first to complete clinical trials (at time period 10). The trials have four possible outcomes, each of which will imply a different demand profile for the product. The trials for product C3 are complete at time period 20 and so on. A total of 64 scenarios are identified, as shown in Figure 12.5.

The optimal solution is obtained with the scenario-based approach, as outlined in Section 12.2.1.2 above. This is arrived at by maximizing the expected value of NPV, subject to some constraints on the worst possible downside. It is interesting to consider the distribution of NPVs for the 64 scenarios that were considered (see Figure 12.6). First, we observe that the distribution is not particularly symmetrical or of a Gaussian form; this is quite typical and provides part of the reason why simple risk measures such as variance are not appropriate. For example, it can be seen that:

- The expected NPV is $120 million.
- The slight chance of losses of up to $110 million on an NPV basis (the leftmost bar) is the downside of the proposed investment plan.
- The reasonable chance of an NPV above $250 million is the upside of the proposed investment plan.

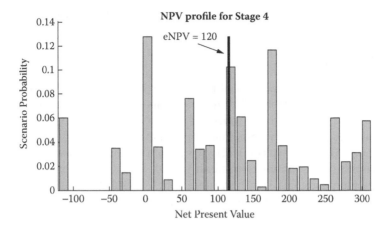

FIGURE 12.6 Distribution of NPVs in optimal solution.

12.2.2 AN INDUSTRIAL APPLICATION OF NETWORK DESIGN

An industrial application was described by Camm et al.,[13] who worked on the restructuring of Procter and Gamble's North American supply chain. A year-long project involving integer programming, network optimization, and geographical information systems (GISs) was responsible for streamlining the U.S. manufacturing and distribution operations with annual savings of $200 million. The initial network was comprised of 50 product lines, 60 plants, 10 distribution centers, and hundreds of customer zones. A number of factors made this initiative particularly timely, including deregulation, brand globalization for production economies, higher plant reliabilities and throughputs, and excess capacity from a series of acquisitions.

Product *sourcing* (i.e., the allocation of products to manufacturing sites) was the focus of the study by Camm et al., who also addressed distribution network design. Rather than develop a single comprehensive production–distribution optimization model, they decomposed the problem into a product–plant allocation problem and a distribution network design problem. Raw material and manufacturing costs tended to dominate, so the product sourcing problem was the more important of the two and was relatively independent of the distribution network design because 80 to 90% of production is shipped directly to customers rather than passing through P&G's distribution network. A family of solutions to the distribution network design problem is then made available to the product sourcing model. This simply allocates production to plants to minimize overall costs. The problem is solved as a capacitated network flow problem, with a very crude production model (each plant simply constrained in terms of total annual production across all products). The authors made the point that being able to visualize the outputs of large-scale models (via GISs, in this case) is important for their credibility. Even with such a simple representation of site capacity, large savings (particularly in terms of manufacturing costs and the removal of excess capacity) were identified.

12.2.3 REMARKS

It is clear that a very large amount of work has been undertaken to address the infrastructure design problem. The results of this work have made their way into software tools provided by a variety of vendors, including Advanced Process Combinatorics (www.combination.com), Aspen Technology (www.aspen-tech.com), SAP (www.sap.com), i2 (www.i2.com), Gensym (www.gensym.com), Manugistics (www.manugistics.com), and Process Systems Enterprise (www.psenterprise.com). However, outstanding issues that provide challenges for ongoing research and technology development include:

- It has not really been concluded what an adequate description of manufacturing processes is at this level and what the potential benefits of including more detail on the manufacturing process might be. In the case study above, significant benefits were achieved with a low level of resolution; subsequent studies may require more detail.
- The issue of complexity vs. productivity has not been dealt with effectively. Generally speaking, as a manufacturing site becomes more complex in terms of numbers of products per unit of manufacturing resource, the productivity of the manufacturing resources decreases.
- The international nature of many supply chains provides additional opportunities for optimization, especially when considering features such as transfer prices, taxes, royalties, and duties. Combined financial and production–distribution models should be considered (see Shapiro's[14] review of strategic planning).
- Most research still has the enterprise envelope as the boundary conditions. Coordinated optimization across the extended supply chain should result in significant benefits (see, for example, Lin et al.[15]).
- The full range of uncertainty is not explored (e.g., raw material availabilities and prices, product prices, international aspects)
- Perhaps most importantly, from the process engineering perspective, is that no connection has been made between process design and supply-chain operation. We have seen many examples where process design has compromised supply-chain operation (see, for example, Shah[16]). Backx et al.[17] concur, and have introduced the concept of *supply-chain conscious process operation*. Process design for supply-chain efficiency will be an important future research area. For example, attempting to operate processes at or near intrinsic rates will increase manufacturing velocities significantly and improve supply-chain responsiveness. Some of the methods described in Section II of this book can be used to support this activity.

12.3 PLANNING PROBLEMS

Supply-chain planning is concerned with making sure that the right product is available at the right place at the right time with the right quality and in an economic fashion. It considers a fixed infrastructure over the short to medium term and is therefore concerned with the coordinated planning of the manufacture of intermediate and final products, distribution of material, and management of inventories.

12.3.1 ELEMENTS OF SUPPLY-CHAIN PLANNING

The main elements of supply-chain planning are:

- *Demand management* — Estimating and even influencing the demands for products in each region where the firm operates
- *Inventory management* — Determining strategies to supply customers and replenish stocks
- *Production and distribution planning* — Making products in the right place and time and sending them to the right place at the right time

The objectives of planning are quantified through *metrics*. These are usually economic or service based. Typical economic metrics include the following goals:

- Maximize profit or economic value added (EVA) throughout the network. This requires considerable data; supply-chain-wide, activity-based costing can support this (see, for example, Shapiro[18]).
- Maximize inventory turns (measured as annual turnover/average inventory). This is a measure that focuses on working capital. Values of, say, 2 to 8 would be considered poor and values greater than, say, 16 would be considered good.
- Maximize return on assets (combining both fixed and working capital).

Typical customer service-level measures include:

- Minimizing the percentage of due dates missed (or maximizing the number of orders met on time in full)
- Maximizing the fill rate, where customer orders are filled on time, but the amounts may fall short (total amounts filled divided by the total amounts requested)
- Maximizing the availability of the product in a particular location (days available divided by total number of days)

12.3.2 DEMAND MANAGEMENT

Demand management means the planning and execution of activities related to getting final products to the customers. The two different ways of operating the demand management process are:

- Order driven — Products are made or customized according to actual orders and the demand management process is all about a quick response to actual orders.
- Forecast driven — When it would take too long to make things in response to orders they must be made according to some forecasts and placed in storage; here, the demand management process is all about accurate forecasts and clever inventory management policies.

The order-driven mode is primarily relevant toward the final customer end of the chain in the process industries and is probably the less common of the two in this sector. The latter tends to dominate, either due to the lack of visible order data or long in-process times requiring demand satisfaction from stock to avoid long lead times. Demand forecasting is therefore a very important component of supply-chain planning in the process industries and is briefly covered next.

12.3.2.1 Forecasting

The accurate forecasting of future demands is a critical element of supply-chain planning. It does not matter how efficient production and distribution plans are with respect to their use of resources if they are aligned against forecasts that are inaccurate. Demand forecasting is not an exact science, but the application of systematic methods can give rise to forecasts with an accuracy of greater than 90% in many sectors if applied wisely. Quantitative forecasting methods are divided into autoprojection and explanatory methods. Autoprojection assumes that the past is a good predictor of the future and uses some function of past demands to predict future demands. Explanatory methods assume that a model of the demand process can be built and its parameters estimated. The details of these methods are beyond the scope of this chapter. Texts such as that by Makridakis and Wheelwright[19] provide excellent details. Most enterprise resource planning (ERP) systems include demand management modules. Specialist providers include Manugistics (www.manugistics.com), Mercia (www.mercia.com), Prescient (www.prescient.com), and txt (www.txt.com).

12.3.2.2 Performance Measurement for Demand Management

Because demand forecasting represents one of the most important and central activities in the process industries, key elements that must be present in a successful business process are accountability for the quality of the forecast, a robust process for measurement of forecast accuracy, and active monitoring of performance in order to drive continuous improvement. It is surprising to observe how often these basic building blocks are weak or missing in industrial practice, especially given the critical influence forecasting has on a large number of business decisions. Many different methods are used to measure forecast accuracy, and some of them employed in practice are flawed. A plot of the forecast vs. shipments should ideally show points scattered along the 45° line. Consider an example where we have shipped 100 units of a single product in a given time period, and we are measuring percentage forecast error. A represents the actual quantity shipped equal to 100, and F is the forecast for that period. Here are some measures of percentage forecast error used in practice:

$$\frac{|F - A|}{A}$$

$$\frac{|F - A|}{F}$$

$$\frac{1 - \min\left(F, A\right)}{1 - \max\left(F, A\right)}$$

$$\frac{A - F}{F}$$

$$\frac{F - A}{F}$$

As can be seen, a very common error is to use the units forecasted in the denominator. Those advocating this approach desire a measure bounded within ±100%; however, we are then measuring forecast error against a variable within the control of the business. Also, for situations where the forecast is far greater than the quantity shipped, the error will be understated and will not scale well as the error increases.

To illustrate these concepts, observe Figure 12.7, which plots different measure of forecast error for a single product where the actual shipments in a period were 100 units. It can be seen that the first measure above, the mean absolute percentage error (MAPE), is linear over a wide range of forecasts, and punishes over-forecasting in a uniform manner. The other measures do not significantly punish severe over-forecasting (forecasts above twice the level of shipments). For example, the error measure is 67% when 300 units are forecast and 75% for 400 units, using the first equation. If we plot the forecast accuracy (100% − weighted

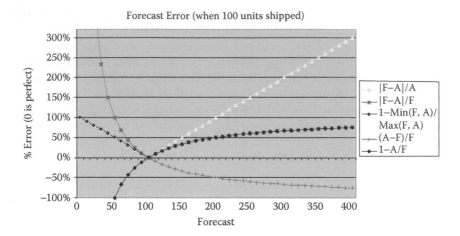

FIGURE 12.7 Forecast errors using different measures.

forecast error), then we can see that, for significant over-forecasting situations, the MAPE measures are –100% and –200% when we have over-forecasted by twice and three times the level and shipments, respectively. This is one of the reasons why the other measures are preferred over MAPE in practice, as a conceptual struggle can arise with regard to accepting a negative forecast accuracy measurement, even though it means merely that we are over-forecasting by at least a factor of 2.

The above was developed for a single product. Now, we will focus on a measure suitable for an entire business with many products, markets, and customers. First, we must determine the attributes for which forecast accuracy will be measured. An attribute might be a product group or a stock-keeping unit (SKU); a country or a continent; a market segment or a corporate division. Second, we must determine how far forward the forecast will be made. The total supply-chain lead time is the most important element in this decision; therefore, for an agrochemical with a year of manufacturing lead time, an appropriate measure would be the forecast made one year in advance, but, for the manager of a drug distribution center, the best measure may be the SKU forecast made one week in the future. Obviously, many combinations of attributes and look-ahead periods could generate a large number of measures of forecast accuracy. A disciplined process of pruning should be enforced so the metrics are not devalued due to their excess.

The error for a forecast made $t - n$ periods ago for period t with a defined set of attributes (a) is:

$$E_{a,t-n} = F_{a,t-n} - A_{a,t}$$

For a business with a wide portfolio of products and markets, an appropriate measure of performance is the *weighted mean absolute percentage error* (wMAPE):

$$wMAPE_t = \frac{\sum_a \left| E_{a,t-n} \right|}{\sum_a A_{a,t}}$$

where a represents any attribute such as product, geography, market segment. Note that a can easily include a set of time periods, allowing meaningful measures to be taken over some time horizon. For example, we may select month, product, and region as our set of attributes to measure forecast accuracy. As long as we measure the error at this level of attributes, we can measure forecast accuracy at higher levels of aggregation; therefore, if we wish to know how accurate our monthly forecasts were last year, the error at the monthly level is measured, and the error for each month is summed over the entire year.

Weighted MAPE, therefore, represents the best measure of forecast accuracy. Forecast accuracy can then be calculated as:

$$FA_{t-n} = 100\% - wMAPE_{t-n}$$

Once the measurement has been decided, then the metric must be made a highly visible one, and approaches developed to continuously improve it.

Other key measures are forecast bias and percentage forecast bias. These bias measures reveal if we are consistently over- or under-forecasting. For businesses without an adequate measurement and monitoring process, the probability is greater that bias will be present. It is defined for a set of selected attributes (a) for the unscaled and scaled case, respectively, as:

$$BIAS_a = \frac{\sum_{t=1}^{n} E_{a,t}}{n} \quad \text{and} \quad \%BIAS_a = \frac{\sum_{t=1}^{n} E_{a,t}}{\sum_{t=1}^{n} A_{a,t}}$$

Because bias may be negative or positive, it is difficult to produce meaningful measures for a business in total, given that the bias at the attribute level may cancel out when added up across all levels. In this case, a useful plot is that of percentage bias by item plotted in order of decreasing volume.

Another key indicator of difficulty in forecasting is the *coefficient of variability*, defined as:

$$cv = \frac{\sigma}{\mu}$$

We can calculate this in two ways. The first is simply to take counts of actual shipments in each time period and evaluate the mean (μ) and standard deviation (σ) of the data. This measure of cv reflects demand volatility. As a rough rule of thumb, values of cv for actual shipments greater than 1 indicate highly variable shipment patterns that will represent a challenge to forecast using either naïve or causal models.

The other calculation involves the same value for the mean (i.e., the average demand), while σ is taken to be the standard deviation of the forecast errors (i.e., E_{at} above). Generally speaking, values of 0.2 and lower for this measure indicate reasonably accurate forecasts.

Another measure is the *tracking signal*, which helps track the quality of the forecast we are making. It is defined as:

$$TS_a = \frac{BIAS_a}{\sigma_a}$$

Values for the tracking signal that are outside some arbitrarily chosen range then indicate that forecasting must improve for a particular item. Some authors use the *mean absolute deviation* instead of the standard deviation. In a real-world business environment, it may be difficult to tell when a tracking signal is giving useful alarm information or whether data artifacts have impacted its applicability. In principle, it does provide the impetus to explore further what is driving the signal.

12.3.2.3   Beyond "Passive" Demand Management

Traditionally, demand forecasts have been based on historical patterns. These are the best estimate of what would happen to the products when the firm does not take any additional action ("baseline forecasts"). The trend today is toward active demand management. The details of this approach are beyond the scope of this chapter but relate closely to marketing and economics. If, for example, it is found that the baseline forecasts result in an optimal balance of supply and demand from a fixed and variable cost perspective, it makes sense to use these. If, on the other hand, there is an anticipated oversupply, then a strategy involving either promotion or price reduction or both might be employed to optimize the overall performance of the business. Similarly, for excess demand, an optimal pricing strategy might be employed that in turn relies on a reasonable model of demand elasticity. Of course, these tactics can then be deployed at a regional level as well and might have to consider the activities of competitors. Some firms (e.g., those operating petrol forecourts or retailers of consumer goods) employ such dynamic pricing strategies almost on a daily basis. Additionally, leading-edge supply-chain partners now undertake collaborative forecasting, where real-time data on the latest forecasts at the customer-facing end of the chain are shared throughout the chain, removing much of the guesswork involved on the part of the nodes at the upstream part of the chain.

12.3.3   Inventory Management

The next stage of demand management is to identify how to relate the forecasts to the amounts of material available. For finished goods, this means identifying how demand forecasts will affect future amounts of stock and whether production is required. For intermediates it means working back from products to identify production requirements. For raw materials, it means working back to identify purchasing requirements.

12.3.3.1   Finished Goods Inventory Management

Assuming that demand is met from stock, a process of finished-goods inventory management would take place at the storage locations (warehouses/distribution centers) and would usually be along the following lines:

- First set safety stock levels and review periods.
- At each review period, consider current stock levels, forward demand forecasts, and current manufacturing plans to update the demands placed on the plant for future manufacturing plans.

12.3.3.2   Production Campaign Optimization

The simple economic order quantity (EOQ) concept is commonly used to determine the optimal batch size in equipment run semicontinuously by balancing changeover and inventory holding costs. While it can be a powerful tool, it is typically controversial when it comes to actual application. One of the primary issues in this regard is that it is trying to find the optimal trade-off between the balance sheet (working capital) and the income statement (plant transition costs). In this section, we show how this trade-off can be highlighted for management so appropriate decisions can be made for the corporation's specific needs at a specific point in time. Some have advocated abandoning the EOQ approach, and it is agreed that it never should be used in isolation but should only be viewed as a guide to production planning.

Higher changeover costs mean the product wheels run faster and inventory levels are lower. On the other hand, if we make fewer product transitions, then average inventory levels will rise. The EOQ formula is given as:

$$EOQ = \sqrt{\frac{2dC}{h(1-d/R)}}$$

where:

    EOQ   = economic order quantity, the number of units to produce during each run (i.e., lot-size).

    d = forecasted period (e.g., 6 months) demand by SKU, expressed as units per unit time.

    R = rate of product on a machine expressed in the same per time unit as d (for example, for demand per day, R = rate per day).

The critical parameters from a financial perspective are:

- Changeover cost (C) and any incremental costs that can be associated with the order, made up of such items as:
  - Foregone contribution margin (revenue–waste revenue), if capacity is constrained.
  - Production scrap (variable–waste revenue), directly associated with the machine setup if capacity is not constrained.
- A common error here might be to allocate the cost of fixed labor to a transition. Only incremental costs relevant to the decision at hand should be included.
- Carrying costs per unit per day (h):
  - Interest paid on loans (which could be repaid if inventory was reduced to free up cash) or, if the corporation is debt free, the estimated return that could be obtained if the money tied up in inventory was invested; however, many practitioners believe that the proper cost is the return on investment that management expects on all forms of capital (i.e., the firm's cost of capital).
  - Costs of insurance directly related to the total value of the inventory.
  - Any taxes that must be paid on the value of the inventory.
  - Storage costs, which include only costs that are variable based on inventory levels; if storage space reductions would not result in further inventory decreases, then all storage costs would be included in the EOQ calculation.
  - Risk factors associated with obsolescence, damage, and theft if they are incremental or the result of an increase in inventory levels.

An issue with EOQ is that it determines a single point at which a machine should be run; however, the cost curves for EOQ are often very flat in the region of the optimum. If we decide we wish to run within, say, 5% of the optimum, then we can see that we now have a very wide range in which we can run our campaigns without leading to excessive business costs. In fact, plant transition costs and working capital costs are split equally at the EOQ. The only difference as we move in either direction from the optimum is the relative makeup of these two costs on the objective function. It, therefore, can be presented to the decision maker as a trade-off curve. At different points in time, either objective will match better or worse with the corporate one; for example, at year end, a business may lean toward reducing working capital levels. The trade-off curve can help decide how far we can cut inventory before plant transition costs will start to ramp up unacceptably.

12.3.3.2.1 Valuation of Inventory for EOQ

Other potential benefits that could result from inventory reduction include:

- Reduced warehouse fixed cost
- Reduced waste due to reductions in aging out of inventory or inventory no longer needed
- Reduced freight costs
- Reduced plant manufacturing cost, if any

12.3.3.3 Setting Safety Stock Levels

Safety stock of a product is held at a location for three reasons:

1. To cover the time between an order for a replenishment being raised and the order being received (often referred to as *cycle stock*)
2. To buffer against uncertainties in the forecast
3. To buffer against supply unreliability

Items (1) and (3) can be considered together. Suppose that it notionally takes n days to replenish a warehouse with a product when an order is raised by the warehouse. Suppose also that the average demand is d, and that sometimes a delay occurs of about a day over and above the value of n. Then, the amount of cycle stock held may be $(n + 1) \times d$. This is the amount that would normally be held for a type A product/customer (see Section 1.2). Lower stocks might be held for other products and customers for which one or two days of stockout might be tolerated from time to time.

The amount of stock to hold to buffer against demand uncertainty again depends on the class of product. It might well be zero for a type C customer. For a type B customer, it might be zero if the uncertainty, which can be estimated either from the random component of the forecast (see Section 12.3.1.2) or from historical errors between forecasts and actual demands. For example, for a type A customer, the following relationship between forecast error and safety stock is recommended:

$$\text{safety stock} = nd + k\sigma_d n^{0.5}$$

where the value of k is chosen to guarantee a certain level of service (see Lewis[20]) and σ_d is the standard deviation of the forecast errors. It should be noted that a common error in practice is to use the standard deviation of actual shipments to calculate safety stock. This should only be used in situations where we are not able to calculate the forecast error. When variability in the lead time is significant, safety stock is determined as follows.

For the case of a supply chain with long lead times, an elementary inventory model can be developed. Let d represent the demand per day at a distribution center (DC); f is the frequency of replenishment per planning period from the supply node to the DC; L is the lead time in days from the supply node to DC; and P is the length of the planning period in days. The average DC stock is then given as:

$$I_{DC}^{av} = \frac{1}{2} \frac{P}{f} d$$

The more often we replenish, the lower the average inventory level will be. The average pipeline stock, which only depends on the lead time, is also given as:

$$I_{PIPE}^{av} = LT \ d$$

Lower and upper control limits can be developed for this scenario. If we define the time between replenishments (*TBR*) as *P/f* and the number of times a shipment arrives through the lead time as $\phi = LT/TBR$, then the lower control limit (LCL) and upper control limit (UCL) are:

$$LCL = d\left\{ LT + TBR\left(\lfloor \phi \rfloor - \lfloor \phi \rfloor_{TBR \leq LT} \right) \right\}$$

$$UCL = d\left\{ LT + TBR\left(\lfloor \phi \rfloor - \lceil \phi \rceil_{TBR \leq LT} + 1\big|_{\lceil \phi \rceil = \lfloor \phi \rfloor} \right) \right\}$$

12.3.3.3.1 Estimating Forward Demands on the Plant

A very common way of estimating forward demands comes under the general heading of *distribution requirements planning* (DRP). First, the time horizon is discretized into a number of equally spaced intervals (e.g., weeks). Basically, at each storage location on the supply chain, the following material balance calculation is then performed:

$$I_{pt} = I_{p,t-1} + R_{pt} - F_{pt} \ \forall \ p, t$$

where

I_{pt}	= inventory of product p over interval t.
R_{pt}	= Planned receipts at storage location from upstream node of product p over interval t.
F_{pt}	= Demand forecast for product p over interval t.

It can be seen that, unless there are already many planned receipts (which are essentially orders placed by the storage node on an upstream storage or manufacturing node), I_{pt} will progressively fall. Eventually, it will fall below the safety stock (I_{*p}). A new planned receipt must be triggered to arrive at this point, thus placing a new order on the upstream node. The size of this order will be based on three factors: the amount below the safety stock that I_{pt} has fallen to, a minimum order quantity (MOQ), or the EOQ as calculated in Section 12.3.3.2 and an order increment (OI). If MOQ $> I_{*p} - I_{pt}$ then the order size is the MOQ; otherwise, the order size is MOQ $+ n \times$ OI, where n is the smallest integer such that the order size $+ I_{pt} > I_{*p}$.

This new order is inserted as a planned receipt, and the material balance calculation proceeds from this time period until the stock falls below the safety stock, when a new order is generated. When the process is complete, a complete set of orders is ready for the upstream node. Sometimes the planning horizon is large and the orders are divided into *firm* (in the near future and fixed) and *planned* (in the more distant future and flexible; may be updated as demand forecasts improve).

A less sophisticated approach to managing inventories is based on reorder points and reorder quantities. Here, there is no forward prediction of material inventories; rather, material inventories are monitored periodically and if any is found to be below a set level (the reorder point), a new order is generated. The size of the order is determined by other parameters, typically either a standard reorder quantity or an "order-up to" level. These might be adjusted to fit order increments as above. The next step in the supply-chain planning process is to make sure the orders placed by the customer-facing nodes on the upstream nodes are actually met.

12.3.4 PRODUCTION AND DISTRIBUTION PLANNING

We now know what must be shipped to the extreme nodes in the supply chain. We must now determine what is produced, in what amounts, and how it will be shipped to storage locations. At this stage, we use planning models (which are essentially simplifications of the detailed single-site scheduling models such as those already described in Chapter 10 of this book).

Optimization methods have found considerable application here. A feature of these problems is that the representation of the production process depends on the gross margin of the business. Businesses with reasonable to large gross margins (e.g., consumer goods, specialties) tend to use recipe-based representations, where processes are operated at fixed conditions and to fixed recipes. Recipes may also be fixed by regulation (e.g., pharmaceuticals) or because of poor process knowledge (e.g., food processing). This is typical of the batch processing sector.

In academic research, process descriptions based on fixed recipes have been used to optimize production, distribution, and storage across multiple sites, normally using mixed-integer linear program (MILP) models. Wilkinson et al.[21]

described a continent-wide industrial case study that involved optimally planning the production and distribution of a system with three factories, 14 market warehouses, and over 100 products. It was found that the ability of the model to capture effects such as multipurpose operation, intermediate storage, and changeovers gave rise to counter-intuitive results, such as producing materials farther away from demand points than would be expected. Such an approach balances the complexity associated with producing many products in each factory with the extra distribution costs incurred by concentrating the manufacture of specific products at specific sites.

McDonald and Karimi[22] described a similar problem for multiple facilities that produce products on single-stage continuous lines for a number of geographically distributed customers. Their model is of multiperiod form and takes account of capacity constraints, transportation costs, and shortage costs. An approximation is used for the inventory costs, and product transitions are not modeled. They include a number of additional supply-chain-related constraints such as single sourcing, internal sourcing, and transportation times.

Kallrath[23] presented a comprehensive review on planning and scheduling in the process industry. He identified the need for careful model formulation for the solution of complex problems in reasonable computational times and described briefly how careful modeling and algorithm design can enable the solution of a 30-day integrated refinery scheduling problem.

Berning et al.[24] described a multisite planning–scheduling application that uses genetic algorithms for detailed scheduling at each site and a collaborative planning tool to coordinate plans across sites. The plants all operate batchwise and may supply each other with intermediates, thus creating interdependencies in the plan. The scale of the problem is large, involving about 600 different process recipes and 1000 resources.

Timpe and Kallrath[25] presented a mixed-integer, optimization-based multisite planning model that aims to give accurate representations of production capacity. It is a multiperiod model, where (as in Kallrath[8]) each unit is assumed to be in one mode per period, which enables the formulation of tight changeover constraints. An interesting feature of the model is that the grid spacings are shorter at the start of the horizon (closer to scheduling) and longer later on (closer to planning). The problem solved involved four sites in three geographical regions.

The approaches above assume deterministic demands. Gupta and Maranas[26] and Gupta et al.[27] considered the problem of mid-term supply-chain planning under demand uncertainty. Gupta and Maranas[26] utilized a two-stage stochastic programming approach, where production is chosen here and now while distribution decisions are optimized in a wait-and-see fashion. This makes sense, as production tends to be the main contributor to lead times. Gupta et al.[27] investigated the trade-offs between customer demand satisfaction and production costs by using a chance-constrained approach applied to the problem of McDonald and Karimi.[22]

12.3.4.1 Planning Models for Production and Distribution

As discussed above, many different planning models are available to solve this type of problem. Here, we will use a typical example.

12.3.4.1.1 Time Discretization

The time is discretized into a number of intervals of even duration. These are much longer than those used at the scheduling level (e.g., one week).

12.3.4.1.2 Degrees of Freedom

The key degrees of freedom are:

- The amount of each product produced at each location over each period
- The amount of each product shipped from one location to another over each period

When these have been fixed, the other variables can be deduced and the performance of the plan assessed.

12.3.4.1.3 Resource Constraints and Availability

Of course, the ability to produce and ship products is constrained by the resources available. Detailed scheduling models use the concept of equipment items that are used one at a time and utilities that are used up to a certain level to make sure that processing tasks do not consume more resource than available. At the planning level, a simpler concept is used. This is based on a unit amount of product being produced requiring a number of resource-hours of a set of resources.

To illustrate this further, consider the simple state–task network (STN) in Figure 12.8. The circles represent materials and the rectangles tasks. The times represent the processing time required to produce each state, and the fractions reflect the relative amounts of each material produced. A reactor of 10 tonnes is available for the React task, and a column with a capacity of 15 tonnes is available for the Distil task. Adequate storage is available for B. To produce, say, 100 tonnes of C would require $100/0.9 = 111$ tonnes of B, which implies 12 batches of React, which in turn will occupy the reactor for 36 hours. Similarly, 111 tonnes must be processed in the column, which means that 8 batches of Distil must be produced. The column will be occupied for 32 hours.

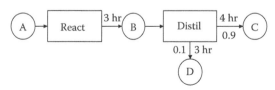

FIGURE 12.8 Sample state–task network.

Inverting this information, we get the following parameters:

$$u_{C,reactor} = 36/100 = 0.36 \text{ hr/t}$$

$$u_{C,column} = 32/100 = 0.32 \text{ hr/t}$$

In general, u_{plr} is the time utilized of resource r to produce a unit amount of product p at location l. To ensure that resource availability is accounted for, the number of hours a resource is available over a discrete time interval must be specified. For a piece of equipment with 100% availability and which operates all the time, the hours available will be equal to the duration of the interval. In other cases, the available hours will be less than this value.

Transportation resources may also be constrained in a similar fashion:

$$v_{pll'r} = \text{time utilized of resource } r \text{ to ship a unit amount of } p \text{ from } l \text{ to } l'$$

12.3.4.1.4   Cost Elements
The cost models can be more or less sophisticated. Here, we will assume that the main costs are variable and correspond to the cost of producing a unit amount of p at location l (CP_{pl}), the cost of storing p at location l (CS_{pl}), and the cost of transporting p from l to l' ($CT_{pll'}$). The latter may depend on the amount shipped, reflecting economies of scale in transportation.

12.3.4.1.5   Overall Model
The overall model is then formulated as follows:

- *Variables:*
  P_{plt} = amount of product p produced at location l over period t
  S_{plt} = amount of product p stored at location l over period t
  $T_{pll't}$ = amount of product p transported from l to l' over period t
- *Manufacturing resource constraints:* The amounts of each product produced must not cause more resources to be used than are available. We write constraints for each resource and time period as follows:

$$\sum_{p} u_{plr} P_{plr} \leq R_{rl} \quad \forall r,l,t$$

where R_{rl} is the availability (usually in terms of hours) of resource r at location l.

- *Inventory balance constraints:* The amount of p stored at location l over interval t is equal to that from the previous interval, plus any amounts shipped from other locations less that shipped to other locations. Because locations may be far apart, a transportation lead time from location l' to l, ($\tau_{l'l}$) must be taken into account:

$$S_{plt} = S_{pl,t-1} + P_{plt} + \sum_{l'} T_{pl'lt-\tau_{l'l}} - \sum_{l'} T_{pll't} - D_{plt} \quad \forall p,l,t$$

- where D_{plt} is the demand (actual order or forecast) of product p at location l and time t. It is likely that many of the locations in L_c do not actually have production capabilities, so it is the inbound transportation terms that will ensure that the correct amounts of material are available.
- *Objective function:* The cost elements have been described above. The objective is to meet the demands at the lowest possible cost. The objective is therefore to minimize:

$$z = \sum_{p,l,t} CP_{pl} P_{plt} + \sum_{p,l,l',t} CT_{pll'} T_{pll't} + \sum_{p,l,t} CS_{pl} S_{plt}$$

12.3.5 SUPPLY-CHAIN PLANNING: AN INDUSTRIAL EXAMPLE

Brown et al.[28] described a planning system used at the Kellogg Company that has two components:

- An operational one, which runs at a weekly level of detail and allocates production products to plants and optimizes distribution of materials between plants and from plants to distribution centers
- A tactical one, which operates at a monthly level of resolution and which sets plant budgets and guides capacity expansion and consolidation decisions

The Kellogg Planning System (KPS) covers the United States and Canada. It models the production of all the major products at Kellogg sites (five plants) and third-party sites (15 plants), as well as distribution to and storage at seven distribution centers (DCs). The system requires accurate manufacturing and transportation costs. Demands for products are met from DCs and plants. Complex rules govern the allocation of products to plants and the flow of products between plants. A large-scale linear program (LP) of the form shown in Section 12.3.4.1 is used for operational planning; constraints that balance the flows between the "making" and the "packing" parts of the process are also required. The forecasts are assumed to be deterministic, and safety stocks (usually 2 weeks' cover at the SKU level) are used to cope with demand uncertainty. A rolling horizon scheme is used whereby a 28-week horizon model is solved, with week 1 decisions already fixed, week 2 decisions being optimized for implementation, and weeks 3 to 28 subject to revision after a week of operation. This avoids the typical end-effects of cost-based optimization whereby production and stocks can fall to unsustainably low levels at the end of the planning horizon. Overall, the operational system is estimated to save $4.5m *per annum* in production, distribution, and inventory costs, and the tactical system has supported a capacity consolidation exercise that will lead to savings of about $35m *per annum*.

12.3.6 REMARKS

Supply-chain planning is maturing, and a number of tools are available from such process-industry-focused companies as Advanced Process Combinatorics (www.combination.com), Aspen Technology (www.aspentech.com), Finmatica (www.finmatica.com), and Process Systems Enterprise (www.psenterprise.com), as well as generic solution providers such as i2 (www.i2.com), Manugistics (www.manugistics.com), PeopleSoft (www.peoplesoft.com), and SAP (www.sap.com). Challenges for the future include:

- Financial and supply-chain planning must be integrated, especially in a global context.
- Most planning activities are still intra-enterprise. More work should be undertaken on multiple-enterprise (extended) supply-chain planning. For illustration, Figure 12.9 shows an order profile for a product of one of our collaborators. The dynamics are generated by their customer's reordering policy. What would be better: an optimized plan trying to meet hundreds of order profiles like this or a collaborative plan driven by smoother end-user demands?

12.4 SUPPLY-CHAIN OPERATION: EXECUTION AND CONTROL

The execution and control layer of the hierarchy is all about establishing effective systems and policies to ensure that the supply-chain operates effectively in real time and that the right product is indeed in the right place at the right time. Most companies now have in place *transactional* information technology (IT) systems (typically, ERP systems). These ensure that, once the key SCM decisions have been made, they are implemented smoothly and all relevant data are visible.

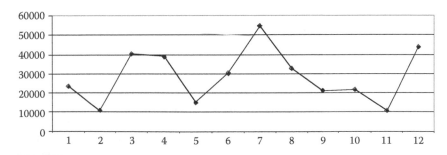

FIGURE 12.9 Order profile for one product.

However, the availability and quality of the decision-making components (so-called *analytical* IT) is much more variable. For example, it is very likely that the most commonly used SCM decision-making tool is Microsoft's Excel®, which may be used for making forecasting, inventory management, production, and distribution decisions. Issues such as organizational design and operational practice are important. Some ongoing debates in this sector include:

- *Leanness vs. agility and robustness* — The concept of leanness comes from the discrete manufacturing world and arises through pressures to reduce non-value-adding activities and keep working capital low. The objective is to aim for standard practices. This causes conflicts with another objective common in batch processes, which is to retain a high degree of flexibility so any product could in principle be made in short order. Hence, leanness can reduce agility and flexibility and compromise robustness. An emerging trend in the sector is to try to be lean at the front end, where significant volumes of intermediate materials are made, and to be agile at the back end, where lower individual volumes of a wide range of products are made.
- *Centralized vs. decentralized control* — Some supply chains (e.g., that of CISCO) have been successfully very tightly integrated and effectively centralized through technology; however, it is generally accepted in the process industries that robust, adaptive, real-time responses require a degree of decentralization with local decision making and control. Policies must be designed to control the storage, release, and replenishment of materials (based on the replenishment policies) and the sequence and timing of manufacturing batches at some local level, with periodic central supervision of the entire chain.

It is necessary to gain an understanding of how these local control actions affect overall supply-chain performance through the study of supply-chain dynamics and policy assessment, discussed below.

12.4.1 SUPPLY-CHAIN DYNAMICS, SIMULATION, AND POLICY ANALYSIS

Supply-chain dynamics have become part of industrial lore. As illustrated as far back as 1958 by Forrester,[29] the processes used at different nodes of the supply chain result in a variety of different dynamic behaviors, often to the detriment of overall performance. Typically, standard control actions taken at a node (respond to demand signals by supplying from stock and then sending new demand signals upstream as stocks become depleted) cause amplification of the original demand signal from the market to a much larger demand signal at the upstream parts of the supply chain. This approach results in excessive stocks in the chain, operation under stress, and overcapacity of manufacturing and logistics.

Dynamic process simulation has long been recognized as a useful tool for understanding and improving processes. Similarly, supply-chain simulation is becoming a popular tool to formulate policy. Simulation is useful in identifying the potential dynamic performance of the supply chain as a function of various operating policies, ahead of actual implementation of any one policy. In most cases, the simulations are stochastic in that they repetitively sample from distributions of uncertain parameters to build up distributions of performance measures, rather than point values.

Many companies have undertaken some form of simulation study to understand the extent to which internal and external policies generate unnecessarily volatile behavior. The tools used for such analyses are normally commercially available, discrete-event simulation tools such as, for example, Witness (www.lanner.com), Extend™ (www.imaginethatinc.com), and Arena® (www.arenasimulation.com). With such tools, users can initially build up high-level models where warehouses store materials and are depleted by demand signals (which can be stochastic). The warehouses request replenishments from factories based on their inventory-control policies, and the factories are represented by their key resources and the rules for processing the queues of orders that build up behind the resources. Often, models of this nature can be used to study and resolve the high level dynamics. In many cases, sharing of demand data and making demand signals visible across the chain (which allows dynamic responses to real-time data across the chain through reforecasting) are the single most important steps for reducing unintentional supply-chain dynamics.

Dynamic supply-chain models can be also be used to identify operational improvements. Sensitivity analyses can locate the reasons for poor performance. Examples of such reasons from our studies include long changeover times, long and poorly managed quality control activities, highly variable processing times, and poor forecast accuracies. Projects such as changeover reductions achieved through single-minute exchange of die (SMED) techniques can then be implemented to improve performance. Another use of such models is to test the robustness of the system — for example, evaluating its ability to deal with demand surges or equipment failures.

A considerable amount of research in academia has also taken place in this field. Beamon[30] presented a review of supply-chain models and partitioned them into *analytical* (i.e., purely declarative) and *simulation* (i.e., including procedural elements). Analytical models are used to optimize high-level decisions involving unknown configurations by taking an aggregate view of the dynamics and details of operation (e.g., supply-chain network design). On the other hand, simulation models can be used to study the detailed dynamic operation of a fixed configuration under operational uncertainty and can be used to evaluate expected performance measures for the fixed configuration to a high level of accuracy. Although the field of industrial dynamics is very large, it tends to concentrate on logistics and inventory planning and normally ignores production or has a very simplistic representation of production; therefore, we will concentrate on research with a significant production element here.

Bose and Pekny[31] used a model predictive control (MPC) framework to understand the dynamic behavior of a consumer-goods supply chain. They studied various levels of coordination between the supply and demand entities. They also considered forecasting techniques, particularly for promotional demands. The forecasting model sets desired inventory targets that the scheduling model (based on MILP optimization) tries to meet. This is performed in a repetitive, rolling horizon approach. This model allows clear conclusions to be drawn regarding promotion and inventory management and the benefits and drawbacks of different degrees of coordination.

Perea-Lopez et al.[32] studied a polymer supply chain in which the manufacturing process was a single-stage batch multiproduct reactor supplying a warehouse, distribution network, and retailers. They captured the supply-chain dynamics by the balance of inventories and the balance of orders in terms of ordinary differential equations, together with the definition of shipping rates to the downstream product nodes, subject to some physical bounds and initial conditions for the inventory and order values. The model therefore assumes that the material and order flows are continuous. They evaluated a variety of different supply-chain control policies based on a decentralized decision-making framework and identified the policies that best mitigate perturbations. They extended this work[33] to include MILP-based scheduling within an MPC framework, whereby regular solutions are generated based on the current state and portions of the solution implemented. They contrasted a centralized approach, where all decisions are made simultaneously by a coordinator, with a decentralized approach, where each entity makes decisions independently. The benefits of central coordination are clear, as increases in profits of up to 15% were observed in the case study presented.

Supply chains can be thought of as distributed systems with somewhat decentralized decision making (especially for short-term decisions). The multiple-agent-based approach is a powerful technique for simulating this sort of system. Agent-based simulation techniques have been reported by Gjerdrum et al.[34] and García-Flores and Wang.[35] In both cases, the different players in the supply chain are represented by agents who are able to make autonomous decisions based on the information they have available and messages they receive. The agents include warehouses, customers, plants, and logistics functions. In the work of Gjerdrum et al.[34] and García-Flores and Wang,[35] the plant decision making involved production scheduling; the plant agent used a commercial schedule optimization package (agent-based systems have the advantage of being able to provide wrappers to existing software). The other agents used a variety of rules (e.g., to generate orders or to manage inventory). Agents are able to negotiate solutions from different starting points. García-Flores and Wang presented a single plant supply chain and evaluated various inventory management policies, while Gjerdrum et al. evaluated two plants and the effect of different product sourcing rules. Overall, the agent-based approach has proven to be a good framework for the abstraction and modular development of supply-chain models and is supported by some good software development tools that have been widely used in other sectors (e.g., telecoms).

Hung et al.[36] developed a flexible, object-oriented approach to the modeling of dynamic supply chains that is based on a generic node that has inbound material management, material conversion, and outbound material management capabilities and can be specialized to describe plants, warehouses, etc. Both physical processes (e.g., manufacturing, distribution, warehousing) and business processes are modeled. By the latter, we mean how decisions are taken at the different nodes of the chain, who makes them, what tools and methods are used, etc. This means that the logic of software tools used for decision making at various nodes (e.g., DRP and MRP) is replicated in the simulation tool. The aim of this approach is to suggest noninvasive improvements to the operation of the supply chain. Such improvements may come about through changes in parameters (e.g., safety stocks) or business processes (e.g., relationships between agents). In order to assess future performance, uncertainties must be taken into account. These include product demands, process yields, processing times, and transportation lead times. A stochastic simulation approach that samples from the uncertain parameters is a useful way of determining expected future performance as well as confidence limits on future performance measures. Because the uncertainty space is very large and uncertainties are time varying, Hung et al.[37] developed a very efficient (quasi–Monte Carlo) sampling procedure.

Shah[16] described two pharmaceutical studies based on this dynamic modeling approach. In the first study, a peculiar dynamic behavior was seen in the market warehouse. Although the background demand for the product was very stable, the manufacturer's warehouse experienced highly fluctuating demands and needed to hold considerable inventories to buffer against this. Upon some investigation, the reason related to a pricing cycle that caused wholesalers to try to anticipate price increases and request large preemptive orders. Of course, when these are received, the wholesaler will not order material again for some time. We used singular-value decomposition techniques to extract the historical dynamics and used them to generate forward forecasts. We compared the future supply-chain performance using this model against a model that used collaborative planning between manufacturer and wholesaler. The key metric was the amount of finished goods safety stock cover the manufacturer required to meet a certain customer service level (defined in this case as the fill rate, which is equal to the amount shipped divided by the amount requested). The results may be compared in Figure 12.10 and Figure 12.11.

The "weeks cover" dimension indicates how many weeks' worth of stock of final product is used as the safety stock figure. To achieve the target service level, the finished goods stock can be approximately halved in the collaborative case. It is clear that significant benefits are possible through an alternative way of running the supply chain. Conservative estimates would put these at $30 million in one-off inventory savings and $3.6m *per annum* savings. Of course, all the relevant reasons for holding stock (e.g., cycle times, manufacturing facility reliability, forecast accuracies) must be included in such models.

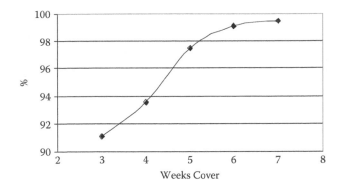

FIGURE 12.10 Variation of service level with finished goods safety stock for the non-collaborative case.

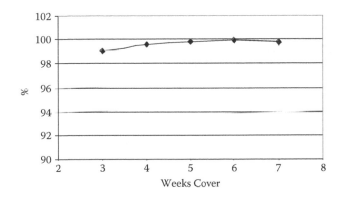

FIGURE 12.11 Variation of service level with finished goods safety stock for the collaborative case.

Another study considered the effects not of the production or inventory aspects but of the quality control (QC) procedures. As mentioned earlier, a prevalence of QC activities can be found in the industry, although they are not really necessary at all the points where they are currently used. These steps account for significant dead time in the process, often of the order of 1 to 2 weeks, when all the intervening processes are considered. We developed a model of a process that has five primary synthesis stages and two secondary manufacturing sites. The as-is (AI) process has QC activities at the end of each primary stage and for the final product. The modified process has a QC step for the AI and a QC step for the final product. The results for one of the products are compared below.

In Figure 12.12, the forward prediction of finished goods inventory of pack C is quite smooth. The lower confidence limit (95%) on the profile is still positive, giving confidence that stockouts are very unlikely. In Figure 12.13, which shows

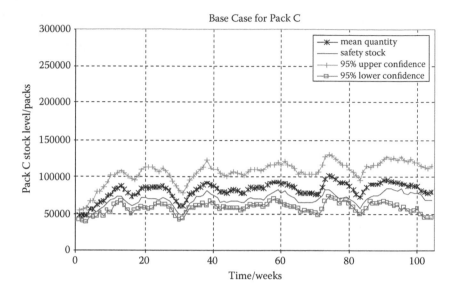

FIGURE 12.12 Time profile of expected inventory of finished goods, including confidence intervals, for quality control at only two points.

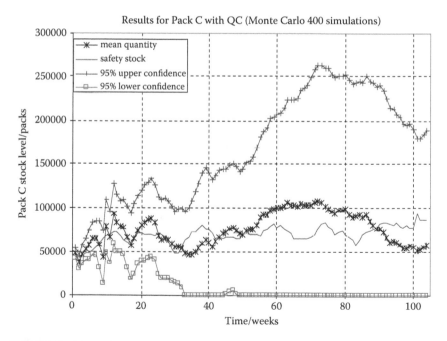

FIGURE 12.13 Time profile of expected inventory of finished goods, including confidence intervals, for quality control at all stages.

QC at all stages, there is much less certainty in the inventory (the variance grows significantly with time) and the lower confidence level goes to zero. In terms of customer service performance measures, over a 2-year period the average service level in the case with QC is 91% and the probability of a stockout in any week is 5%. On the other hand, in the low QC case, these figures are 100% and 0% respectively. Clearly, as process development and design advance, the comfort provided by QC at so many stages in the supply chain will not be required, and the dynamic behavior will improve markedly.

An area where stochastic simulation is finding increased use is in refining the results of relatively coarse optimization models. In this case, optimization models are used to determine important structural and parametric decisions, and simulation is used to evaluate the distributions of performance measures and constraints more accurately. This has been reported by Karabakal et al.,[38] who studied the Volkswagen (VW) distribution network in the United States, and Gnoni et al.,[39] who developed a robust planning procedure for a multisite automotive components facility.

Blau et al.[40] considered the value-chain problem of risk management at the development stage in the pharmaceutical industry. This is a long, costly, and inherently risky process with a large up-front commitment. The aim of their work was to support the process of product selection and test planning while managing risk effectively. The development activities are modeled as a probabilistic activity network, where each activity has a time, precedence relations, resource requirements, and probability of success. The risk of a set of decisions must be balanced against the potential reward. The risk/reward ratio can then be used to compare different drug candidates. A screening process removes any obviously unpromising candidates, and the remainder must be sequenced through the development pipeline. A heuristic approach using simulation with local rules in response to trigger events (e.g., failure of a test) is employed. This approach aims to process tasks as quickly as possible and, although there is no guarantee of not violating resource constraints, these violations are usually not large.

Subramanian et al.[41,42] extended this work to take explicit account of the resource requirements of the problem. They make the point that a single-level mathematical programming problem cannot hope to capture all these features. On the other hand, simulation techniques cope well with the stochastic elements but require local, myopic rules to resolve conflicts or make choices as they arise. Subramanian et al. therefore developed an integrated optimization-simulation framework (SIM-OPT) in which a simulator reverts to an optimization layer (with different degrees of optimization) to resolve conflicts or make choices such as task sequencing. The results show that using optimization far outperforms the typical local rules used in classical simulation. By repetitive simulation, the statistical trends can be tracked and corporate policy (particularly in relation to risk and resourcing) can be analyzed. Also, data from the inner simulation loop can be used to update parameters in the optimization loop.

12.4.2 AN INDUSTRIAL APPLICATION

The polymers and resins business of Rohm and Haas was being squeezed by powerful customers and suppliers, and they had not been able to increase the prices of key products between 1992 and 1997.[43] An ERP system was rolled out between 1992 and 1995, but because underlying processes did not change the expected productivity improvements did not materialize. The division therefore undertook a study to try to improve supply-chain margins. Prior to the study, the policy was quite chaotic, aiming to serve all customers equally with constant disruptions to production plans. The study involved: (1) a review of customer service policy, (2) a review of product demand management, and (3) a review of production planning and manufacturing management.

The review of customer service policy recognized that treating all customers uniformly was not a good idea and placed unnecessary stress on the supply chain. The customer base was then arranged into four tiers, where the first tier reflected very important customers responsible for a significant proportion of demand and the fourth tier represented the long tail of very low-volume customers with erratic demands. This fourth tier was then not serviced directly but rather through distributors who managed stock themselves.

No formal demand management policy existed prior to the study, and most products were made to stock with a view to supplying on short lead times. In the study, products were categorized into four quadrants based on demand volume and demand variability. The contribution of products in each quadrant to the prevailing inventory costs was found to be very different. This resulted in a new strategy, whereby some capacity was dedicated to high-volume, low-variability products, which were made to stock for low lead times. This resulted in far fewer changeovers. The low-volume, high-variability products were to be made to order, so these customers would have to expect longer lead times and would be expected to order in production batch multiples.

In order to identify how to allocate products to production capacity and to estimate the new lead times, a discrete-event supply-chain simulation model was developed. Various rules for make-to-stock and make-to-order products were evaluated, and it was found that segregating the resources for these classes of products was beneficial. Overall, an estimated improvement in throughput of 15% was achieved, and millions of dollars were saved while operating a more predictable, less stressful system. Again, the simulation model is not very complicated but still identifies significant benefits.

12.4.3 REMARKS

The use of dynamic supply-chain simulation is very much an emerging area, and one which is expected to expand rapidly. One key issue is the integration of business process modeling with the physical aspects (e.g., recipes, resources). No consensus has yet been reached on frameworks for addressing this. A simulation engine must be able to replicate or incorporate algorithms used at certain parts of the supply chain. The emerging frameworks appear to be agent based and object oriented, attributes that are suited to modeling complex systems with degrees of distributed decision making. These complex, stochastic, discrete-event models contain adjustable parameters. The application of optimization procedures (probably gradient free) to select good values for these is another interesting avenue to pursue. In the meantime, some real-time supply-chain management solutions are being made available by some of the larger vendors (e.g., www.sap.com, www.i2.com; www.manugistics.com; www.peoplesoft.com).

12.5 ORGANIZATIONAL SYSTEMS AND IT ISSUES

The role of the supply-chain function in the organization differs widely between enterprises. In some cases, it is simply seen as a vehicle for demand fulfillment, while in others it is tightly integrated into the overall business strategy. As highlighted by Kavanaugh and Matthews,[44] the role of the supply-chain organization in the company is partly dictated by the external environment. They identified five levels of involvement:

- *Stable supplier* — In an environment of stable demand of simple products, the supply-chain function is normally limited to cost-effective supply of product.
- *Reactive supplier* — Here, significant market dynamics exist over which the enterprise has no control. The supply-chain function chases demand, without necessarily understanding the cost implications of attempting to fulfill all orders. Little coordination with other related functions (e.g., marketing, product innovation) occurs.
- *Efficient reactive supplier* — The supply-chain function is still effectively reactive to the external environment but is better integrated with other functions in the enterprise (e.g., manufacturing and inventory planning are coordinated).
- *Efficient proactive supplier* — The supply-chain function is also involved in sales and marketing and in influencing and managing demand to make it more predictable and therefore supporting planning of operations, rather than simply responding to external events. Companies employing vendor-managed inventory policies fit into this category.
- *Revenue and margin drivers* — The enterprise is built around the supply-chain function, which is integral to the heart of its strategy. The supply-chain function often operates on a inter-company basis with visibility of information across supply-chain partners. Dell's build-to-order personal computer supply chain with its tiers of integrated suppliers is a good example of such a chain. A key feature of these chains is that they are truly demand driven.

A sensible approach, therefore, is first to decide on what role the company sees for its supply-chain function, then design business processes that support this, and then finally invest in appropriate IT systems. Historically, investments in IT systems have not always followed such a rigorous analysis. Many books and articles are available that provide advice on designing the business process and supporting organizational elements.

Rogers[45] described six elements of superior design for an effective supply-chain organization:

- *Tasks/process* — Define the key internal-facing and external-facing tasks; organize and integrate these.
- *People/skills* — Develop the workforce to undertake the tasks successfully; recognize that different skills are needed for internal- and external-facing activities.
- *Information/information flow* — The three main flows are external flows between companies, internal flows across internal boundaries, and intra-flows within divisions; people should have easy access to all the relevant information to support decision making.
- *Decision making* — Determine what decisions are to be made where in the system and by whom; decide on the methods and metrics that will guide decision making.
- *Rewards* — Keep rewards in line with corporate strategy and encourage supply-chain performance; functional elements must operate in the best interests of the enterprise and indeed the chain as a whole.
- *Structure* — Design the "extra" structure that interfaces with the external world, the "inter" structure that works across functions, and the "intra" structure.

12.5.1 Tools and Models

Because of its importance and the huge financial potential benefits, numerous research activities on supply-chain modeling have been reported in the literature, and abundant software packages present themselves as supply-chain solutions on the commercial market. Enterprise resource planning (ERP) systems and e-commerce software packages improve supply-chain management by communicating real-time data regarding companies' supply-chain activities and providing more efficient transactions between supply-chain members. These tools are effective in coordinating diverse business processes across the supply chain, from raw material purchasing to issuing invoices to customers. They therefore support the execution part of supply-chain management. These tools are not, however, necessarily sufficient for decision making to support significant supply-chain improvements because they only deal with current transactions and compile historical data. Models that guide decision making by considering supply-chain-wide activities in a forward-looking context are needed. Besides models developed in academia, certain commercial supply-chain software packages, marketed by vendors such as i2, SAP, Oracle, and Manugistics, also provide this type of functionality. To achieve the purported benefits and substantially better decisions, it is crucial that underneath the user-friendly features the models used are accurate (i.e., represent the supply chain realistically. According to Ingalls,[46] the supply-chain algorithms in commercial software are not much different from the techniques available in academia.

The decision-guiding models can be classified into three main types: (1) analytical models, (2) simulation models, and (3) combined simulation and optimization models. Analytical models are optimization models that maximize certain benefits by choosing the best set of decision variables of a supply chain subject to the constraints that represent the features of the supply chain and management decisions in the form of mathematical descriptions. In contrast, simulation models emulate supply-chain activities and predict performance measures. The third type of supply-chain models attempt to combine the strengths of simulation and optimization.

In addition to the above classification, supply-chain models can be described according to several other aspects as well:

- *Static vs. dynamic* — This aspect denotes whether time contributes to the dimensionality of the model concerned. In a static model, the passage of time is not considered in the evaluation, whereas in a dynamic model the supply chain is represented as it evolves over time.
- *Discrete vs. continuous* — This aspect can have two meanings. First, it often describes the type of numerical values that the variables can take in the models. In discrete-variable models, all variables are integers or limited to some specified integral values, such as (0,1) (binary) variables, whereas in continuous-variable models all variables can take any values within the permitted ranges. Hybrid models contain both discrete and continuous variables and are also referred to as *mixed integer*. This first differentiation is often required for analytical models for selecting the most efficient optimization solution method. Second, for dynamic models this aspect also signifies how the variables change with respect to time. In discrete-time models, variables change their values at certain fixed points in time, such as when an event occurs, whereas in continuous-time models variables can change their values at any time. Models in which variables change their value only when events occur are also referred to as *discrete-event models*.
- *Deterministic vs. stochastic* — This aspect indicates the absence or presence of uncertain factors in the models. In deterministic models, no uncertainty is considered. Stochastic models evaluate the effects of uncertain factors on the supply-chain performance. The uncertain factors are usually represented in terms of probability distributions to give the expected performance measures. Alternatively, fuzzy-set theory is used in some simulation models where statistics are unreliable.[47] For

analytical models, scenario analysis[48] or polytope integration[49] can also be used for approximation of the effects of uncertainties.

- *Strategic vs. operational* — This aspect determines the scope, time scale, and level of details of the models, which are chosen according to the objective of the particular problems at hand. Strategic models can involve the determination of suppliers and market segments to serve; the number, location, capacity, and type of a company's facilities; the amount of various materials to produce and hold at different supply-chain members and to be shipped among them; and the routing of material flows. With regard to operational models, the aspects concerned can be inventory management at various points along the supply-chain network or production planning and scheduling. In addition, some models are tactical if their scopes are somewhere between strategic and operational.

12.6 PERFORMANCE MEASURES AND BENCHMARKING

Supply-chain benchmarking has become a popular activity. It helps identify gaps between a company's performance and that of its best peers. These gaps may be translated into financial opportunities and identify improvement projects. Some examples of benchmark figures are given below:

PRTM 2001 Study[50]

Item	Median (days)	Best in Class (days)
Inventory	74	35
Cash-to-cash cycle	84	36
Response time for 20% rise in demand	20	9

Note: The inventory is the number of days' worth of average demand which exists in the chain.

CIO[51] Benchmarks

Item	Median	Best in Class
SCM costs (e.g., costs of planning, IT, acquiring materials, managing orders and inventory)	8–12% of sales	4–5% of sales
Cash-to-cash cycle	100 days	30 days
OTIF	69–81%	94%

Note: OTIF (on time in full) reflects the percentage of time that orders are met on time in full.

The UK Process Industries Center for Manufacturing Excellence (PICME) has published the following benchmarks:

Metric	U.K. Plants	Britain's Best Process Factory	World Class
OTIF (%)	93.1	94.5	>99.5
Complaints (%)	3.1	0.29	<0.001
Equipment utilization (%)	71.3	80	>85
Changeovers (% of capacity)	11.3	0	<0.05
Finished goods days of cover	—	16	0.03
Raw material days of cover	33.9	—	5

Data obtained from www.picme.org.

Ahmad and Benson[52] have published the following manufacturing-oriented benchmarks:

Measure	Average of Worst Five	Average of Others	Average of Best Five	World Class
OTIF (%)	40	90	99.9	>99.9
Customer complaints (%)	6	1	0.01	<0.01
Availability (%)	70	85	96	>97
Equipment utilization (%)	20	60	94	—
Stock turns	4	12	19	>25
Simple changeover (min)	480	240	5	<5

Note: Stock turns are essentially equal to (annual sales volume)/(average stock levels).

Supply-chain benchmarks should be used with some caution. The types of benchmarks listed above, according to Lambert and Pohlen,[53] are really "internal logistics" metrics and do not reflect channel-spanning measures. These authors recommend monitoring the sorts of metrics that help determine the relationship between corporate and supply-chain performance. They proposed a framework that initially focuses on each supplier–customer link in the chain and developed metrics (e.g., profit and loss) for these. The framework can then be extended to develop channel-spanning measures.

12.6.1 BENEFITS OF MODELING APPROACH TO SUPPLY-CHAIN MANAGEMENT

In our view, modeling based approaches to supply-chain management should realize the following benefits:

- Throughput should be increased by 3 to 15%.
- Stock should be reduced by 15 to 30%.
- Value-added time percentage should be increased by 30%.
- Chain costs should be reduced by at least 7%.

All of the above should be achievable without capital investment. The following authors have made the following claims associated with supply-chain studies:

- Evans et al.[54] reported a savings of £50,000 per day due to a reduction in the chain cycle time for United Health.
- Davis[55] reported a 25 to 50% reduction in stocks for Hewlett Packard by strategic redistribution of stocks.
- Schmidt[56] reported that, according to Phillips (USA), 10 to 15% of annual revenues can be released through better operation.
- Child et al.[57] claim that 10 to 40% of total supply-chain costs are due to unnecessary complexity that can be eliminated through streamlining of associated processes.
- Arntzen et al.[58] claim to have saved DEC $100m *per annum* through a mixed-integer optimization modeling project.
- Worthen[50] described how ChevronTexaco's downstream profits increased from $290 million to $662 million without any investment in refining or retail capacity but with a strategic redesign of their supply-chain processes.

12.7 CONCLUSIONS AND FUTURE PERSPECTIVES

Supply-chain management is the glue that ensures that all the benefits of good plant design and effective plant operation are realized in operational practice. A plant that is well designed and operated will produce the right products at the right time. Supply-chain management can be organized along a three level hierarchy:

- *Infrastructure* — Establish an effective asset base, including key suppliers and partners.
- *Planning* — Looking ahead, plan the production and distribution activities effectively.
- *Operations* — Manage the supply chain effectively in real time; monitor relevant signals and react accordingly.

A wide range of technology has recently been developed to support the effectively deployment of supply-chain solutions. In addition, much research has been undertaken in academia toward the development of the next generation of technology. A number of challenges have already been posed in Sections 12.2.3, 12.3.2, and 12.4.4, above. We see two generic important future challenges:

- *Improved design for existing processes* — A distinguishing feature of process industry supply chains is that supply-chain performance is very strongly affected by the flexibility and responsiveness of the production process. This is not the case to the same extent in other industries. For example, consider the multimedia products supply chain. Here, efficient forecasting, flexible warehouses, and real-time downstream supply-chain management and adaptation are critical; production is very straightforward (stamping out CDs and DVDs) and often a lead time of one day can be assumed for a product. We believe "process design for supply-chain responsiveness" is an important area that has not received much attention so far. The process industries have not fully grasped the concept of mass customization. For example, instead of using a single reactor to produce different complete polymers from monomers, why not try to develop building blocks of medium molecular weights and combine them as appropriate? To what extent can intermediates be made at world-scale centralized facilities and specialized products be configured at flexible, near-market facilities?
- *Effective design of "new" supply chains* — It is evident that the process industry supply chains of the future will be quite different from those of the past. In addition, a number of new supply chains (parts of which may already be present) will emerge. There exists a relatively short window of opportunity to explore the optimal configuration of such supply chains before they develop organically; this may be of vital importance in informing national and international policy as well as strategic decisions in industry. Examples of such supply chains of the future include:
  - Hydrogen and, more generally, supply chains to support fuel cells
  - Water
  - Fast response therapeutics (particularly vaccines) for civilian and homeland security uses
  - Energy (the provision of the energy needs for a country can be viewed as a supply chain subject to significant decarbonization pressures)
  - Life-science products
  - Crops for non-food use and biorefineries
  - Gas-to-value (i.e., generating high-value products such as very-low-sulfur diesel from natural gas *in situ*)
  - Waste-to-value (i.e., effective recycling of materials at the end-of-use) and reverse (closed-loop) production systems

Although research in the basic sciences related to emerging industries is very topical, supply-chain research as applied to these will be important. Wang[59] noted that enablers for emerging industries (e.g., micro-nano technology, biotechnology, and advanced material technology) are information technology, supply-chain management, modeling and simulation, human development, and knowledge management.

REFERENCES

1. CLM, *What's It All About?*, Council of Logistics Management, Oak Brook, IL, 1986.
2. Anderson, D.L., Britt, F.E., and Favre, D.J., The seven principles of supply-chain management, *Supply-Chain Manage. Rev.*, 1, 31, 1997.
3. Lambert, D.M. and Cooper, M., Issues in supply-chain management, *Ind. Mktg. Mgt.*, 29, 65, 2000.
4. Geoffrion, A.M. and Graves, G., Multicommodity distribution system design by Benders decomposition, *Manage. Sci.*, 20, 822, 1974.
5. Geoffrion, A.M. and van Roy, T.J., Caution: common sense planning methods can be hazardous to your corporate health, *Sloan Manage. Rev.*, Summer, 31, 1979.
6. Brown, G.G., Graves, G.W., and Honczarenko, M.D., Design and operation of a multicommodity production/distribution system using primal goal decomposition, *Manage. Sci.*, 33, 1469, 1987.
7. Vidal, C.J. and Goetschalckx, M., Strategic production-distribution models: a critical review with emphasis on global supply-chain models, *EJOR*, 98, 1, 1997.
8. Kallrath, J., Combined strategic and operational planning: an MILP success story in chemical industry, *OR Spectrum*, 24, 315, 2002.
9. Sabri, E.H. and Beamon, B.M., A multi-objective approach to simultaneous strategic and operational planning in supply-chain design, *Omega*, 28, 581, 2000.
10. Tsiakis, P., Shah, N., and Pantelides, C.C., Design of multi-echelon supply-chain networks under demand uncertainty, *Ind. Eng. Chem. Res.*, 40, 3585, 2001.
12. Tsiakis, P., Shah, N., and Pantelides, C.C., Optimal Structures for Supply-Chain Networks, AIChE Annual Meeting, November 2001.
13. Schoemaker, P.J.H., Multiple scenario development: its conceptual and behavioural foundation, *Strategic Manage. J.*, 14, 193, 1993.
14. Camm, J.D. et al., Blending OR/MS, judgment, and GIS: restructuring P&G's supply chain, *Interfaces*, 27, 128, 1997.
15. Shapiro, J.F., Challenges of strategic supply-chain planning and modelling, *Comput. Chem. Eng.*, 28, 855, 2004.
16. Lin, G. et al., Extended-enterprise supply-chain management at IBM personal systems group and other divisions, *Interfaces*, 30, 7, 2000.
17. Shah, N., Pharmaceutical supply chains: key issues and strategies for optimisation, *Comput. Chem. Eng.*, 28, 929, 2004.
18. Backx, T., Bosgra, O., and Marquardt, W., Towards intentional dynamics in supply-chain conscious process operations, *AIChE Symp. Ser.*, 320(94), 5, 1998.
19. Shapiro, J.F., On the connections among activity-based costing, mathematical programming models for analyzing strategic decisions, and the resource-based view of the firm, *EJOR*, 118(2), 295, 1999.
20. Makridakis, S. and Wheelwright, S.C., *Forecasting Methods for Management*, John Wiley & Sons, New York, 1989.
21. Lewis, C.D., *Scientific Inventory Control*, Elsevier, Amsterdam, 1970.

22. Wilkinson, S.J. et al., Integrated production and distribution scheduling on a Europe-wide basis, *Comput. Chem. Eng.,* S20, 1275, 1996.

23. McDonald, C.M. and Karimi, I.M., Planning and scheduling of parallel semicontinuous processes. 1. Production planning, *Ind. Eng. Chem. Res.,* 36, 2691, 1997.

24. Kallrath, J., Planning and scheduling in the process industry, *OR Spectrum,* 24, 219, 2002.

25. Berning, G. et al., An integrated system for supply-chain optimisation in the chemical process industry, *OR Spectrum,* 24, 371, 2002.

26. Timpe, C.H. and Kallrath, J., Optimal planning in large multi-site production networks, *EJOR,* 126, 422, 2000.

27. Gupta, A. and Maranas, C.D., A two-stage modeling and solution framework for multisite midterm planning under demand uncertainty, *Ind. Eng. Chem. Res.,* 39, 3799, 2000.

28. Gupta, A., Maranas, C.D., and McDonald, C.M., Mid-term supply-chain planning under demand uncertainty: customer demand satisfaction and inventory management, *Comput. Chem. Eng.,* 24, 2613, 2000.

29. Brown, G. et al., The Kellogg Company optimizes production, inventory and distribution, *Interfaces,* 31(6), 1, 2001.

30. Forrester, J., Industrial dynamics: a major breakthrough for decision makers, *Harvard Bus. Rev.,* July–August, 37, 1958.

31. Beamon, B.M., Supply-chain design and analysis: models and methods, *Int. J. Prod. Econ.,* 55, 281, 1998.

32. Bose, S. and Pekny, J.F., A model predictive framework for planning and scheduling problems: a case study of consumer goods supply chain, *Comput. Chem. Eng.,* 24, 329, 2000.

33. Perea-Lopez, E., Ydstie, B.E., and Grossmann, I.E., Dynamic modelling and decentralised control of supply chains, *Ind. Eng. Chem. Res.,* 40, 3369, 2001.

34. Perea-Lopez, E., Ydstie, B.E., and Grossmann, I.E., A model predictive control strategy for supply-chain optimisation, *Comput. Chem. Eng.,* 27, 1201, 2003.

35. Gjerdrum, J., Shah, N., and Papageorgiou, L.G., A combined optimisation and agent-based approach for supply-chain modelling and performance assessment, *Prod. Plan. Control,* 12, 81, 2000.

36. García-Flores, R. and Wang, X.Z., A multi-agent system for chemical supply-chain simulation and management support, *OR Spectrum,* 24, 343, 2002.

37. Hung, W.Y., Samsatli, N., and Shah, N., Object-oriented dynamic supply-chain modelling incorporated with production scheduling, *EJOR,* in press, 2005.

38. Hung, W.Y. et al., An efficient sampling technique for stochastic supply-chain simulations, in *Proc. of the 2003 Summer Computer Simulation Conference (SCSC 2003),* Montreal, July 2003.

39. Karabakal, N., Gunal, A., and Ritchie, W., Supply-chain analysis at Volkswagen of America, *Interfaces,* 30(4), 46, 2000.

40. Gnoni, M.G. et al., Production planning of a multi-site manufacturing system by hybrid modelling: a case study from the automotive industry, *Int. J. Prod. Econ.,* 85, 251, 2003.

41. Blau, G. et al., Risk management in the development of new products in highly regulated industries, *Comput. Chem. Eng.,* 22, 1005, 2000.

42. Subramanian, D., Pekny, J.F., and Reklaitis, G.V., A simulation-optimization framework for research and development pipeline management, *AIChE J.,* 47, 2226, 2001.

42. Subramanian, D. et al., Simulation-optimization framework for stochastic optimization of R&D pipeline management, *AIChE J.*, 49, 96, 2003.
43. D'Alessandro, A.J. and Baveja, A., Divide and conquer: Rohm and Haas' response to a changing specialty chemicals market, *Interfaces*, 30(6), 1, 2000.
44. Kavanaugh, K.M. and Matthews, P., Maximising supply-chain value, *ASCET*, 2, 2000 (http://www.ascet.com/documents.asp?d_ID=281).
45. Rogers, S., Supply management: six elements of superior design, *Supply-Chain Manage. Rev.*, April 2004 (http://www.manufacturing.net/scm/).
46. Ingalls, R.G., The value of simulation in modelling supply chains, in *Proc. 1998 Winter Simulation Conference*, Medeiros, D.J., Watson, E.F., Carson, J.S., and Manivannan, M.S., Eds., Institute of Electrical and Electronic Engineers, Piscataway, NJ, 1998, 900–906.
47. Giannoccaro, I. et al., A fuzzy echelon approach for inventory management in supply chains, *EJOR*, 149(1), 185, 2003.
48. Escudero, L.F., Quintana, J.F., and Salmeron, J., CORO, a modeling and an algorithmic framework for oil supply, transformation and distribution optimization under uncertainty, *EJOR*, 114(3), 638, 1999.
49. Applequist, G.E., Pekny, J.F., and Reklaitis, G.V., Risk and uncertainty in managing chemical manufacturing supply chains, *Comput. Chem. Eng.*, 24, 2211, 2000.
50. Worthen, B., Drilling for every drop of value, *CIO Mag.*, June 2002 (http://www.cio.com/archive/060102/drilling.html).
51. CIO, Benchmarks, *CIO Mag.*, October 2000 (http://www.cio.com/archive/100100/payoff_sidebar2.html).
52. Ahmad, M.M. and Benson, R., *Benchmarking in the Process Industries*, Institute of Chemical Engineers, Rugby, U.K., 1999.
53. Lambert, D. M. and Pohlen, T. L., Supply-chain metrics, *Int. J. Logistics Manage.*, 12, 1, 2001.
54. Evans, G.N., Towilll, D.R., and Naim, M.M., Business process re-engineering the supply chain, *Prod. Plan. Control*, 6, 38, 1995.
55. Davis, T., Effective supply-chain management, *Sloane Manage. Rev.*, Summer, 35, 1993.
56. Schmidt, J.D., Achieving the elusive integrated supply chain, *Proc. 2nd Ind. Eng. Res. Conf.*, 138, 1993.
57. Child, P. et al., The management of complexity, *Sloane Manage. Rev.*, Fall, 73, 1991.
58. Arntzen, B.C. et al., Global supply-chain management at Digital Equipment Corporation, *Interfaces*, 25, 69, 1997.
59. Wang, B., Advanced manufacturing in the 21st century, in *Proc. of the 5th Annual International Conference on Industrial Engineering: Theory, Applications and Practice*, Taiwan, 2000.

Part IV

Future of Batch Processing

13 Concluding Remarks and Future Prospects

Ekaterini Korovessi and Andreas A. Linninger

CONTENTS

A number of key characteristics and trends have had an impact on the chemical industry and are fueling a sustained interest in the technologies used for the design and management of the batch processing industries. The most important are globalization and strong competition, continuing demand for differentiated specialty chemicals and new pharmaceuticals and biochemicals, environmental health and safety concerns, and growing regulatory requirements. This chapter briefly discusses some of the technological and operational challenges that these factors introduce to the business. It compiles a few of the most common and promising approaches to solving the problems that the industry faces and focuses on the multiple opportunities for technological advancements and areas of arising research and development. Discussions on most of these technologies, how they are currently being implemented, and potential future improvements are provided in the respective chapters throughout this book.

13.1 GLOBALIZATION AND INCREASED COMPETITION

Globalization has strongly affected the entire chemical industry in recent years. The effects have been twofold:

- Rapid economic development along with an increase in research is taking place in developing countries. Manufacturing development in these countries is steadily growing, resulting in an increase in competition in all sectors. Currently, specialty chemical products are being produced proficiently and at lower cost in developing countries and exported to developed countries.

• Leading corporations traditionally considered the backbone of the U.S. or European chemical industry now identify themselves as global companies, deploying their resources to serve global markets.

The specialty chemicals and pharmaceutical sectors are characterized by the production of products in relatively small volumes, and in general they enjoy profit margins higher than those for commodity or bulk chemicals. But, they have been equally vulnerable to competition from Asia. The industry has reacted to the increased competitive pressures resulting from globalization with various cost-reduction approaches: reorganizations, restructuring, consolidation, job cutting, and plant closures. Some companies have taken advantage of the competition by sourcing raw materials from Asia and outsourcing the manufacture of some of their intermediates to China, India, or Latin America.

Long-term viability requires more fundamental competitive advantages than merely cost cutting. Reduced time to market, lower production costs, and improved flexibility are known to be critical success factors for batch processes. In the current environment, multiple differentiating factors are necessary for companies to survive and excel. The industry is realizing the important role that continuous innovation plays in maintaining an advantage in the market place and is investing in novel technologies and quality and operations management.

Great opportunities for cost reduction lie in the conceptual process design phase during the early stages of process development. Pharmaceutical companies used to initiate major process research and development efforts only after drug approval. Under the current competitive pressures of decreased drug life-cycle times and cost reduction, this is no longer the case. The main challenge is now the coordination of new drug development and its manufacturing process development workflows so as to bring the product to market quickly and cost effectively. This means that pilot-plant and process development must be launched with very little lead time. Competitive efforts, therefore, try to scale-up from the lab directly to manufacturing levels whenever possible. After Food and Drug Administration (FDA) approval, neither the drug nor its operating procedures may be altered without a new drug approval process, leaving very little room for optimization of the manufacturing recipe. When a drug and its manufacturing recipe receive FDA approval, the manufacturer is restricted in the types of process improvements that it can initiate to improve the efficiency of the chemistry or operations. Identification of potential alternative chemical routes to the desired product and selection of the one that will lead to the highest yields and improved productivity and waste reduction early in the process development lifetime can result in significant savings. In addition to the original design, it is also important to establish a technical plan for necessary increases in product throughput so as to successfully satisfy future increased demand without the need for additional regulatory approvals. Concurrent product development and process development are necessary to achieve both of these goals. The insight gained during the early process development effort can result in timely identification and elimination of potential problems and continuous gains in efficiency and reductions in cost as the product moves through development, launch, and maturity. Systematic methods for the synthesis and design of batch processes are a central point of interest in these activities. The role that computer-aided approaches can play in shortening the time to market are discussed in several sections of this book.

Traditional approaches to the design of batch unit operations include shortcut methods, rules of thumb, and design by analogy. Recent research activities try to come up with more fundamental and rigorous design methodologies for unit operation design improvement. Increased process efficiencies can be achieved through improved reactor design, better or new technologies for separations and purifications, cost-effective engineering, and low-cost manufacturing equipment. Following are a few representative examples, some of which have been discussed in detail in the book.

New advances in batch distillation, a well-established mature technology, include novel complex column configurations, azeotropic, extractive and reactive batch distillations, optimal designs, optimal operation policies, and new methods of analysis. These new advances can increase the possibility of using batch distillation profitably for a much wider variety of separations, but they also give rise to complicated problems of selection of the proper configuration, correct operating mode, and optimal design parameters.

Laboratory-scale tests are extremely important and can be excellent tools for evaluating critical parameters for the selection, design, and optimization of batch equipment when computational models are not available. For example, despite the requirements for crystal product control across a range of industries, we still have much to learn about designing batch crystallizers to meet the desired product specifications. A better fundamental understanding of seeding, nucleation, and the effects of impurities on solubility and kinetics is the subject of academic research. Laboratory feasibility testing can be used to test the efficacy of crystallization for purity specifications and particle size distribution. It should be noted that feasibility testing for the screening of alternative separation and purification technologies should also be deployed early in the process development stage. Developmental testing of crystallization processes at the pilot scale is a good practice to establish the robustness of the process, determine the principal scale-up parameters, and generate product samples for evaluation.

In many cases, even when computer-aided first-principles models for unit operation design are available, empirical techniques still predominate for initial design purposes due to a lack of essential physicochemical and engineering property data required for the fundamental approaches. Quite often the time and cost required to measure the necessary property data for process design are prohibitive, and an empirical approach is more commonly used. Various physical property prediction models are available, especially for liquids and gases, but they have not been applied successfully to the final active products and intermediates of the specialty chemical, pharmaceutical, and food industries which are complex, multifunctional molecules. New fundamental property estimation models and improved prediction models for large organic molecules are an active area of research.

Identification of new applications for known technologies and unit operations offers other technological innovations. Spray drying is a well-known, already mature technology with many applications in the food and chemical process industries for heat-sensitive powdered, granulated, and agglomerated products of various quality specifications. It has recently been finding new applications in the pharmaceutical industry for active pharmaceutical ingredients with stability and special formulation requirements. Preparative chromatography has found many applications in fields ranging from pure chemistry to biotechnology oriented. Protein purification, peptide production, and the resolution of enantiomers are some challenging separation problems where chromatographic methods are already being used or have been developed successfully.

Pharmaceutical and fine chemical companies are looking closely at microreactors, a new but noncommercially available technology, as an alternative to batch processing. Continuous operation, good mixing, efficient heat transfer, and precise control of the reaction parameters are just some of the characteristics that make microreactor technology promising for increased yields and throughputs. This technology also has the advantage of minimizing scale-up time. Problems associated with solids handling in microdevices, the lack of available microdevices for unit operations other than reactors, and the lack of commercial manufacturing of microdevices are all challenges that still must be overcome.

Today, most aspects of model-based batch process design and optimization are by no means mature. Many open questions and points of improvement in this area are the subject of research in academia. Among these are aspects of model development and validation, rigorous approaches to considering the effect of model and parameter uncertainty, and the challenges of robust and expeditious numerical solution techniques, especially for dynamic optimization problems in batch operations. Much research is currently being done in improving the efficiency and increasing the scope of current algorithms. Simulation and optimization under uncertainty, dynamic or distributed quantitative–qualitative tools, improved real-time modeling, improved tools to represent heuristics, and more information management and data-mining tools are still under development. Rigorous mathematical models and optimization-based strategies have a high potential for the future and will remain a central focus of research. Powerful optimization techniques and solvers are opening up new opportunities for planning and scheduling and supply-chain management, areas of utmost interest for large manufacturing businesses.

Batch process planning and scheduling are important for maximization of facility utilization and production rates while meeting product market demands. This is a mature yet still active area of research, and multiple commercial tools are available to companies. Even though planning and scheduling tools continue to develop, many industrial-scale plants are so complex that automated solutions are quite large and cumbersome. It is expected that as more efficient algorithms are introduced the solution of more realistic problems will be possible, with results being implemented automatically in the batch plant.

Advances in information technology and instrumentation make possible improved collection and analysis of manufacturing data. Instrumentation can now be adapted to manufacturing for in-process feedback in real time. Automated batch process monitoring and control, although still in the very early stages of development, are areas of research that have the potential to produce substantial improvements in batch operations and their optimization.

Effective supply-chain management is a challenging and important activity, particularly given the external changes imposed by globalization of the industry. Several commercial software tools that are available to the industry improve supply-chain management by communicating real-time data regarding companies' supply-chain activities and allowing for more efficient transactions between supply-chain members. Supply-chain models to guide decisions in a forward-looking context and thus result in significant financial improvements are still under development. It has also been argued in the pages of this book that, although the supply-chain performance is very strongly affected by the flexibility and responsiveness of the manufacturing process, process design for supply-chain responsiveness is a research area that has not yet received the attention it deserves.

13.2 DEMAND FOR DIFFERENTIATED PRODUCTS OR NEW ROUTES TO KNOWN PRODUCTS

The market demand for new differentiated products as well as the demand for new routes to existing products can also be driving forces for the development of innovative technologies. The increasing demand for commodity chemicals (e.g., ethanol, which is promoted as a fuel additive in the United States) and for new specialty chemicals or pharmaceuticals has played an important role in the continuing development of advanced biotechnologies in recent years. Biotechnology provides a superior alternative to other methods of producing chemicals through less expensive, cleaner, and higher yield processes.

Differentiated products that can offer special performance at a price premium provide opportunities for the creation of new synergies at the interfaces of chemistry with biology, physics, materials science, and engineering. The change in the product mix of commodities and specialties brings up the need for constant reevaluation and use of existing facilities for new products. The introduction of new products in the market increases the complexity of both the scheduling and planning problem and the longer term supply-chain problem due to the consideration of greater market-related uncertainties (demand levels, timing, product pricing) as well as new production uncertainties (cycle times, yields). Software systems or modeling tools are required that can deal with many of these uncertainties and allow appropriate strategic planning.

13.3 ENVIRONMENT, HEALTH, AND SAFETY

The industry is under increasing pressure to operate in a way that does not diminish an increasingly limited supply of resources and that reduces the environmental impact of old process technologies. New processes must be efficient in energy and raw material consumption and, just as importantly, produce minimal waste that does not harm living beings or the environment. The ecological promise of the science has prompted governmental agencies to sponsor research, and private biotech companies are seeking to improve their existing process and product lines or offer entirely new products.

In the United States, the Environmental Protection Agency's Presidential Green Chemistry Challenge Awards Program recognizes and promotes fundamental and innovative methods that accomplish pollution prevention through source reduction and which can have broad applicability in the industry. An important focus area for green chemistry is the use of alternative synthetic pathways, such as catalysis or biocatalysis, accompanied by the use of alternative feedstocks that are renewable (e.g., biomass) and more innocuous than current petrochemically derived feedstocks.

In the manufacture of commodity chemicals, older traditional and environmentally unacceptable processes have largely been replaced by cleaner, catalytic alternatives. This has not been the case, though, in the specialty chemical and pharmaceutical industries, whose manufacturing processes are still characterized by stoichiometric synthetic technologies that generate large amounts of waste. Reductions using metals and metal hydrides, stoichiometric oxidations, halogenations, and many reactions that require stoichiometric use of inorganic acids or bases are just a few examples. A significant opportunity exists for the development of catalytic, environmentally benign processes with minimal waste generation. The environmental friendliness of biotechnology has also prompted research into new applications. Biosynthetic processes such as fermentation are naturally green, because they typically require fewer harsh chemical agents than purely chemical processing. For instance, instead of extracting fuel from mineral sources, crude petroleum ethanol can be obtained by fermenting sugars from corn. Plastic manufacturers that previously relied on petrochemical processing are now investing time and money in the production of polymers from sugar-based lactic acids derived from corn and other renewable sources.

The use of alternative reaction conditions is a second important area of development that encompasses the use of solvents that have a reduced impact on human health and the environment and the use of alternative chemistries with increased selectivities that result in reduced wastes and emissions. Pollution prevention at the source is obviously the most responsible and preferred approach to minimizing impact on the environment and also minimizing the potential health effects on workers using toxic or hazardous substances or handling wastes, thus reducing compliance vulnerabilities and saving money otherwise spent on waste management. An additional challenge specific to batch processes is having to deploy solvents in the complex organic synthesis routes. The use of specific solvents may be required in order to improve the selectivity, expedite the conversion, or aid in the separation of products from unwanted byproducts. Ongoing efforts are geared toward recovering valuable solvents via distillative solvent recovery in central facilities. Recent progress has been reported with regard to conditioning effluents for alternative uses with inferior quality demands, such as the use of spent solvents as alternative fuel in the cement industry, as paint additives, or as drilling agents.

Finally, a third area of focus is the design of safer chemicals that can serve as replacements for more toxic current alternatives and the application of inherently safer process (ISP) principles with regard to accident potential during the design of new processes or the retrofit of existing processes.

13.4 REGULATORY ISSUES

Most manufacturers are subject to a wide range of environmental regulations related to air emissions and solid, organic, and aqueous waste. They are also subject to premanufacture notification (PMN) requirements of the Toxic Substances Control Act (TSCA), the Inventory Update Rule under TSCA, Toxics Release Inventory (TRI) reporting, and Occupational Safety and Health Administration (OSHA) regulations. Companies in the pharmaceutical industry are subject to the FDA's current Good Manufacturing Practice regulations. Government safety and environmental regulations have been implemented in many areas, and increasing regulatory requirements continue to put a lot of pressure on the industry. New regulations are being discussed globally to address several contentious issues: environment and sustainability, global climate change, energy use, carbon dioxide emissions, and health effects of chemicals.

It has been observed that, at least in developed countries, the chemical industry has adopted strong voluntary codes of behavior. Companies engage in activities that improve their environmental compliance and performance, and not just because they want to avoid the legal actions and repercussions of noncompliance. A good track record with regulatory compliance is an important competitive advantage for companies manufacturing custom chemicals. In addition, companies want to maintain a good public image of being environmentally responsible, good neighbors, and examples for other companies. Raised awareness of environmental issues among end-use consumers can result in a competitive marketing advantage for manufacturers that demonstrate increased product stewardship.

Batch processing generates many different types of emissions, discharges, and wastes. Air emissions are probably the most difficult to control among all environmental releases. The cost of meeting regulations can be a serious financial challenge. Additional challenges and opportunities pertain to new types of environmental regulations aimed at controlling absolute emission levels rather than ensuring merely regulatory compliance. These new models, known as *emission trading*, already affect volatile organic emissions from batch industries in Connecticut, Illinois, Florida, Maine, Michigan, New Jersey, and Virginia. If emission trading is expanded as an instrument for environmental regulation, total emissions will have an effect on direct costs, thus creating competitive incentives for implementing pollution prevention efforts. All phases of pollution prevention require a great deal of time and effort by experienced engineers. Time constraints, lack of information, and limited in-house expertise may lead to suboptimal decisions, especially for small manufacturers. A number of computer-aided pollution prevention tools have been developed in recent years to assist plant managers in identifying pollution prevention opportunities in a consistent way. They are originating from either government agencies or academic research programs and are aimed at assisting designers in quantifying environmental implications and generating suggestions for process modifications. Computer-aided design methodologies can ultimately guide the selection of pollution prevention strategies in anticipation of regulatory changes.

Index